Springer Series in Synergetics

Springer Series in Synergetics

Founding Editor: H. Haken

The Springer Series in Synergetics was founded by Herman Haken in 1977. Since then, the series has evolved into a substantial reference library for the quantitative, theoretical and methodological foundations of the science of complex systems.

Through many enduring classic texts, such as Haken's *Synergetics and Information and Self-Organization*, Gardiner's *Handbook of Stochastic Methods*, Risken's *The Fokker Planck-Equation* or Haake's *Quantum Signatures of Chaos*, the series has made, and continues to make, important contributions to shaping the foundations of the field.

The series publishes monographs and graduate-level textbooks of broad and general interest, with a pronounced emphasis on the physico-mathematical approach.

More information about this series at http://www.springer.com/series/712

R. I. Sujith • Samadhan A. Pawar

Thermoacoustic Instability

A Complex Systems Perspective

Springer

R. I. Sujith
Aerospace Engineering
Indian Institute of Technology Madras
Chennai, India

Samadhan A. Pawar
Aerospace Engineering
Indian Institute of Technology Madras
Chennai, India

ISSN 0172-7389 ISSN 2198-333X (electronic)
Springer Series in Synergetics
ISBN 978-3-030-81134-1 ISBN 978-3-030-81135-8 (eBook)
https://doi.org/10.1007/978-3-030-81135-8

This Springer imprint is published by the registered company Springer Nature Switzerland AG
The registered company address is: Gewerbestrasse 11, 6330 Cham, Switzerland

To our families and well wishers

Foreword by Jürgen Kurths

Combustors are main elements in various important applications, such as rockets, aeroengines, or power generating gas turbines. Several unwanted and dangerous regimes of operation have been reported in these highly complex systems. In particular, thermoacoustic instabilities resulting in acoustic oscillations having ruinously large amplitudes occurring due to manifold interactions between sound waves and heat release rate fluctuations in the flame are typical problems there and can lead to severe damages. Their first observations go back to 1777 and were explained physically by Baron Rayleigh in 1878. Thermoacoustic instabilities were investigated using linear stability analysis in the 1950s. However, later observations have shown that this kind of analysis is often insufficient because perturbations in even linear stable regimes can lead to thermoacoustic instabilities, often referred to as "triggering instabilities." In the last decade, a bunch of nonlinear phenomena have been found in such thermoacoustic systems, and methods based on nonlinear dynamics and complex systems theory have been applied and further developed to identify them and explain their underlying mechanisms.

In this book, the first systematic treatment of this class of phenomena is presented by comprehensively using classical as well as modern concepts of nonlinear dynamics and complex systems theory. Both authors are very active researchers in this field, and R. I. Sujith has pioneered a deep complex systems-based study of thermoacoustic instabilities in the last decade.

This book gives first a clear and concise mathematical introduction to the basics and how to use them for analyzing laminar thermoacoustic systems. Subsequently, fundamental phenomena in turbulent regimes of such systems regarded as complex spatio-temporal dynamics are discussed and analyzed from multifractal, synchronization, and even complex networks perspectives. It is important to emphasize that the authors also describe strategies to mitigate and forewarn thermoacoustic instability using methodologies derived from complex systems theory. Further, they highlight the generality of the intermittency route to oscillatory instability in turbulent flows, by examining instabilities in other systems including aeroacoustic and aeroelastic systems.

The authors provide excellent guidance on how to treat thermoacoustic instabilities. This book is a very useful introduction for starters in the field, but it provides also important information, tools, and suggestions for researchers and practitioners working on combustion systems as well as also for users from various other fields related to instabilities in spatio-temporal dynamics.

Berlin and Potsdam, Germany
May 2021

Jürgen Kurths

Foreword by Timothy C. Lieuwen

The most pressing applied combustion problems for the last 50 years have been motivated by unsteady challenges. For example, thermoacoustic instabilities have served as one of the most prominent risks in the development of rockets, aircraft engines, or ultra-low emissions power generation systems, often leading to expensive failures, delays, and operational problems. Similarly, engine operational limits, particularly blowoff, have significantly impacted how aircraft engines or power plants are designed and how they are operated. Slowly but surely, steady progress by the community in terms of experiments, theory, and computations has moved these phenomena from being utter mysteries to behaviors firmly rooted in physical understanding (although still very difficult to predict). A major part of this work has been improved understanding and taxonomies for classifying, organizing, and understanding based upon the growing understanding of nonlinear systems in physics, and the overlay of these more general approaches to the intersection of combustion, acoustics, and fluid mechanics problems.

Prof. Sujith has been at the forefront of this work, providing frameworks based on dynamical systems theory and complex systems theory for understanding thermoacoustic systems, and the integration of narrowband, deterministic dynamics with chaotic and/or higher degree of freedom system dynamics. He has shown how these frameworks and analysis tools can be used to deconstruct very complex time series and provide insights into underlying processes, while also shedding light on important practical questions, such as the overall system damping or stability margin. He has pioneered the application of various complex systems approaches, such as synchronization, multifractality, and complex networks, to reveal useful information from the spatial field of the combustor. Prof. Sujith has similarly developed frameworks for the interpretation of transient combustion phenomena, such as blowoff. He has generalized various linearized signal analysis approaches to rigorous, nonlinear time series analysis to predict the amplitude of thermoacoustic instability, and to detect precursors for both thermoacoustic instability and flame blowoff. Prof. Sujith has developed various novel methodologies, based on amplitude death and asynchronous quenching phenomena and the detection of critical regions in the reaction field to mitigate thermoacoustic instability. He has also uncovered some underlying commonality between thermoacoustic instabilities and other oscillatory instabilities that occur in turbulent flows such as aeroacoustic and aeroelastic instabilities and the existence of a universal scaling behavior, during their onset.

This work has spanned the range from fundamental and rigorous to providing practical, actionable insight for designers and engineers on the field. I was delighted to see that Prof. Sujith has compiled and integrated this body of work into a single volume, providing an overarching framework for the application of dynamical systems theory and complex systems theory into unsteady combustor problems. This book will provide a valuable reference and set of case studies, showing how modern dynamical systems concepts can be used to understand and classify unsteady behaviors in modern combustion systems.

Georgia Tech,
Atlanta, GA, USA
May 2021

Timothy C. Lieuwen

Preface

Thermoacoustic instability, also known as combustion instability, continues to be a problem in the development of modern gas turbine combustors and rocket engines. Such instability is often observed in the form of large amplitude, self-sustained oscillations in the acoustic field of the combustor. The phenomenon of thermoacoustic instability has been a lingering thought in the minds of several scientists and engineers in the last century. However, the subject came to the forefront in the 1960s during the development of rocket engines for the Space Race and the subsequent lunar missions. Since then, thermoacoustic instability has been an active area of research due to the high interdisciplinary nature of the problem and the insufficient understanding of the interaction of multiple subsystems involved in the occurrence of such instabilities in combustion systems. Traditionally, the analyses pertaining to thermoacoustic instability and its onset from the state of stable operation were limited to the usage of tools from linear theory. However, the last decade saw tremendous developments with the application of dynamical systems and complex systems theories to understand the occurrence of thermoacoustic instability, through both temporal and spatiotemporal analyses.

In this book, we guide the reader through a fascinating journey where we examine the nonlinear behavior of thermoacoustic systems using different approaches based on dynamical systems theory and complex systems theory. Dynamical systems theory deals with the understanding of the time evolution of a system and characterizes its behavior using various nonlinear measures. Complex systems approach treats the system as a whole and helps us understand the interaction between subsystems at local scales that lead to collective behavior at a global scale. Over the last few decades, these approaches have progressed significantly and have been used in various fields of science and engineering. These mathematical concepts developed using simple nonlinear models can explain the dynamical behavior of many complex practical systems and have far-reaching implications.

Traditional approaches describe thermoacoustic instability as limit cycle oscillations. However, using dynamical systems theory, different manifestations of thermoacoustic instability, beyond limit cycle oscillations, have been revealed in various laminar and turbulent thermoacoustic systems. Various kinds of bifurcations and routes to chaos are also discovered. Methodologies based on synchronization, multifractality, pattern formation, and complex networks have advanced the analysis of temporal and spatiotemporal fields of turbulent thermoacoustic systems. Many control and prediction strategies have been proposed to mitigate and forewarn thermoacoustic instabilities. Overall, new powerful directions to study and mitigate thermoacoustic instability have emerged with tremendous potential for improving future propulsion systems.

For enjoyable reading and better understanding, the book is organized in such a way that we first present the bifurcation characteristics of laminar thermoacoustic systems, then we show the temporal and spatiotemporal analyses of turbulent thermoacoustic systems, and finally we highlight the recent developments in the control and prediction strategies of thermoacoustic instability. In the introductory chapter (Chap. 1), we discuss the traditional viewpoint of thermoacoustic instability and the need for dynamical systems and complex systems approaches for studying such instabilities. Chapter 2 is dedicated to the basics of dynamical systems theory, which covers several topics such as linear stability analysis, fixed points, bifurcations, signals and their classification, phase space reconstruction, Poincaré section, recurrence plots, and different measures to characterize the dynamical behavior of a phase space trajectory. In Chap. 3, we present the bifurcation characteristics of laminar thermoacoustic systems, where we focus on the characterization of subcritical and supercritical Hopf bifurcations, bistability, noise-induced transitions, and triggering. Chapter 4 focuses on characterizing the dynamical regimes of thermoacoustic instability other than limit cycle oscillations. We uncover the different dynamical states, such as quasiperiodicity, frequency-locked, intermittency, and chaos, observed during thermoacoustic instability, discuss different routes to chaos, and present the nonlinear nature of flame-acoustic interaction.

After discussing the nonlinear properties of laminar thermoacoustic systems, we move on to describe the recent developments in both temporal and spatiotemporal analyses of turbulent thermoacoustic systems, during the transition from stable operation (combustion noise) to unstable operation (thermoacoustic instability). In Chap. 5, we introduce the fundamental properties of complex systems such as self-organization, pattern formation, and the emergence of order from chaos. Different approaches developed to analyze such complexities in the system dynamics such as multifractality, synchronization, and complex networks are also briefly introduced here. Here, we make an effort to convince the reader about the necessity to apply complex systems approach to study the transition to thermoacoustic instability in turbulent systems. Chapter 6 describes the temporal dynamics of turbulent combustors during the intermittency route to thermoacoustic instability and the intermittency route to flame blowout. We also uncover the chaotic dynamics hidden in the aperiodic oscillations observed during the state of combustion noise using different tests of deterministic chaos. The spatiotemporal interactions between the flow, flame, and acoustic fields of turbulent combustors are described in terms of pattern formation in Chap. 7. In this chapter, we also discuss the phenomenon of collective interaction of vortices observed in turbulent combustors during the onset of thermoacoustic instability.

The coupled behavior of the acoustic pressure and the heat release rate fluctuations during the transition to thermoacoustic instability is examined in the context of synchronization theory, in Chap. 8. Both temporal and spatiotemporal analyses of these fluctuations are performed using several tools, and a new description of the intermittent route to thermoacoustic instability is provided. Chapter 9 provides details of a reduced-order model, based on vortex-acoustic interaction, to study synchronization transition of the acoustic field and the heat release rate fluctuations during the intermittency route to thermoacoustic instability. In Chap. 10, we introduce the fractal and multifractal analyses of the temporal and spatial fluctuations observed in turbulent systems. We study the multifractal features of low amplitude aperiodic fluctuations observed in the state of combustion noise and characterize the route to thermoacoustic instability and the route to flame blowout in this context. The complex network approach to study the transition to thermoacoustic instability in turbulent systems is presented in Chap. 11. Using the visibility algorithm, we reveal the scale-free feature of the state of combustion noise. Further, using recurrence network analysis, the phase space topologies of different dynamics are investigated. Spatial networks of the reacting field are constructed, and the variation of different network properties is analyzed during the onset of thermoacoustic instability.

Chapter 12 describes different methodologies devised from dynamical systems theory and complex system theory to mitigate thermoacoustic instability and measures that can help predict the occurrence of impending thermoacoustic instability in turbulent combustors. Chapter 13 presents the universal nature of the dynamical transitions observed during the onset of oscillatory instability in different turbulent systems such as thermoacoustic, aeroacoustic, and aeroelastic systems. Finally, Chap. 14 provides a summary of the book and discusses future directions to resolve several unsolved issues pertaining to thermoacoustic instability.

Chennai, India R. I. Sujith

Chennai, India Samadhan A. Pawar
May 2011

Acknowledgments

Writing this book has been challenging and stimulating. The completion of this book would not have been possible without the guidance, support, and valuable contribution of many talented individuals.

We deeply appreciate the invaluable support of our families for cheerfully bearing with us over the years it has taken to bring this book to fruition. Sujith is very grateful to his dear wife Indu and her family, daughters Kalyani and Lakshmi, parents Jalaja and Indusekharan, and brother Rajith for their love, encouragement, and support. Sujith also wishes to acknowledge Silke Kraus, Sujatha Chakravarthy, and G. Jayakumar who have been a constant source of inspiration and support. Samadhan would like to thank his beloved, Akshata; parents Ananda, Dropada, and Jayashree; and brothers Santosh, Sagar, and Chaitanya for their constant love and support.

This book would not have been possible without the inspiration and influence of Koushik Balasubramanian, Lipika Kabiraj, Vineeth Nair, and Vishnu Unni. Due to the never-ending discussions with them and their valuable suggestions, a new perspective to study thermoacoustic instability using complex systems theory has bloomed.

We are indeed very fortunate to have strong collaborators who supported us in our endeavor to develop this perspective. Juergen Kurths, Norbert Marwan, and Elena Surovyatkina from Potsdam Institute for Climate Impact Research; Wolfgang Polifke from TU Munich; Tim Lieuwen, Ben Emerson, and Vishal Acharya from Georgia Tech; William Anderson and Rohan Gejji from Purdue; Matthew Juniper from Cambridge; Swetaprovo Chaudhuri from Toronto; Abhishek Saha from the University of California at San Diego; and Isaac Boxx and Ianko Chterev from DLR Stuttgart have played a major role in shaping the ideas presented in this book.

The work and more importantly the viewpoint presented here have been supported constantly over the last decade by the Office of Naval Research Global, the Science and Engineering Research Board of the Department of Science and Technology, Air Force Office of Scientific Research, Marie Skłodowska-Curie Actions (European Commission), Liquid Propulsion Systems Centre and Vikram Sarabhai Space Centre of the Indian Space Research Organization, and Gas Turbine Research

Establishment and Research and Innovation Centre of the Defense Research and Development Organization. This book would not have been possible without the staunch support of Ramesh Kolar (ONRG), who supported our research in the last decade. We are deeply grateful for his inspiration and support, which made this book possible. We are also very grateful to Venke Sankaran (AFRL) whose probing questions during our intense interactions were instrumental in driving our research in the last few years, and also for providing the inspiration to write this book.

We thank our friends Larry Li, Hiroshi Gotoda, Tim Lieuwen, Ben Emerson, Vineeth Nair, Vishal Acharya, and Achintya Mukhopadhyay for taking the time to review the entire book. We have greatly benefitted from their comments and suggestions, which helped in making the presentation more lucid.

This book would not have been possible without the enormous amount of help provided by the members of our group here at IIT Madras. They went through the draft multiple times and provided invaluable suggestions to improve it. In particular, thanks to Krishna Manoj, Amitesh Roy, Praveen Kasthuri, Vedasri Godavarthi, Shruti Tandon, Induja Pavithran, Sneha Srikanth, Manikandan Raghunathan, Alan Varghese, Ankit Sahay, Samarjeet Singh, Siva Kumar, Ramesh Bhavi, and Premraj D.

We are very grateful to many previous students and postdocs; it is through their work that the complex systems perspective to thermoacoustic instability took shape. Special thanks to Gopalakrishnan E. A., Gireesh Thampi, Meenatchi Murugesan, Abin Krishnan, Sirshendu Mondal, Nitin B. George, S. Manikandan, Tony John, Akshay Seshadri, Suraj Dange, Nevin Thomas, Sreelekha Etikyala, Priya Subramaniyan, Satheesh Mariappan, and Aditya Saurabh.

Sujith also wishes to thank IIT Madras for granting him a sabbatical year, providing him time to research and write.

Sujith sincerely thanks his mentors and colleagues Ben Zinn, Jeff Jagoda, Bob Daniel, Greg Waldherr, Markus Raffel, Krishnan Balasubramanian, Sarith K. Das, S. Narayanan, Kalyana Raman, T. Pradeep, Maria Heckl, Rama Govindarajan, Pankaj Wahi, Thomas Sattelmayer, Christoph Hirsch, Bruno Schuermans, Nicolas Noiray, Peter Schmid, Santosh Shanbhogue, Santosh Hemchandra, Venkat Narra, Renith Richardson, G. Ambika, Partha Dutta, Joseph Mathew, Niket Kaisare, Preeti Aghalayam, Mahesh Panchagnula, Bala Pesala, S. R. Chakravarthy, Sunetra Sarkar, Sayan Gupta, Arul Laksminarayan, Neelima Gupte, Muthusamy Lakshmanan, M. S. Padmanabhan, T. Jayachandran, Lazar Chittilapilly, V. Siddhartha, S. Krishnan, and the late K. A. Damodaran for their help, support, and inspiration.

Contents

About the Authors

R. I. Sujith received his Ph. D. adorned with the "top graduate student in the college of engineering" award from Georgia Institute of Technology in 1994 under the supervision of Prof. Ben T. Zinn. He is currently the D. Srinivasan Chair Professor in the Department of Aerospace Engineering at the Indian Institute of Technology Madras. He is a recipient of the prestigious Alexander von Humboldt Fellowship and the Hans Fischer Senior Fellowship of the Institute for Advanced Study (IAS) at the Technical University of Munich. Prof. Sujith was the founding editor-in-chief of the *International Journal of Spray and Combustion Dynamics* from 2009 to 2015 and is at present a member of the editorial advisory board of the interdisciplinary journal Chaos. He won the Young Engineer Award of the Indian National Academy of Engineering. He has also been awarded the Swarnajayanti Fellowship and the J. C. Bose Fellowship by the Department of Science and Technology, India. He is a Distinguished Fellow of the International Institute of Vibration and Acoustics (IIAV). He is a fellow of the Indian National Academy of Engineering and the Indian Academy of Sciences, and has been conferred the title of "TUM Ambassador" by the Technical University of Munich. Prof. Sujith currently works on the application of dynamical systems and complex systems theory to study and mitigate thermoacoustic instability.

Samadhan A. Pawar received his Ph.D. from the Indian Institute of Technology Madras in 2018 under the supervision of Prof. R. I. Sujith and Prof. Mahesh V. Panchagnula. He has received the INAE Young Engineer Award 2021 of the Indian National Academy of Engineering. He was conferred the Institute Research Award in 2018 from IIT Madras in recognition of his doctoral research work on the application of synchronization theory to thermoacoustic systems. He has also been awarded the Young Scientist Award 2020 by the International Society for Energy, Environment and Sustainability (ISEES). He is an INAE Young Associates. His work highlights the maiden experimental and theoretical characterization of the onset of thermoacoustic instability and its control using dynamical systems theory and complex systems theory. He has authored 26 papers in reputed journals and filed one patent.

Chapter 1
Introduction

Thermoacoustic instability has plagued the development of combustors for gas turbine and rocket engines. In this chapter, we discuss the origin and consequences of thermoacoustic instability. We will first discuss the mechanisms that cause thermoacoustic instability in generic systems and then focus on the specific case of turbulent combustors and their unique challenges. Finally, we will discuss the importance of dynamical systems theory and complex systems theory to study thermoacoustic instability in turbulent combustors.

1.1 Introduction to Thermoacoustic Instability and its Consequences

Thermoacoustic instability (also known as combustion instability) refers to self-sustained, large amplitude oscillations of pressure and velocity occurring in reacting flows. It arises primarily through an interaction of sound waves in a confined space with the unsteady heat release rate caused by combustion. The flame acts as an acoustic actuator and the combustion chamber as an acoustic resonator. If the rate at which heat released by the flame responds to the pressure and velocity fluctuations, a feedback loop is established, which can destabilize the system (see Fig. 1.1).

The occurrence of thermoacoustic instability depends on the phase difference between the heat release rate fluctuations and the acoustic pressure fluctuations at the flame. Fluctuating heat release rate will amplify the acoustic oscillations, if the maximum heat addition or removal occurs during the compression and rarefaction phases of the pressure oscillation, respectively; this is known as the Rayleigh criteria [124]. The fluctuations grow exponentially until nonlinearities in the system become significant, resulting in a saturation of thermoacoustic oscillations when the total energy losses from the chamber balance the energy input from combustion.

R. I. Sujith, S. A. Pawar, *Thermoacoustic Instability*, Springer Series in Synergetics,
https://doi.org/10.1007/978-3-030-81135-8_1

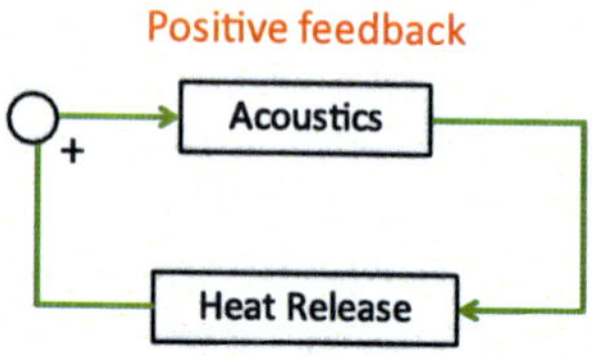

Fig. 1.1 The heat release rate, if dependent on the acoustic pressure and velocity fluctuations in the system, results in a feedback loop that can destabilize the thermoacoustic system

Predicting and controlling thermoacoustic instability requires an understanding of the interactions between the combustion process and the acoustic waves.

Rockets, aeroengines, and power generating gas turbines are particularly susceptible to the coupling between the heat release rate and the acoustic field because they have high energy densities, on the order of 10 GW/m^3 for liquid rockets, 1 GW/m^3 for solid rockets, and 0.1 GW/m^3 for gas turbine engines, as well as low acoustic damping [143]. To put these numbers in perspective, the first stage of the Saturn rocket has a power of 60 GW, which is similar to the peak electricity demand of the United Kingdom [161]. Even if only a small fraction of this energy is converted into acoustic (mechanical) energy, large amplitude acoustic oscillations can be sustained in the system.

Thermoacoustic instabilities involve ruinous large amplitude pressure fluctuations. In rockets, they are responsible for excessive vibrations, leading to structural damage of components and failure of electronic systems in the vehicle and the satellite (payload) [33]. These large amplitude oscillations cause excessive heat transfer leading to thermal failure [33, 129]. They can also increase the mass, momentum and energy transport, changing the thrust levels of engines from their design values [53]. Even low amplitude vibrations can cause failure of engine components because of fatigue, and wear and tear when exposed over a long period of time [146]. In industrial gas turbine engines used for power production, stringent emission norms require the combustor to operate in lean equivalence ratios where thermoacoustic instabilities are more prone to occur [4]. Thermoacoustic instabilities in such conditions can even induce flame blowout [93]. In rockets, the problem is more severe, as catastrophic mission failures can occur [14, 33]. Thus, identifying the possibility of thermoacoustic instabilities in the early phase of the design stage is an important but formidable task.

Thermoacoustic instability (then referred to as singing flames) was first reported in 1777 [54]. In 1858, John LeConte [81] described the interaction between sound and flames when he observed synchronous response of flames to music: "*Soon after the music commenced, I observed that the flame exhibited pulsations exactly synchronous with the audible beats. This phenomenon was very striking to everyone in the room, and especially so when the strong notes of the violoncello came in.... A deaf man might have seen the harmony...*". With the advent of high intensity combustion systems, thermoacoustic instability moved on from what was originally an academic curiosity to a pressing problem. Detrimental thermoacoustic oscillations were observed in boilers, blast furnaces and oil, coal and gas fired heating systems [123]. The cold war and the space race led to the development of a large number of high-power density solid and liquid rockets, turbojets and their afterburners, and ramjet engines, most of which encountered thermoacoustic instability during their development [34]. For example, the F-1 engines used in the Apollo moon rocket had severe problems with combustion instability, which were overcome by the installation of baffles in the injector plates [114]. Thermoacoustic instability was also common in rockets from Soviet Union; famous examples being RD-170 and RD-171, where they solved the problem by fitting four nozzles to the engine, giving the visual appearance of four separate engines [38]. Thermoacoustic instabilities did not spare missiles either; they have been reported in Standard missile, Sidewinder, Harm, Trident, Hellfire, and Minuteman [13]. Combustors of land-based gas turbine engines for power production that operate close to the lean limit to satisfy the stringent emission norms are particularly prone to thermoacoustic instability, with the expected losses reportedly exceeding a billion dollars annually [167].

Although thermoacoustic instability arises primarily from the interaction between acoustic waves and heat release rate fluctuations, this coupling may be affected significantly by additional factors such as hydrodynamic structures, mixture strength fluctuations, or entropy waves. The various effects are linked in feedback loops as shown in Fig. 1.2.

Rayleigh [124] presented the first scientific explanation for the occurrence of thermoacoustic instability. Later, Putnam [123] formulated it mathematically and is now popularly known as the Rayleigh criterion. The Rayleigh criterion states

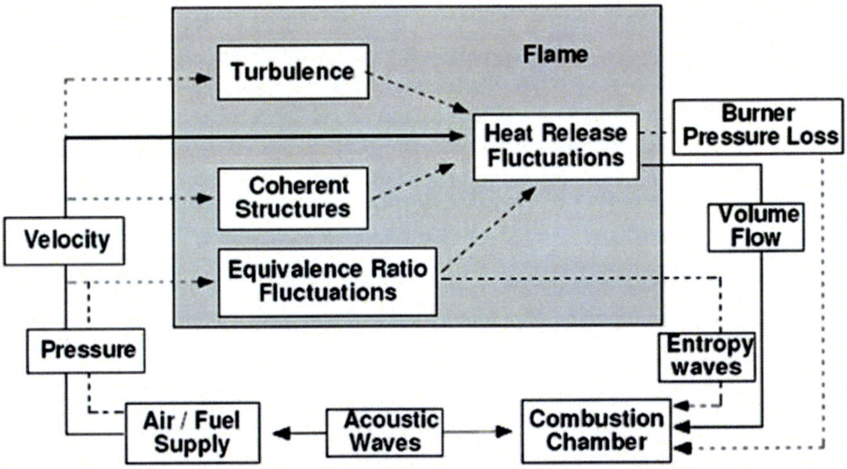

Fig. 1.2 Interactions between the flow, acoustics, and heat release fluctuations in a combustion system. Reproduced with permission from Polifke [120]

that during the growth of thermoacoustic instability, unsteady heat release rate and acoustic pressure oscillations are in phase with each other.

When pulsations start spontaneously in a thermoacoustic system, the system is said to be linearly unstable; i.e., the system is unstable to any small amplitude disturbance that may be present in the combustor. This scenario has been successfully investigated for various thermoacoustic systems by using a linear stability analysis that models the system as a network model. In a network model, each element is modeled using a linear transfer function [59, 60]. The stability of the thermoacoustic system can then be determined easily by examining the eigenvalues of the system, which will give the growth/decay rate of the system.

It is also possible, however, that a linearly stable combustor; i.e., one that does not pulse spontaneously, could be "triggered" into pulsating operation by the introduction of a finite amplitude disturbance [33]. Such a disturbance may be caused by a spark plug ignition or a small explosion, or nozzle ejection such as the expulsion of an igniter or insulation fragment. The system will then be stable with respect to all disturbances whose amplitudes are below a certain threshold value, but will transition into pulsating operation when the amplitude of the disturbance exceeds this threshold value. Such nonlinear behavior of the system is referred to as triggering and can occur only in systems with a subcritical bifurcation (subcritical bifurcation will be explained in detail in Chap. 2.). Triggering cannot be explained through classical linear stability analysis [62, 92].

Despite decades of active research, the appearance of thermoacoustic instability has remained a serious problem in the design and development of combustors for rockets, ramjets, and gas turbines [65, 99, 167]. Although much work has been done over the last 50 years, most of this work is in the framework of classical linear stability analysis. A comprehensive prediction of the conditions for the onset of thermoacoustic instabilities is a difficult task, which is not yet mastered. In particular, "*predicting the conditions under which finite-amplitude disturbances destabilize a linearly stable system and predicting the limit-cycle amplitude of the thermoacoustic instability remain a key challenge, as little is known, even in a qualitative sense, about the key parameters controlling nonlinear flame dynamics*" (reproduced from [167]). In order to predict the onset of combustion instabilities, a nonlinear approach to studying thermoacoustic oscillations is necessary.

1.2 Mechanisms that Cause Thermoacoustic Instability

Many physical processes are involved during the occurrence of thermoacoustic instability. Among these, the propagation of sound waves is fairly well understood. However, the interaction between the unsteady heat release rate from the flames and the sound waves is very complex, and not yet completely understood. There are a number of mechanisms that can contribute to the positive feedback between the unsteady heat release rate and the acoustic waves. These include flame surface area modulations, equivalence ratio fluctuations, coherent structures in the flow, droplet and spray dynamics, and entropy waves (refer to Fig. 1.2). In the following subsections, we give a brief description of some of these mechanisms; for a comprehensive treatment, the reader may refer to [92].

1.2.1 Flame Surface Area Modulations

Modulations in the flame surface area that arise from the interactions between acoustic waves and the flame is an important mechanism driving thermoacoustic instability in premixed flames [84]. For a steady flame, there is a balance between the flame front propagation velocity and the flow velocity perpendicular to the flame front. Acoustic oscillations can disturb this balance, causing a disturbance in the flame that originates at its base. These disturbances are then convected along the flame front from the base of the flame to the tip by the tangential flow component along the flame, causing the flame to wrinkle (Fig. 1.3). As a consequence, the effective surface area of the flame is modulated. The heat released by a premixed flame is proportional to its surface area. Therefore, when the flame surface area modulates, the heat release rate fluctuates. When this unsteady heat release rate is coupled to the acoustic field in the confinement, it produces acoustic power. If this acoustic driving is able to overcome the losses in the system,

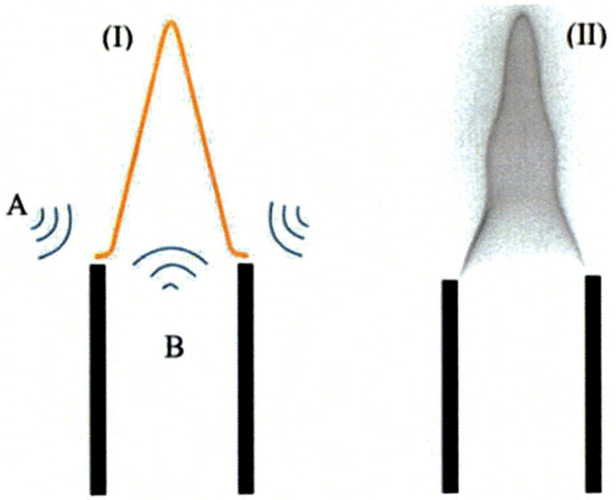

Fig. 1.3 (I) Schematic of the possible external disturbances that can cause flame perturbations. (II) Experimental observation of flame wrinkling caused by inherent acoustic oscillations. In the case of self-excited thermoacoustic instability in a confined flame system, the downstream acoustics velocity fluctuations (A) cause modulations of the flame surface. Flame wrinkling can also be caused by the upstream acoustic field (B) when the coupling is established between the flame and burner tube acoustics or in the presence of upstream forcing. Reproduced with permission from Kabiraj [66]

thermoacoustic instability occurs. The role of flame area modulations has been investigated thoroughly in [18, 71, 84].

1.2.2 Equivalence Ratio Fluctuations

Having perfectly premixed reactants upstream of the injector poses the risk of the flame flashing back, causing autoignition; i.e., spontaneous ignition without an external source of ignition such as a flame or spark [93]. To avoid this risk, in industrial systems, fuel is injected into the airflow, a short distance upstream of the combustor. Such systems are called technically premixed (or industrially premixed) systems.

In such systems, the fuel and air supplies are prone to pressure and velocity fluctuations traveling upstream from the combustion zone. As illustrated in Fig. 1.4, the acoustic waves propagate upstream of the burner. The pressure waves cause an oscillatory pressure drop across the fuel supply, leading to the modulation of the mass flow rate of fuel entering into the combustor. The velocity fluctuations cause the flow rate of air to fluctuate. Thus, the equivalence ratio (i.e., the ratio between the actual fuel-air ratio to the stoichiometric fuel-air ratio) of the fuel-air mixture fluctuates. These equivalence ratio fluctuations are convected to the flame by the mean flow, and at the flame, they cause heat release rate fluctuations. Lieuwen and Zinn [88] and Lieuwen et al. [87] established the sensitivity of lean premixed combustion to equivalence ratio fluctuations.

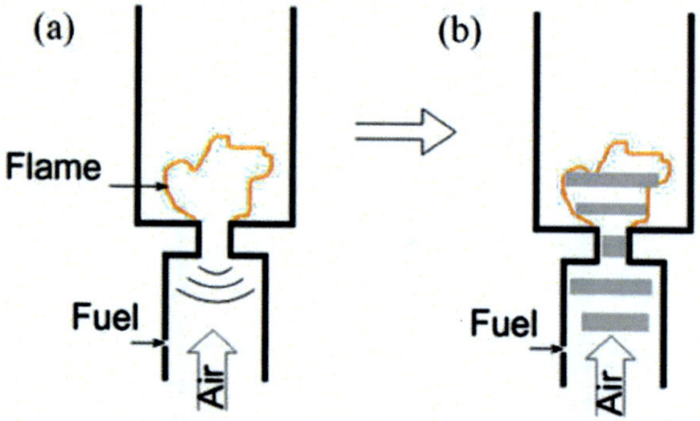

Fig. 1.4 (**a**) Acoustic perturbations move upstream of the burner into the fuel and air supply lines and lead to oscillations in the air and fuel supplied to the burner. This leads to oscillations in the equivalence ratio. (**b**) Equivalence ratio fluctuations lead to heat release rate fluctuations from the flame, which could couple with the acoustic field in the combustor to cause thermoacoustic instability. Reproduced with permission from Kabiraj [66]

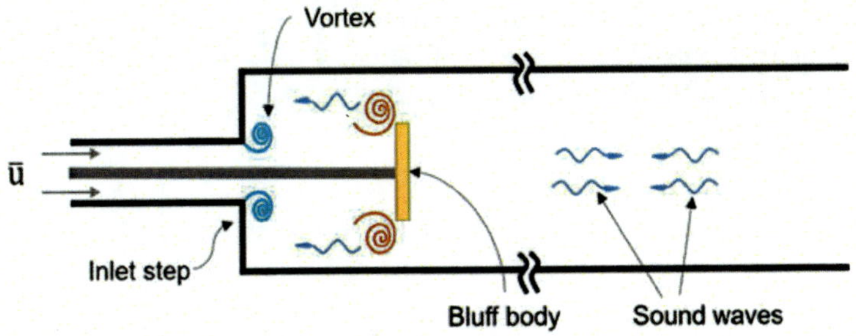

Fig. 1.5 In a backward facing step (dump plane in the combustion terminology) combustor, vortices are shed from the inlet step. Periodic vortex shedding in such systems leads to periodic heat release rate fluctuations. Reproduced with permission from Pawar et al. [116]

1.2.3 Coherent Structures in the Flow

Regions of high velocity gradient such as shear layers (there are other high velocity gradient regions as well, other than shear layers) are used to stabilize flames in most practical combustion systems, such as swirl-stabilized or bluff body stabilized combustors. Periodic vortex shedding in such systems leads to heat release rate fluctuations. Figure 1.5 shows vortex shedding in a dump combustor.

Schadow and Gutmark [127] and Venkataraman [151] showed that this flame-vortex interaction is a dominant mechanism for combustion instability in lean premixed combustors. During the occurrence of thermoacoustic instability, a feedback loop is formed, thereby causing the vortex shedding frequency to match the acoustic frequency. In swirl stabilized burners, a precessing vortex core (PVC) that arises at the boundary of the direct and reverse flows plays an important role in causing thermoacoustic instability [58].

1.2.4 Entropy Waves

Entropy waves are hotspots (regions of non-uniform temperature) in the flow originating from the flame. These entropy waves are convected by the flow towards the combustor exit. When the entropy waves travel through a region of non-uniform base flow (e.g. a nozzle), acoustic waves are formed [95]. These acoustic waves cause thermoacoustic instability [40, 41].

These are only a few mechanisms that cause thermoacoustic instabilities; there are many other mechanisms and we do not attempt to provide a comprehensive description. In liquid rockets, thermoacoustic instabilities can arise from the coupling between the combustion processes and the feed system [160]. Atomization, vaporization, and mixing of liquid fuel can respond to pressure and velocity disturbances [79, 142], and can lead to thermoacoustic instability [57, 119]. In solid rocket motors, the response of the gasification of the solid propellant to acoustic oscillations can lead to thermoacoustic instability [13].

1.3 Mechanisms that Damp Thermoacoustic Instability

Although often considered to be the less glamorous aspect of thermoacoustic instability, damping is just as important as driving [1, 76]. Viscous and thermal diffusion contributes to damping. These effects manifest in the form of boundary layer losses and flow separation losses. Boundary layer losses occur close to the walls or surfaces, where viscous and thermal dissipation are significant. Regions of strong velocity gradients such as flow over a backward facing step cause flow separation losses, where acoustic energy is transferred to the vorticity fluctuations. Acoustic energy can be radiated out from the end of the combustors to the ambient, or lost through the nozzle. Further, the vibration of the combustor and the associated structures will lead to loss. Another mechanism involves the transfer of energy to higher modes that are not driven by the combustion. Acoustic energy is more rapidly dissipated from modes that have a higher frequency.

1.4 Current Approaches: Acoustic Oscillations Driven by Unsteady Combustion, Network Modelling, and Eigenvalues

Traditionally, a thermoacoustic system has been modelled as an acoustic system, which is driven by unsteady combustion. For gas turbine combustors, this is done using the so-called network model. In a network model, the thermoacoustic system is modelled as a series of network elements with specified boundary conditions. The stability of the network is then analysed by performing a linear stability analysis of the resulting system of equations. In a network model, the analysis is performed in the frequency domain, wherein each element is described using a linear transfer matrix [59, 60]. These transfer matrices describe the relationship between the acoustic pressure and the acoustic velocity across the element and also provide the necessary boundary conditions across the element interfaces. The flame, which itself is the source of acoustic energy, is described using a transfer matrix or a transfer function (frequency response function). These descriptions yield a system of coupled linear equations. The stability analysis is performed by determining the eigenvalues of the system and examination of the resulting growth/decay rates (obtained from the imaginary part of the eigenvalue.[1]) Network models can be established for situations involving entropy waves as well [40, 41]. For rocket motors, the propellant response is traditionally described by a pressure coupled or velocity coupled response function [33].

Modelling acoustic wave propagation is relatively straightforward and well established. However, there is no simple way to predict the flame transfer function. This needs to be measured either from experiments [39, 130] or determined from computational fluid mechanics (CFD) simulations [59, 60].

1.5 Why Do We Need a Nonlinear Description?

When thermoacoustic oscillations start spontaneously at an operating condition, the combustor is said to be linearly unstable at that condition. It means the combustor is prone to thermoacoustic instability for arbitrarily small disturbances that may arise at that operating condition. As described in the previous section, such a system can be analyzed with a network model, and its stability is easily determined by examining the eigenvalues of a matrix composed of transfer functions.

However, even if such a system initially has a positive growth rate, its oscillations do not grow indefinitely, as predicted by linear theory. As the amplitude grows, nonlinearities in the system become increasingly important, limiting further growth in amplitude, leading to the establishment of limit cycle oscillations. Note that after the first bifurcation to limit cycle oscillations, further bifurcations are possible, leading to quasiperiodic and chaotic oscillations. Further, during triggering, a linearly stable thermoacoustic system can be excited to limit cycle oscillations by disturbances above a critical amplitude.

In summary, thermoacoustic systems are nonlinear. They can be influenced by global nonlinearities such as nonlinear gas dynamics, which are present in the entire flow field, or by local nonlinearities such as the nonlinearities in the flow and flame responses, which are present in the region where the interaction between the reacting flow and acoustic waves occurs. Nonlinearities can be present in the losses as well. Thus, it is necessary to adopt a nonlinear approach towards characterizing and predicting the nonlinear states (e.g., limit cycles) that are observed in thermoacoustic systems. Dynamical systems theory provides the language and tools necessary to meet those challenges [65, 145].

At this point, it is helpful to introduce some terminology from dynamical systems theory. The qualitative change in the behavior of a system due to a change in the control parameter from stable operation to unstable operation is termed a bifurcation [139]. A transition from a steady behavior to periodic oscillations is termed an Andronov-Hopf bifurcation, or more commonly as a Hopf bifurcation. During a

[1] $p' = \hat{p}e^{i(\omega_r + i\omega_{im})t} = \hat{p}e^{i\omega_r t}e^{-\omega_{im}t}$. Thus, the imaginary part of the eigenvalue, i.e., $-\omega_{im}$, is the growth rate.

Hopf bifurcation, the growth rate (the imaginary part of the eigenvalue) changes sign; i.e., the system transitions from one with a decay rate to one with a growth rate.

A transition to periodic oscillations due to a change in the initial disturbances in the linearly stable regime is a characteristic of subcritical Hopf bifurcations. For a system exhibiting subcritical Hopf bifurcations, there exists a range of operating conditions in a control parameter (such as the airflow rate) prior to the Hopf point (i.e., where the Hopf bifurcation occurs) for which two stable solutions are possible in the asymptotic state (i.e., the final state occurring after a large time, when the transients have died down). The solution at which the system settles depends on the magnitude and type of the initial perturbation and the higher order nonlinearities (specifically, 5th order and above [139]). It must be emphasized that the techniques of classical linear stability analysis do not include the effects of initial conditions in the analysis and hence cannot explain triggering.

In summary, to describe the onset of thermoacoustic instability, to estimate the amplitude of the resultant self-sustained oscillations, and to describe features such as triggering and limit cycles, we need to adopt a framework of nonlinear theory.

In the context of gas turbine engines, Zinn and Lieuwen [167] remarked that a linearly stable system may be nonlinearly unstable at low-amplitude disturbances that are of the order of the background noise level, in a scenario analogous to the hydrodynamic stability of a laminar Poiseuille flow. Practical thermoacoustic systems operate in turbulent environments. Lieuwen and Banaszuk [86] suggested that these turbulent fluctuations can be thought of as additive and parametric sources of noise. Jagadesan and Sujith [62] and Waugh and Juniper [154] have shown that transition to thermoacoustic instability can occur even when the noise level is significantly below the triggering amplitude. Understanding noise-induced transition, therefore, requires accounting for nonlinearities in the system.

1.6 Nonlinearities in a Thermoacoustic System

Motivated by the observation of thermoacoustic instability in both solid and liquid rocket motors, much work in the 1970s and 1980s was focused on identifying the role of nonlinear gas dynamic processes in establishing thermoacoustic instability. In such cases, the resultant amplitude of acoustic pressure oscillations can reach a significant proportion of the mean pressure value; typical values correspond to $p'/\bar{p} = 20 \sim 50\%$. Culick [29–31] studied the nonlinearities of acoustic waves in the combustion chamber of a rocket motor using an one-dimensional wave equation and by treating the combustion response as linear. He predicted the limit cycle amplitudes by numerically solving a set of coupled nonlinear wave equations. His work paved the first steps in tackling the general problem of analyzing the nonlinear growth and saturation of the amplitude of acoustic waves in a combustion chamber.

In contrast to the model proposed by Culick, Sterling [136] reasoned that the only nonlinearities of significance in a combustion chamber originated from the combustion process itself and the acoustic dynamics could be treated as essentially linear. His ideas were "*motivated by experimental results from a laboratory combustor that demonstrated that heat-addition and fluctuations are dominated by the Rayleigh mechanism*". He concluded that in order to adequately represent the flame dynamics, a nonlinear expression for the heat release rate is necessary.[2] Culick [32] and Wicker et al. [156] showed that the combustion response needs to be nonlinear to be able to trigger spontaneous pulsations in a combustor. Nevertheless, we remark that this suggestion is not consistent with the pioneering work on nonlinear thermoacoustic instability done at Princeton in the 1960s using asymptotic theory [100, 132, 166] or the recent CFD findings at University of California, Irvine [133]. We further remark that the nonlinear models adopted in the studies of Culick [32] and Wicker et al. [156] seems to be ad hoc and not derived from first principles.

Nonlinear gas dynamical processes are thought to be insignificant in premixed gas turbine combustors, where the reported acoustic pressure amplitudes are typically around 1–5% of the mean pressure value [35, 90, 117]. Therefore, according to these studies, the acoustic (i.e., gas dynamic) processes remain in the linear regime, even during thermoacoustic instability. These studies also highlight the fact that the dominant nonlinear contribution arises from the relationship between the flow and heat release oscillations, which in the frequency domain can be represented as a flame transfer function or a flame describing function. Since the typical flow Mach numbers in a premixed combustor are low, the amplitude of pressure fluctuations tends to be quite small compared to the mean pressure, which allows a linear analysis of acoustic processes, even if the combustion response turns out to be nonlinear [33].

Later studies have confirmed that the heat release rate response in premixed combustors is nonlinear [56, 85, 121, 122]. Typically, we need to worry about nonlinear effects related to heat release rate response only when the amplitude of the velocity fluctuations is comparable to the mean flow velocity. The response of bluff-body stabilized lean premixed flames to acoustic forcing was investigated by Balachandran et al. [7] to understand the combustion response during the occurrence of limit cycles. They found that the nonlinearity in the heat release rate response becomes significant only when the amplitude of the inlet velocity fluctuations reaches around 15% of the mean value, a value that depends on the forcing frequency and the equivalence ratio. Tyagi et al. [150] have shown that the combustion response of diffusion flames also becomes nonlinear, even for very low amplitudes of acoustic excitation.

For the type of flow-fields that exist in a combustor, hydrodynamic instabilities play an important role in determining the overall stability of the thermoacoustic system. A large class of thermoacoustic instabilities is driven by the interaction of vortices with flames and the acoustic field. Such interactions also—to a great extent—dictate the structure of turbulent flames and the corresponding rates of reaction inside the combustion chamber [125]. These interactions are inevitably highly nonlinear.

[2]For a contrasting view point, see Sirignano [133].

The subject of nonlinear dynamics started out in the 1960s with the pioneering study of Lorenz [94]. Around the same time, the Princeton group spear headed by Crocco [28, 100, 132, 165, 166] worked on the nonlinear nature of thermoacoustic oscillations; in particular, they examined nonlinear behavior such as limit cycles and triggering. It is worth mentioning that while Crocco and his students were studying bistability, nonlinear dynamics was at its infancy.

Dynamical systems theory (also popularly referred to as nonlinear dynamics) flowered in the 1970s with an explosive growth of research and found applications not only in natural sciences such as physics, chemistry, biology, and medicine, but also in subjects such as economics and even social sciences [46]. In the 1980s, complex systems theory, which is a related subject started flowering, and then had an explosive, unfettered growth starting from the 1990s [21]. In contrast, after the impressive beginning in the 1960s, unfortunately, nonlinear thermoacoustics did not move forward rapidly much till recent years.

1.7 Thermoacoustic Stability Analysis: Acoustic Versus Dynamical Systems Approach

1.7.1 Traditional Acoustic Approach

The traditional approach involves modelling a thermoacoustic system as an acoustic system driven by unsteady fluctuations. The combustion response is traditionally modelled using a response function that models the response of the combustion process to excitation by the velocity fluctuations (or pressure fluctuations) caused by the acoustic field. The stability is then determined by examining the eigenvalues of the system. Nonlinear extensions of this approach have been developed. A major share of the literature on the nonlinear analysis of combustion instabilities has been devoted to obtaining nonlinear describing functions, either experimentally or theoretically [36, 111]. The procedure involves decoupling the flame from the acoustic field, and forcing the flame over all the possible ranges of frequencies and amplitudes. The nonlinear flame describing function (or the nonlinear flame transfer function) thus obtained is a function of both the amplitude and frequency, which is then substituted in the wave equation (transformed to the frequency domain) to model the acoustic driving. The dispersion relation (relation connecting the frequency and the wave number) is then solved for its eigenvalues at different amplitudes of the acoustic velocity. These eigenvalues indicate the growth or decay in the system at any amplitude. The point of zero growth or decay rate is interpreted as a limit cycle. The stability of a limit cycle is then determined by the change in the sign of the growth rate. A change in the sign of the growth rate from a positive value to a negative value with increasing amplitude of perturbations indicates that the limit cycle is stable. This is because perturbations with an amplitude smaller than the limit cycle amplitude grow with time, whereas perturbations with amplitude larger than that at limit cycle decay to the limit cycle amplitude. Conversely, a change in the growth rate from negative to positive with increasing amplitude of perturbations indicates that the limit cycle is unstable.

At its core, the technique still relies on a quasi-linear analysis that involves measuring the flame describing functions (which is essentially the flame transfer function as a function of the forcing amplitude) at different amplitudes and using eigenvalues to infer stability. Although a large body of recent literature exists on this topic and success is reported by various groups in predicting the amplitude of oscillations in this manner, such an analysis requires costly experiments to characterize the response of a flame to different types of external forcing at a range of amplitudes and frequencies. Further, designing actuators that can produce high amplitude oscillations comparable to the amplitude of limit cycle oscillations is very difficult. Also, it is not desirable to excite such high amplitudes in high pressure rigs or in gas turbine combustors. As the acoustic field is assumed to be harmonic during thermoacoustic instability, the work done by the heat release rate fluctuations at higher harmonics when integrated over a cycle is zero. Thus, the flame describing function approach needs to consider only the heat release rate at the forcing frequency. Although this is a drastic simplification, it embeds a questionable assumption that the acoustics are harmonic in time [72, 73].

Further, the dynamics of a forced system is different from that of a self-excited system. If the system is unstable, a disturbance grows exponentially in time without limit, provided all the processes stay linear. This growth of amplitude can be expressed as $A_0 e^{\alpha t}$, where A_0 represents the initial amplitude and α indicates a growth rate [33]. This is represented by a curve that is concave upward and will evolve into a constant limiting value only if nonlinear processes are active in the system. In contrast, the initial transient response of a linear system forced externally grows according to the form $1 - e^{\beta t}$, a curve that is concave downward, and approaches a limiting value for a linear system because the driving agent supplies only a finite amount of power [33]. Further, the phase on a limit cycle is free to evolve; in contrast, the phase on a stable closed curve of the forced system is unambiguously related to the phase of the external force [118].

1.7.1.1 Dynamical Systems Approach

An alternative viewpoint based on dynamical systems approach has emerged recently. A dynamical system is a system that evolves in time. Dynamical systems theory provides well-established tools for analyzing the evolution of both linear and nonlinear systems [139]. It is therefore sensible to adopt the tools from dynamical systems theory to study the evolution of thermoacoustic systems.

The time evolution of a pressure measurement acquired from a combustor can give information about its stability and the nature of the resulting asymptotic state. However, this information is highly dependent on the system parameters, operating conditions, and initial conditions (e.g., the levels of flow noise, ignition of a spark plug or an explosion). Hence, trying to assess the stability of a combustor by following the evolution in time of pressure oscillations individually, for all the possible ranges of parameter values and initial conditions, is not only expensive but also impractical. Tools from dynamical systems theory can provide a systematic and efficient framework to investigate both the linear [42, 43, 128, 149] and nonlinear [139] behavior of the system.

Jahnke and Culick [61] discussed the operating conditions under which stable limit cycles can exist in a system that is linearly unstable, and conditions under which bifurcations to a limit cycle can occur. Culick and co-workers applied some concepts from dynamical systems theory such as continuation methods to calculate the limit cycle amplitudes [3, 16]. These were isolated efforts, and did not catch on, until the approach saw its revival in recent years wherein the onset and triggering of combustion instability was successfully investigated in a number of scenarios by Sujith and co-workers [8, 9, 96, 98, 140], Juniper and co-workers [63, 64, 154] and others [157, 164].

1.8 Beyond Limit Cycles

Pressure oscillations more complex than a limit cycle have also been reported previously in the context of combustion instability, by a few authors. Jahnke and Culick [61] reported the possibility of quasiperiodic oscillations (oscillations characterized by two incommensurate dominant frequencies and their multiples). Using a numerical bifurcation analysis, Sterling [136] and later Lei and Turan [83] reported the presence of chaotic oscillations in models of premixed combustors. In experiments, Keanini et al. [75] and Fichera et al. [44] reported chaotic dynamics in a laboratory ramjet combustor and a lean gas turbine combustor, respectively.

Kabiraj et al. [67, 69, 70] and Kabiraj and Sujith [68] performed a systematic bifurcation analysis on the acoustic pressure and CH* chemiluminescence time traces obtained from simple setups comprising ducted, laminar premixed flames to investigate the features of nonlinear thermoacoustic oscillations. They reported that as the bifurcation parameter is varied, the system undergoes a series of bifurcations leading to characteristically different states of nonlinear oscillations. Through the application of techniques from nonlinear time series analysis, these oscillations were characterized as periodic, quasiperiodic, aperiodic or chaotic oscillations; subsequently, the nature of the obtained bifurcations was explained based on dynamical systems theory. Vishnu et al. [152] investigated the role of flame dynamics on the dynamical states of a ducted V-flame by using high-speed imaging. They highlighted processes such as wrinkling, roll-up, and mutual annihilation of the flame elements and their importance for the dynamical states.

These recent findings indicate that limit cycle is just one of the possible end states in a combustor. A thermoacoustic system can undergo further bifurcations and attain states characterized by quasiperiodicity, period-doubling, frequency-locking, intermittency, strange nonchaos, and chaos. Sujith's group has observed both the Ruelle–Takens [67] and the frequency-locking quasiperiodic route to chaos [69] in their experiments. Using high-speed flame images acquired simultaneously with the pressure time trace, they showed that the source of nonlinearity is in the interaction between the flame and the acoustic field. These findings were later echoed by the numerical studies of Kashinath [72] and Orchini et al. [115]. More recently, the intermittency route to chaos has been shown by Guan et al. [52].

1.9 Thermoacoustic Instability in Turbulent Combustors

In turbulent combustors, the stable operation is not a completely silent state. The turbulent reacting flow generates sound, which has a broadband characteristic; this sound is referred to as combustion noise.

Combustion noise is a well-studied topic (see [37, 131, 137] for extensive reviews on the topics). Lieuwen [90] showed that the inherent noise in a thermoacoustic system can strongly affect the limit cycles. Further, he showed that the inherent noise, under certain operating conditions, may even be responsible for causing the combustor to become unstable under linearly stable conditions. He also investigated the characteristics of the fluctuating pressure in an unstable combustor [89] and concluded that the phase drift characteristics are caused mainly by random processes (noise) and do not reflect the underlying low-dimensional dynamics of thermoacoustic instability. The statistical characteristics of self-excited and noise-driven pressure oscillations in a premixed combustor were investigated by Lieuwen [91]. Using experimental data, he showed that the probability density function of the amplitude of these oscillations transitions from Gaussian to a Rayleigh-type distribution as the combustor moves from stable to unstable operation. Lieuwen and Banaszuk [86] considered the effect of background turbulent fluctuations on the stability boundaries of a combustor. They showed that additive noise sources change only the quantitative aspects of thermoacoustic oscillations. However, parametric noise sources can affect the dynamics qualitatively as well; in particular, parametric noise can destabilize a system that is stable in the absence of these noise sources.

Most studies focus on individually assessing and contrasting stable and unstable thermoacoustic systems. There are not many studies that focus on the transition itself by varying the operating parameters smoothly, starting from the stable operation, towards thermoacoustic instability (unstable operation). Even though various physical mechanisms responsible for thermoacoustic instability have been identified in the prior studies, not much attention has been paid on the exact nature of the transition, or the pathways (routes) through which thermoacoustic instability is established. Chakravarthy and co-workers [23, 24] were the first to perform a systematic variation of the control parameters in bluff-body and backward-facing step combustors from stable to unstable operation. They highlighted the low-amplitude broadband nature of the pressure fluctuations in what they called the non-lock-on regime (where vortex shedding is not locked to the duct acoustic field). However, as the vortex shedding locks to the acoustic oscillations, the acoustic field has high-amplitude discrete tones resulting from limit cycle oscillations. Recently, the groups of Gotoda and co-workers [50, 51, 74, 102] and Sujith and co-workers [103, 105, 107, 116, 144, 145, 147] shed light on the transition from combustion noise to thermoacoustic instability using novel methods of nonlinear time series analysis.

1.10 Transition to Thermoacoustic Instability in Turbulent Reacting Flow Systems

1.10.1 Is Combustion Noise Deterministic or Stochastic?

Combustion noise originates from fluid dynamic and combustion processes such as flame roll-up, vortex coalescence and impingement, fluid dilatation, etc. [26]. These processes are physical processes which are governed by deterministic set of equations; thus, the term "combustion noise" may be misleading.

Often, combustion noise is analyzed as an acoustic problem, decoupling the hydrodynamics. Candel et al. [19] clearly describes the formulation and its drawbacks as: '*Studies of combustion noise generally focus on situations where the flow dynamics can be considered to be independent of the radiated sound. It is implicitly assumed that the flow dynamics is decoupled from the induced wave motion and the sound emission from unstable flames is generally not considered when dealing with combustion noise.*' They further note that such decoupling, although could ease computations, cannot be justified because practical systems are confined and boundaries do reflect sound towards the combustion zone.

In summary, it is critical that we take a fresh look at combustion noise and identify whether it is deterministic or not.

1.10.2 Studying the Transition to Thermoacoustic Instability in "Noisy" Systems

The current practice is to treat thermoacoustic instability fundamentally as an acoustic problem and the effects of turbulence are often decoupled or neglected [86, 90, 91, 112]. Traditionally, unsteady measurements acquired from combustors are treated as signals modulated by random perturbations. In models, turbulence is introduced as a random perturbation to the wave equation; the properties of the perturbations being determined by the measured power spectrum [17, 25, 86]. In such a mean-field description, the possible dynamical states are restricted to fixed points and limit cycle oscillations with the observed amplitude modulations in the measured data being described as the effects of background noise. The goal of all such analysis is to identify the linear instability boundaries of the system which form the margins of operability.

In turbulent combustors, the transition to self-sustained oscillations from regimes of stable operation can often be triggered due to unsteadiness in the flow and combustion [90, 167]. The distributions of the pressure measurements acquired from combustors well before conditions of thermoacoustic instability have a characteristic Gaussian distribution [90], suggestive of the dynamics dictated by random processes in these regimes. However, bursts of pressure oscillations have been reported close to the critical transition to thermoacoustic instability in liquid-propellant rocket engines [25]. This erratic behavior of the pressure fluctuations was then incorporated as a multiplicative noise term in the wave equation. For an unchoked fuel flow at the injector in a swirl combustor, Hong et al. [55] reported the presence of pressure oscillations that alternated between a '*noisy period of 200 Hz fluctuation and a silent period with a small pressure fluctuation*'. Arndt et al. [5] reported a transition in the flame dynamics between a state of stable combustion and self-excited oscillations in a premixed gas turbine model combustor using simultaneous OH* chemiluminescence, OH* PLIF, and stereoscopic measurements.

Noiray and Schuermans [112] explained that during stable operation, thermoacoustic systems are driven by strong forcing produced by the highly turbulent reactive flow. They used stochastic differential equations to describe such a situation and were able to extract the deterministic quantities such as the growth rate. They applied their methods to engine data. Noiray and Schuermans [113] studied the dynamics of standing and rotating azimuthal thermoacoustic modes in annular gas turbine combustion chambers. Using a detailed statistical analysis of the spatial Fourier amplitudes extracted from this data, they showed that the limit cycles intermittently switch between spinning and standing modes. They modeled this phenomenon using a system of three coupled stochastic differential equations and were able to capture the properties of the dynamic pressure field, which results from combined nonlinear thermoacoustic coupling and random perturbations by the non-coherent combustion noise. This methodology was improved upon in subsequent publications [15, 108].

1.10.3 Noise-Induced Transition, Stochastic Bifurcation and Fokker–Planck Equation

Experimental and numerical studies on prototypical thermoacoustic systems have shown that the transition to thermoacoustic instability can happen through a subcritical Hopf bifurcation [9, 47, 63, 97, 141]. A system exhibiting subcritical transition can be triggered to thermoacoustic instability even if the system is linearly stable, in the bistable zone. Zinn and Lieuwen [167] reported that perturbations of the order of background noise levels are sufficient to trigger a thermoacoustic system. The presence of noise can transition the thermoacoustic system from a non-oscillatory state to an oscillatory state when the system is in the bistable regime [62, 153, 154].

Clavin et al. [25] introduced the framework of stochastic differential equations to study turbulence induced noise effects in high frequency thermoacoustic instabilities in liquid rockets. They derived a Fokker–Planck equation for the evolution of the transition probability density function for the amplitude. Noiray and Schuermans [112], Noiray and Denosov [109] used stochastic differential equations to model thermoacoustic systems phenomenologically. They then performed an output-only model-based system identification to extract deterministic quantities such as the growth rate from the measurements of the acoustic pressure. They validated their methodology against numerical simulations [110]. Further, by measuring OH* chemiluminescence fluctuations from the flame simultaneously, they were able to determine the acoustic damping of the system, and thus able to determine the linear growth rate due to the flame [15]. They were successful in applying their methods to data from real engines. Using the same approach, they [113] studied the dynamics of thermoacoustic modes intermittently switching between spinning and standing modes, in annular gas turbine combustion chambers. More recently, Lee et al. [82] used a similar Fokker–Planck framework to perform output-only system identification on a model gas turbine combustor operating near a supercritical Hopf bifurcation. They showed that the noise-induced dynamics produced by turbulence in the pre-bifurcation regime can aid in predicting the Hopf point (growth rate) and the resultant limit cycle amplitude.

Gopalakrishnan and Sujith [48] performed a study on horizontal Rijke tube which underwent a subcritical Hopf bifurcation in order to understand the effect of noise fluctuations on the hysteresis characteristics of subcritical bifurcation. They reported a complete suppression of the hysteresis zone and the inability to ascertain a critical point of transition in the presence of high intensity noise. This is because the measured observable is a stochastic variable. Thus, a single realization is incapable of providing the complete information. We need to calculate the probability density function of the observable rather than its absolute value in the presence of noise. The probability density function of a stochastic variable can be obtained by solving the Fokker-Plank equation associated with the stochastic differential equations (SDE) [45, 126, 138]. The qualitative changes observed in the probability distribution of the observable is termed as phenomenological bifurcation (P-bifurcation). The bifurcation associated with the absolute value of the order parameter is known as dynamic bifurcation (D-bifurcation). In a stochastic system, both phenomenological and dynamical bifurcations are observed [2, 6, 12, 162, 163].

Thus, we need to adopt the concept of stochastic bifurcation to study the effect of noise in a thermoacoustic system. Using such an analysis, Gopalakrishnan et al. [49] identified stochastic P-bifurcation in a horizontal Rijke tube system in the presence of low intensity noise through changes in the stationary probability distribution. In contrast, they showed that P-bifurcation is not observed in the presence of high intensity noise.

1.10.4 Is 'Signal Plus Noise' Paradigm the Right Way to Go About?

Although a number of researchers have studied thermoacoustic instability in the presence of noise and noisy limit cycles, it is not clear if the effects of turbulence are the same as the effect of forcing with noise. Predicting the amplitude and frequency of such triggered oscillations, or even the stability margins of combustors with turbulent reacting flow remain yet a challenge for the researchers in the field due to the complex nature of the interactions in combustors amongst the flow, the heat release rate, and the chamber acoustics [167]. An understanding of the universal features of such transitions is limited and operators often rely on heuristic measures to prevent thermoacoustic instability in fielded combustors. This limitation may possibly arise from the inappropriate use of the 'signal plus noise' paradigm where the signal is thought of as consisting of a deterministic signal and noisy fluctuations. However, the irregular fluctuations seen in measurements in a thermoacoustic systems are a direct result of the inherent complexity of turbulent combustion dynamics. Thus, a separation of the measurements into a signal and noise is questionable [105].

1.11 Alternate Perspectives

Engineers and technicians in the field often can tell well in advance just by listening to the sounds and feeling the vibrations; i.e., well before the root-mean-square (rms) of the acoustic pressure fluctuations goes up, if an impending thermoacoustic instability is going to occur. They often speak about "the combustor is trying to become unstable", "the system is bursting", etc. In the scientific literature, Clavin et al. [25] mentioned that bursts of pressure oscillations have been reported close to the critical transition to thermoacoustic instability in liquid-propellant rocket engines. This makes one guess that there is a "bursting state" before the onset of "full-blown" thermoacoustic instability.

1.11.1 Combustion Noise Is Chaos

Nair et al. [106] suggested that contrary to traditional beliefs, combustion noise is not stochastic and is in fact chaos of moderately high dimension (of the order of 8 to 10). Further, they suggested that the onset of thermoacoustic instability is essentially a transition from chaos to order. They introduced a test for chaos, known as the 0–1 test for chaos, as a measure of the proximity of the combustor to an impending instability. Using surrogate analysis Tony et al. [148] proved rigorously that combustion noise is indeed high-dimensional deterministic chaos.

1.11.2 Intermittency Presages the Onset of Thermoacoustic Instability

Nair et al. [107] investigated the transition from combustion noise to thermoacoustic instability in two laboratory-scale turbulent combustors (namely, swirl-stabilized and bluff-body-stabilized backward-facing-step combustors). By systematically varying the Reynolds number of the reactant flow, they observed that the onset of thermoacoustic oscillations is always presaged by intermittent bursts of high amplitude periodic oscillations that appear in a near-random fashion amidst regions of low-amplitude aperiodic fluctuations. These excursions to periodic oscillations last longer in time as the operating conditions approach thermoacoustic instability, until the system transitions completely into a state of periodic oscillations. The traditional paradigm for thermoacoustic instability involved a shift from fixed point (stable operation) to a limit cycle (unstable operation). Nair et al. [107] proposed an alternate paradigm for the transition; i.e., chaos $\rightarrow$ intermittency $\rightarrow$ limit cycle.

Thermoacoustic instability results from the positive coupling between the heat release rate fluctuations and the acoustic field in the combustor. Pawar et al. [116] showed that during the occurrence of combustion noise, these signals are desynchronized chaos. During intermittency, they are intermittently synchronized. During thermoacoustic instability, these oscillations show phase synchronization (low amplitude thermoacoustic instability), and generalized synchronization (high amplitude thermoacoustic instability) [116].

The recurrence properties of intermittent bursts in a signal were quantified using recurrence plots [107] and the distribution of the aperiodic phases was examined. From the statistics of these aperiodic phases, robust early warning signals of an impending thermoacoustic instability could be obtained.

1.11.3 Multifractal Description of Combustion Noise and its Transition to Thermoacoustic Instability

In the traditional approaches for studying combustion noise and its transition to high amplitude thermoacoustic instability, the irregular fluctuations observed in the measured data are often averaged out or neglected, or treated as stochastic background. This is unfortunate as these irregular fluctuations contain valuable information for diagnosis or prognosis [105]. As shown by Tony et al. [148] and Nair et al. [106], combustion "noise" is "deterministic chaos" resulting from the nonlinear interaction between combustion, turbulent flow, and the acoustic field in the confinement. In other words, the measurements do not display features that one expects from a stochastic random process, and the use of the term "noise" is misleading [105, 148]. Most importantly, these fluctuations contain a gold mine of useful information that can provide early warning to the onset of an impending thermoacoustic instability. These low amplitude aperiodic pressure fluctuations that are observed during the occurrence of combustion noise display scale-invariance [103] and have a multifractal signature[3] that disappears at the onset of thermoacoustic instability [105]. These irregular fluctuations contain useful information of prognostic value, which can be quantified using measures such as Hurst exponents[4] that can act as early warning signals to the onset of an impending transition to thermoacoustic instability.

1.11.4 Complex Networks

Thermoacoustic systems are complex, as they operate over a wide range of scales—from large scale acoustic waves to turbulent transport to molecular mixing. Complexity, which arises due to the interaction between a large number of elements, is a characteristic of many physical, biological, and social systems. The complex network approach which studies the connectivity patterns between the interacting elements is a popular approach to study complex systems [11].

Using complex networks constructed using a visibility algorithm [80], Murugesan and Sujith [103] investigated the scale invariance of a time series obtained during the occurrence of combustion noise. The complex networks obtained during the occurrence of combustion noise can be represented as a scale-free network [10] which have a heavy-tailed distribution of connectivity, with no clear characteristic scales. Scale-free networks are ubiquitous in nature; examples include scientific collaboration networks, the World Wide Web, protein networks, and social networks to name a few. The scale-free behaviour of networks obtained from combustion noise is related to the scale invariance of combustion noise. Further, Murugesan and Sujith [103] showed that as the thermoacoustic system transitions from combustion noise to thermoacoustic instability, the scale-free nature of combustion noise is lost, leading to the occurrence of regular networks, thus characterizing the emergence of order.

The statistical theory of complex networks has been used to devise precursors to the onset of impending thermoacoustic instabilities [104]. Further, spatiotemporal analysis using complex networks have thrown light on the fundamental processes occurring during the thermoacoustic transition. From a practical perspective, the most sensitive points in the complex networks were used to implement smart passive controls [77, 78, 101].

[3]A fractal object displays self-similarity over an extended, but finite, scale range (Chap. 10). A multifractal system is a generalization of a fractal system in which a single exponent (the fractal dimension) is not enough to describe its dynamics; instead, a continuous spectrum of exponents (the so-called singularity spectrum) is needed.

[4]Hurst exponent is a measure that quantifies the fractal nature of a time series.

1.11.5 On the Importance of Being Nonlinear

The previous sections suggest that nonlinearities are of critical importance in the study of thermoacoustic instabilities.[5] Cook [27] says that "*nonlinear is not a hypothesis but the lack of a hypothesis*". This was more dramatically stated by Ulam, who says "*using a term like nonlinear science is like referring to the bulk of zoology as the study of non-elephant animals*" [46].

Nature is nonlinear, with nonlinearity being the rule rather than the exception. The behavior of nonlinear systems can be very different (even qualitatively) from the behavior of linear systems, and therefore one cannot use the solutions of linear equations to understand the behavior of real systems. Nonlinear systems display many interesting behaviors such as limit cycles, quasiperiodicity, intermittency, strange nonchaos, and chaos. This very interesting fact has deep consequences for all sciences and has altered our view on the principle of predictability and determinism [46].

But then, why do we always study linear systems? This is because we can solve all linear problems! Therefore, we do *what we can do*, rather than *what we need to do*! We often approximate nonlinear systems as linear, to the effect of small perturbations about points of equilibrium. Nonlinear problems are very rarely exactly solvable. Before the era of computers, nearly nothing could be said about the behavior of nonlinear systems. The scenario has changed dramatically with the advent of high-speed computers, high-speed data acquisition, and high-speed lasers, and high-speed digital imaging systems, each of which shed light on a vast world out there, which is primarily nonlinear.

1.11.6 Reductionist vs Complex Systems Approach

Traditionally physics and engineering science have progressed by adopting a reductionist perspective. We adopt a "divide and conquer" approach, where we split the full system into a number of components, and study the component systems in great detail. Such approach has indeed been followed in thermoacoustics as well [59, 60].

Combustion noise has a broadband spectrum. This highlights the absence of a single characteristic scale in thermoacoustic systems and reflects the complex nature of the dynamics involved. Complexity is indeed a characteristic of real physical systems. Everyday examples include economic systems, biological systems, language, airline traffic network, mobile phone network, social systems, spread of epidemics, atmosphere and geosphere of our planet earth, etc. [20–22]. These systems are complex because a great number of interacting elements are involved.

[5] The title of this section is inspired by the book title "An essay on the importance of being nonlinear" by Bruce J. West [155]"

As can be seen in Fig. 1.2, the behavior of thermoacoustic system arises from a number of factors such as molecular mixing, turbulent transport involving a wide range of scales, acoustic waves, and chemical reaction. This leads to a rich variety of dynamics ranging from periodic oscillations to chaos.

In recent years, a new approach to science called complex systems approach has emerged that investigates how the interaction between the constituent elements gives rise to the collective behavior of the system. In the complex systems approach, a complex system is viewed as a system with "*multiple interacting components whose behavior cannot be simply inferred from the behavior of the components*" [134, 135, 159]. Emergence is an important aspect of complex systems, whereby larger entities, patterns, and regularities arise through interactions among constituent entities that themselves could not exhibit such properties. For example, the phenomena of "life" as studied in biology is often perceived as an emergent property of interacting molecules as studied in chemistry [20, 158]. Reductionism does not preclude the existence of emergent phenomena. However, we may not be able to understand emergence in terms of the processes from which they are composed.

In the rest of the book, we take you through a fascinating journey where we will examine thermoacoustic systems using dynamical systems theory and complex systems theory. We start with the basics of dynamical systems theory in the next chapter, where we will cover several topics such as linear stability analysis, fixed points, bifurcations, signals and their classification, phase space reconstruction, Poincaré section, recurrence plots, and different measures to characterize the behaviour of the phase space trajectory.

References

1. Alemela, P.R.: Measurement and scaling of acoustic transfer matrices of premixed swirl flames. Ph.D. Thesis, Technical University of Munich, Germany (2009)
2. Altares, V., Nicolis, G.: Stochastically forced Hopf bifurcation: approximate Fokker-Planck equation in the limit of short correlation times. Phys. Rev. A **37**(9), 3630–3633 (1988)
3. Ananthkrishnan, N., Deo, S., Culick, F.E.C.: Reduced-order modeling and dynamics of nonlinear acoustic waves in a combustion chamber. Combust. Sci. Tech. **177**(2), 221–248 (2005)
4. Annaswamy, A.M., Fleifil, M., Hathout, J.P., Ghoniem, A.F.: Impact of linear coupling on the design of active controllers for the thermoacoustic instability. Combust. Sci. Tech. **128**(1–6), 131–180 (1997)
5. Arndt, C.M., Steinberg, A.M., Boxx, I.G., Meier, W., Aigner, M., Carter, C.D.: Flow-field and flame dynamics of a gas turbine model combustor during transition between thermo-acoustically stable and unstable states. In: Proceedings of the ASME Turbo Expo 2010: Power for Land, Sea, and Air, GT2010-22830, pp. 677–687 (2010)
6. Arnold, L.: Random dynamical systems. In: Dynamical Systems, pp. 1–43. Springer (1995)
7. Balachandran, R., Ayoola, B.O., Kaminski, C.F., Dowling, A.P., Mastorakos, E.: Experimental investigation of the nonlinear response of turbulent premixed flames to imposed inlet velocity oscillations. Combust. Flame **143**(1–2), 37–55 (2005)
8. Balasubramanian, K., Sujith, R.I.: Non-normality and nonlinearity in combustion–acoustic interaction in diffusion flames. J. Fluid Mech. **594**, 29–57 (2008)

9. Balasubramanian, K., Sujith, R.I.: Thermoacoustic instability in a Rijke tube: Non-normality and nonlinearity. Phys. Fluids **20**(4), 044103 (2008)
10. Barabási, A.L., Albert, R.: Emergence of scaling in random networks. Science **286**(5439), 509–512 (1999)
11. Barabási, A.L., et al.: Network Science. Cambridge University Press, Cambridge (2016)
12. Bashkirtseva, I., Ryazanova, T., Ryashko, L.: Stochastic bifurcations caused by multiplicative noise in systems with hard excitement of auto-oscillations. Phys. Rev. E **92**(4), 042908 (2015)
13. Blomshield, F.: Historical perspective of combustion instability in motors-case studies. In: 37th Joint Propulsion Conference and Exhibit, AIAA 2001-3875 (2001)
14. Blomshield, F.S., Crump, J.E., Mathes, H.B., Stalnaker, R.A., Beckstead, M.W.: Stability testing of full-scale tactical motors. J. Propuls. Power **13**(3), 349–355 (1997)
15. Boujo, E., Denisov, A., Schuermans, B., Noiray, N.: Quantifying acoustic damping using flame chemiluminescence. J. Fluid Mech. **808**, 245–257 (2016)
16. Burnley, V.S.: Nonlinear combustion instabilities and stochastic sources. Ph.D. Thesis, California Institute of Technology (1996)
17. Burnley, V.S., Culick, F.E.C.: Influence of random excitations on acoustic instabilities in combustion chambers. AIAA J. **38**(8), 1403–1410 (2000)
18. Candel, S.: Combustion dynamics and control: progress and challenges. Proc. Combust. Inst. **29**(1), 1–28 (2002)
19. Candel, S., Durox, D., Ducruix, S., Birbaud, A.L., Noiray, N., Schuller, T.: Flame dynamics and combustion noise: progress and challenges. Int. J. Aeroacoust. **8**(1), 1–56 (2009)
20. Capra, F.: The Web of Life: A New Scientific Understanding of Living Systems. Anchor, London (1996)
21. Capra, F.: The Hidden Connections: A Science for Sustainable Living. Anchor, London (2004)
22. Capra, F., Luisi, P.L.: The Systems View of Life: A Unifying Vision. Cambridge University Press, Cambridge (2014)
23. Chakravarthy, S.R., Shreenivasan, O.J., Boehm, B., Dreizler, A., Janicka, J.: Experimental characterization of onset of acoustic instability in a nonpremixed half-dump combustor. J. Acoust. Soc. Am. **122**(1), 120–127 (2007)
24. Chakravarthy, S.R., Sivakumar, R., Shreenivasan, O.J.: Vortex-acoustic lock-on in bluff-body and backward-facing step combustors. Sadhana **32**(1), 145–154 (2007)
25. Clavin, P., Kim, J.S., Williams, F.A.: Turbulence-induced noise effects on high-frequency combustion instabilities. Combust. Sci. Tech. **96**(1–3), 61–84 (1994)
26. Coats, C.M.: Coherent structures in combustion. Prog. Energy Combust. Sci. **22**(5), 427–509 (1996)
27. Cook, D.: Nonlinear is not a hypothesis (2012). https://www.johndcook.com/blog/2012/01/04/nonlinear-is-not-a-hypothesis/
28. Crocco, L., Mitchell, C.E.: Nonlinear periodic oscillations in rocket motors with distributed combustion. Combust. Sci. Tech. **1**(2), 146–169 (1969)
29. Culick, F.E.C.: Stability of longitudinal oscillations with pressure and velocity coupling in a solid propellant rocket. Combust. Sci. Tech. **2**(4), 179–201 (1970)
30. Culick, F.E.C.: Nonlinear behavior of acoustic waves in combustion chambers—II. Acta Astronaut. **3**(9–10), 735–757 (1976)
31. Culick, F.E.C.: Combustion instabilities in liquid-fuelled propulsion systems - an overview. In: AGARD Conference Proceedings, vol. 450, pp. 1–73 (1988)
32. Culick, F.E.C.: Some recent results for nonlinear acoustics in combustion chambers. AIAA J. **32**(1), 146–169 (1994)
33. Culick, F.E.C.: Unsteady motions in combustion chambers for propulsion systems. Tech. Rep., AGARDograph, NATO/RTO-AG-AVT-039 (2006)
34. Culick, F.E.C., Yang, V.: Overview of combustion instabilities in liquid-propellant rocket engines. In: Liquid Rocket Engine Combustion Instability, vol. 169, pp. 3–37. Progress in Astronautics and Aeronautics, AIAA, Washington (1995)

35. Dowling, A.P.: Nonlinear self-excited oscillations of a ducted flame. J. Fluid Mech. **346**, 271–290 (1997)
36. Dowling, A.P.: A kinematic model of a ducted flame. J. Fluid Mech. **394**, 51–72 (1999)
37. Dowling, A.P., Mahmoudi, Y.: Combustion noise. Proc. Combust. Inst. **35**(1), 65–100 (2015)
38. Dranovsky, M.L.: Combustion Instabilities in Liquid Rocket Engines: Testing and Development Practices in Russia. Progress in Astronautics and Aeronautics. AIAA, Washington (2007)
39. Ducruix, S., Durox, D., Candel, S.: Theoretical and experimental determinations of the transfer function of a laminar premixed flame. Proc. Combust. Inst. **28**(1), 765–773 (2000)
40. Eckstein, J.: On the mechanisms of combustion driven low-frequency oscillations in aero-engines. Ph.D. Thesis, Technical University of Munich, Germany (2004)
41. Eckstein, J., Freitag, E., Hirsch, C., Sattelmayer, T.: Experimental study on the role of entropy waves in low-frequency oscillations in a RQL combustor. J. Eng. Gas Turbines Power **128**(2), 264–270 (2006)
42. Farrell, B.F., Ioannou, P.J.: Generalized stability theory. Part I: autonomous operators. J. Atmos. Sci. **53**(14), 2025–2040 (1996)
43. Farrell, B.F., Ioannou, P.J.: Generalized stability theory. Part II: nonautonomous operators. J. Atmos. Sci. **53**(14), 2041–2053 (1996)
44. Fichera, A., Losenno, C., Pagano, A.: Experimental analysis of thermo-acoustic combustion instability. Appl. Energy **70**(2), 179–191 (2001)
45. Gardiner, C.W., et al.: Handbook of Stochastic Methods, vol. 3. Springer, Berlin (1985)
46. Gleick, J.: Chaos: Making a New Science. Penguin Books, New York (1987)
47. Gopalakrishnan, E.A., Sujith, R.I.: Influence of system parameters on the hysteresis characteristics of a horizontal Rijke tube. J. Spray Combust. Dyn. **6**(3), 293–316 (2014)
48. Gopalakrishnan, E.A., Sujith, R.I.: Effect of external noise on the hysteresis characteristics of a thermoacoustic system. J. Fluid Mech. **776**, 334–353 (2015)
49. Gopalakrishnan, E.A., Tony, J., Sreelekha, E., Sujith, R.I.: Stochastic bifurcations in a prototypical thermoacoustic system. Phys. Rev. E **94**(2), 022203 (2016)
50. Gotoda, H., Amano, M., Miyano, T., Ikawa, T., Maki, K., Tachibana, S.: Characterization of complexities in combustion instability in a lean premixed gas-turbine model combustor. Chaos **22**(4), 043128 (2012)
51. Gotoda, H., Nikimoto, H., Miyano, T., Tachibana, S.: Dynamic properties of combustion instability in a lean premixed gas-turbine combustor. Chaos **21**(1), 013124 (2011)
52. Guan, Y., Gupta, V., Li, L.K.B.: Intermittency route to self-excited chaotic thermoacoustic oscillations. J. Fluid Mech. **894**, R3 (2020)
53. Harrje, D.T., Reardon, F.H.: Liquid propellant rocket combustion instability. Scientific and Technical Information Office, NASA, SP-194 (1972)
54. Higgins, B.: On the sound produced by a current of hydrogen gas passing through a tube. J. Nat. Phil. Chem. Arts **1**(2), 129–131 (1802)
55. Hong, J.G., Oh, K.C., Lee, U.D., Shin, H.D.: Generation of low-frequency alternative flame behaviors in a lean premixed combustor. Energy Fuel **22**(5), 3016–3021 (2008)
56. Hosseini, S.M.R., Lawn, C.J.: Non-linearities in the thermo-acoustic response of swirled premixed flames. In: In 12th International Congress on Sound and Vibration (ICSV12), vol. 389 (2005)
57. Huang, C., Yoon, C., Gejji, R.M., Anderson, W., Sankaran, V.: Computational study of combustion dynamics in a single-element lean direct injection gas turbine combustor. In: 52nd Aerospace Sciences Meeting, AIAA 2014-0620 (2014)
58. Huang, Y., Yang, V.: Dynamics and stability of lean-premixed swirl-stabilized combustion. Prog. Energy Combust. Sci. **35**(4), 293–364 (2009)
59. Huber, A., Polifke, W.: Dynamics of practical premixed flames, part I: model structure and identification. J. Spray Combust. Dyn. **1**(2), 199–228 (2009)
60. Huber, A., Polifke, W.: Dynamics of practical premixed flames, part II: identification and interpretation of CFD data. J. Spray Combust. Dyn. **1**(2), 229–249 (2009)

61. Jahnke, C.C., Culick, F.E.C.: Application of dynamical systems theory to nonlinear combustion instabilities. J. Propuls. Power **10**(4), 508–517 (1994)
62. Jegadeesan, V., Sujith, R.I.: Experimental investigation of noise induced triggering in thermoacoustic systems. Proc. Combust. Inst. **34**(2), 3175–3183 (2013)
63. Juniper, M.P.: Triggering in the horizontal Rijke tube: non-normality, transient growth and bypass transition. J. Fluid Mech. **667**, 272–308 (2011)
64. Juniper, M.P.: Triggering in thermoacoustics. J. Spray Combust. Dyn. **4**(3), 217–237 (2012)
65. Juniper, M.P., Sujith, R.I.: Sensitivity and nonlinearity of thermoacoustic oscillations. Annu. Rev. Fluid Mech. **50**, 661–689 (2018)
66. Kabiraj, L.: Intermittency and route to chaos in thermoacoustic oscillations. Ph.D. Thesis, Indian Institute of Technology Madras, India (2012)
67. Kabiraj, L., Saurabh, A., Wahi, P., Sujith, R.I.: Route to chaos for combustion instability in ducted laminar premixed flames. Chaos **22**(2), 023129 (2012)
68. Kabiraj, L., Sujith, R.I.: Nonlinear self-excited thermoacoustic oscillations: intermittency and flame blowout. J. Fluid Mech. **713**, 376–397 (2012)
69. Kabiraj, L., Sujith, R.I., Wahi, P.: Bifurcations of self-excited ducted laminar premixed flames. J. Eng. Gas Turbines Power **134**(3), 031502 (2012)
70. Kabiraj, L., Sujith, R.I., Wahi, P.: Investigating the dynamics of combustion-driven oscillations leading to lean blowout. Fluid Dyn. Res. **44**(3), 031408 (2012)
71. Karimi, N., Brear, M.J., Jin, S.H., Monty, J.P.: Linear and non-linear forced response of a conical, ducted, laminar premixed flame. Combust. Flame **156**(11), 2201–2212 (2009)
72. Kashinath, K.: Nonlinear thermoacoustic oscillations of a ducted laminar premixed flame. Ph.D. Thesis, University of Cambridge (2013)
73. Kashinath, K., Waugh, I.C., Juniper, M.P.: Nonlinear self-excited thermoacoustic oscillations of a ducted premixed flame: bifurcations and routes to chaos. J. Fluid Mech. **761**, 399–430 (2014)
74. Kasuya, H., Gotoda, H., Yoshida, S., Tachibana, S.: Dynamic behavior of combustion instability in a cylindrical combustor with an off-center installed coaxial injector. Chaos **28**(3), 033111 (2018)
75. Keanini, R., Yu, K., Daily, J.: Evidence of a strange attractor in ramjet combustion. In: 27th Aerospace Sciences Meeting, AIAA 1989-624 (1989)
76. Keller, J.J.: Thermoacoustic oscillations in combustion chambers of gas turbines. AIAA J. **33**(12), 2280–2287 (1995)
77. Krishnan, A., Manikandan, R., Midhun, P.R., Reeja, K.V., Unni, V.R., Sujith, R.I., Marwan, N., Kurths, J.: Mitigation of oscillatory instability in turbulent reactive flows: a novel approach using complex networks. Europhys. Lett. **128**(1), 14003 (2019)
78. Krishnan, A., Sujith, R.I., Marwan, N., Kurths, J.: Suppression of thermoacoustic instability by targeting the hubs of the turbulent networks in a bluff body stabilized combustor. J. Fluid Mech. **916**, A20 (2021)
79. Kumara Gurubaran, R., Sujith, R.I.: Dynamics of spray–swirl–acoustics interactions. Int. J. Spray Combust. Dyn. **3**(1), 1–22 (2011)
80. Lacasa, L., Luque, B., Ballesteros, F., Luque, J., Nuno, J.C.: From time series to complex networks: the visibility graph. Proc. Natl. Acad. Sci. **105**(13), 4972–4975 (2008)
81. Leconte, J.: On the influence of musical sounds on the flame of a jet of coal-gas. Phil. Mag. **15**(99), 235–239 (1858)
82. Lee, M., Kim, K.T., Gupta, V., Li, L.K.B.: System identification and early warning detection of thermoacoustic oscillations in a turbulent combustor using its noise-induced dynamics. Proc. Combust. Inst. **38**(4), 6025–6033 (2021)
83. Lei, S., Turan, A.: Nonlinear/chaotic behaviour in thermo-acoustic instability. Combust. Theor. Model. **13**(3), 541–557 (2009)
84. Lieuwen, T.: Modeling premixed combustion-acoustic wave interactions: a review. J. Propuls. Power **19**(5), 765–781 (2003)
85. Lieuwen, T.: Nonlinear kinematic response of premixed flames to harmonic velocity disturbances. Proc. Combust. Inst. **30**(2), 1725–1732 (2005)

86. Lieuwen, T., Banaszuk, A.: Background noise effects on combustor stability. J. Propuls. Power **21**(1), 25–31 (2005)
87. Lieuwen, T., Torres, H., Johnson, C., Zinn, B.T.: A mechanism of combustion instability in lean premixed gas turbine combustors. J. Eng. Gas Turbines Power **123**(1), 182–189 (2001)
88. Lieuwen, T., Zinn, B.T.: The role of equivalence ratio oscillations in driving combustion instabilities in low NOx gas turbines. In: Symposium (International) on Combustion, vol. 27, pp. 1809–1816. Elsevier, Amsterdam (1998)
89. Lieuwen, T.C.: Phase drift characteristics of self-excited, combustion-driven oscillations. J. Sound Vib. **242**(5), 893–905 (2001)
90. Lieuwen, T.C.: Experimental investigation of limit-cycle oscillations in an unstable gas turbine combustor. J. Propuls. Power **18**(1), 61–67 (2002)
91. Lieuwen, T.C.: Statistical characteristics of pressure oscillations in a premixed combustor. J. Sound Vib. **260**(1), 3–17 (2003)
92. Lieuwen, T.C.: Unsteady Combustor Physics. Cambridge University Press, Cambridge (2012)
93. Lieuwen, T.C., Yang, V.: Combustion Instabilities in Gas Turbine Engines (Operational Experience, Fundamental Mechanisms and Modeling). Progress in Astronautics and Aeronautics, vol. 210. AIAA (2005)
94. Lorenz, E.N.: Deterministic nonperiodic flow. J. Atmos. Sci. **20**(2), 130–141 (1963)
95. Marble, F.E., Candel, S.M.: Acoustic disturbance from gas non-uniformities convected through a nozzle. J. Sound Vib. **55**(2), 225–243 (1977)
96. Mariappan, S., Sujith, R.I.: Thermoacoustic instability in solid rocket motor: non-normality and nonlinear instabilities. In: 48th AIAA Aerospace Sciences Meeting Including the New Horizons Forum and Aerospace Exposition, AIAA 2010-1517 (2010)
97. Mariappan, S., Sujith, R.I.: Modelling nonlinear thermoacoustic instability in an electrically heated Rijke tube. J. Fluid Mech. **680**, 511–533 (2011)
98. Mariappan, S., Sujith, R.I., Schmid, P.J.: Experimental investigation of non-normality of thermoacoustic interaction in an electrically heated Rijke tube. J. Spray Combust. Dyn. **7**(4), 315–352 (2015)
99. McManus, K.R., Poinsot, T., Candel, S.M.: A review of active control of combustion instabilities. Prog. Energy Combust. Sci. **19**(1), 1–29 (1993)
100. Mitchell, C.E., Crocco, L., Sirignano, W.A.: Nonlinear longitudinal instability in rocket motors with concentrated combustion. Combust. Sci. Tech. **1**(1), 35–64 (1969)
101. Murayama, S., Gotoda, H.: Attenuation behavior of thermoacoustic combustion instability analyzed by a complex-network-and synchronization-based approach. Phys. Rev. E **99**(5), 052222 (2019)
102. Murayama, S., Kinugawa, H., Tokuda, I.T., Gotoda, H.: Characterization and detection of thermoacoustic combustion oscillations based on statistical complexity and complex-network theory. Phys. Rev. E **97**(2), 022223 (2018)
103. Murugesan, M., Sujith, R.I.: Combustion noise is scale-free: transition from scale-free to order at the onset of thermoacoustic instability. J. Fluid Mech. **772**, 225–245 (2015)
104. Murugesan, M., Sujith, R.I.: Detecting the onset of an impending thermoacoustic instability using complex networks. J. Propuls. Power **32**(3), 707–712 (2016)
105. Nair, V., Sujith, R.I.: Multifractality in combustion noise: predicting an impending combustion instability. J. Fluid Mech. **747**, 635–655 (2014)
106. Nair, V., Thampi, G., Karuppusamy, S., Gopalan, S., Sujith, R.I.: Loss of chaos in combustion noise as a precursor of impending combustion instability. J. Spray Combust. Dyn. **5**(4), 273–290 (2013)
107. Nair, V., Thampi, G., Sujith, R.I.: Intermittency route to thermoacoustic instability in turbulent combustors. J. Fluid Mech. **756**, 470–487 (2014)
108. Noiray, N.: Linear growth rate estimation from dynamics and statistics of acoustic signal envelope in turbulent combustors. J. Eng. Gas Turbines Power **139**(4), 041503 (2017)
109. Noiray, N., Denisov, A.: Stochastic aspects of thermoacoustic instabilities in combustion chambers. In: Proceedings of the 24th International Congress of Theoretical and Applied Mechanics (ICTAM 2016). ETH-Zürich (2016)

110. Noiray, N., Denisov, A.: A method to identify thermoacoustic growth rates in combustion chambers from dynamic pressure time series. Proc. Combust. Inst. **36**(3), 3843–3850 (2017)
111. Noiray, N., Durox, D., Schuller, T., Candel, S.: A unified framework for nonlinear combustion instability analysis based on the flame describing function. J. Fluid Mech. **615**, 139–167 (2008)
112. Noiray, N., Schuermans, B.: Deterministic quantities characterizing noise driven Hopf bifurcations in gas turbine combustors. Int. J. Nonlinear Mech. **50**, 152–163 (2013)
113. Noiray, N., Schuermans, B.: On the dynamic nature of azimuthal thermoacoustic modes in annular gas turbine combustion chambers. Proc. R. Soc. A **469**(2151), 20120535 (2013)
114. Oefelein, J.C., Yang, V.: Comprehensive review of liquid-propellant combustion instabilities in F-1 engines. J. Propuls. Power **9**(5), 657–677 (1993)
115. Orchini, A., Illingworth, S.J., Juniper, M.P.: Frequency domain and time domain analysis of thermoacoustic oscillations with wave-based acoustics. J. Fluid Mech. **775**, 387–414 (2015)
116. Pawar, S.A., Seshadri, A., Unni, V.R., Sujith, R.I.: Thermoacoustic instability as mutual synchronization between the acoustic field of the confinement and turbulent reactive flow. J. Fluid Mech. **827**, 664–693 (2017)
117. Peracchio, A.A., Proscia, W.M.: Nonlinear heat-release/acoustic model for thermoacoustic instability in lean premixed combustors. J. Eng. Gas Turbines Power **121**, 415 (1999)
118. Pikovsky, A., Rosenblum, M., Kurths, J.: Synchronization: A Universal Concept in Nonlinear Sciences, vol. 12. Cambridge University Press, Cambridge (2003)
119. Pillai, A.L., Nagao, J., Awane, R., Kurose, R.: Influences of liquid fuel atomization and flow rate fluctuations on spray combustion instabilities in a backward-facing step combustor. Combust. Flame **220**, 337–356 (2020)
120. Polifke, W.: Combustion instabilities. In: Advances in Aeroacoustics and Applications, vol. 5. Von Karman Institute Rhode, St-Genèse (2004)
121. Preetham, P., Lieuwen, T.: Nonlinear flame-flow transfer function calculations: flow disturbance celerity effects part 2. In: 43rd AIAA Aerospace Sciences Meeting and Exhibit, AIAA 2005-543 (2005)
122. Preetham, S.H., Lieuwen, T.C.: Response of turbulent premixed flames to harmonic acoustic forcing. Proc. Combust. Inst. **31**(1), 1427–1434 (2007)
123. Putnam, A.A.: Combustion Driven Oscillations in Industry. Elsevier, Amsterdam (1971)
124. Rayleigh, J.W.S.: The explanation of certain acoustical phenomena. Nature **18**(455), 319–321 (1878)
125. Renard, P.H., Thevenin, D., Rolon, J.C., Candel, S.: Dynamics of flame/vortex interactions. Prog. Energy Combust. Sci. **26**(3), 225–282 (2000)
126. Risken, H.: Fokker-Planck equation. In: The Fokker-Planck Equation, vol. 18, pp. 63–95. Springer, Berlin (1996)
127. Schadow, K.C., Gutmark, E.: Combustion instability related to vortex shedding in dump combustors and their passive control. Prog. Energy Combust. Sci. **18**(2), 117–132 (1992)
128. Schmid, P.J.: Nonmodal stability theory. Annu. Rev. Fluid Mech. **39**, 129–162 (2007)
129. Schob, Jr., W.J.: An experimental investigation of heat transfer and pressure effects on longitudinal combustion instability in a rocket motor using premixed gaseous propellants. Ph.D. Thesis, Princeton University (1963)
130. Schuller, T., Ducruix, S., Durox, D., Candel, S.: Modeling tools for the prediction of premixed flame transfer functions. Proc. Combust. Inst. **29**(1), 107–113 (2002)
131. Schwarz, A., Janicka, J.: Combustion Noise, vol. 102. Springer, Berlin (2009)
132. Sirignano, W.A.: A theoretical study of nonlinear combustion instability: longitudinal mode. Ph.D. Thesis, Princeton University (1964)
133. Sirignano, W.A.: Driving mechanisms for combustion instability. Combust. Sci. Tech. **187**(1–2), 162–205 (2015)
134. Smith, K., Brighton, H., Kirby, S.: Complex systems in language evolution: the cultural emergence of compositional structure. Adv. Complex Syst. **6**(4), 537–558 (2003)
135. Steinmetz, R., Wehrle, K.: Peer-to-peer Systems and Applications, vol. 3485. Springer, Berlin (2005)

136. Sterling, J.D.: Nonlinear analysis and modelling of combustion instabilities in a laboratory combustor. Combust. Sci. Tech. **89**(1–4), 167–179 (1993)
137. Strahle, W.C.: Combustion noise. Prog. Energy Combust. Sci. **4**(3), 157–176 (1978)
138. Stratonovich, R.: Topics in the Theory of Random Noise Vol. I: General Theory of Random Processes Nonlinear Transformations of Signals and Noise. Gordon & Breach, London (1963)
139. Strogatz, S.H.: Nonlinear Dynamics and Chaos: With Applications to Physics, Biology, Chemistry, and Engineering. CRC Press, Boca Raton (1994)
140. Subramanian, P.: Dynamical systems approach to the investigation of thermoacoustic instabilities. Ph.D. Thesis, Indian Institute of Technology Madras, India (2011)
141. Subramanian, P., Mariappan, S., Sujith, R.I., Wahi, P.: Bifurcation analysis of thermoacoustic instability in a horizontal Rijke tube. J. Spray Combust. Dyn. **2**(4), 325–355 (2010)
142. Sujith, R.I.: An experimental investigation of interaction of sprays with acoustic fields. Exp. Fluids **38**(5), 576–587 (2005)
143. Sujith, R.I., Juniper, M.P., Schmid, P.J.: Non-normality and nonlinearity in thermoacoustic instabilities. J. Spray Combust. Dyn. **8**(2), 119–146 (2016)
144. Sujith, R.I., Unni, V.R.: Complex system approach to investigate and mitigate thermoacoustic instability in turbulent combustors. Phys. Fluids **32**(6), 061401 (2020)
145. Sujith, R.I., Unni, V.R.: Dynamical systems and complex systems theory to study unsteady combustion. Proc. Combust. Inst. **38**(3), 3445–3462 (2021)
146. Suresh, S.: Fatigue of Materials. Cambridge University Press, Cambridge (1998)
147. Tandon, S., Sujith, R.I.: Condensation in the phase space and network topology during transition from chaos to order in turbulent thermoacoustic systems. Chaos **31**(4), 043126 (2021)
148. Tony, J., Gopalakrishnan, E.A., Sreelekha, E., Sujith, R.I.: Detecting deterministic nature of pressure measurements from a turbulent combustor. Phys. Rev. E **92**(6), 062902 (2015)
149. Trefethen, L.N., Embree, M.: Spectra and Pseudospectra: The Behavior of Nonnormal Matrices and Operators. Princeton University Press, Princeton (2005)
150. Tyagi, M., Chakravarthy, S.R., Sujith, R.I.: Unsteady combustion response of a ducted non-premixed flame and acoustic coupling. Combust. Theor. Model. **11**(2), 205–226 (2007)
151. Venkataraman, K.K.: An investigation of the acoustic instability mechanism in lean premixed dump combustors. Ph.D. Thesis, Pennsylvania State University, University Park
152. Vishnu, R., Sujith, R.I., Aghalayam, P.: Role of flame dynamics on the bifurcation characteristics of a ducted V-flame. Combust. Sci. Tech. **187**(6), 894–905 (2015)
153. Waugh, I., Geuß, M., Juniper, M.: Triggering, bypass transition and the effect of noise on a linearly stable thermoacoustic system. Proc. Combust. Inst. **33**(2), 2945–2952 (2011)
154. Waugh, I.C., Juniper, M.P.: Triggering in a thermoacoustic system with stochastic noise. J. Spray Combust. Dyn. **3**(3), 225–241 (2011)
155. West, B.J.: An Essay on the Importance of Being Nonlinear, vol. 62. Springer, Berlin (1985)
156. Wicker, J.M., Greene, W.D., Kim, S.I., Yang, V.: Triggering of longitudinal combustion instabilities in rocket motors-nonlinear combustion response. J. Propuls. Power **12**(6), 1148–1158 (1996)
157. Wieczorek, K., Sensiau, C., Polifke, W., Nicoud, F.: Assessing non-normal effects in thermoacoustic systems with mean flow. Phys. Fluids **23**(10), 107103 (2011)
158. Wikipedia: Emergence. https://en.wikipedia.org/wiki/Emergence
159. Wikipedia: New england complex systems institute. https://en.wikipedia.org/wiki/New_England_Complex_Systems_Institute
160. Wood, D.J., Dorsch, R.G.: Effect of propellant-feed-system coupling and hydraulic parameters on analysis of chugging. Tech. Rep., Lewis Research Center, Cleveland, Ohio (1967) NASA -TN D-3896 (1967)
161. Woods, W.D.: How Apollo Flew to the Moon. Springer, Berlin (2011)
162. Xu, Y., Gu, R., Zhang, H., Xu, W., Duan, J.: Stochastic bifurcations in a bistable Duffing–Van der Pol oscillator with colored noise. Phys. Rev. E **83**(5), 056215 (2011)
163. Zakharova, A., Vadivasova, T., Anishchenko, V., Koseska, A., Kurths, J.: Stochastic bifurcations and coherence–like resonance in a self-sustained bistable noisy oscillator. Phys. Rev. E **81**(1), 011106 (2010)

164. Zhao, D.: Transient growth of flow disturbances in triggering a Rijke tube combustion instability. Combust. Flame **159**(6), 2126–2137 (2012)
165. Zinn, B.T.: A theoretical study of nonlinear transverse combustion instability in liquid propellant rocket motors. Ph.D. Thesis, Princeton University (1966)
166. Zinn, B.T.: A theoretical study of nonlinear combustion instability in liquid-propellant rocket engines. AIAA J. **6**(10), 1966–1972 (1968)
167. Zinn, B.T., Lieuwen, T.C.: Combustion instabilities: basic concepts. In: Combustion Instabilities in Gas Turbine Engines: Operational Experience, Fundamental Mechanisms, and Modeling, vol. 210, pp. 3–26. AIAA, Reston (2005)

Chapter 2
An Introduction to Dynamical Systems Theory

As we discussed in Chap. 1, the onset of thermoacoustic instability is a nonlinear phenomenon resulting from the coupled interaction between the acoustic field in a combustor and the heat release rate field of the flame. A thermoacoustic system can undergo different dynamical transitions (called bifurcations), due to a change in any system (control) parameter. We observe that the characteristics of such dynamical transitions are governed primarily by the type of the underlying flow field, i.e., laminar or turbulent flows, present in a combustor. As a result, we notice the occurrence of different dynamical states during the onset of thermoacoustic instability in laminar and turbulent thermoacoustic systems. Furthermore, tools based on linear time series analysis are not sufficient to reveal many features exhibited by nonlinear thermoacoustic systems. In order to overcome the limitations of linear approaches in analyzing the data from experiments or numerical simulations, we need different methodologies based on dynamical systems theory to accurately characterize the nonlinear behaviors of thermoacoustic systems. In this chapter, we will introduce the basics of dynamical systems theory and cover different topics such as stability analysis, bifurcations, attractors, nonlinear measures, phase space reconstruction, Poincaré section, and recurrence plot. In the subsequent chapters, we will explore the application of these methodologies to perform time series analysis of the data obtained from experiments and numerical simulations performed on both laminar and turbulent thermoacoustic systems.

2.1 Dynamical System

The properties of most natural and man-made systems vary with time. Some examples include the displacement of a swinging pendulum, the growth of a particular species in forest, the temperature of the earth's surface in a day, heartbeats, the local velocity of a turbulent flow, etc. A system whose behavior changes as

R. I. Sujith, S. A. Pawar, *Thermoacoustic Instability*, Springer Series in Synergetics,
https://doi.org/10.1007/978-3-030-81135-8_2

a function of time is referred to as a *dynamical system* [12, 20, 36, 92, 105], and the mathematical description used to define the evolution of such systems is called dynamical systems theory [6, 25, 44]. Using dynamical systems theory, we can analyze and also predict the behavior of a system when its control parameters or initial conditions are systematically varied. Thus, dynamical systems theory models the temporal information of the state of a system under specific predefined rules to describe its behavior in an abstract phase space. Here, the phase space is a multidimensional space obtained by plotting the independent variables of a system as its coordinates. A point in the phase space, indicating the instantaneous value of all coordinates, is called a state. The evolution of such a state in time specifies the global behavior of the system. Thus, the phase space provides information about the dynamical behavior of a system. Dynamical systems theory has been successfully applied in many fields of science and engineering [37, 39, 41, 100, 103, 114].

2.1.1 Conservative and Dissipative Dynamical Systems

Dynamical systems are basically classified into two parts: *conservative* and *dissipative* [48, 81]. For conservative systems (also referred to as Hamiltonian systems), due to the absence of any dissipative mechanism, the total energy (or Hamiltonian) of the system is always preserved. This means that a trajectory of a particular initial condition and energy stays with the same energy level for all times.

Conservative systems do not possess any *attractor*. Here, the attractor refers to a set of points in the phase space that asymptotically (i.e., as the time $t \to \infty$) attract or repel all neighboring trajectories encountered in their basin of attraction. The attractor can be stable or unstable [92]. For a stable attractor, a small perturbation to the trajectories exponentially decays and approaches the basin boundary of the attractor. In contrast, for an unstable attractor, a small perturbation to the trajectories exponentially diverge from the boundary of the unstable attractor. In the absence of any attractor for conservative systems, trajectories in the phase space do not get pulled or pushed towards any attracting or repelling region. Therefore, the volume of the phase space for conservative systems is preserved; however, the shape of the phase space may change in time. Examples of conservative (or Hamiltonian) systems are planetary systems (n-body problem), electromagnetic field, spring-mass system, pendulum, dynamical billiards, etc. [18, 54].

Dissipative systems, on the other hand, do not conserve energy. They dissipate energy due to the interaction with the surroundings [53, 66]. As a result, the volume of the phase space is not conserved for dissipative systems, while it contracts/expands depending on the choice of control parameters. Dissipative systems represent the most of the practical dynamical systems where energy dissipation is an inherent phenomenon. In such systems, the transient dynamics associated with different initial conditions die down in a short interval of time and their long term (asymptotic) behavior is constrained by boundaries of nearest attractors in the phase space. Once the phase space trajectory reaches a particular attractor, it stays on the attractor for all future evolution; thus, it reveals the shape of the attractor. The evolution of the phase space trajectory of dissipative systems, therefore, provides information on whether the system is moving towards or away from an equilibrium point, whether it is revolving around a regular orbit, or whether it is moving in a complex manner in time.

For deterministic dynamical systems with lower dimensions, as the future states can be uniquely defined from the present and the past information, their trajectory in the phase space is bounded by a definite shaped attractor (e.g., limit cycle or quasiperiodic orbit). However, for deterministic systems with dimensions greater than or equal to three, the trajectories can be complex, highly unstable and bounded by a fractal set, also called a strange attractor (e.g., chaos or strange nonchaos). A few real life examples of dissipative dynamical systems include climate, solar system, human body, ecosystem, finance market, turbulent flow system, thermoacoustic system, etc. In this chapter, we restrict our discussion only to dissipative dynamical systems. The description of conservative/Hamiltonian systems is beyond the scope of this book and for further information on this, interested readers may refer to [21, 52, 60, 62, 110].

2.1.2 Modeling Dynamical Systems as Discrete and Continuous Functions of Time

We can mathematically express the dynamical behavior of a system in terms of either a discrete or a continuous function of time [75]. A dynamical system with a discrete variation of time is referred to as a *map*, and a system which consists of a continuous variation of time is called a *flow* [36, 92]. In the case of maps, for a given initial condition x_0, we can obtain the future states $(x_1, x_2, \ldots.)$ of the system through an iterative process governed by a particular functional relation (f) as follows,

$$
\begin{aligned}
x_1 &= f(x_0) \\
x_2 &= f(x_1) = f(f(x_0)) = f^{(2)}(x_0) \\
&\cdot \\
&\cdot \\
&\cdot \\
x_n &= f(x_{n-1}) = f^{(n)}(x_0)
\end{aligned}
\tag{2.1}
$$

Examples of discrete dynamical systems are logistic map, Henon map, population growth model, etc. [39, 75, 92].

In the case of a continuous dynamical system (flow), we can capture the evolution of a system appropriately using a set of first-order coupled ordinary differential equations [50]. Let us consider an example of a dynamical system with three degrees of freedom, i.e., the system requires at least three time-dependent variables to describe its state uniquely. Let these variables be u, v, and w. Then, we can express the time derivative of these variables in terms of a set of first-order linearly coupled ordinary differential equations as,

$$\dot{u} = f(u, v, w), \quad \dot{v} = g(u, v, w), \quad \text{and} \quad \dot{w} = h(u, v, w) \tag{2.2}$$

where f, g, and h are functions that govern the dynamics of variables u, v, and w, respectively. Further, the dynamics of these variables are dependent on each other. The dot represents the derivative with respect to time, i.e., $\dot{u} = du/dt$, $\dot{v} = dv/dt$ and $\dot{w} = dw/dt$.

Note that the time variable t does not occur explicitly in (2.2). Such dynamical systems where the functions f, g, and h are not explicitly dependent on time are referred to as '*autonomous systems*' [92]. If functions f, g, and h explicitly involve a time-dependent variable, i.e., $\dot{u} = f(u, v, w, t)$, then such dynamical systems are called '*non-autonomous systems*'. Therefore, for an autonomous system, the behavior of the phase space trajectory appears as a frozen state which, in turn, eases the understanding of a geometric picture of the system behavior. In the next section, to facilitate ease of understanding, we deal with one and two-dimensional autonomous systems. Examples of continuous dynamical systems are van der Pol oscillator, Stuart-Landau oscillator, Lorenz model, and spring-mass system [36, 39, 75, 92].

2.2 Linear Approximation of One-Dimensional Systems

In this section, we explore the local stability analysis of fixed points of a linear system using the method of linear approximation. In local stability analysis, we mostly deal with what happens in the neighborhood of a fixed point and not in the entire (global) phase space [36, 92]. According to the Hartman-Grobman theorem [35, 38], we can apply linear stability analysis to determine the stability of nonlinear dynamical systems as well [67]. This theorem suggests that near the fixed point, a nonlinear system behaves like a linear system, where all the nonlinear terms in the governing equations become negligibly small and, hence, can be ignored. With this linear approximation, we can study the behavior of trajectories around the fixed points in a phase space and also predict the global behavior of a nonlinear system to some extent without explicitly solving its nonlinear equations.

Let us consider $\dot{x} = f(x)$ to be the linear differential equation of a one-dimensional continuous dynamical system. To perform a linear approximation of this equation, we first find Taylor's series expansion of function $f(x)$. Letting ϵ be a small perturbation of x about the fixed point x^*, i.e., $x = x^* + \epsilon$, we obtain Taylor's series expansion of $f(x)$ as,

$$f(x^* + \epsilon) = f(x^*) + \epsilon \frac{df(x^*)}{dx} + \frac{\epsilon^2}{2!}\frac{d^2 f(x^*)}{dx^2} + \frac{\epsilon^3}{3!}\frac{d^3 f(x^*)}{dx^3} + \dots \tag{2.3}$$

where $\frac{df(x)}{dx}|_{x^*} = \frac{df(x^*)}{dx}$.

As x^* is a fixed point, we have $f(x^*) = 0$. On neglecting the second and higher-order terms, we obtain the linearized equation of a system about the fixed point as,

$$f(x^* + \epsilon) = \epsilon \frac{df(x^*)}{dx} \tag{2.4}$$

Since $x = x^* + \epsilon$, we can obtain the time derivative of x as $\dot{x} = \dot{\epsilon}$. Therefore, we can write $\dot{x} = f(x) = \dot{\epsilon}$ and modify the left hand side of (2.4) as,

$$f(x^* + \epsilon) = f(x) = \dot{\epsilon} \tag{2.5}$$

Substituting (2.5) in the left hand side of (2.4), we get,

$$\dot{\epsilon} = \epsilon \frac{df(x^*)}{dx} \tag{2.6}$$

If we assume $\lambda = \frac{df(x^*)}{dx}$, then we can modify (2.6) as follows,

$$\dot{\epsilon} = \frac{d\epsilon}{dt} = \epsilon\lambda \tag{2.7}$$

We can obtain the solution of equation (2.7) by using the method of variable separation,

$$\frac{1}{\epsilon} d\epsilon = \lambda dt \tag{2.8}$$

On integration,

$$\log_e \epsilon = \lambda t + C \tag{2.9}$$

where C is an integration constant.

By taking antilog on both sides of above equation, we obtain,

$$\epsilon = \epsilon_0 e^{\lambda t} \tag{2.10}$$

where $\epsilon_0 = e^C =$ constant.

Equation (2.10) suggests that the evolution of a perturbation trajectory from the fixed point has an exponential dependence on time. Here, the exponent λ can also be called eigenvalue of the system. From the sign of λ, we can determine the stability of the fixed point, while from the magnitude of λ, we can determine the rate of exponential decay or growth of the perturbation towards or away from the fixed point, respectively. Furthermore, we can obtain the characteristic timescales associated with the perturbation trajectory as $1/\lambda$. If λ is complex, then we refer to the real part of λ as the Lyapunov exponent [5].

For $\lambda > 0$, the fixed point is unstable; hence, any perturbation, $\epsilon(t)$, away from the fixed point grows in time. In contrast, for $\lambda < 0$, the fixed point is stable. As a result, any small perturbation decays in time and asymptotically settles down at the fixed point. In the case where $\lambda = 0$, the linear approximation is insufficient to detect the stability of the perturbation and, hence, we need to consider higher-order terms in the stability analysis of the system [92].

2.2.1 Two-Dimensional Linear Systems

In a one-dimensional dynamical system, the movement of the phase space trajectory is bounded about a line or a fixed point. However, in a higher-dimensional system, as the degrees of freedom for the movement of the phase trajectory increases, we observe more complex dynamics. In this section, we will briefly discuss the dynamical properties of a two-dimensional linear system and characterize the distinct properties of its fixed points [29, 92, 93].

Let us consider a set of first-order linear differential equations of a two-dimensional system,

$$\dot{x} = ax + by \tag{2.11a}$$

$$\dot{y} = cx + dy \tag{2.11b}$$

Here, x and y are the system variables and, a, b, c and d are constants. We can represent these equations in a compact matrix form as follows,

$$\begin{bmatrix} \dot{x} \\ \dot{y} \end{bmatrix} = \begin{bmatrix} a & b \\ c & d \end{bmatrix} \begin{bmatrix} x \\ y \end{bmatrix} \tag{2.12}$$

Consider,

$$\dot{X} = \begin{bmatrix} \dot{x} \\ \dot{y} \end{bmatrix}, \quad A = \begin{bmatrix} a & b \\ c & d \end{bmatrix}, \quad X = \begin{bmatrix} x \\ y \end{bmatrix}$$

Then, we can write (2.12) as, $\dot{X} = AX$, where A is a coefficient matrix, also called the Jacobian matrix [92].

For the two-dimensional system shown in (2.11), we always observe the origin $(x^*, y^*) = (0, 0)$ as a fixed point for any value of A. As a result, trajectories in the two-dimensional (x, y) phase plane always approach or diverge from this fixed point at an exponential rate when $t \to \pm\infty$.

Next, we can detect the stability of this fixed point by finding the eigenvalues (λ) of the system. Eigenvalues are obtained from the solution of a characteristic equation as follows,

$$\mathrm{Det}(A - \lambda I) = 0 \tag{2.13}$$

where I is the identity matrix.

On substituting the values of A and I on the left side of (2.13), we get,

$$A - \lambda I = \begin{bmatrix} a & b \\ c & d \end{bmatrix} - \lambda \begin{bmatrix} 1 & 0 \\ 0 & 1 \end{bmatrix} = \begin{bmatrix} a - \lambda & b \\ c & d - \lambda \end{bmatrix} \tag{2.14}$$

We can calculate the determinant of the above equation as,

$$\begin{aligned} \mathrm{Det}(A - \lambda I) &= \begin{vmatrix} a - \lambda & b \\ c & d - \lambda \end{vmatrix} \\ &= (a - \lambda)(d - \lambda) - bc \\ &= \lambda^2 - (a + d)\lambda + ad - bc \end{aligned} \tag{2.15}$$

Now, we substitute (2.15) into (2.13) and get,

$$\lambda^2 - (a + d)\lambda + ad - bc = 0 \quad \text{Or,} \quad \lambda^2 - \tau\lambda + \Delta = 0 \tag{2.16}$$

where $\tau = \mathrm{trace}(A) = a + d$, and $\Delta = \mathrm{Det}(A) = ad - bc$. We refer to (2.16) as the characteristic equation of the system. By solving this equation, we obtain its roots $\lambda_{1,2}$ as,

$$\lambda_1 = \frac{\tau + \sqrt{\tau^2 - 4\Delta}}{2}, \quad \lambda_2 = \frac{\tau - \sqrt{\tau^2 - 4\Delta}}{2} \tag{2.17}$$

Thus, we note that the roots of the characteristic equation are dependent only on the values of τ and Δ.

Depending on the value (real or complex) of $\lambda_{1,2}$, a linearized two-dimensional system can possess different types of fixed points (Fig. 2.1). The sign (positive or negative) of the real part of $\lambda_{1,2}$ aids us in deciding the stability of fixed points [98]. When the real part of $\lambda_{1,2}$ is negative, the fixed point is stable, and when the real part of $\lambda_{1,2}$ is positive, the fixed point is unstable. If the real part of all eigenvalues ($\lambda_{1,2}$) is non-zero, we refer to the corresponding fixed point as a *hyperbolic* fixed point. In contrast, if the real part of at least one eigenvalue of $\lambda_{1,2}$ is zero, we refer to such a fixed point as a *non-hyperbolic* fixed point.

As we mentioned earlier, according to the Hartman-Grobman theorem, every nonlinear dynamical system behaves like a linear system in the vicinity of hyperbolic fixed points [67]. However, such a description is not valid for non-hyperbolic

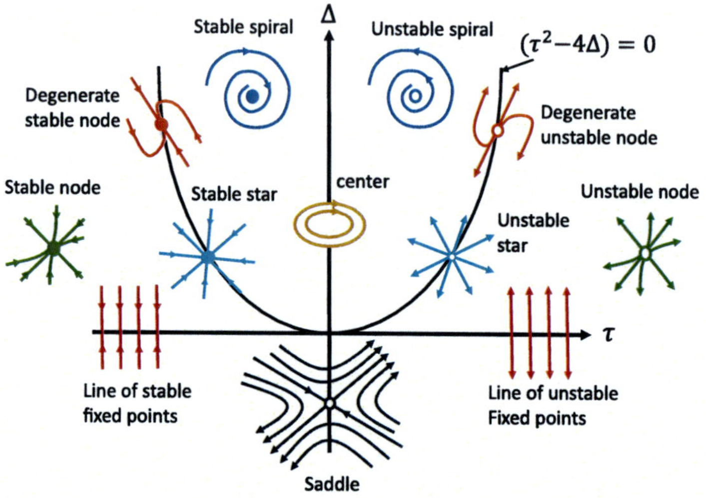

Fig. 2.1 A summary of the classification of all fixed points and their stability observed in a two-dimensional linear dynamical system in the $\tau - \Delta$ plane. The classification depends primarily on whether the eigenvalues are real or complex, while their stability depends on the sign of the real part of the eigenvalues. Arrows in the plot indicate the phase portraits of a particular fixed point. For a stable fixed point, arrows point inward, while for an unstable fixed point, arrows point outward in the two-dimensional phase space. Stable fixed points are indicated by filled circles and unstable fixed points are denoted by open circles

fixed points [41]. Furthermore, hyperbolic fixed points are structurally stable against small perturbations, while non-hyperbolic fixed points are structurally unstable for any perturbations. This means that, contrary to the case of hyperbolic fixed points, small perturbations can lead to changes in the local phase portraits of non-hyperbolic fixed points of a system.

Next, we classify different types of fixed points observed in a linearized two-dimensional system [29, 92, 93], as shown in Fig. 2.1.

1. **Saddle point (λ_1, λ_2 are real and $\lambda_1 < 0 < \lambda_2$):**
   When $\Delta < 0$ and $-\infty < \tau < \infty$, the eigenvalues are real but carry different signs. Here, the trajectory corresponding to $\lambda_1 < 0$ has a decaying solution, while the trajectory corresponding to $\lambda_2 > 0$ has a growing solution. Such a fixed point is referred to as "saddle point". A saddle point consists of two manifolds: one is stable and the other is unstable. The saddle point is always considered unstable, in spite of having both stable and unstable solutions [92].
2. **Node (λ_1, λ_2 are real and have the same sign):**
   When $\Delta > 0$ and $\tau^2 - 4\Delta > 0$, the eigenvalues are real, distinct, and have the same sign. In this case when $\tau < 0$, since both eigenvalues are negative ($\lambda_1 < \lambda_2 < 0$), we refer to the corresponding fixed point as a "stable node". In contrast, for $\tau > 0$, since the values of both $\lambda_{1,2}$ are positive ($\lambda_1 > \lambda_2 > 0$), we call the corresponding fixed point as an "unstable node". As the values of

λ_1 and λ_2 are different, the corresponding rates at which trajectories approach the fixed point are also different. A smaller magnitude of λ corresponds to a slow rate along the corresponding eigendirection, while a larger magnitude of λ corresponds to a fast rate along the corresponding eigendirection.

3. **Spiral and center ($\lambda_1 = \alpha + i\beta$ and $\lambda_2 = \alpha - i\beta$):**
   For $\Delta > 0$ and $\tau^2 - 4\Delta < 0$, the eigenvalues are complex. The sign of α decides the stability of the phase space trajectory. If α is positive, we refer to the fixed point as an "unstable spiral". In contrast, if α is negative, we refer to the fixed point as a "stable spiral". Furthermore, α indicates the decay rate and β indicates the rotation rate of the trajectory around the fixed point. The example of a stable spiral is a damped harmonic oscillator, wherein the trajectory approaches the fixed point at an exponential rate after making several revolutions around it. In every revolution, the trajectory loses a small portion of its energy, resulting in a decay of the oscillation. An opposite argument can be made for the unstable spiral.
   When $\Delta > 0$ and $\tau = 0$, the eigenvalues are purely imaginary, i.e., $\lambda_{1,2} = \pm i\beta$. In this case, the fixed point is neutrally stable. As a result, the trajectory does not approach or move away from the fixed point, but revolves at a constant rate in closed orbits around the fixed point. We call such a fixed point as a "center".
4. **Star, degenerate node, and non-isolated fixed points ($\lambda = \lambda_1 = \lambda_2$):**
   For $\Delta > 0$ and $\tau^2 - 4\Delta = 0$, we observe a single eigenvalue (i.e., $\lambda = \lambda_1 = \lambda_2$). In this case, when $\lambda < 0$, we call the fixed point as a "stable star" node. In contrast, for $\lambda > 0$, we refer to the fixed point as an "unstable star node". For star nodes, we do not have a unique eigenvector and, hence, every direction is considered as an eigendirection. For such fixed points, the trajectories always approach to (or repel from) the fixed point in a straight line.
   In some cases, although the eigenvalues are the same, the characteristic equation possesses only one eigenvector. In such a case, we refer to the fixed point as "a degenerate node". For this node, the trajectory is parallel to the eigendirection at $t \to -\infty$; however, instead of approaching the fixed point directly, it overshoots, then turns and approaches the fixed point tangentially in the reverse direction.
   In the case where one of the eigenvalues is zero (i.e., $\Delta = 0$, $\tau \neq 0$), the system possesses a set of "non-isolated fixed points". Depending on the sign of τ, the system consists of a line of unstable non-isolated fixed points ($\tau > 0$ or $\lambda_2 > \lambda_1 = 0$) or a line of stable non-isolated fixed points ($\tau < 0$ or $\lambda_1 < \lambda_2 = 0$). If both the eigenvalues are zero $\lambda_1 = \lambda_2 = 0$ (i.e., for $\Delta = 0$, $\tau = 0$), the system consists of a plane of non-isolated fixed points.
   Note that, for non-isolated fixed points, several fixed points are observed arbitrarily close to each other. In contrast, for isolated fixed points (e.g., saddle point, node, etc.), a region around them do not possess any other fixed point. Further, for a conservative system, due to the absence of attracting equilibrium points in the phase space, only two fixed points are usually observed, which are saddle point and center. In contrast, a dissipative system can exhibit all the aforementioned fixed points except centers [48].

Having gained a basic understanding of the stability analysis and the detection of fixed points for one and two-dimensional dissipative dynamical systems, we will next study different types of bifurcations observed in such systems. In addition, we will also discuss the properties of different signals and attractors that emerge due to bifurcation in such systems when we systematically change a control parameter.

2.3 Bifurcations and Their Classification

The dynamical properties of a dissipative system depend primarily on the control parameter that is varied. Examples of control parameters include the temperature in phase transition of water, the concentration of a reactant in chemical reactions, or the equivalence ratio in thermoacoustic systems. The increase of control parameter above a critical value results in a sudden change in the qualitative behavior of the system. We refer to this qualitative change in the system dynamics as a '*bifurcation*', and the value of the control parameter leading to this change in the system as a '*bifurcation point*' [80, 92]. During bifurcation, the stability of attractors (e.g., fixed point, limit cycle) undergoes a dramatic change due to small perturbations (i.e., the system is structurally unstable), which leads to the formation of new attractors and the destruction of old attractors. Thus, bifurcation causes a topological change in the phase space of the system.

Let us consider a simple example of bifurcation characteristics of a beam subjected to an external force (Fig. 2.2). On increasing magnitude of the external force on the beam, we notice a corresponding change in the structural properties (especially, deflection) of the beam [92, 98]. Here, we consider the magnitude of the force as a control parameter and the deflection induced in the beam as a system variable. If the magnitude of the force applied on the top of the beam is small, the beam does not bend. However, when the magnitude of the force crosses a critical value, we notice the occurrence of a symmetry-breaking phenomenon in the beam. Here, symmetry-breaking indicates the fact that the beam can deform either to the left or the right side with an equal probability without having any specific preference. During this symmetry-breaking, the beam undergoes a permanent deformation and does not restore to its original position even after the force is removed. We call this state of the beam as buckled state. Further, we refer to a critical value of the magnitude of the force at which the beam starts showing buckling as a bifurcation point.

A simple method to detect the occurrence of bifurcations (or bifurcation points) in a system is to track the change in the sign of the real part of its eigenvalues [4, 47, 88]. If the eigenvalues of a system depend on the control parameter (r), the variation of the parameter affects the stability of the system (Fig. 2.3). At the bifurcation point, the real part of the eigenvalues becomes zero.

For a one-dimensional system, all eigenvalues are always real [89]. In such systems, during the occurrence of the bifurcation of steady (stationary) states, the eigenvalues exhibit a zero-crossing (i.e., the transition from negative to positive), see

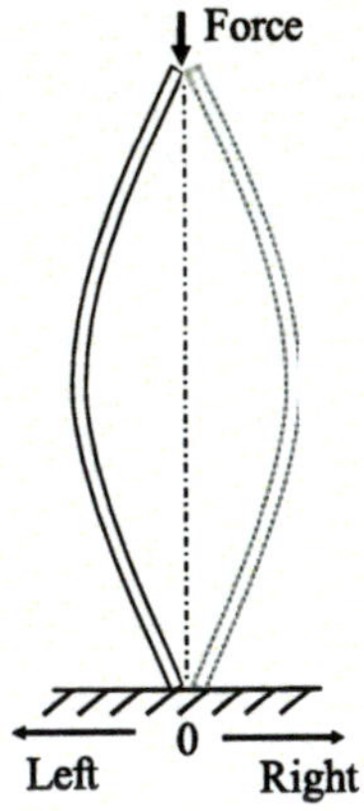

Fig. 2.2 Schematic showing bifurcation of a cantilever beam when an the magnitude of an axial force exerted on its tip is varied as a control parameter. The increase of magnitude of the force above a critical value leads to a bifurcation, observed in the form of buckling of the beam either to its left or right side

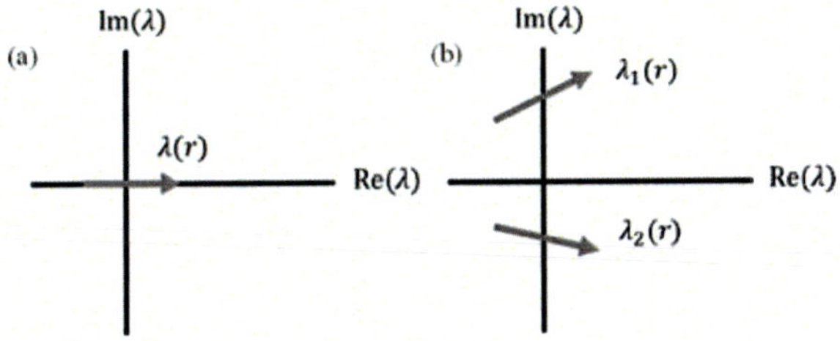

Fig. 2.3 Bifurcation characteristics of (**a**) one-dimensional and (**b**) two-dimensional systems. In a one-dimensional system, bifurcation occurs when Re(λ) crosses zero. In a two-dimensional system, bifurcation occurs when complex conjugate pairs of $\lambda = \lambda_1 + i\lambda_2$ cross the imaginary axis. Here, λ is the eigenvalue, Re and Im indicate the real and imaginary parts of λ, respectively, and r is the control parameter

Fig. 2.3a. Therefore, the bifurcation in one-dimensional systems is also referred to as *zero-eigenvalue bifurcation* or *stationary bifurcation* [88]. Examples of this type of bifurcations are saddle-node, transcritical, and pitchfork bifurcation (as explained in the following subsection). In contrast, bifurcations observed in higher-dimensional systems consist of more than one eigenvalue, which could be real or complex. For example, a two-dimensional system exhibiting periodic oscillations has a complex conjugate pair of eigenvalues (i.e., $\lambda = \lambda_1 + i\lambda_2$). In this system, the crossing of the imaginary axis by a complex conjugate pair of eigenvalues is considered as an indicator for the occurrence of Hopf bifurcation (Fig. 2.3b) as discussed in Sect. 2.3.4.

In the following subsections, we describe basic types of bifurcations, such as saddle-node, transcritical, pitchfork, and Hopf bifurcation, commonly observed in physical systems [48, 50, 92]. We can observe the type of bifurcations witnessed in one-dimensional systems also in two or higher dimensional systems. Note that Hopf bifurcation cannot be observed in one-dimensional systems.

Bifurcations are basically classified into two types: *local bifurcations* and *global bifurcations* [92]. Local bifurcations usually occur in the neighborhood of fixed points, periodic orbits, or any other invariant sets. We can investigate the occurrence of these bifurcations from the linear stability analysis of the fixed points when a control parameter crosses a critical value. On the other hand, for global bifurcations, large invariant sets (e.g., periodic orbits) of the system 'collide' with each other or with any fixed point. As a result, during the occurrence of such bifurcations, the changes in the behavior of phase trajectories are not restricted to the vicinity of the fixed points but can affect a larger region (global) of the phase space. Therefore, we cannot detect the occurrence of these bifurcations from linear stability analysis. Examples of local bifurcation are saddle-node, pitchfork, transcritical, Hopf, etc., while that of global bifurcations are homoclinic bifurcation, heteroclinic bifurcation, etc. [48, 92]. In the following section, we restrict our discussion to local bifurcations alone, and for more details on global bifurcations, we request the reader to refer to [67, 86, 106].

2.3.1 *Saddle-Node Bifurcation*

In saddle-node bifurcation, due to a change in the control parameter, the fixed points collide at the onset of bifurcation and vanish after the bifurcation. This type of bifurcation is also referred to as fold bifurcation or tangent bifurcation. The analysis of the bifurcation in the reverse direction would picturize a sudden emergence of the fixed point after the bifurcation out of nowhere. Hence, this type of bifurcation is also referred to as *blue-sky bifurcation* [92]. Thus, the saddle-node bifurcation provides a way to create or destroy the fixed points in the system. We can use a normal form equation to study saddle-node bifurcation as,

$$f(x) = \dot{x} = \mu \pm x^2 \tag{2.18}$$

where $x = x(t)$ is the single dynamic variable and μ is the control (or bifurcation) parameter.

To elaborate on the local stability analysis of the saddle-node bifurcation, let us consider a normal form of $\dot{x} = \mu + x^2$. We can obtain the fixed points of this dynamical equation by equating $\dot{x} = 0$ as $x^* = \pm\sqrt{-\mu}$. This suggests that there are two fixed points for $\mu < 0$ (Fig. 2.4a), and no fixed point for $\mu > 0$ (Fig. 2.4c). At the bifurcation point (i.e., $\mu = 0$, $f'(x^*) = 0$), both the fixed points coalesce to form a half stable and a half unstable fixed point (Fig. 2.4b). To find the stability of these fixed points, we need to obtain the derivatives of $f(x)$ with respect to x so that, we get $f'(x) = 2x$. Hence, for $\mu < 0$, we have $x^* = -\sqrt{-\mu}$ is the stable fixed point (i.e., $f'(x^*) < 0$) and $x^* = +\sqrt{-\mu}$ is the unstable fixed point (i.e., $f'(x^*) > 0$), as shown in Fig. 2.4d.

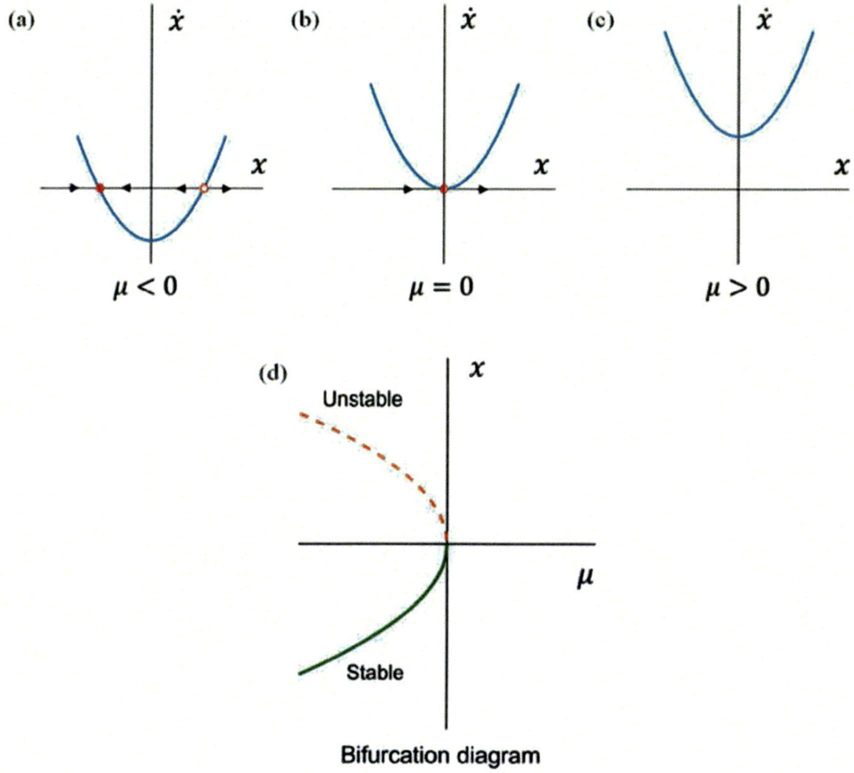

Fig. 2.4 **(a)**–**(c)** Flow diagrams and **(d)** a bifurcation diagram for a dynamic variable x undergoing a saddle-node bifurcation when the control parameter μ is smoothly varied across $\mu = 0$. During this bifurcation, two fixed points: one stable (indicated by a filled circle) and one unstable (indicated by an open circle) observed for $\mu < 0$ approach each other, merge at $\mu = 0$ (indicated by a half filled circle), and vanish for $\mu > 0$. Here, $\mu = 0$ corresponds to a bifurcation point. Arrows indicate the direction of the flow trajectories. In **(d)**, a dotted orange line indicates an unstable branch and a solid green line indicates a stable branch

2.3.2 *Transcritical Bifurcation*

During transcritical bifurcation, the fixed points exchange their stability after the bifurcation. In other words, as we change the bifurcation parameter, two fixed points (stable and unstable) approach each other, cross at the bifurcation point, and change their stability after the bifurcation. Thus, the fixed points always exist in the parameter space, unlike the saddle-node bifurcation where the fixed points disappear after the bifurcation. We represent transcritical bifurcation using the following normal form,

$$f(x) = \dot{x} = \mu x - x^2 \tag{2.19}$$

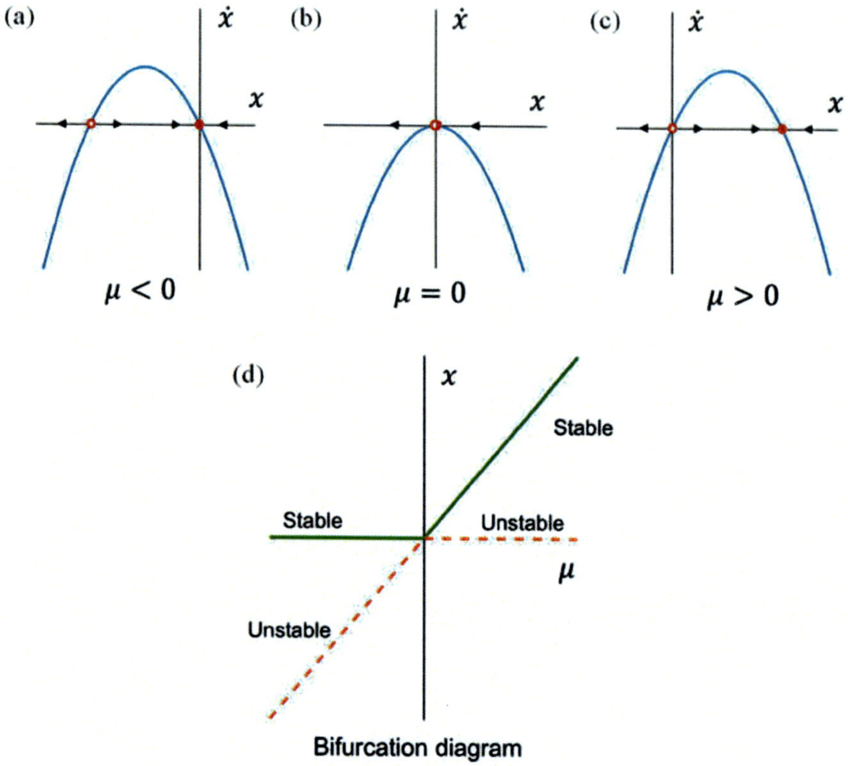

Fig. 2.5 (**a**)–(**c**) Flow diagrams and (**d**) a bifurcation diagram for a variable x undergoing a transcritical bifurcation at different values of the control parameter μ. During this bifurcation, the fixed points are preserved; however, they exchange their stability after the bifurcation point ($\mu = 0$)

The fixed points for this equation are $x^* = 0$ and μ. We can obtain the stability of these fixed points from the sign of $f'(x) = \mu - 2x$. For $\mu < 0$ (Fig. 2.5a), $x^* = 0$ behaves as a stable fixed point (i.e., $f'(x^*) < 0$) and $x^* = \mu$ behaves as an unstable fixed point (i.e., $f'(x^*) > 0$). As μ is increased, both the fixed points approach closer to each other and merge when $\mu = 0$ (Fig. 2.5b). For $\mu > 0$ (Fig. 2.5c), $x^* = 0$ becomes an unstable fixed point (i.e., $f'(x^*) > 0$) and $x^* = \mu$ becomes a stable fixed point (i.e., $f'(x^*) < 0$). Thus, the exchange of stability of these fixed points happens across $\mu = 0$ (Fig. 2.5d).

2.3.3 *Pitchfork Bifurcation*

Pitchfork bifurcation is common for systems that exhibit spatial symmetry. For example, in the case of a vertical beam (as discussed in Fig. 2.2), for the application of a small weight on the top of the beam, the beam is stable and vertically straight. However, when the weight is increased beyond a critical value, the beam could buckle to the left or right with an equal probability. Thus, a pitchfork bifurcation is associated with a symmetric generation of fixed points so that the system can attain any of these fixed points after the bifurcation. A pitchfork bifurcation is classified into two types: *supercritical* and *subcritical* pitchfork bifurcation.

2.3.3.1 Supercritical Pitchfork Bifurcation

We can express the normal form of a supercritical pitchfork bifurcation using the following equation,

$$f(x) = \dot{x} = \mu x - x^3 \tag{2.20}$$

Note that this equation is invariant for the transformation of x to $-x$. It means, if we replace x by $-x$ in (2.20), we get $f(-x) = \mu(-x) - (-x)^3 = -(\mu x - x^3) = -f(x)$. Thus, we obtain the original equation with a negative sign. This property also indicates the symmetric nature exhibited by the bifurcation.

The fixed points of this equation are $x^* = 0$ and $\pm\sqrt{\mu}$. The first derivative of the equation is $f'(x) = \mu - 3x^2$. Therefore, for $\mu \leq 0$, we have only one fixed point at $x^* = 0$ (Fig. 2.6a, b), and for $\mu > 0$, we have three fixed points (Fig. 2.6c). The stability of the fixed points is such that for $\mu \leq 0$, the point $x^* = 0$ is always stable, and for $\mu > 0$, the points $x^* = \pm\sqrt{\mu}$ are stable and the point $x^* = 0$ is unstable (Fig. 2.6d). For $\mu = 0$, the stability of the fixed point at $x^* = 0$ is weakly stable (Fig. 2.6b), indicating a slower decay than an exponential decay of the perturbed phase space trajectory. We can refer to this behavior of the system as *critical slowing down* that occurs when the system approaches the bifurcation point [92].

2.3.3.2 Subcritical Pitchfork Bifurcation

In the case of a subcritical pitchfork bifurcation, we can write the normal form equation as,

$$f(x) = \dot{x} = \mu x + x^3. \tag{2.21}$$

The fixed points of this equation are $x^* = 0$ and $\pm\sqrt{-\mu}$. Thus, we have three fixed points for $\mu < 0$ (Fig. 2.7a) and one fixed point at $x^* = 0$ for $\mu \geq 0$ (Fig. 2.7b,

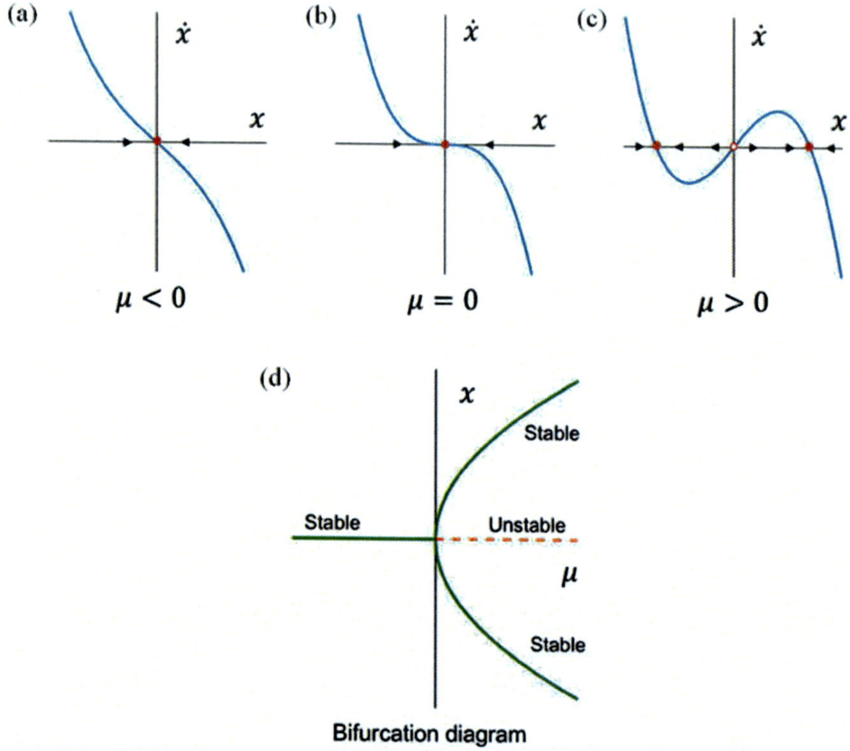

Fig. 2.6 **(a)**–**(c)** Flow diagrams and **(d)** a bifurcation diagram for a variable x undergoing a supercritical pitchfork bifurcation at different values of the control parameter μ. During this bifurcation, a single stable fixed point observed for $\mu \leq 0$ undergoes a symmetry-breaking bifurcation for $\mu > 0$, leading to the formation of two new stable fixed points along with the change of stability of the fixed point at the origin

c). The stability of fixed points is such that, for $\mu < 0$, the fixed points $\pm\sqrt{-\mu}$ are unstable, while the fixed point at 0 is stable (Fig. 2.7d). For $\mu > 0$, the fixed point at 0 is always unstable. Thus, due to a lack of stability boundary for $\mu > 0$, the solution of the equation blows-up to infinity in a finite time for a small value of the perturbation.

In order to have a stabilizing boundary for $\mu > 0$ in the case of subcritical pitchfork bifurcation, which is preferable in practical applications, we need to incorporate a higher-order nonlinearity in the normal form equation. Thus, we can revise the normal form of the subcritical pitchfork bifurcation as,

$$f(x) = \dot{x} = \mu x + x^3 - x^5 \tag{2.22}$$

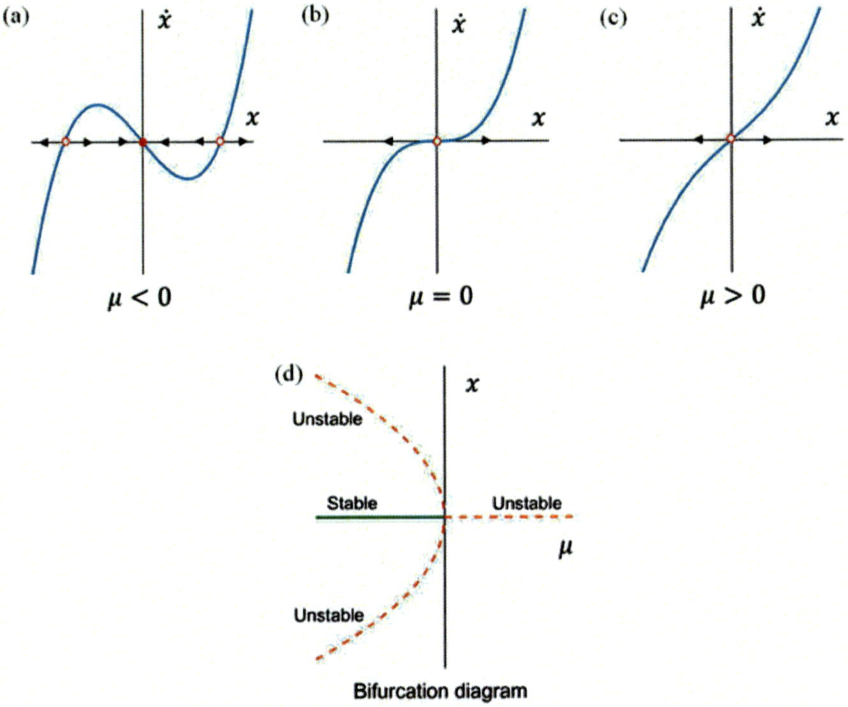

Fig. 2.7 (**a**)–(**c**) Flow diagrams and (**d**) a bifurcation diagram for a variable x undergoing a subcritical pitchfork bifurcation at different values of the control parameter μ. Three fixed points (two unstable and one stable) observed for $\mu < 0$ disappear and form a single unstable fixed point at the origin during the bifurcation for $\mu \geq 0$

With the addition of the fifth-order term, we can preserve the invariant property ($f(x) = -f(x)$) of the pitchfork bifurcation. In this case, we notice that the fixed points of the dynamical equation are different in different regions of μ (Fig. 2.8d). For $\mu < -1/4$, we have a single fixed point, which is always stable (Fig. 2.8a). For $-1/4 < \mu < 0$, we have five fixed points: three are stable and two are unstable (Fig. 2.8b). For $\mu > 0$, we have three fixed points: two are stable and one is unstable (see Fig. 2.8c). Thus, the addition of fifth-order nonlinearity leads to the introduction of a stabilizing solution at high amplitude for all unstable branches observed for $\mu > -1/4$. For the range $-1/4 < \mu < 0$, a system with subcritical bifurcation exhibits hysteresis and also a bistable region with multiple stable solutions (Fig. 2.8d, e).

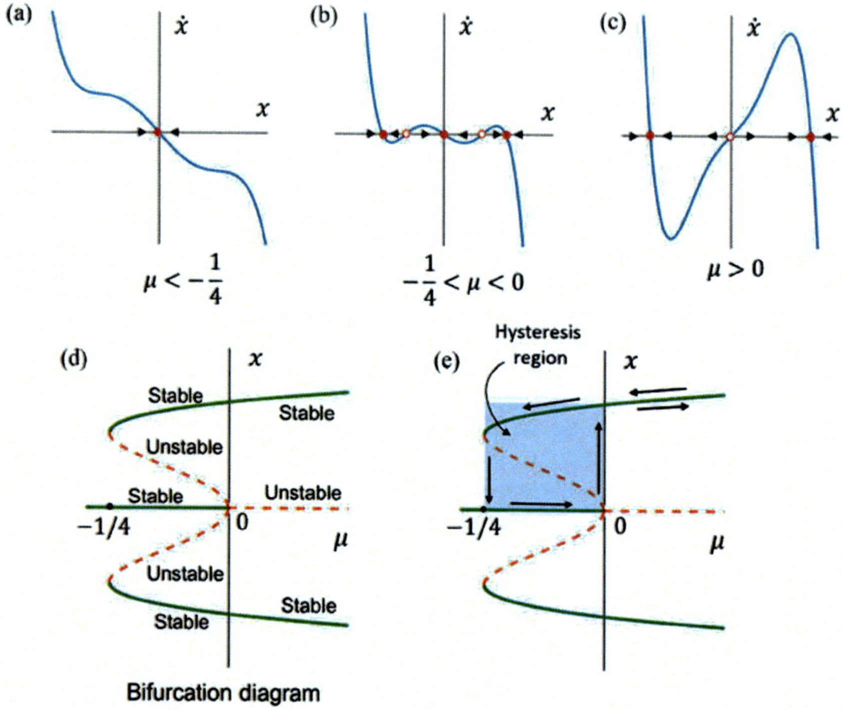

Fig. 2.8 (**a**)–(**c**) Flow diagrams and (**d**) a bifurcation diagram for a variable x undergoing a subcritical pitchfork bifurcation for a higher-order nonlinearity when μ is varied as the control parameter. The increase in nonlinearity increases the number of fixed points in the system. A single stable fixed point observed for $\mu < -1/4$ bifurcates into five fixed points (three stable and two unstable) in the bistable region ($-1/4 < \mu < 0$). For $\mu > 0$, we notice three fixed points (two stable and one unstable). Thus, a stable solution is observed for any value of μ due to the presence of higher-order nonlinearity in the system. The presence of hysteresis region in the subcritical pitchfork bifurcation is highlighted in (**e**). Arrows in (**e**) indicate the variation of x for increasing and decreasing values of μ

2.3.4 Hopf Bifurcation

A Hopf bifurcation, also referred to as a Poincaré–Andronov–Hopf bifurcation [56, 61], is commonly observed in two or higher dimensional dynamical systems [92]. During this bifurcation, a stable fixed point loses its stability and simultaneously, a limit cycle attractor is born. This happens when a pair of eigenvalues that are complex conjugate cross over the imaginary axis. Before the bifurcation, any perturbation of a finite amplitude dampens out through spiraling of the trajectory toward the fixed point (Fig. 2.9a). The decay rate of these damped oscillations decreases as the control parameter approaches the bifurcation point. After the bifurcation point, the fixed point becomes an unstable spiral and the trajectory in the

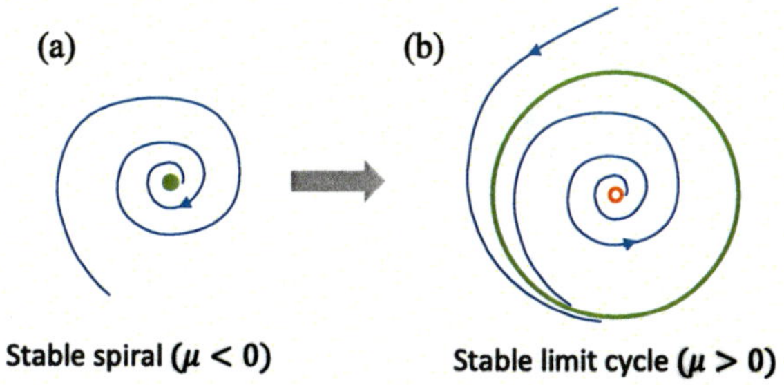

Fig. 2.9 Schematic showing the process of Hopf bifurcation, where the fixed point of a stable spiral ($\mu < 0$) in (**a**) becomes unstable during the bifurcation and leads to the occurrence of a stable limit cycle attractor ($\mu > 0$) in (**b**)

phase space saturates to a closed-loop orbit of the limit cycle attractor (Fig. 2.9b). Thus, Hopf bifurcation leads to the generation of periodic oscillations in the system.

Similar to the pitchfork bifurcation, Hopf bifurcation is also classified into two types: *supercritical Hopf* and *subcritical Hopf*. However, the key difference between these bifurcations is that pitchfork bifurcation results in the formation of fixed points, whereas Hopf bifurcation leads to an emergence of oscillatory solutions (or limit cycle oscillations) in a system.

2.3.4.1 Supercritical Hopf Bifurcation

In supercritical Hopf bifurcation, a stable fixed point loses its stability, while a low amplitude stable limit cycle solution is created. We can express the normal form of supercritical Hopf bifurcation in polar coordinates as follows,

$$\dot{r} = \mu r - r^3 \quad \text{and} \quad \dot{\theta} = \omega + br^2 \tag{2.23}$$

where μ is the control parameter, ω is the angular frequency whose sign decides the direction of rotation of the spiral, and b is a constant that determines the dependence of frequency on the amplitude of oscillations. The solutions of equation (2.23) are $r^* = 0, \pm\sqrt{\mu}$. The bifurcation occurs at $\mu = \mu_c$. For $\mu < 0$, the fixed point $r^* = 0$ is a stable spiral, and for $\mu > 0$, this fixed point is an unstable spiral. The trajectories of the unstable spiral attract towards a finite amplitude stable limit cycle attractor (see Fig. 2.10a). Close to the Hopf point (i.e., when μ is close to μ_c), the amplitude of this stable limit cycle attractor gradually increases as $\sqrt{(\mu - \mu_c)}$ [92].

2.3.4.2 Subcritical Hopf Bifurcation

A subcritical Hopf bifurcation results in a sudden emergence of large amplitude periodic oscillations after the bifurcation. This is in contrast to a supercritical Hopf bifurcation where the amplitude of the limit cycle generated after the bifurcation is of comparatively very low. Thus, during subcritical Hopf bifurcation, the phase space trajectory evolves to a distant limit cycle attractor (Fig. 2.11). We can write the normal form of a subcritical Hopf bifurcation in polar coordinates as,

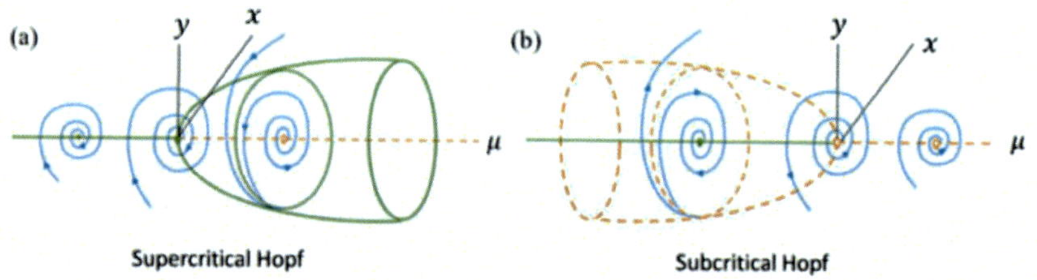

Fig. 2.10 Bifurcation diagrams of (**a**) supercritical Hopf bifurcation and (**b**) subcritical Hopf bifurcation, where the dotted line indicates the unstable branch and solid line indicates the stable branch. During supercritical Hopf bifurcation, a stable spiral becomes unstable, leading to the formation of a stable limit cycle attractor. During subcritical Hopf bifurcation, a stable spiral bounded by an unstable limit cycle attractor becomes unstable

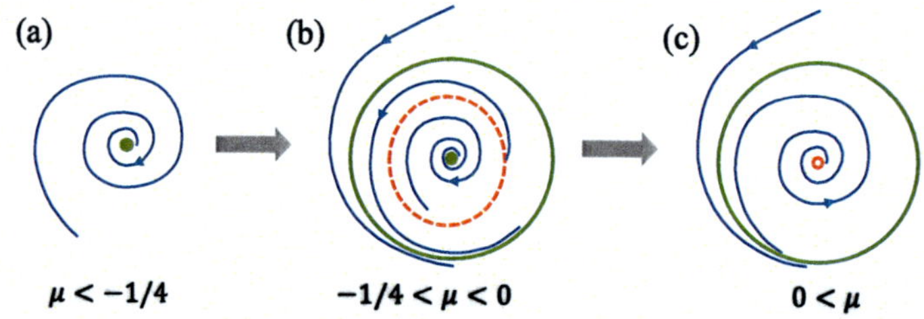

Fig. 2.11 Schematic showing the process of subcritical Hopf bifurcation observed for the presence of higher order nonlinearity. During this bifurcation, (**a**) a stable spiral first transitions to (**b**) a multi-attractor phase portrait, consisting of stable and unstable limit cycle attractors that coexists with a stable fixed point, and then a final transition happens to (**c**) where a stable limit cycle attractor coexists with an unstable fixed point

$$\dot{r} = \mu r + r^3 \quad \text{and} \quad \dot{\theta} = \omega + br^2 \tag{2.24}$$

The solutions of the above equation are $r^* = 0$ and $\pm\sqrt{-\mu}$. For $\mu < 0$, we have three possible solutions, where $r^* = 0$ is a stable fixed point that attracts all the trajectories repelled from an unstable limit cycle attractor located at $r^* = \pm\sqrt{-\mu}$. When $\mu > 0$, we have a single solution at $r^* = 0$, where the unstable fixed point spirals out the trajectories till infinity (see Fig. 2.10b).

In order to have a stabilizing boundary for the unstable solutions of subcritical Hopf bifurcation observed for both $\mu < 0$ and $\mu > 0$, a term of higher-order nonlinearity is added in (2.24). The introduction of such a term provides a bounded solution to the evolution of the phase space trajectories that, in turn, closely matches real situations, as the blow-out of the unstable trajectories to infinity is impractical [92]. Thus, we can write a modified normal form equation for the subcritical Hopf bifurcation as,

$$\dot{r} = \mu r + r^3 - r^5 \quad \text{and} \quad \dot{\theta} = \omega + br^2 \tag{2.25}$$

The stability calculation of the above equation shows that, for $\mu < -1/4$, we have a single stable solution for a fixed point $r^* = 0$ (see Fig. 2.11a). For the range $-1/4 < \mu < 0$, we observe bistability in the bifurcation diagram, that consists of an unstable limit cycle attractor located in between a distant large amplitude stable limit cycle attractor and a stable fixed point centered at $r^* = 0$ (see Fig. 2.11b). As μ approaches the Hopf bifurcation point (i.e., $\mu_c = 0$), the size of this unstable attractor gradually shrinks, causing the formation of an unstable fixed point at $r^* = 0$ coexisting with a stable limit cycle attractor (see Fig. 2.11c). Thus, the trajectories emanating from $r^* = 0$ spiral outwards and asymptotically approach the large amplitude limit cycle attractor for all $\mu > 0$.

Although both subcritical and supercritical Hopf bifurcations exhibit the transition from steady state to limit cycle oscillations, there are a few key differences between them. The distinguishing feature is that a hysteresis region is observed in a subcritical Hopf bifurcation and not in a supercritical Hopf bifurcation. Hysteresis occurs when the transition of the system dynamics from steady state to oscillatory state (say, the forward path) and from oscillatory state to steady state (say, the reverse path) do not occur at the same value of the control parameter; however, there is a finite difference in the values of control parameter at which the forward and reverse transitions occur. In the absence of hysteresis, these transitions happen at the same value of the control parameter, which is the behavior seen in a supercritical Hopf bifurcation. Furthermore, the amplitude of limit cycle oscillations generated during supercritical Hopf bifurcation is very low, whereas the amplitude of limit cycle oscillations generated during subcritical Hopf bifurcation is very high. A sudden generation of such large amplitude periodic oscillations during subcritical Hopf bifurcation can have catastrophic consequences in many engineering and natural systems. Hence, we usually try to avoid the occurrence of this type of bifurcation in a system.

Having discussed the different types of bifurcations, we next present the classification of different types of signals that occur during the bifurcation in physical systems.

2.4 Signals and Their Classification

Many practical systems display complex dynamical states. Appropriate classification of these states is important for understanding the behavior of a system. This, in turn, helps in either controlling the occurrence of such dynamics or modeling their features. We discuss some of the most widely observed system dynamics and their characterization in the following subsections [39, 49, 64].

2.4.1 Limit Cycle Oscillations

Limit cycle oscillations indicate constant amplitude, discrete frequency, periodic oscillations. The phase space trajectory of such oscillations repeats its behavior after

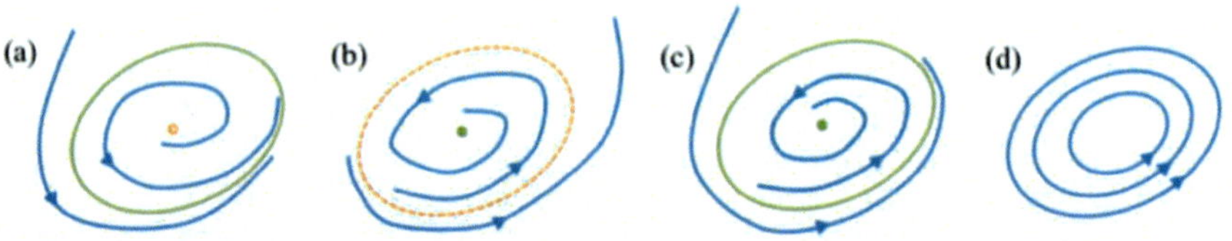

Fig. 2.12 **(a)–(d)** The classification of different types of limit cycles: stable limit cycle, unstable limit cycle, semi-stable limit cycle, and neutrally stable limit cycle (center), respectively. In these plots, an open circle at the center indicates an unstable fixed point, while a filled circle at the center represents a stable fixed point. Arrows signify the direction of the evolution of the phase space trajectory. A closed-loop in the solid green curve represents a stable attractor and that in the dotted orange curve represents an unstable attractor

a time period T, i.e., $x(t + T) = x(t)$ for all t [48]. Such oscillations are self-sustained in nature, as they occur due to the inherent nonlinearity associated with the system and not due to external forcing. The attractor of the limit cycle exhibits an isolated closed curve and is usually projected onto a two or higher dimensional phase space [98]. The term isolated curve implies the absence of another closed curve in its immediate neighborhood [92]. Nevertheless, there is a possibility for the existence of a fixed point inside or outside the closed curve [98].

From a stability perspective, a limit cycle could be stable, unstable, or semi-stable [31, 92]. For a stable limit cycle (Fig. 2.12a), any small perturbation to the phase space trajectory would get attracted back towards the attractor at an exponential rate. In the case of an unstable limit cycle (Fig. 2.12b), a small perturbation would lead to an exponential divergence of the phase space trajectory away from the attractor. For a semi-stable limit cycle (Fig. 2.12c), the phase space trajectories are attracted from one side of the attractor and repelled from the other side. The formation of the limit cycle takes place through Hopf bifurcation, as discussed in the previous section. Note that limit cycles exist only for nonlinear systems.

Similar to the limit cycle, another type of closed orbit observed in both linear and nonlinear systems is called a *center* [17, 84], as illustrated in Fig. 2.1. Centers in linear systems are neutrally stable orbits (also called neutrally stable limit cycles), where small perturbations lead to the formation of other closed orbits in the phase space (Fig. 2.12d). The amplitude of centers in linear systems is dependent on the initial condition, while centers in conservative nonlinear systems are more robust against any perturbations [84]. Examples of centers are unforced undamped Duffing oscillator, Lotka-Volterra system, Lienard system, etc. [84]. In contrast to centers, the amplitude of stable limit cycle oscillations is independent of the initial condition and all trajectories from different initial conditions asymptotically approach the same attractor. Note that centers are observed in conservative (Hamiltonian) systems and limit cycles are observed in dissipative systems. Nevertheless, conservative systems can possess periodic, quasiperiodic, and chaotic solutions, but not attractors [48].

The presence of stable limit cycle oscillations is desirable in some systems such as circadian rhythm [9], hormone secretion [32], chemical reaction [27], etc. However, in some physical systems, the existence of limit cycle oscillations in the form of oscillatory instabilities such as wing flutter [97], thermoacoustic instability [19], flow-induced vibration [115], etc., is completely undesirable. The large amplitude oscillations during oscillatory instabilities affect the structural integrity of the system. Examples of well-known limit cycle oscillators are van der Pol oscillator, Stuart-Landau oscillator, Hodgkin–Huxley model, etc. [41, 85].

2.4.2 *Period-n Oscillations*

A period-n signal (or mode-locked signal) consists of two or more discrete frequencies that are rationally related to each other (here, n is an integer number). For example, let f_1 and f_2 be the two frequencies arising either due to nonlinear interaction of harmonics of the same system or due to nonlinear coupling of two subsystems having different frequencies. The relation among these frequencies can be expressed as $f_2/f_1 = p/q$, where p and q are integer numbers. For a period-n signal, this frequency relation is rational (commensurate); hence, we can call such signals as *frequency-locked* or *mode-locked* signals [39]. For example, a period-2 signal consists of a frequency relation of $f_1 = 2f_2$, where f_1 is the dominant frequency and f_2 is its subharmonic frequency.

We can detect the properties of a period-n signal using a Poincaré section. A Poincaré section is a transversely cut portion of an attractor, which represents the projection of the attractor into a lower-dimensional phase space (as detailed in Sect. 2.7). A Poincaré section of a period-n signal shows n discrete points (Fig. 2.13a), indicating the type of frequency-locking observed in the signal. The motion of the phase space trajectory for these signals repeats after n revolutions and forms a closed-loop orbit on the surface of a torus in the phase space.

2.4.3 *Quasiperiodic Oscillations*

A quasiperiodic signal consists of two or more frequencies that are irrationally related (incommensurate) with each other [39]. If f_1 and f_2 are two frequencies in a quasiperiodic signal, then the frequency relation between them is such that $f_1/f_2 \neq p/q$, where p and q are integer numbers. Due to the presence of multiple irrationally related frequencies, quasiperiodic signals exhibit aperiodic oscillations, i.e., $x(t) \neq x(t+T)$, where T is the time period of the oscillation.

The quasiperiodic behavior of a signal can be characterized using different methods [39, 64]. One of the methods is to compute a frequency ratio amongst two dominant frequency peaks in the power spectrum of a signal. If the ratio of these dominant frequencies cannot be expressed in terms of integer numbers, then the signal may be quasiperiodic. For such signals, contributing frequencies are always present at non-harmonics of the dominant frequency. In practical situations, it is

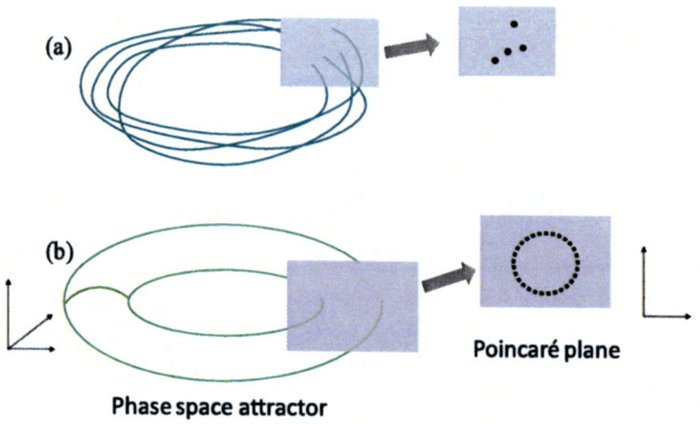

Fig. 2.13 Diagrammatic representation of the phase portrait and the Poincaré section of (**a**) mode-locked oscillations and (**b**) quasiperiodic oscillations. As the trajectory in the phase space repeats its behavior after a few rotations for mode-locked oscillations, the Poincaré section of this state shows only discrete points. For quasiperiodic oscillations, as the non-repeating trajectories fill the surface of a torus in the phase space, the Poincaré section of this state shows a continuous distribution of points along a closed-orbit

common to observe multiple frequencies in the spectrum of a signal. However, it is not always easy to identify the contributing frequencies that exhibit incommensurate frequency relationship. In such a situation, the use of the phase space and the Poincaré section are alternative methods to detect the quasiperiodicity underlying to a given signal.

For a two-frequency quasiperiodic signal, the trajectory in the phase space evolves on the surface of a two-dimensional torus attractor (or a doughnut-like structure), see Fig. 2.13b. For such attractors, one component of the trajectory revolves around the large circumference of the torus (say, with frequency f_1) and the other revolves around its cross-section (say, with frequency f_2). Since these frequencies are irrationally related to each other, the trajectories corresponding to them never meet each other in the phase space and wander on the entire surface of the torus. The Poincaré section of a quasiperiodic signal therefore displays a dense ring-like structure (Fig. 2.13b), which is not necessarily a circle.

2.4.4 *Chaotic Oscillations*

A chaotic signal consists of aperiodic oscillations with a broadband amplitude spectrum [87]. Chaos is the property of a deterministic system. Therefore, aperiodicity, i.e., the irregular oscillations that never repeat as $t \rightarrow \infty$, primarily arises due to nonlinearity and not due to stochastic forcing in the system [39]. An important characteristic of a chaotic signal is its sensitive dependence on initial conditions [51]. A small change in the initial condition can cause parallelly running phase space trajectories to diverge exponentially after a short time (Fig. 2.14).

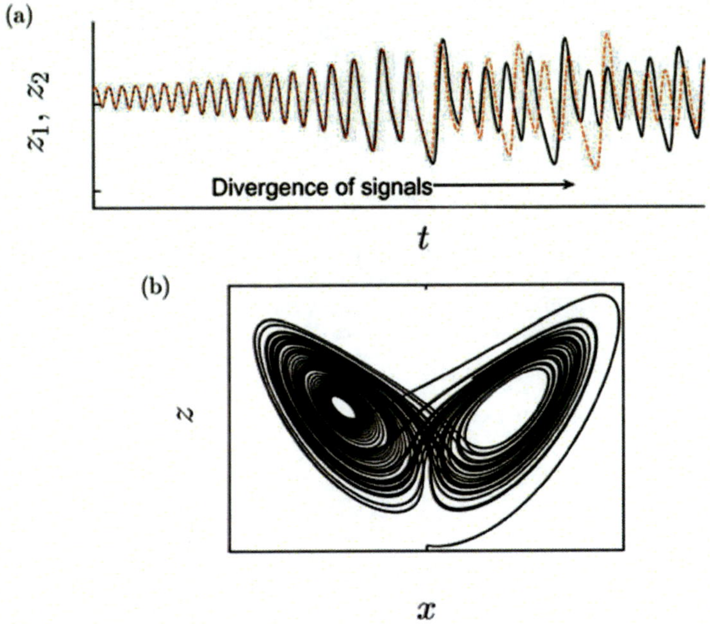

Fig. 2.14 (**a**) The time series of a Lorenz oscillator for two different initial conditions (z_1 and z_2) show the drift after a few cycles of oscillations; thus, indicating the sensitivity to a change in the initial condition. (**b**) The attractor of the Lorenz oscillator exhibits a butterfly shape, where the phase space trajectory alternately switches between two lobes of the attractor

Chaotic signals have short-predictability (determinism). This is in contrast to the property of a random signal where the predictability of the signal at the next time instant is zero. Edward Lorenz [51] coined such behavior of chaotic signals as the '*butterfly effect*', while he was studying atmospheric convection in the climate system. Due to the butterfly effect, small fluctuations resulting from the flapping of a butterfly in Brazil could possibly cause a torrential rainstorm in Texas. Although the motion of a chaotic signal is erratic (i.e., complex), it is always bounded in the phase space. The attractor of a chaotic signal is sometimes referred to as a *strange attractor* [82]. The minimum number of dimensions needed for chaotic motion is three [64]. As the movement of the trajectory in the phase space is fractal, the chaotic attractors mostly possess a non-integer dimension [95], refer to Chap. 10 for more discussion on fractals. Chaotic oscillations are widely observed in diverse fields such as meteorology, biology, physiology, astronomy, telecommunications, economics, sociology, and engineering [7, 16, 43, 78, 90, 91].

Distinguishing chaotic behavior from random (noisy) and regular behaviors is important. We will next discuss some of the commonly used quantitative measures for detecting chaos in the system dynamics [39, 42, 74, 92].

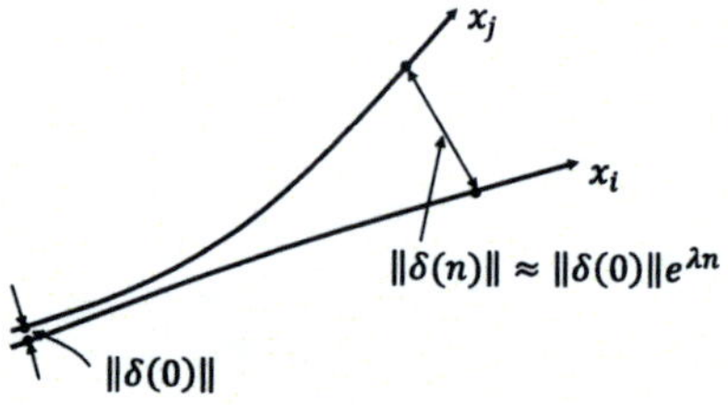

Fig. 2.15 Schematic showing the divergence property of neighboring trajectories x_i and x_j in the phase space, where $\|\delta(0)\|$ is the initial Euclidean distance between the trajectories and $\|\delta(n)\|$ is their distance after the nth time instant. For chaotic oscillations, the trajectories diverge exponentially in time, where the rate of divergence (i.e., Lyapunov exponent) is indicated by λ

2.4.4.1 Lyapunov Exponent

Lyapunov exponent measures the average divergence property of neighboring trajectories in the phase space [5, 23]. A simple way to evaluate this divergence is to calculate a temporal evolution of the distance between these trajectories (Fig. 2.15). In this manner, Lyapunov exponent quantifies the rate at which neighboring points on the chaotic attractor approach, repel, or stay parallel to each other in time [108]. The number of Lyapunov exponents observed for a signal depends on the dimension of the phase space [68]. Thus, a range of Lyapunov exponents can be observed for a signal. For a chaotic signal, the largest value of the Lyapunov exponent (also called maximal Lyapunov exponent) is always positive. A positive Lyapunov exponent indicates that the neighboring trajectories diverge exponentially after a short distance in the phase space.

Let us consider $x(t)$ to be a vector of state points in the phase space such that $x(t) = [x_1, x_2, x_3, \ldots\ldots., x_N]$. Further, we assume x_i and x_j be two portions of $x(t)$ such that points in these vectors are close to each other in the phase space. Then, we can calculate the evolution of the distance between the neighboring state points of x_i and x_j as follows [39],

$$
\begin{aligned}
\delta(0) &= \|x_i - x_j\| \\
\delta(1) &= \|x_{i+1} - x_{j+1}\| \\
&\;\;\vdots \\
\delta(n) &= \|x_{i+n} - x_{j+n}\| \qquad (2.26)
\end{aligned}
$$

If we assume the divergence of the neighboring phase space trajectories with an increase in n to be exponential, then we can write $\delta(n) = \delta(0)e^{\lambda n}$. Here, λ is called the Lyapunov exponent. We can write this relation alternately as,

$$\lambda = \frac{1}{n} \ln \frac{\delta(n)}{\delta(0)} \tag{2.27}$$

Thus, if the signal is chaotic, the value of λ will be a positive number, indicating the unstable nature of neighboring trajectories in the phase space.

Furthermore, the value of the Lyapunov exponent also provides important information about the identification of different dynamical states present in the system [8, 68]. As discussed earlier, a chaotic signal has at least one positive Lyapunov exponent. When a signal has at least two positive Lyapunov exponents, it is called hyper-chaos [79]. For a fixed point, all Lyapunov exponents need to be negative. In contrast, for limit cycle (period-1) oscillations, at least one Lyapunov exponent need to be zero and the remaining exponents need to be negative. For a quasiperiodic signal, at least two Lyapunov exponents should be zero and the remaining exponents should be negative.

2.4.4.2 Correlation Dimension

As we mentioned before, chaotic attractors possess a fractal structure; hence, the dimension of such attractors is a non-integer number. Thus, finding the fractal dimension of the attractor through some means would indirectly help in detecting the chaotic behavior of a signal. There are several methods available in the literature to calculate the fractal dimension of an attractor. These methods include the Hausdorff dimension, the box-counting dimension, the information dimension, and the correlation dimension [95]. Here, we will briefly discuss a procedure for measuring the fractal dimension of a signal through the calculation of the correlation dimension. Towards this purpose, we use a method based on the algorithm of Grassberger and Procaccia [33]. In this method, we first reconstruct the signal into an appropriate dimensional phase space using Takens' delay embedding theorem [94], as discussed later in Sect. 2.6. Then, we measure the closeness of the density of points in the phase space as a correlation sum.

Let us consider x_i and x_j to be two state points on the phase space trajectory of a signal, where $i, j = 1, 2, 3, \ldots, N$ and N is the total number of state points in the phase space. We need to calculate the pairwise Euclidean distances between the state points x_i and x_j. When the distance between these state points falls within a threshold ϵ (see Fig. 2.16a), we consider the points to be correlated points in the phase space. We denote the number of correlated points in the phase space as $C(\epsilon)$. With the help of Heaviside theta function (Θ), we then mark the correlated points as 1 and uncorrelated points as 0, i.e., $\Theta(\epsilon > X) = 1$ and $\Theta(\epsilon < X) = 0$, where X is the distance between the state points.

Thus, the correlation sum of an attractor for a particular choice of ϵ is calculated as the mean of all correlated points as follows,

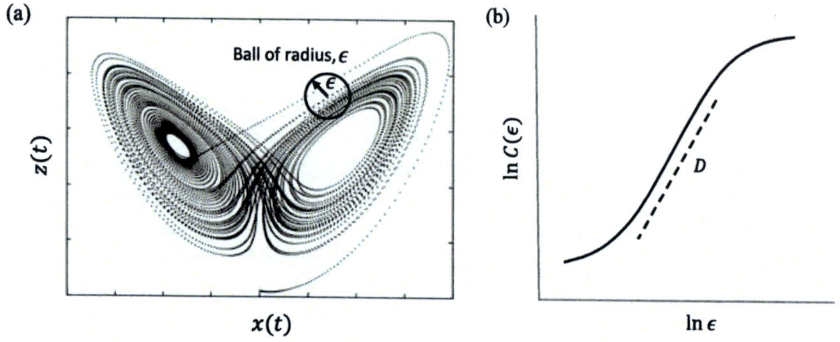

Fig. 2.16 (**a**) A scatter of state points in the phase space of a Lorenz attractor where a threshold ball of radius ϵ is shown to surround a state point of the trajectory. The points inside the ball are considered as correlated points. (**b**) Variation of the correlation sum $C(\epsilon)$ versus the threshold size ϵ on a log-log plot, where the slope of a linear variation in the plot is referred to as the correlation dimension (D)

$$C(\epsilon) = \frac{\text{Number of instances when the distance is less than } \epsilon}{\text{Total number of distances between all pair of state points}}$$
$$= \frac{1}{N(N-1)} \sum_{i,j=1,i\neq j}^{N} \Theta(\epsilon - \|x_i - x_j\|) \tag{2.28}$$

Here, the correlation sum counts the number of instances when the distance between state points is smaller than the threshold ϵ. The increase in ϵ would correspondingly increase the number of correlated points in the phase space. We assume that in the limit of $N \to \infty$ and $\epsilon \to 0$, the variation of the correlation sum exhibits a scaling behavior with ϵ such that,

$$C(\epsilon) \sim \epsilon^D \quad \text{or} \quad D = \lim_{N\to\infty} \lim_{\epsilon\to 0} \frac{\log C(\epsilon)}{\log \epsilon} \tag{2.29}$$

Here, D is called the correlation dimension, which is the slope of the log-log plot between $C(\epsilon)$ and ϵ.

In a practical situation, the scaling (power-law) behavior in the plot of $C(\epsilon)$ and ϵ is valid only for an intermediate region of ϵ (see Fig. 2.16b). If ϵ is very small, very few points are present around x_i within ϵ, resulting in the value of $C(\epsilon)$ to approach zero. In contrast, if ϵ is very large (say, close to the size of the attractor), the variation of $C(\epsilon)$ saturates with a further increase in ϵ [42]. When the value of D is a non-integer number greater than 2, the corresponding attractor may be chaotic. When the value of D is an integer number, the corresponding attractor is regular (i.e., for periodic attractor $D = 1$ and for quasiperiodic attractor $D = 2$) [4, 10]. For a random signal, the value of D tends to infinity.

2.4.5 *Difference Between Strange Chaotic, Strange Nonchaotic, and Chaotic Nonstrange Signals*

As we discussed in the previous section, the attractor of strange chaotic signals is bounded into a finite phase space [82]. Due to the aperiodic nature of strange chaotic signals, the evolution of trajectories in the phase space is always unstable (i.e., neighboring trajectories diverge exponentially for a small perturbation); hence, the shape of the attractor is not regular (i.e., no closed curves). The important property of the strange chaotic signal is its sensitive dependence on the initial condition, which leads to an exponential divergence of the nearby trajectories in the phase space. Thus, the value of Lyapunov exponent for such signals is always a positive number. The spectrum of such signals is continuous (broadband) for a certain range of frequencies. Further, the geometric dimension of such an attractor is a fractional number. The autocorrelation function for such signals decay suddenly to zero. The strange chaotic signals are also called *deterministic chaos* [87].

In some cases, however, the attractor of a signal is strange (i.e., fractal), but it does not exhibit the property of chaotic oscillations, i.e., the neighboring trajectories do not diverge exponentially in phase space. For such attractors, all Lyapunov exponents are negative, which means the asymptotic dynamics of the phase space trajectories are stable. The attractors with such dynamical properties are called *strange nonchaotic attractors* (SNA). Grebogi et al. [34], for the first time, identified SNA in a quasiperiodically driven dynamical system. The first experimental evidence of SNA in a self-excited system was reported by Premraj et al. [72] in the self-excited dynamics of a laminar thermoacoustic system. Strange nonchaotic oscillations are mostly observed between the states of quasiperiodic and chaotic oscillations. The spectrum of SNA contains both discrete and continuous frequencies; therefore, such a spectrum is also called a singular-continuous spectrum. The autocorrelation function of SNA does not decay to zero as $t \to \infty$ [69].

We also observe another type of chaotic oscillations where the attractor is regular (not strange) but the phase space trajectories are exponentially unstable, i.e., the system consists of a positive Lyapunov exponent. Since the geometric structure of the attractor is regular and not fractal, its dimension is an integer number. The attractor with such properties is called a *chaotic nonstrange attractor* (CNA) [2].

2.4.6 *Intermittency*

Intermittency refers to the switching of the system dynamics between periodic (laminar phase) and chaotic (turbulent phase) oscillations in an apparently random manner [55], see Fig. 2.17. This random switching property of the phase space trajectory between two distinct attractors is completely deterministic and is not caused by external noisy fluctuations. The transition of the system behavior from periodic to chaotic oscillations via intermittency displays a sequence of events due

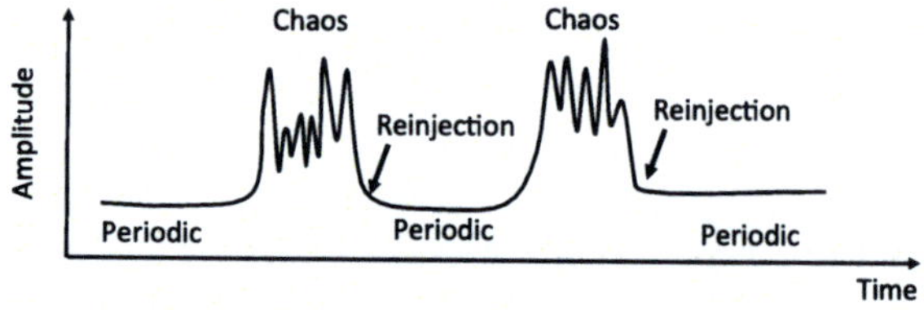

Fig. 2.17 The intermittency signal consisting of alternate occurrence of laminar (periodic) phases and turbulent (chaotic) phases. Adapted with permission from Schuster and Just [87]

to variation in the control parameter. We will discuss such intermittency route to chaos in detail in Sect. 2.5.2.

Pomeau and Manneville [71] characterized the intermittency route to chaos in three main types, such as type-I, type-II, and type-III intermittency. In addition to these basic types of intermittency, several other types of intermittency including on-off, type-V, type-X, eyelet, crisis-induced, spatiotemporal intermittency also exist in the literature [24, 87]. The phenomenon of intermittency has been observed in several physical systems such as Rayleigh–Bénard convection system [11], turbulent flows [101], lasers [83], electronic circuits [109], thermoacoustic and aeroacoustic [63] and aeroelastic systems [102]. In the following subsection, we elaborate on the properties of type-I, type-II, and type-III intermittency.

2.4.6.1 Types of Intermittency

Manneville and Pomeau [55] studied the Lorenz model to identify the mechanism for the occurrence of intermittency. They showed that when the control parameter is less than a critical value, the system exhibits a stable fixed point for limit cycle oscillations in the Poincaré section. The stability of such fixed points could change in three different ways which, in turn, decide the type of intermittency associated with the system. When the real part of eigenvalues (or Floquet multipliers) of the unstable fixed point crosses the unit circle at $+1$, such a bifurcation is called saddle-node or inverse-tangent bifurcation. We refer to the corresponding type of intermittency as type-I. In another case, complex conjugate eigenvalues simultaneously cross the unit circle, leading the system to undergo subcritical Hopf bifurcation. We call the intermittency associated with this route to chaos as type-II. Finally, the crossing of the real part of eigenvalues at -1 on the unit circle indicates the presence of period-doubling bifurcation. We refer to the intermittency corresponding to this route as type-III. As the transition from periodic to chaotic state is not sudden due to the presence of intermittency, identification of the type of intermittency may help in knowing the underlying bifurcation that the system undergoes during this transition.

In addition to the aforementioned measures, there are several other qualitative and quantitative tools for identifying the type of intermittency observed in experimental systems, as the exact dynamical equations for such systems might not always be known [24, 87]. These methods include the probability distribution of laminar phases near the onset of intermittency, the variation of the average laminar phase with the control parameter, the Poincaré section, and the recurrence plot, to name a few of them.

The calculation of the duration for which the system dynamics remains in the laminar phase for a particular signal aids in distinguishing different types of intermittency. The distribution of such laminar lengths obtained from an intermittency signal in the neighborhood of the critical point (i.e. the onset of intermittency) shows a bimodal distribution for type-I intermittency (Fig. 2.18a) and an exponential distribution for type-II and type-III intermittency (Fig. 2.18b, c, respectively). The power-law exponents (or scaling relation) for such distributions are $-1/2$ for type-I intermittency, -2 for type-II intermittency, and $-3/2$ for type-III intermittency (Fig. 2.18a–c). The variation of the mean duration of the laminar phase in the intermittency signal $\langle l \rangle$ with the control parameter (ϵ) shows a slope of $-1/2$ for type-I intermittency, and -1 for both type-II and type-III intermittency (Fig. 2.18a–c).

A Poincaré section or a first return map (discussed later in Sect. 2.7) of type-I intermittency shows a small opened channel between the map and the bisector line, wherein the trajectory re-injected from the chaotic state to the laminar state gets trapped and exhibits numerous iterations before exiting the channel (Fig. 2.18d). For type-II intermittency, the trajectory in the Poincaré section shows a spiraling behavior about the fixed point (Fig. 2.18e). For type-III intermittency, the Poincaré section shows a two-fold trajectory arising from the fixed point along the diagonal line, wherein one trajectory displays an increasing trend for the subharmonic

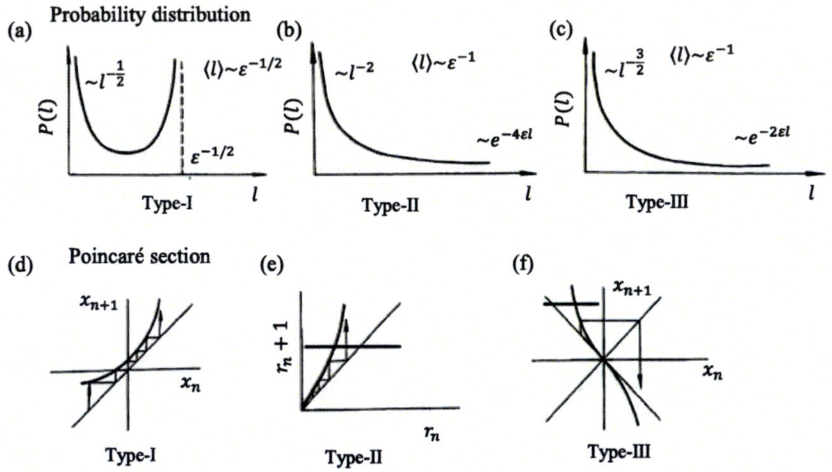

Fig. 2.18 Distinguishing properties of three types of intermittency. **(a)–(c)** Probability distribution $P(l)$ of the length of the laminar phase (l) near the onset of intermittency for type-I, type-II and type-III intermittency, respectively. Scaling of the average length of the laminar phase $\langle l \rangle$ with the control parameter ε shows various exponents for these types of intermittency. **(d)–(f)** The Poincaré sections (i.e., first return maps) for type-I, type-II and type-III intermittency, respectively. Adapted with permission from Schuster and Just [87]

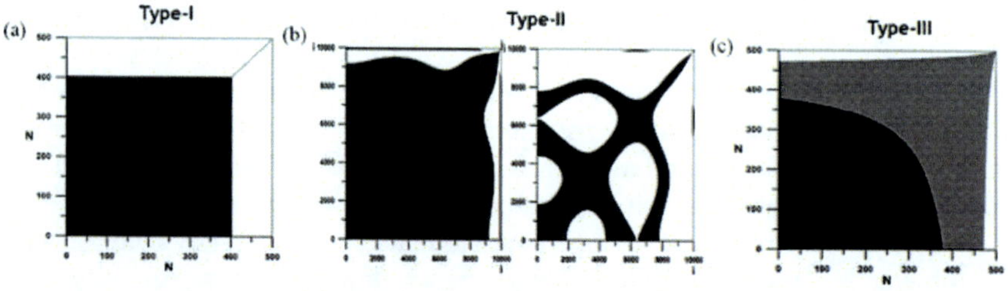

Fig. 2.19 (a)–(c) Recurrence plots for type-I, type-II, and type-III intermittency, respectively. Each type of intermittency consists of specific patterns in the corresponding recurrence plot. Reproduced with permission from Klimaszewska and Zebrowski [46]

amplitude and another trajectory shows a decreasing trend for the fundamental amplitude (Fig. 2.18f). At the end of the laminar phase, both trajectories are reinjected back to the fixed point with a uniform probability distribution.

Statistical methods such as the probability distribution of the laminar phase and the variation of the average laminar phase with the control parameter, as mentioned above, need long time series data. Acquiring such lengthy data in practical situations is sometimes difficult. Therefore, Klimaszewska and Zebrowski [46] developed a method based on a recurrence plot (discussed later in detail in Sect. 2.8) to distinguish the three types of intermittency, as this method requires a shorter time series to characterize the features of the intermittency signal. According to them, each type of intermittency is associated with a specific type of pattern in the recurrence plot. For type-I intermittency, the recurrence plot shows a black square box located along the main diagonal line (Fig. 2.19a). In the case of type-II intermittency, the recurrence plot shows an elongated right upper corner (kite-like structure) with perforations (Fig. 2.19b). For type-III intermittency, the recurrence plot shows a kite-like structure whose bottom left corner is uniformly black, while the upper right corner is round (Fig. 2.19c). Outside the round corner, the presence of parallel diagonal lines gives a non-uniform black structure to the kite.

2.5 Routes to Chaos

The presence of chaos has fascinated various researchers due to its inherent property of sensitivity to initial conditions leading to irregularity and, in turn, the long term unpredictability of the system behavior. There are several routes through which regular dynamics can transition to chaos as the control parameter is changed [39, 48, 50, 87]. In the following subsections, we briefly discuss the main routes to chaos that happens via period-doubling, quasiperiodicity, and intermittency states.

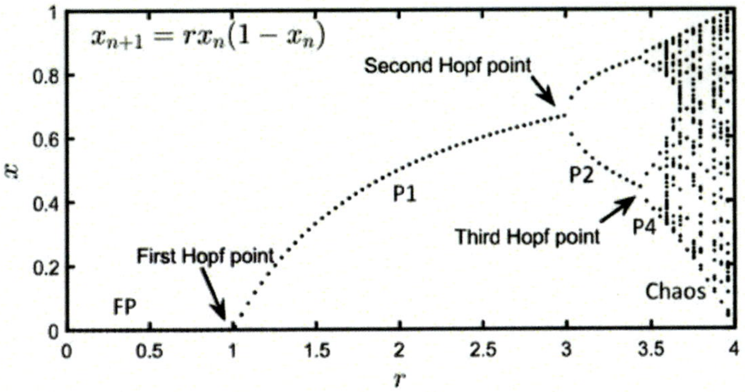

Fig. 2.20 Bifurcation diagram of a logistic map, showing the transition to chaos via a period-doubling cascade, where r is the control parameter and x is the measure of a signal. As the value of r is varied, the system exhibits a transition from a fixed point (FP), to a period-1 limit cycle (P1), to a period-2 limit cycle (P2), to a period-4 limit cycle (P4), and so on until chaos is observed

2.5.1 *Period-Doubling Route to Chaos*

Period-doubling route to chaos begins with the state of limit cycle oscillations having a single frequency. These oscillations are born out of the unstable fixed point through an Andronov-Hopf bifurcation when a control parameter crosses a particular value on the bifurcation diagram [39]. The limit cycle oscillations are indicated as a point on the Poincaré section. A further change in the control parameter causes the limit cycle to become unstable, leading to the birth of the period-2 limit cycle oscillations. These period-2 oscillations are indicated by two distinct points in the Poincaré section. The power spectrum of such a signal shows the presence of a new peak at the subharmonic of the natural frequency of the oscillations. With a further increase in the control parameter, the period-2 limit cycle bifurcates to period-4 limit cycle oscillations (Fig. 2.20). Thus, a cascade of period-doubling bifurcation continues as the control parameter is changed, resulting in the generation of period-8, period-16, ..., period-2^n limit cycle oscillations, which ultimately ends up in having infinite period oscillations with a continuous power spectrum. Such signals are called chaotic oscillations. This route to the chaos was initially proposed by Feigenbaum [26] in a one-dimensional map; hence, it is referred to as the '*Feigenbaum scenario*'.

2.5.2 *Quasiperiodic Route to Chaos*

During quasiperiodic route to chaos, the system is initially in a state of a fixed point. As the control parameter crosses a critical value, the fixed point becomes

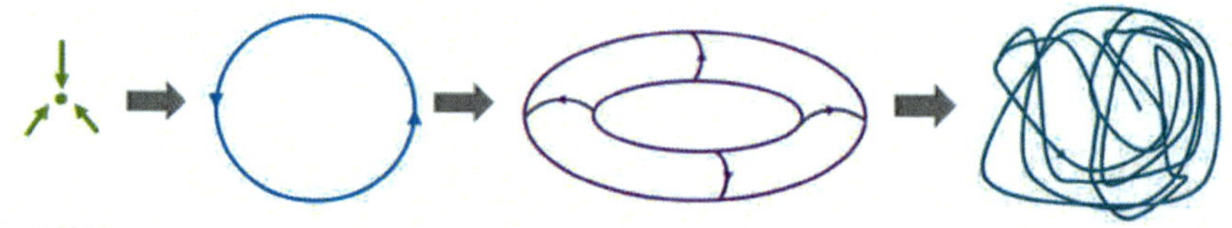

Fig. 2.21 Attractors observed during the quasiperiodic route to chaos, where the system behavior changes from a fixed point, to a limit cycle, to a quasiperiodicity, and finally to a chaos

unstable, resulting in the generation of period-1 limit cycle oscillations through Andronov-Hopf bifurcation (Fig. 2.21). A further increase in the control parameter above another critical point leads to the birth of a new frequency in the spectrum. If this newly generated frequency is incommensurate with the existing dominant frequency, the corresponding system dynamics exhibits quasiperiodic oscillations. During quasiperiodic oscillations, the trajectory in the phase space revolves around an attractor of a two-dimensional torus. Further increase in the control parameter leads to the occurrence of torus-breaking bifurcation, wherein the introduction of a third new frequency leads to the transition of the system behavior to chaotic oscillations. This route to the chaos was first discovered by Ruelle and Takens [82] and Newhouse et al. [65] in independent theoretical studies. Since then, the quasiperiodicity route to chaos is also referred to as the '*Ruelle–Takens–Newhouse scenario*'.

2.5.3 *Intermittency Route to Chaos*

As we mentioned before, the intermittency route to the chaos was initially proposed by Pomeau and Manneville [71] in one-dimensional maps. They observed that for a particular value of the control parameter, the asymptotic dynamics of the system is periodic (limit cycle), see Fig. 2.22a. Once the control parameter crosses a critical value, the periodic oscillations are interrupted by bursts of irregular (chaotic) oscillations for relatively shorter duration (Fig. 2.22b). The occurrence of these bursts is apparently random in the signal. During the onset of a burst, the trajectory first moves away from the neighborhood of a periodic orbit, performs irregular motions in the phase space, and then reinjected back in the vicinity of the periodic orbit at the end of the burst. A further change in the control parameter increases the duration for which bursts of irregular dynamics persist in the signal (Fig. 2.22c). This, in turn, leads to a corresponding decrease in the duration of epochs of periodic oscillations in the signal. After a critical value of the control parameter, the signal becomes fully chaotic (Fig. 2.22d).

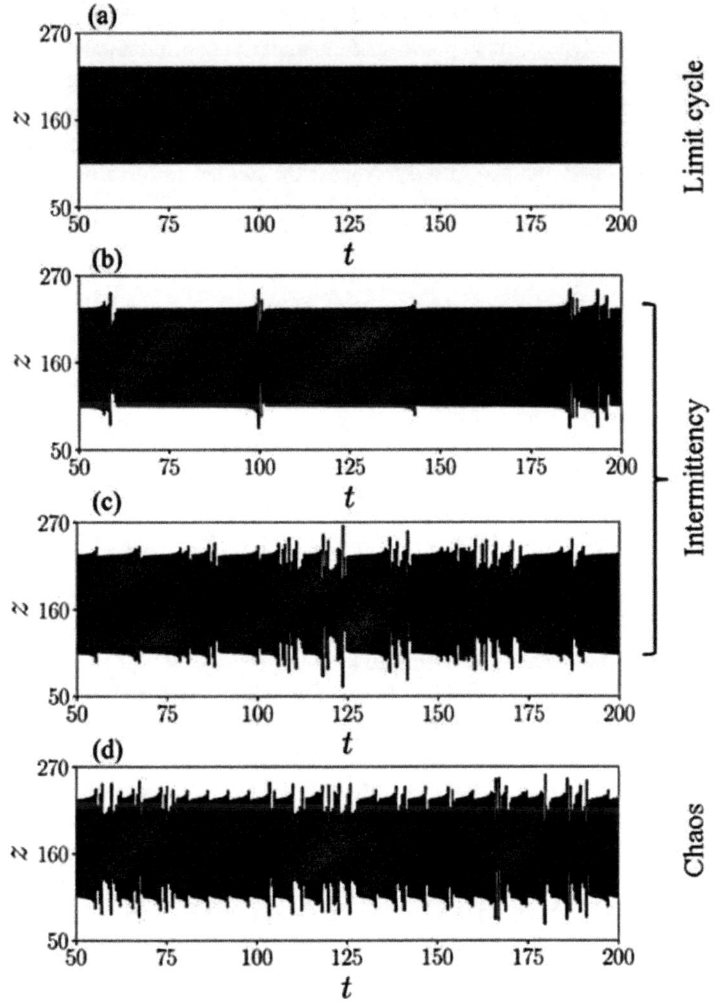

Fig. 2.22 Time series of z variable of a Lorenz model: $\dot{x} = \sigma(y - z)$, $\dot{y} = \rho x - y - xz$, and $\dot{z} = -\beta z + xy$, where $\sigma = 10$, $\beta = 8/3$, and ρ is varied as the control parameter. For different values of ρ, the system behavior transitions from the state of (**a**) limit cycle oscillations ($\rho = 166$) and (**b**, **c**) intermittency ($\rho = 166.335, 166.5$) to (**d**) chaotic oscillations ($\rho = 167$), where $\rho_c \approx 166.06$ is the critical value corresponding to the onset of intermittency in the Lorenz model

Until now, we discussed the basics of dynamical systems theory, required for characterizing the type of fixed points and the underlying bifurcations that a system undergoes in response to a smooth change in the control parameter. We also classified the features of different signals and presented several tools necessary for their quantitative analysis. Next, we discuss different methods from dynamical systems theory that have been used to reveal the hidden features of experimental data.

2.6 Phase Space Reconstruction

We can represent any nonlinear dynamical system in the form of a set of first-order differential equations [64] as follows,

$$\frac{dX(t)}{dt} = F(X(t)) \tag{2.30}$$

Here, $X(t)$ is a vector of state variables $[X_1(t), X_2(t), \ldots\ldots., X_d(t)]$ in a d-dimensional phase space and F is vector of functions $[F_1, F_2, \ldots\ldots., F_d]$ that explicitly or implicitly couple these variables. The value of these state variables at a particular time instant t is indicated as a 'state point' in the phase space. The evolution of these state points in time forms a 'trajectory' in the phase space which, in turn, provides information about the underlying dynamics of the system. Such dynamics could be periodic, chaotic, steady state, or something else.

In numerical studies, all dynamical variables and governing equations are known; hence, the state of the system is accurately defined. However, in many practical situations, we do not have knowledge of all the state variables, either due to our incomplete knowledge of the system or due to limitations in our measurements. In such a situation, Takens [94] proposed a method to derive information about all unknown dynamic variables of a physical system from the observations of its single dynamic variable. This procedure of reconstruction of the phase space from a single variable of a system is known as *Takens' delay embedding theorem*.

In experiments, let us consider x_t to be a measured signal of a system variable sampled at intervals of Δt such that $x_t = [x_1, x_{1+\Delta t}, \ldots., x_{1+(N-1)\Delta t}]$, where N is the total number of data points in the signal (Fig. 2.23). In order to get information about the unknown variables, we construct a number of delayed vectors using the Takens' delay embedding theorem. In this method, we first delay the sampled signal by an appropriate value of the delay time ($\tau = n\Delta t$, where $n = 1, 2, \ldots.., N-1$) and then identify the optimum dimension (say, d) required for the phase reconstruction of the signal. Thus, from the knowledge of τ and d, we can obtain the delayed vectors of a signal as follows,

$$Z_i(d) = [x_i, x_{i+\tau}, \ldots\ldots., x_{i+(d-1)\tau}] \tag{2.31}$$

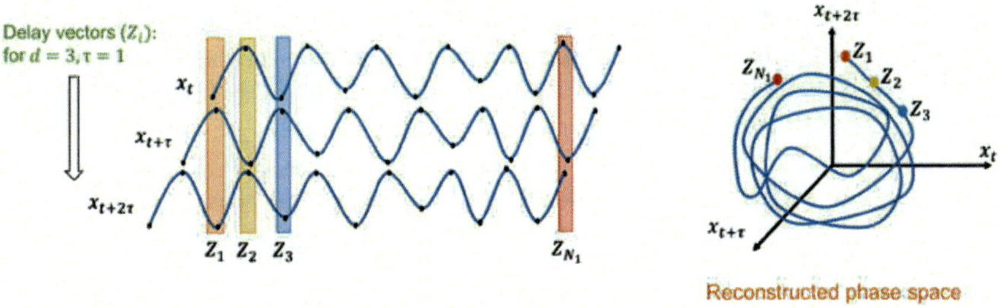

Fig. 2.23 Diagrammatic representation of a procedure to construct delayed vectors and reconstructed phase space. The delayed vectors (Z_i) are first constructed from a signal x_t for the appropriate values of τ and d. Subsequently, these vectors are projected in the d-dimensional phase space

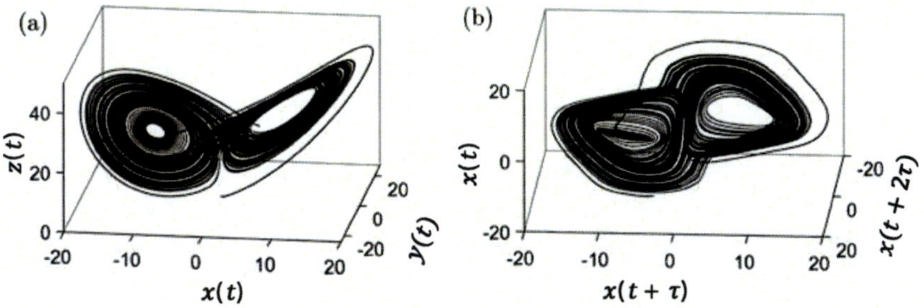

Fig. 2.24 (**a**) Phase space of a Lorenz attractor constructed using actual variables (x, y, z) of the model. (**b**) The reconstructed phase space of the Lorenz attractor obtained from the delayed vectors of the variable x. Although both the phase portraits look different, the reconstructed attractor preserves the key property (i.e., two-lobes) of the original Lorenz attractor

where $i = 1, 2, \ldots, N_1$; $N_1 = N - (d-1)\tau$.

Note that the topology of the reconstructed phase space of a signal is not the same as that of its actual attractor. Nevertheless, the phase space reconstruction of successfully preserves the values of dynamical invariants, such as Lyapunov exponent, correlation dimension, etc., of the original attractor [42]. Further, the state points on the reconstructed phase space and those on the original attractor preserve a one-to-one mapping. Figure 2.24 shows a three-dimensional phase space constructed from the actual variables of a well-known Lorenz oscillator and that constructed from the application of delay embedding theorem to variable x. We notice that although both the attractors do not appear the same, the reconstructed phase space preserves the key feature of the two-lobe structure observed in the original Lorenz attractor. According to Takens' embedding theorem [94], an appropriate embedding dimension (d) required for the phase space reconstruction of a signal is much higher than the actual dimension of its attractor (say, d_A) such that $d \geq (2d_A + 1)$ [13, 42, 64]. The value of d_A can be measured from several fractal-based methods [14, 95].

From the procedure of the phase space reconstruction of a single time series data, we found that the appropriate reconstruction of the phase space depends on two key parameters, i.e., the time delay (τ) and the embedding dimension (d). Therefore, the selection of the optimum value of these parameters is an important task. Towards this purpose, several methods have been proposed in the past [1]. We will present a brief discussion of these methods in the following subsections.

2.6.1 *Selection of Optimum Time Delay (τ)*

One of the key parameters required for the reconstruction of the phase space is the time delay (τ). The choice of delay should be such that it creates statistically independent vectors. If the delay is too small, the points on the trajectory would be very close, resulting in highly correlated vectors. On the contrary, if the delay is too large, the vectors would become completely uncorrelated, giving rise to a random attractor [1]. The values of the delay will always be in integer multiples of the sampling time. In the case of noise-free data, an arbitrary choice of delay ($\tau \geq 1$) would be acceptable [64]; however, for experimental signals that are contaminated with noise, the choice of delay through an appropriate method is crucial. In the literature, there are two common methods available for the selection of delay, which are based on the autocorrelation function and the average mutual information [1].

2.6.1.1 Autocorrelation Function

Autocorrelation function (r) quantifies a linear dependence between a signal and its delayed version at different values of delay (τ). The expression to calculate autocorrelation function is given as,

$$r(\tau) = \frac{\sum_{i=1}^{N-\tau}[x(t_i + \tau) - \overline{x}(t)][x(t_i) - \overline{x}(t)]}{\sum_{i=1}^{N}[x(t_i) - \overline{x}(t)]^2} \tag{2.32}$$

where $x(t)$ is a signal sampled at equal intervals, $\overline{x} = (1/N)\sum_{i=1}^{N}[x(t_i)]$ is its temporal mean, and N is the total number of data points in the signal.

If the variation of the autocorrelation function with delay τ has a zero-crossing point, then we can use this value of τ as an optimum delay for the phase space reconstruction. In the absence of a distinct zero-crossing point for the variation of $r(\tau)$, we can consider the time instant corresponding to the first local minima of the autocorrelation function as a delay time for phase space reconstruction [1].

2.6.1.2 Average Mutual Information

The method of average mutual information was proposed by Fraser and Swinney [28] to calculate the optimum value of delay time for phase space reconstruction. This method is based on information theory. Mutual information tells us about the information of some measurement say, b, drawn from a set B, from the known information of measurement a in another A. Thus, we can obtain the average mutual information in these two sets as follows,

$$I_{AB} = \sum_{a,b} P_{AB}(a,b) \log_2 \left[\frac{P_{AB}(a,b)}{P_A(a) P_B(b)} \right] \tag{2.33}$$

where $P_{AB}(a, b)$ is the joint probability density, and $P_A(a)$ and $P_B(b)$ are the individual probability densities of measurements a and b from sets A and B, respectively. Thus, I_{AB} measures the degree of mutual correlation between two measurement sets A and B. If the measurements are completely independent, i.e., $P_{AB}(a, b) = P_A(a)P_B(b)$, then $I_{AB} = 0$.

Similarly, we can obtain the average mutual information (I) between a signal $x(t)$ and its delayed version $x(t + \tau)$. This method provides information about $x(t + \tau)$ if we know the details of $x(t)$. If we consider $x(t_i)$ as set A and $x(t_i + \tau)$ as set B, then we can calculate the quantity $I(\tau)$ for different values of delay (τ) as,

$$I(\tau) = \sum_{x(t_i),x(t_i+\tau);i=1}^{N-\tau} P[x(t_i), x(t_i + \tau)] \log_2 \left[\frac{P[x(t_i), x(t_i + \tau)]}{P[x(t_i)]P[x(t_i + \tau)]} \right] \tag{2.34}$$

In the plot of $I(\tau)$ versus τ, we consider the value of τ corresponding to the first local minima of $I(\tau)$ as an optimum time delay for phase space reconstruction. If the decrease of $I(\tau)$ is monotonous without any distinct minimum point, then we can consider the value of τ corresponding to $I(\tau)/I(0) = 0.2$ or $I(\tau)/I(0) = 1/e$ as an optimum delay for phase space reconstruction [1, 40].

2.6.2 *Selection of the Minimum Embedding Dimension (d)*

The choice of minimum embedding dimension for phase space reconstruction is based on the concept of removal of false neighbors of any state point in the phase space [45]. The false nearest neighbors are the neighbors of a state point in the reconstructed state space that could be a result of the projection of a high-dimensional attractor onto a lower-dimensional subspace [1]. Such a projection brings the trajectories that are otherwise far apart in the phase space closer, resulting in the presence of false neighbors.

Let us understand this concept through a simple example. Consider A, B, and C to be three points on the neighboring vertices of a cube projected in three-

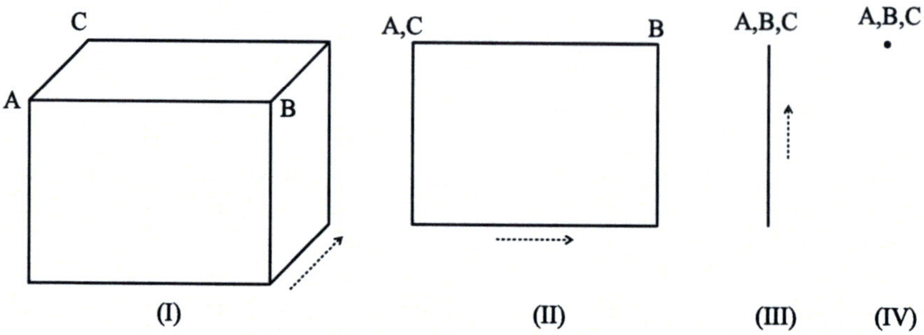

Fig. 2.25 Diagrammatic illustration of the concept of false nearest neighbors, where vertices A, B, and C of a cube (I) can appear to be false neighbors of each other when the cube is projected onto two (II), one (III), and zero (IV) dimensional subspaces. A dotted arrow indicates the direction of the projection

dimensional phase space (Fig. 2.25I). When this cube is projected on its front view in a two-dimensional plane, we notice that points A and B remain far apart, while points A and C appear to collapse onto each other and, thus, appear to be neighbors of each other (Fig. 2.25II). It means that, in this projection view of the cube, points A and C are false neighbors. Now, if the cube is further projected onto a one-dimensional and zero dimensional spaces (i.e., a line and a point, respectively), all three points appear to be located at the same position (Fig. 2.25III, IV). Thus, all these points appears to be false neighbors of each other. Similarly, if we do the same exercise by increasing the dimension from 1 to 3 (from Fig. 2.25III to I), we observe the removal of false neighbors at every iteration of an increase in the dimension. Based on this simple idea of the removal of the false neighbors of a state point, we next discuss a systematic procedure to obtain the minimum embedding dimension required for the phase space reconstruction of a signal.

2.6.2.1 False Nearest Neighbor Method

Kennel et al. [45] proposed a novel method to remove the false neighbors of a state point in the phase space. This method, in turn, helps in revealing the actual dimension of the attractor where neighbors of any state point are the result of the geometric nature of the attractor and not due to their projection onto a lower-dimensional space. In this case, we can recognize false neighbors from the self-crossings of trajectories in the phase space.

In the method of false nearest neighbor, we calculate the Euclidean distance of any state point on the trajectory with every other state point for a particular dimension. Based on the minimum distance between two state points of the trajectory, we identify a nearest neighbor for every state point. We then repeat the procedure for a higher dimension and verify the existence of the same neighbor for the given state point. If the nearest neighbor for a state point remains the same as that observed in the lower-dimensional space, we stop the process of identifying false neighbors; otherwise, we continue the process for the next dimension. We consider an appropriate embedding dimension for phase space reconstruction as the minimum dimension at which we do not observe any false neighbors in the phase space. We refer to the phase space wherein the presence of false neighbors is completely resolved as the *embedded phase space.*

Let us consider $Z_i(d) = [x_i, x_{i+\tau}, x_{i+2\tau}, \ldots.., x_{i+(d-1)\tau}]$ to be a set of delayed vectors constructed from the optimum value of time delay (τ) in a d-dimensional phase space, where $i = 1, 2, \ldots.., N-(d-1)\tau$. We can obtain the value of the delay from the methods such as autocorrelation function or average mutual information, as discussed in the previous section. We then increase the dimension of an attractor in steps until all false neighbors are completely removed.

We now consider $\hat{Z}_i(d) = [\hat{x}_i, \hat{x}_{i+\tau}, \hat{x}_{i+2\tau}, \ldots.., \hat{x}_{i+(d-1)\tau}]$ to be a vector of the nearest neighbors of Z_i in the d-dimensional phase space. We can calculate the Euclidean distance between these nearest neighbors as follows,

$$R_i^2(d) = \sum_{m=1}^{d}[\hat{Z}_i(m) - Z_i(m)]^2 = \sum_{m=1}^{d}[\hat{x}_{i+(m-1)\tau} - x_{i+(m-1)\tau}]^2 \tag{2.35}$$

If the dimension of the attractor is increased to $d+1$, we get the distance between the neighbors as,

$$R_i^2(d+1) = \sum_{m=1}^{d+1}[\hat{x}_{i+(m-1)\tau} - x_{i+(m-1)\tau}]^2 = R_i^2(d) + [\hat{x}_{i+d\tau} - x_{i+d\tau}]^2 \tag{2.36}$$

If the value of $R_i^2(d+1)$ is larger than $R_i^2(d)$, we consider the neighbors observed in the d-dimensional space to be false neighbors. In this case, we repeat the calculation for the next dimension. In the method of false nearest neighbors, to consider whether the given state points on the trajectory are the nearest neighbors of each other or not, we need to use a distance threshold R_T. When the Euclidean distance is less than the threshold, we consider the neighbors $\hat{Z}_i$ of a state point Z_i as actual neighbors; otherwise, they are false neighbors. Thus, we use the criterion for the detection of false neighbors in the case of an infinite amount of noise-free data as follows,

$$\frac{|\hat{x}_{i+d\tau} - x_{i+d\tau}|}{R_i(d)} > R_T \tag{2.37}$$

Here, the value of R_T ranges from 10 to 50 [1].

However, in practical situations, we know the signals are always contaminated with noise. For such type of signals, Abarbanel et al. [1] suggested the following criterion for the detection of false neighbors,

$$\frac{R_i(d+1)}{R_A} \geq 2 \tag{2.38}$$

where $R_A^2 = (1/N)\sum_{i=1}^{N}(x_i - x_{av})^2$ is a measure of the size of the attractor and $x_{av} = (1/N)\sum_{i=1}^{N} x_i$.

2.6.2.2 Cao's Method

As discussed in the method of false nearest neighbors, the criterion to detect a false nearest neighbor of a state point on the phase space trajectory depends on the distance threshold R_T. Depending on the value of R_T, the number of false nearest neighbors observed at a particular d would be different. There is no robust framework developed so far to choose an appropriate value of R_T that can be used for different types of signals and is independent of dimension d. Therefore, to avoid such problem of the choice of threshold, Cao [15] modified the algorithm of false nearest neighbor that is independent of R_T, and provided a new quantity (E_1) which can be used to detect the appropriate embedding dimension (d) required for the phase space reconstruction.

In a similar manner to the method of false nearest neighbors, we consider a quantity $a(i, d)$ to compare the distance of a state point in the phase space with its nearest neighbor for two consecutive dimensions as follows,

$$a(i,d) = \frac{\|Z_i(d+1) - \hat{Z}_{(i,d)}(d+1)\|}{\|Z_i(d) - \hat{Z}_{(i,d)}(d)\|} \tag{2.39}$$

where $i = 1, 2, \ldots\ldots, N - (d-1)\tau$, d is the phase space dimension, τ is the optimum time delay, and $\|.\|$ is the Euclidean distance calculated with the maximum norm. Further, $Z_i(d)$ is the ith reconstructed delay vector and $\hat{Z}_{(i,d)}(d)$ is its nearest neighbor in the d-dimension phase space.

Based on this distance ratio, Cao [15] proposed a new measure $E1$ that quantifies the ratio of mean values of $a(i, d)$ in d and $d+1$ dimensional phase spaces.

$$E1(d) = \frac{E(d+1)}{E(d)} \tag{2.40}$$

where $E(d) = \frac{1}{N-d\tau}\sum_{i=1}^{N-d\tau} a(i,d)$.

The above (2.40) suggests that $E(d)$ depends only on d and τ and not any arbitrary threshold (say, R_T). As we increase d, the value of $E_1(d)$ exhibits an initial growth for lower values of d and saturates when d is above a particular value (say, d_0). We can then use the value of $d_0 + 1$ as an embedding dimension required for the phase space reconstruction of a signal.

Furthermore, Cao [15] defined another measure ($E2$) that can help in distinguishing a deterministic signal from a stochastic signal as follows,

$$E2(d) = \frac{E^*(d+1)}{E^*(d)} \tag{2.41}$$

where $E^*(d) = \frac{1}{N-d\tau} \sum_{i=1}^{N-d\tau} |x_{i+d\tau} - \hat{x}_{n(i,d)+d\tau}|$.

For a random signal, the value of $E2(d)$ remains constant near unity for all values of d. However, for a deterministic signal, $E2(d)$ takes a value different from unity for lower values of d and approaches $E1(d)$ for higher values of d. Thus, the calculation of both the measures $E1$ and $E2$ helps in identifying the embedding dimension and also in differentiating deterministic or random properties of a signal.

2.7 Poincaré Section (or First Return Map)

In the previous section, we discussed the procedure for the phase space reconstruction of a univariate time series of a system. We showed that the temporal behavior of the trajectory in the reconstructed phase space displays a particular pattern, which further helps in determining the underlying dynamics of the system. However, in some situations, the method of phase space reconstruction is insufficient to reveal all the hidden features in a signal, usually when the dynamics is highly complex. In such a situation, we can use a simpler method based on the Poincaré section [70] to unravel the hidden dynamical characteristics of the signal.

In order to obtain Poincaré section of a signal, we need to project an attractor from a d-dimensional phase space onto a $(d-1)$-dimensional phase space by making a transverse cut to the attractor [73, 107], see Fig. 2.26. The location of the transverse plane along the attractor is not much important. Mostly, the half-cut of the attractor is preferred for better visualization of the Poincaré section (Fig. 2.26). If the trajectory evolves forward in time, we mark all crossings of the trajectory through the transverse plane (Poincaré section) as discrete points. In this manner, a complete evolution of the phase space trajectory creates several marks on the Poincaré section. The patterns generated by these marks indicate the dynamical features of a particular attractor in the embedded phase space, which are otherwise difficult to detect from just a visual inspection of the phase space [39].

For example, a Poincaré section of a limit cycle attractor in three-dimension shows a single dot or a fixed point (Fig. 2.26a). For a period-2 attractor, the Poincaré section displays two discrete dots. Similarly, we can extend the argument to characterize period-n attractors, where the corresponding Poincaré sections show n distinct dots. For the two-period quasiperiodic attractor, as the motion of the trajectory revolves around a 2-torus, its Poincaré section shows a closed-loop orbit (Fig. 2.26b). In the case of chaos, as the evolution of the phase space trajectories is bounded by a strange attractor, its Poincaré section shows a clutter of points without having any specific pattern (Fig. 2.26c).

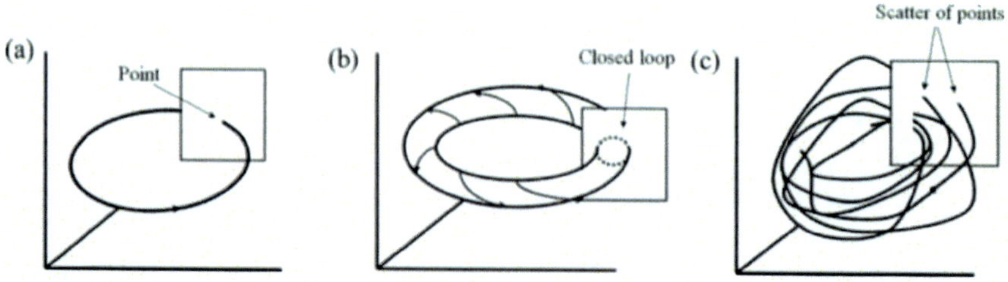

Fig. 2.26 Diagrammatic representation of Poincaré sections of (**a**) a period-1 limit cycle attractor, (**b**) a two-frequency quasiperiodic attractor, and (**c**) a chaotic attractor embedded in a three-dimensional phase space. The Poincaré section shows a point for the limit cycle attractor, a closed loop structure for the quasiperiodic attractor, and a random distribution of points for the chaotic attractor

2.8 Recurrence Plots

In Sect. 2.6, we discussed a method of phase space reconstruction that has been widely used in various disciplines of science and engineering to visualize the dynamical behavior of a signal in the higher-dimensional phase space. However, we have restriction of dimensions up to three, for which the evolution of the phase space trajectory can be practically visualized. To overcome this problem of revealing the dynamical features of higher-dimensional (greater than three) attractors, Eckmann et al. [22] developed a method based on the recurring behavior of the phase space trajectory, popularly known as recurrence plot (RP) analysis. According to him, since attractors of dissipative deterministic systems are bounded, the phase space trajectory of such attractors re-visits the neighborhood of a previous location after some divergence in the phase space. The evidence of such recurrence behavior of the trajectory was first documented by Poincaré in the explanation of his restricted three-body problem of conservative system as, "*In this case, neglecting some exceptional trajectories, the occurrence of which is infinitely improbable, it can be shown, that the system recurs infinitely many times as close as one wishes to its initial state*" [58]. The behavior of the phase space trajectory to arrive very close to its previous location is called recurrence and a plot that visualizes these recurrence features is referred to as a recurrence plot [58] (see Fig. 2.27).

Recurrence is a part of our day-to-day life. Most real-life situations are predictable to some extent as there is some cyclic/seasonal behavior associated with them. Examples include events of sunrise or sunset, respiration, heartbeat, menstrual cycle, bird migration, weather, etc.

In order to construct the recurrence plot from a univariate time series, we first need to project a signal onto its actual or embedded phase space. Then we need to calculate the pairwise Euclidean distances between all state points of the phase space trajectory, for building a distance matrix. Since the trajectories do not recur exactly at the same location but approach in the neighborhood of its previous location, we need to use an appropriate threshold (ϵ) to convert the distance matrix into a

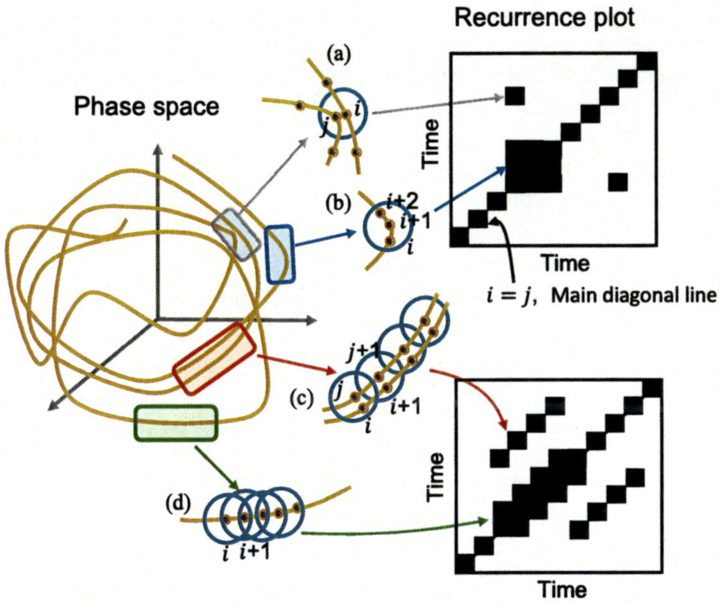

Fig. 2.27 Diagrammatic illustration of the construction of a recurrence plot (RP) from the phase space information of a signal. Depending on the recurrence behavior of the phase space trajectories inside the threshold (indicated by circles around the state points of the trajectory), the recurrence plot shows different arrangements of black points. (**a**) The recurrence of only two state points of neighboring trajectories is observed as an isolated (single) point in RP. (**b**) The trapping of more than one state points of the trajectory inside the threshold is observed as vertical and horizontal lines in RP. The recurrence of (**c**) two parallel running trajectories and (**d**) consecutive points of the same trajectory are visualized as diagonal lines in RP

recurrence matrix. Thus, we can construct the recurrence matrix from the time series as follows [58],

$$R_{i,j} = \Theta(\epsilon - \|x_i - x_j\|) \tag{2.42}$$

Here, $R_{i,j}$ is the recurrence matrix such that $R_{i,j} \in \mathbb{R}^d$ and $i, j = 1, 2, \ldots\ldots, N - (d-1)\tau$. N is the number of data points in the signal, d is the embedding dimension, and τ is the optimum time delay required for the phase space reconstruction. x_i and x_j are delayed vectors, which correspond to state points in the phase space. Θ is the Heaviside step function such that $\Theta(X < 0) = 0$ and $\Theta(X > 0) = 1$, and $\|.\|$ is the distance calculated from Euclidean or maximum norm.

The choice of an optimum threshold (ϵ) for detecting the recurrence behavior of the phase space trajectory is crucial. No single methodology has been established until now that can be universalized to all types of dynamics [57]. The general consideration is to use a smaller size threshold; however, at the same time, the threshold should contain a sufficient number of recurrences of the state points. There are two common ways to fix the threshold and quantify the recurrence behavior of the phase space trajectory [58]. One way is to fix the recurrence threshold based on the percentage of the size of the attractor. Depending on the attractor size, the threshold size may vary from one attractor to other. However, the threshold size is kept the same for all state points of a particular attractor. In this case, whenever any state point and its neighbors fall inside the threshold, the corresponding points are assumed to have recurred in the phase space.

Another way to measure the recurrence behavior of the trajectory is to fix the number of nearest neighbors for a state point, i.e., fixing the recurrence rate. In this analysis, the threshold size would change from one state point to another state point of the same attractor to accommodate a fixed number of neighbors in the threshold. Some of the "rules of thumb" that have been followed for the selection of the threshold are: (1) ϵ should not exceed 10% of the mean or maximum diameter of the phase space [112], (2) it should be such that the recurrence rate (RR) is approximately 1% [113], (3) it corresponds to a maximum value of derivative $d(RR)/d\epsilon$ [30], or (4) in the case of noisy signals, we can choose $\epsilon > 5\sigma$, where σ is the standard deviation of the observational noise [96]. All this discussion suggests that the choice of threshold is not obvious and is mostly signal specific.

Furthermore, we note that the recurrence matrix in Eq. (2.42) is binary and consists of a combination of zeros and ones [22]. The points with ones are marked as black points and that with zeros are marked as white points in the recurrence plot. Thus, the recurrence plot is a two-dimensional projection of the d-dimensional phase space, visualized in the form of black and white points. The advantage of recurrence plots is that we can obtain such plots from a shorter time series and non-stationary signals [58].

The patterns of black points in the recurrence plot aid in identifying different dynamical features governed by a signal [58], see Fig. 2.27. A homogeneous distribution of black points in the recurrence plot indicates the presence of stationarity or randomness in the signal. The presence of continuous diagonal lines in the recurrence plot is suggestive of the periodic dynamics of the signal, where the spacing between such diagonal lines represents the period of the oscillation. For quasiperiodic signals, the diagonal lines are broken and the spacing between them is non-uniform due to the aperiodic nature of the oscillations. For chaotic signals, as the trajectories remain parallel for small durations, their recurrence plots show a distribution of short diagonal lines; thus, indicating low predictability of such signals. For noisy signals, recurrence plots show a rare occurrence of isolated points due to the absence of predictability in the signal. For the intermittency signal, recurrence plots show an alternate occurrence of large sized black patches and white patches, corresponding to epochs of low amplitude and high amplitude oscillations in the signal, respectively.

A close inspection of recurrence plots reveals the possibility of only three types of black structures in it (Fig. 2.27). These structures include single dots, diagonal lines, and vertical and horizontal lines [3, 58]. Isolated points occur mainly for random or uncorrelated signals where the dynamics at the next event in the phase space is completely unpredictable. Diagonal lines parallel to the main diagonal

($R_{i+l,j+l}$, where l is the length of the diagonal line), indicate the presence of determinism or the parallel running of neighboring trajectories in the phase space. Lines perpendicular to the main diagonal suggest the possibility of inappropriate embedding of a signal in the phase space. The vertical ($R_{i,j+v}$, where v is the length of the vertical line) and horizontal lines indicate the trapping of the phase space trajectory for a short duration or the slow variation of the states in time. Such behavior is observed for intermittent oscillations.

In the following subsections, we discuss the extension of the recurrence plot analysis of a single variable system to a multivariable system, and present different measures derived from the simultaneous quantification of black structures in the recurrence plots of two variables.

2.8.1 Cross Recurrence Plots

Zbilut et al. [111] extended the analysis of the recurrence properties of a single time series to capture the recurrence properties of multivariate time series. This method of simultaneous characterization of the recurrence properties of two signals is called cross-recurrence analysis and the corresponding plot is known as cross recurrence plot (CRP). In this method, the time series of both the signals are embedded in the same phase space, and the recurrence behavior of a state point (x_i) on the trajectory of one signal is measured with respect to the neighborhood location of the state points (y_j) on the trajectory of another signal (see Fig. 2.28a). We can compute the cross recurrence matrix of two signals as follows,

$$CR_{i,j} = \Theta(\epsilon - \|x_i - y_j\|), \tag{2.43}$$

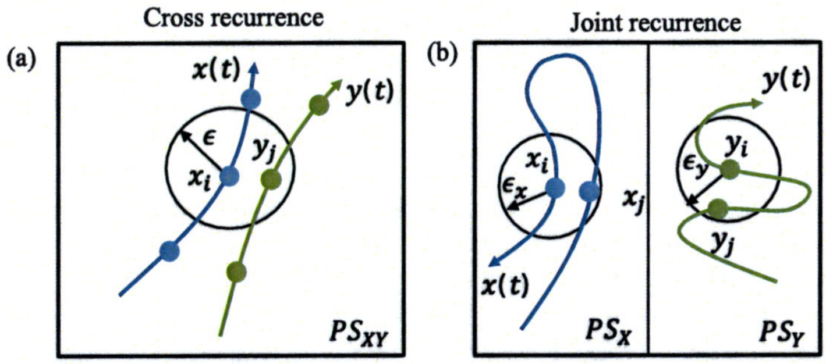

Fig. 2.28 Schematics showing the process of computation of **(a)** cross recurrence and **(b)** joint recurrence of trajectories $x(t)$ and $y(t)$ of systems X and Y in the same phase space PS_{XY} and the individual phase space (PS_X and PS_Y), respectively. Adapted with permission from Zou et al. [116]

where $x_i, y_j \in \mathbb{R}^d$; $i = 1, 2, \ldots, N$ and $j = 1, 2, \ldots, M$. Since the total number of data points in both the signals need not be the same for the calculation of CRP, the recurrence matrix of CRP can be asymmetric $[N \times M]$. Similar to recurrence plots, the long diagonal structures in CRP indicate the identical behavior of both the signals in the phase space.

2.8.2 *Joint Recurrence Plot*

Romano et al. [76, 77] introduced joint recurrence plots (JRP) as another method for recurrence analysis of multivariate signals. In this method, the simultaneous recurrence of phase space trajectories of two signals is measured in their individual phase space (see Fig. 2.28b). Thus, JRP counts the simultaneous occurrence of black points at particular time instances in the recurrence plots of both the signals. As JRP is constructed through the product of recurrence matrices of two signals, the number of data points in these signals must be the same. We can obtain the joint recurrence matrix of two signals using the following expression,

$$JR_{i,j} = \Theta(\epsilon_x - \|x_i - x_j\|)\Theta(\epsilon_y - \|y_i - y_j\|), \tag{2.44}$$

where $x_i \in \mathbb{R}^d, y_i \in \mathbb{R}^m$; $i, j = 1, 2, \ldots, N$. The JRP can also be used for the detection of synchronization properties of two signals as it captures the simultaneous recurrence of the dynamics of different systems in time [77].

2.8.3 *Recurrence Quantification Analysis*

Various measures have been developed over the years to compute the recurrence properties of a signal. These measures account for the duration for which the phase space trajectories recur in the embedded phase space. Such a characterization of black structures in recurrence plots is commonly referred to as recurrence quantification analysis (RQA). As mentioned earlier, the black points are arranged primarily in three manners in the recurrence plot, which are isolated points, diagonal lines, and vertical or horizontal lines. Zbilut and Webber [104, 112] were the first to propose a method of recurrence quantification analysis of diagonal structures in the recurrence plot. Marwan et al. [59] extended this analysis to quantify the vertical structures present in the recurrence plot (note that as the recurrence plot is symmetric, the quantification of vertical or horizontal lines gives the same results). Recurrence quantification analysis helps in detecting the transition from chaotic to periodic oscillations (using measures that quantify diagonal lines [99]) or chaos-chaos transition (using measures that quantify vertical lines [59]) observed in the same signal or in the bifurcation diagram when a control parameter is varied systematically [59].

Next, we present a list of the recurrence quantification measures [58] that are popularly used in nonlinear time series analysis of practical signals.

1. **Recurrence rate** (*RR*)**:** It quantifies the ratio of recurrence (black) points with respect to the total number of state points present in the phase space. The value of this measure can range from 0 (no recurrence) to 1 (complete recurrence). The calculation of RR and the correlation sum (as discussed in Sect. 2.4.4.2) are nearly the same, except that the main diagonal line is excluded in the computation of RR and not in the calculation of the correlation sum [58]. We can obtain RR as follows,

$$RR = \frac{1}{N^2} \sum_{i,j=1}^{N} R_{i,j} \tag{2.45}$$

2. **Determinism** (*DET*)**:** It calculates the fraction of recurrence points that form diagonal lines in the recurrence plot. Long diagonal lines indicates the deterministic feature of a signal. For periodic signals, as the diagonal lines are continuous due to strong determinism, DET takes a value near 1. As white noise signals are highly uncorrelated with mostly isolated black points in the recurrence plot, the value of DET lies close to zero. In contrast, as chaotic signals possess broken diagonal lines due to their low determinism characteristics, the value of DET appears in between 0 and 1. We can calculate DET for a signal using the following expression,

$$DET = \frac{\sum_{l=l_{\min}}^{N} l P(l)}{\sum_{l=1}^{N} l P(l)} \tag{2.46}$$

   where $P(l)$ is the frequency distribution of diagonal lines of length l present in the recurrence plot and $l_{\min}$ is the minimum number of black points considered to form a diagonal line.
3. **Laminarity** (*LAM*)**:** This measure quantifies the number of black points that form vertical lines in the recurrence plot. The vertical or horizontal lines are formed in the recurrence plot only when a signal possesses low amplitude (or slow varying) fluctuations, similar to laminar phases observed in the intermittency signal. We can compute LAM as,

$$LAM = \frac{\sum_{v=v_{\min}}^{N} v P(v)}{\sum_{v=1}^{N} v P(v)} \tag{2.47}$$

   where $P(v)$ is the frequency distribution of vertical lines of length v, and $v_{\min}$ is the minimum number of black points considered to form a vertical line in the recurrence plot.

4. **Ratio ($RATIO$):** It is the ratio of DET and RR. This measure is used in detecting the dynamic transitions where both RR and DET have independent variation. For example, in some cases RR decreases but DET remains unchanged, resulting in the increase in the value of $RATIO$.

$$RATIO = \frac{DET}{RR} = N^2 \frac{\sum_{l=l_{\min}}^{N} lP(l)}{\left(\sum_{l=1}^{N} lP(l)\right)^2} \tag{2.48}$$

5. **Average diagonal line length (L):** This measure computes the average predictability time of a signal. We can obtain L using the following expression,

$$L = \frac{\sum_{l=l_{\min}}^{N} lP(l)}{\sum_{l=l_{\min}}^{N} P(l)} \tag{2.49}$$

6. **Trapping time (TT):** This measure indicates the average laminarity time of a signal, i.e., the average time for which a signal remains in the same state. We can get TT through the following expression,

$$TT = \frac{\sum_{v=v_{\min}}^{N} vP(v)}{\sum_{v=v_{\min}}^{N} P(v)} \tag{2.50}$$

7. **Longest diagonal line ($L_{\max}$):** This indicates the length of the longest diagonal line in the recurrence plot, excluding the main diagonal line (i.e., $i = j$). $L_{\max}$ can also be considered as the inverse of the most positive Lyapunov exponent [22, 99]. A positive Lyapunov exponent indicates exponential divergence of nearby phase trajectories, which is the characteristic of chaotic signals. Thus, for chaotic signals, the value of $L_{\max}$ is small, while for periodic signals, its value is very high. The expression of $L_{\max}$ is given below,

$$L_{max} = \max(l_i; i = 1, 2, \ldots\ldots, N_l) \tag{2.51}$$

where N_l is the number of diagonal lines in the recurrence plot.
8. **Divergence (DIV):** It measures the divergence of the trajectories that run parallel for some time in the phase space. It is expressed as the inverse of $L_{\max}$,

$$DIV = \frac{1}{L_{\max}} \tag{2.52}$$

9. **Entropy ($ENTR$):** This is a measure of the complexity of diagonal structures present in the recurrence plot and it is calculated using the following expression,

$$ENTR = -\sum_{l=l_{\min}}^{N} p(l) \ln p(l) \quad (2.53)$$

where $p(l)$ is the probability distribution of diagonal lines having length l, calculated as $p(l) = P(l)/\sum_{l=l_{\min}}^{N} P(l)$. The value of $ENTR$ is higher for chaotic signals and very low for regular signals.

With this discussion on recurrence measures, we end the chapter on the basics of dynamical systems theory. Next, we move on to the practical application of this acquired knowledge in analyzing the experimental data obtained from thermoacoustic systems having both laminar and turbulent flow fields. In particular, we will discuss the application of bifurcation theory in detecting the different routes to thermoacoustic instability, and the use of phase space reconstruction, Poincaré section, and recurrence plot analyses in identifying the dynamical features of different thermoacoustic oscillations. Furthermore, we will illustrate the application of measures derived from aforementioned methods in detecting the transition between various dynamical states as the control parameter is varied (Chaps. 4 and 6) and in providing early warnings to the onset of thermoacoustic instability (Chap. 12).

References

1. Abarbanel, H.D.I., Brown, R., Sidorowich, J.J., Tsimring, L.S.: The analysis of observed chaotic data in physical systems. Rev. Mod. Phys. **65**(4), 1331 (1993)
2. Anishchenko, V.S., Strelkova, G.I.: Irregular attractors. Discrete Dyn. Nat. Soc. **2** (1998)
3. Aparicio, T., Pozo, E.F., Saura, D.: Detecting determinism using recurrence quantification analysis: Three test procedures. J. Econ. Behav. Organ. **65**(3), 768–787 (2008)
4. Argyris, J.H., Faust, G., Haase, M., Friedrich, R.: An Exploration of Dynamical Systems and Chaos: Completely Revised and Enlarged. Springer, Berlin (2015)
5. Arnold, L., Wihstutz, V.: Lyapunov exponents: a survey. In: Lyapunov Exponents, pp. 1–26. Springer, Berlin (1986)
6. Arrowsmith, D.K., Place, C.M., Place, C., et al.: An Introduction to Dynamical Systems. Cambridge University Press, Cambridge (1990)
7. Awrejcewicz, J.: Bifurcation and Chaos: Theory and Applications. Springer Science & Business Media, Berlin (2012)
8. Awrejcewicz, J., Lamarque, C.H.: Bifurcation and Chaos in Nonsmooth Mechanical Systems, vol. 45. World Scientific, Singapore (2003)
9. Babloyantz, A., Destexhe, A.: Is the normal heart a periodic oscillator? Biol. Cybern. **58**(3), 203–211 (1988)
10. Bader, R.: Nonlinearities and Synchronization in Musical Acoustics and Music Psychology, vol. 2. Springer Science & Business Media, Berlin (2013)
11. Bergé, P., Dubois, M., Mannevillel, P., Pomeau, Y.: Intermittency in Rayleigh-Bénard convection. J. Phys. Lett. **41**(15), 341–345 (1980)
12. Birkhoff, G.D.: Dynamical Systems. American Mathematical Society, New York (1960)
13. Bolotin, Y., Tur, A., Yanovsky, V.: Chaos: Concepts, Control and Constructive Use. Springer, Berlin (2009)

14. Camastra, F.: Data dimensionality estimation methods: a survey. Pattern Recognit. **36**(12), 2945–2954 (2003)
15. Cao, L.: Practical method for determining the minimum embedding dimension of a scalar time series. Physica D **110**(1–2), 43–50 (1997)
16. Chen, G., Yu, X.: Chaos Control: Theory and Applications, vol. 292. Springer Science & Business Media, Berlin (2003)
17. Christopher, C., Li, C.: Limit Cycles of Differential Equations. Springer Science & Business Media, Berlin (2007)
18. Craig, W.: Hamiltonian Dynamical Systems and Applications. Springer Science & Business Media, Berlin (2008)
19. Culick, F.E.C.: Some recent results for nonlinear acoustics in combustion chambers. AIAA J. **32**(1), 146–169 (1994)
20. Devaney, R.L.: A first course in chaotic dynamical systems: theory and experiment. Addison-Wesley, Reading, MA (1992)
21. Dickey, L.A.: Soliton Equations and Hamiltonian Systems. World Scientific, Singapore (2003)
22. Eckmann, J.P., Kamphorst, S.O., Ruelle, D.: Recurrence plots of dynamical systems. Europhys. Lett. **4**(9), 973–977 (1987)
23. Eckmann, J.P., Ruelle, D.: Ergodic theory of chaos and strange attractors. In: The Theory of Chaotic Attractors, pp. 273–312. Springer, Berlin (1985)
24. Elaskar, S., Del Río, E.: New Advances on Chaotic Intermittency and Its Applications. Springer, Berlin (2017)
25. Elhadj, Z.: Dynamical Systems: Theories and Applications. CRC Press, Boca Raton (2019)
26. Feigenbaum, M.J.: The universal metric properties of nonlinear transformations. J. Stat. Phys. **21**(6), 669–706 (1979)
27. Field, R.J., Noyes, R.M.: Oscillations in chemical systems. IV. Limit cycle behavior in a model of a real chemical reaction. J. Chem. Phys. **60**(5), 1877–1884 (1974)
28. Fraser, A.M., Swinney, H.L.: Independent coordinates for strange attractors from mutual information. Phys. Rev. A **33**(2), 1134–1140 (1986)
29. Fuchs, A.: Nonlinear Dynamics in Complex Systems. Springer, Berlin (2014)
30. Gao, Z., Jin, N.: Flow-pattern identification and nonlinear dynamics of gas-liquid two-phase flow in complex networks. Phys. Rev. E **79**(6), 066303 (2009)
31. Ghaffari, A., Tomizuka, M., Soltan, R.A.: The stability of limit cycles in nonlinear systems. Nonlinear Dyn. **56**(3), 269–275 (2009)
32. Glass, L.: Synchronization and rhythmic processes in physiology. Nature **410**(6825), 277—284 (2001)
33. Grassberger, P., Procaccia, I.: Characterization of strange attractors. Phys. Rev. Lett. **50**(5), 346–349 (1983)
34. Grebogi, C., Ott, E., Pelikan, S., Yorke, J.A.: Strange attractors that are not chaotic. Physica D **13**(1–2), 261–268 (1984)
35. Grobman, D.: Homeomorphisms of systems of differential equations (Russian). Dokl. Akad. Nauk. **128**, 880–881 (1959)
36. Guckenheimer, J., Holmes, P.: Nonlinear Oscillations, Dynamical Systems, and Bifurcations of Vector Fields, vol. 42. Springer Science & Business Media, Berlin (2013)
37. Haddad, W.M., Chellaboina, V., Nersesov, S.G.: Thermodynamics: A Dynamical Systems Approach, vol. 23. Princeton University Press, Princeton (2009)
38. Hartman, P.: A Lemma in the theory of structural stability of differential equations. Proc. Am. Math. Soc. **11**, 610–620 (1960)
39. Hilborn, R.C.: Chaos and Nonlinear Dynamics: An Introduction for Scientists and Engineers. Oxford University Press, Oxford (2000)
40. Horak, J., Krlín, L., Raidl, A.: Deterministicky Chaos a Jeho Fyzikalni Aplikace. Academia, Praha (2003)
41. Izhikevich, E.M.: Dynamical Systems in Neuroscience. MIT Press, Cambridge (2007)

42. Kantz, H., Schreiber, T.: Nonlinear Time Series Analysis, vol. 7. Cambridge University Press, Cambridge (2004)
43. Kapitaniak, T.: Chaos for Engineers: Theory, Applications, and Control. Springer Science & Business Media, Berlin (2012)
44. Katok, A., Hasselblatt, B.: Introduction to the Modern Theory of Dynamical Systems, vol. 54. Cambridge University Press, Cambridge (1997)
45. Kennel, M.B., Brown, R., Abarbanel, H.D.I.: Determining embedding dimension for phase-space reconstruction using a geometrical construction. Phys. Rev. A **45**(6), 3403–3411 (1992)
46. Klimaszewska, K., Żebrowski, J.J.: Detection of the type of intermittency using characteristic patterns in recurrence plots. Phys. Rev. E **80**(2), 026214 (2009)
47. Kuznetsov, Y.A.: Elements of Applied Bifurcation Theory, vol. 112. Springer Science & Business Media, New York (2013)
48. Lakshmanan, M., Rajaseekar, S.: Nonlinear Dynamics: Integrability, Chaos and Patterns. Springer Science & Business Media, Berlin (2012)
49. Lam, L., Morris, H.C.: Nonlinear Structures in Physical Systems: Pattern Formation, Chaos, and Waves. Springer Science & Business Media, New York (1990)
50. Layek, G.: An Introduction to Dynamical Systems and Chaos. Springer, Berlin (2015)
51. Lorenz, E.N.: Deterministic nonperiodic flow. J. Atmos. Sci. **20**(2), 130–141 (1963)
52. Lowenstein, J.H.: Essentials of Hamiltonian Dynamics. Cambridge University Press, Cambridge (2012)
53. Lozano, R., Brogliato, B., Egeland, O., Maschke, B.: Dissipative systems. In: Dissipative Systems Analysis and Control, pp. 111–166. Springer, Berlin (2000)
54. MacKay, R.S., Meiss, J.D.: Hamiltonian Dynamical Systems: A Reprint Selection. CRC Press, Boca Raton (2020)
55. Manneville, P., Pomeau, Y.: Intermittency and the Lorenz model. Phys. Lett. A **75**(1–2), 1–2 (1979)
56. Marsden, J.E., McCracken, M.: The Hopf Bifurcation and Its Applications, vol. 19. Springer Science & Business Media, New York (2012)
57. Marwan, N.: How to avoid potential pitfalls in recurrence plot based data analysis. Int. J. Bifurc. Chaos **21**(04), 1003–1017 (2011)
58. Marwan, N., Romano, M.C., Thiel, M., Kurths, J.: Recurrence plots for the analysis of complex systems. Phys. Rep. **438**(5), 237–329 (2007)
59. Marwan, N., Wessel, N., Meyerfeldt, U., Schirdewan, A., Kurths, J.: Recurrence-plot-based measures of complexity and their application to heart-rate-variability data. Phys. Rev. E **66**(2), 026702 (2002)
60. Mawhin, J.: Critical Point Theory and Hamiltonian Systems, vol. 74. Springer Science & Business Media, New York (2013)
61. Moiola, J.L., Chen, G.: Hopf Bifurcation Analysis: A Frequency Domain Approach, vol. 15. World Scientific, Singapore (1996)
62. Moser, J., Kyner, W.T.: Lectures on Hamiltonian systems. In: Hamiltonian Dynamical Systems. CRC Press, Boca Raton (1987)
63. Nair, V.: Role of intermittent dynamics in the onset of combustion instability. Ph.D. thesis, Indian Institute of Technology Madras (2014)
64. Nayfeh, A.H., Balachandran, B.: Applied Nonlinear Dynamics: Analytical, Computational, and Experimental Methods. Wiley, New York (2008)
65. Newhouse, S., Ruelle, D., Takens, F.: Occurrence of strange AxiomA attractors near quasi periodic flows on T^m, $m \geqq 3$. Commun. Math. Phys. **64**(1), 35–40 (1978)
66. Nicolis, G.: Dissipative systems. Rep. Prog. Phys. **49**(8), 873 (1986)
67. Perko, L.: Differential Equations and Dynamical Systems, vol. 7. Springer Science & Business Media, New York (2013)
68. Pikovsky, A., Politi, A.: Lyapunov Exponents: A Tool to Explore Complex Dynamics. Cambridge University Press, Cambridge (2016)
69. Pikovsky, A.S., Feudel, U.: Characterizing strange nonchaotic attractors. Chaos **5**(1), 253–260 (1995)

70. Poincaré, H.: Mémoire sur les courbes définies par une équation différentielle (I). J. Math. Pures Appl. **7**, 375–422 (1881)
71. Pomeau, Y., Manneville, P.: Intermittent transition to turbulence in dissipative dynamical systems. Commun. Math. Phys. **74**(2), 189–197 (1980)
72. Premraj, D., Pawar, S.A., Kabiraj, L., Sujith, R.I.: Strange nonchaos in self-excited singing flames. Europhys. Lett. **128**(5), 54005 (2020)
73. Puu, T.: Attractors, Bifurcations, & Chaos: Nonlinear Phenomena in Economics. Springer Science & Business Media, Berlin (2013)
74. Rasband, S.N.: Chaotic Dynamics of Nonlinear Systems. Wiley, New York (1990)
75. Robinson, R.C.: An Introduction to Dynamical Systems, vol. 19. Prentice-Hall, Englewood Cliffs, NJ (2004)
76. Romano, M.C., Thiel, M., Kurths, J., von Bloh, W.: Multivariate recurrence plots. Phys. Lett. A **330**(3–4), 214–223 (2004)
77. Romano, M.C., Thiel, M., Kurths, J., Kiss, I.Z., Hudson, J.: Detection of synchronization for non-phase-coherent and non-stationary data. Europhys. Lett. **71**(3), 466–472 (2005)
78. Rondoni, L., Mitra, M., Banerjee, S.: Applications of Chaos and Nonlinear Dynamics in Engineering. Springer, Berlin (2011)
79. Rössler, O.E.: An equation for hyperchaos. Phys. Lett. A **71**(2–3), 155–157 (1979)
80. Ruelle, D.: Chaotic Evolution and Strange Attractors, vol. 1. Cambridge University Press, Cambridge (1989)
81. Ruelle, D.: Turbulence, Strange Attractors, and Chaos, vol. 16. World Scientific, Singapore (1995)
82. Ruelle, D., Takens, F.: On the nature of turbulence. Les rencontres physiciens-mathématiciens de Strasbourg-RCP25 **12**, 1–44 (1971)
83. Sacher, J., Elsässer, W., Göbel, E.O.: Intermittency in the coherence collapse of a semiconductor laser with external feedback. Phys. Rev. Lett. **63**(20), 2224–2227 (1989)
84. Sarkar, A.: Renormalization group as a probe for dynamical systems. Ph.D. thesis (2013)
85. Sawicki, J.: Delay Controlled Partial Synchronization in Complex Networks. Springer Nature, Cham (2019)
86. Schaeffer, D.G., Cain, J.W.: Examples of global bifurcation. In: Ordinary Differential Equations: Basics and Beyond, pp. 403–450. Springer, Berlin (2016)
87. Schuster, H.G., Just, W.: Deterministic Chaos: An Introduction. Wiley, New York (2006)
88. Seydel, R.: Practical Bifurcation and Stability Analysis, vol. 5. Springer Science & Business Media, New York (2009)
89. Shivamoggi, B.K.: Nonlinear Dynamics and Chaotic Phenomena: An Introduction, vol. 103. Springer, Berlin (2014)
90. Skiadas, C.H., Dimotikalis, I.: Chaotic Systems: Theory and Applications. World Scientific, Singapore (2010)
91. Stavroulakis, P.: Chaos Applications in Telecommunications. CRC Press, Boca Raton (2005)
92. Strogatz, S.H.: Nonlinear Dynamics and Chaos: With Applications to Physics, Biology, Chemistry, and Engineering. CRC Press, Boca Raton (1994)
93. Sulis, W.H., Combs, A.: Nonlinear Dynamics in Human Behavior, vol. 5. World Scientific, Singapore (1996)
94. Takens, F.: Detecting strange attractors in turbulence. Lect. Notes Math. **898**(1), 366–381 (1981)
95. Theiler, J.: Estimating fractal dimension. J. Opt. Soc. Am. A **7**(6), 1055–1073 (1990)
96. Thiel, M., Romano, M.C., Kurths, J., Meucci, R., Allaria, E., Arecchi, F.T.: Influence of observational noise on the recurrence quantification analysis. Physica D **171**(3), 138–152 (2002)
97. Thomas, J.P., Dowell, E.H., Hall, K.C.: Nonlinear inviscid aerodynamic effects on transonic divergence, flutter, and limit-cycle oscillations. AIAA J. **40**(4), 638–646 (2002)
98. Thompson, J.M.T., Stewart, H.B.: Nonlinear Dynamics and Chaos. Wiley, New York (2002)
99. Trulla, L.L., Giuliani, A., Zbilut, J.P., Webber, C.L.: Recurrence quantification analysis of the logistic equation with transients. Phys. Lett. A **223**(4), 255–260 (1996)

100. Vallacher, R.R., Nowak, A.E.: Dynamical Systems in Social Psychology. Academic, San Diego (1994)
101. Vassilicos, J.C., et al.: Intermittency in Turbulent Flows. Cambridge University Press, Cambridge (2001)
102. Venkatramani, J., Nair, V., Sujith, R.I., Gupta, S., Sarkar, S.: Precursors to flutter instability by an intermittency route: a model free approach. J. Fluids Struct. **61**, 376–391 (2016)
103. Wainwright, J., Ellis, G.F.R.: Dynamical Systems in Cosmology. Cambridge University Press, Cambridge (2005)
104. Webber, C.L., Zbilut, J.P.: Dynamical assessment of physiological systems and states using recurrence plot strategies. J. Appl. Physiol. **76**(2), 965–973 (1994)
105. Wiggins, S.: Introduction to Applied Nonlinear Dynamical Systems and Chaos, vol. 2. Springer Science & Business Media, New York (2003)
106. Wiggins, S.: Global Bifurcations and Chaos: Analytical Methods, vol. 73. Springer Science & Business Media, New York (2013)
107. Williams, G.: Chaos Theory Tamed. CRC Press, Boca Raton (1997)
108. Wolf, A., Swift, J.B., Swinney, H.L., Vastano, J.A.: Determining lyapunov exponents from a time series. Physica D **16**(3), 285–317 (1985)
109. Yamada, T., Fukushima, K., Yazaki, T.: A new type of intermittency in an electronic circuit. Prog. Theor. Phys. Suppl. **99**, 120–130 (1989)
110. Zaslavsky, G.M.: The Physics of Chaos in Hamiltonian Systems. World Scientific, Singapore (2007)
111. Zbilut, J.P., Giuliani, A., Webber Jr, C.L.: Detecting deterministic signals in exceptionally noisy environments using cross-recurrence quantification. Phys. Lett. A **246**(1–2), 122–128 (1998)
112. Zbilut, J.P., Webber, C.L.: Embeddings and delays as derived from quantification of recurrence plots. Phys. Lett. A **171**(3-4), 199–203 (1992)
113. Zbilut, J.P., Zaldivar-Comenges, J.M., Strozzi, F.: Recurrence quantification based Liapunov exponents for monitoring divergence in experimental data. Phys. Lett. A **297**(3–4), 173–181 (2002)
114. Zhao, X.Q.: Dynamical Systems in Population Biology, vol. 16. Springer, Berlin (2003)
115. Zhou, C.Y., So, R.M.C., Lam, K.: Vortex-induced vibrations of an elastic circular cylinder. J. Fluids Struct. **13**(2), 165–189 (1999)
116. Zou, Y., Donner, R.V., Marwan, N., Donges, J.F., Kurths, J.: Complex network approaches to nonlinear time series analysis. Phys. Rep. **787**, 1–97 (2019)

Chapter 3
Bifurcation to Limit Cycle Oscillations in Laminar Thermoacoustic Systems

In this chapter, we study the bifurcation characteristics of a laminar thermoacoustic system through the application of dynamical system theory. We restrict our discussions to two simple prototypes of real thermoacoustic systems (known as Rijke-type thermoacoustic systems): a horizontal Rijke tube with an electrical heater and a ducted premixed flame combustor, both having a laminar flow field. We describe the transition of a thermoacoustic system from a state of stable operation (called stable fixed point) to unstable operation (called stable limit cycle) on variation of any control parameter. First, we will present the dynamical transitions and characteristics of the dynamical states observed in laminar thermoacoustic systems in the absence of any stochastic background fluctuations in the flow field. Subsequently, we will study the effect of external stochastic perturbations on the occurrence of different dynamical transitions in such systems. These investigations performed on simple laminar thermoacoustic systems can be considered as a baseline study before we attempt to understand the complex dynamics exhibited by turbulent thermoacoustic systems in Chap. 6.

As mentioned in Chap. 1, the occurrence of thermoacoustic instability has been a concern for most practical combustion systems used in power generating units and propulsion engines. Traditionally, both linear and nonlinear methodologies have been used to explain the occurrence and the characteristics of thermoacoustic oscillations observed in such systems. In linear analysis, the equations describing flame-acoustic interactions are subjected to small amplitude perturbations for calculating the eigenvalues [12]. When all the eigenvalues are negative, the system is considered to be linearly stable, and when one of the eigenvalues is positive, the system is considered to be linearly unstable. For a linearly unstable system, an infinitesimally small perturbation introduced in the system grows in time and asymptotically approaches another stable state that is different from the initial state. As seen in Chap. 2, for the case of subcritical Hopf bifurcation, a finite amplitude perturbation (or initial condition) in the regime of bistability (hysteresis) can alter

R. I. Sujith, S. A. Pawar, *Thermoacoustic Instability*, Springer Series in Synergetics,
https://doi.org/10.1007/978-3-030-81135-8_3

the stability of the system from being in a linearly stable steady state to a linearly stable oscillatory state (limit cycle) [20, 92, 94, 102].

Using the linear stability analysis with infinitesimally small perturbations, we can predict the bifurcation points and the linearly unstable modes of a thermoacoustic system. However, the effects due to large amplitude perturbations, such as triggering phenomenon (discussed in Sect. 3.3.3) and the amplitude of limit cycle oscillations, cannot be predicted through the linear analysis [25]. The prediction of the amplitude of limit cycle oscillations is crucial from a structural point of view while designing gas turbine and rocket engines [22, 64]. Furthermore, the occurrence of these phenomena in thermoacoustic systems is indicative of nonlinear characteristics of such systems [92]. Therefore, we need to use nonlinear stability analysis or tools from nonlinear time series analysis to study and predict these phenomena in thermoacoustic systems.

Many numerical [2, 21, 25, 38, 39, 43, 47, 48, 59, 66, 68, 71, 76, 94, 95, 104, 105] and experimental [3, 11, 16, 29, 33, 34, 44, 50, 52, 62, 67, 68, 72, 73, 77, 80] studies have been performed over the years to investigate the bifurcation characteristics of thermoacoustic systems. The nonlinear behavior of thermoacoustic systems can be examined using two different methodologies, which include the describing function approach and the dynamical systems approach.

The describing function approach is derived from nonlinear control theory [32]. Using the describing function approach, we can predict the amplitude and frequency of limit cycle oscillations observed in a system. Such a system consists mainly of a single nonlinear subsystem that is coupled (feedback) with other linear subsystems. In a thermoacoustic system, we often assume the reaction field (flame) as a nonlinear subsystem and the acoustic field as a linear subsystem [4, 13, 25, 61]. In the describing function approach, we measure the flame response against the sinusoidal perturbations imposed in the underlying mixture field or the acoustic field of the system. We retain the response of the flame only at the frequency of forcing and discard the contribution of frequencies observed at higher-harmonics and sub-harmonics in the system [42]. This approach is also referred to as the flame describing function (FDF) approach. The FDF approach was originally introduced in a theoretical analysis of thermoacoustic systems by Dowling [25, 26] and later implemented in experimental studies by Noiray et al. [77] and Durox et al. [28].

While implementing the FDF approach, the flame is perturbed in an unconfined configuration, i.e., in a regime without strong coupling between the flame and acoustic subsystems, and measure the flame response for different conditions of forcing amplitudes and forcing frequencies. The flame response is characterized using the following expression,

$$\mathrm{FDF}(\omega, u') = \frac{\dot{Q}'/\overline{\dot{Q}}}{u'/\overline{u}} \tag{3.1}$$

where ω and u' are the frequency and the amplitude of external velocity perturbations, respectively. $\dot{Q}'$ and u' are the measured heat release rate fluctuations and the velocity fluctuations, respectively. Similarly, $\overline{\dot{Q}}$ and $\overline{u}$ are the measured mean values of the heat release rate fluctuations and the velocity fluctuations, respectively. The FDF is usually characterized in terms of its gain and phase. Here, the gain quantifies the degree to which the input perturbations are amplified or dampened by the flame, whereas the phase measures the instant of occurrence of the heat release rate fluctuation with respect to the input velocity perturbation. The FDF is used as an input in lower-order models of a system, which are then solved to detect the stability of the system.

Note that FDF is a modified form of the flame transfer function (FTF). In the FTF approach, we characterize the flame response at different forcing frequencies when the amplitude of perturbations is very small [8, 15, 17, 79]. The FTF is a linear model of the heat release rate response to velocity fluctuations. This approach helps in determining only the growth rates and the frequencies of linearly unstable modes of a combustor. On the other hand, the FDF approach helps to predict the amplitude and the frequency of thermoacoustic instability, and also the nonlinear behaviors such as triggering, hysteresis, and mode-switching of a system [13, 42].

In the FDF approach, since we assume the acoustic field to be linear, we encounter several limitations in predicting the complex nonlinear behaviors exhibited by acoustic fluctuations in a thermoacoustic system. These complex dynamical behaviors occur due to the nonlinear interaction between the flame and multiple acoustic modes of the combustor [49]. Furthermore, since the FDF approach accounts for the heat release rate fluctuations occurring only at the forcing frequency, we always consider the work done by the heat release rate oscillations over a cycle at higher harmonics zero. Therefore, although this approach simplifies the analysis of flame-acoustic interaction, the inherent assumptions of the acoustic field being linear and harmonic in time maybe questionable. In addition, the experimental implementation of this approach is laborious as we need to obtain FDF for a large range of forcing frequencies and forcing amplitudes. Most importantly, it is difficult to design actuators that can produce large amplitude acoustic oscillations comparable to the amplitude of limit cycle oscillations observed in practical gas turbine combustors. Further, it is undesirable to excite large amplitude oscillations on a real engine operating at high pressure, high temperature environments and measure its response.

The approach based on dynamical system theory (as elaborated in Chap. 2) provides an alternative for studying the asymptotic behavior (or attractor) of complex dynamics exhibited by a thermoacoustic system. In contrast with the traditional describing function approach, where we need to model a thermoacoustic system as an acoustic system driven by combustion and measure the system response for different conditions of forcing, the dynamical systems approach helps in analyzing the inherent global behavior of a thermoacoustic system which occur due to the nonlinear interactions between its various subsystems.

Recent studies [36, 51, 53, 55, 82, 103] using dynamical systems approach have shown that, in addition to limit cycle oscillations, the complex flame-acoustic interactions may manifest thermoacoustic instability in the form of rich dynamical states, such as frequency-locking, quasiperiodicity, chaos, strange nonchaos, and intermittency (as we will discuss in Chap. 4). We cannot predict such nonlinear dynamical states (for instance, chaos and intermittency) using traditional describing function approaches, as these approaches consider the system response only at the forcing frequency and neglect the contribution at other frequencies. Hence, approaches based on the flame response characterization do not provide a complete comprehension of the nonlinear thermoacoustic oscillations observed in practical combustion systems. However, to provide efficient strategies required for evading or suppressing thermoacoustic instabilities in a system, we need to understand the actual features of such instabilities in their self-excited state. Dynamical systems approach provides a useful platform to characterize such different nonlinear states of thermoacoustic instability observed in a system.

In the subsequent sections of this chapter, we discuss the bifurcation characteristics of a Rijke-type laminar system during the onset of thermoacoustic instability, i.e., during the transition from stable steady state to stable limit cycle oscillations. We characterize this system response both in the presence and in the absence of stochastic fluctuations (noise) in the system. We now begin our discussion with a short introduction on different Rijke-type thermoacoustic systems.

3.1 A Brief History of Rijke-Type Thermoacoustic Systems

The tonal sound waves generated due to the presence of a flame in a confined system have fascinated various researchers over the last three centuries. Higgins [40] was possibly the first to report the observation of the sound produced by a hydrogen diffusion flame located in an open-ended glass duct. He referred to this oscillatory behavior of the confined flame as the 'singing flame'. From further investigations, he noticed that the intensity of the sound waves, amplified by the flame, depends on the size of the fuel supply tube, flow rates of the fuel, size and boundary conditions of the resonating tube, and the location of the flame inside the resonator. Subsequently, Sondhauss [90] performed experiments on another system having one end open to the atmosphere and the other end closed by a hot glass bulb. He observed that the acoustic oscillations are produced in such systems due to the existence of a temperature gradient between the hot and cold regions of the tube. He also proposed a relationship between the frequency of the oscillations and the dimensions of the tube.

Later on, Rijke [85] simplified the experiments of Higgins, where he placed a piece of fine metallic wire gauze in the lower half of a vertical tube that is opened at both ends. He then heated the gauze using a flame until it got red hot. He found that when the heat source is located near the quarter length of a resonating tube, the strength of the acoustic oscillations generated in the system is the highest. Rijke conjectured that a convecting air current established by the heated gauze in the upward direction is the main driving source for the acoustic oscillations in the tube. Since then such simple systems consisting of a flame or an electrically heated wire-gauze embedded in a duct have been referred to as 'Rijke-type thermoacoustic systems'.

Rayleigh [84] provided the first physical explanation for the occurrence of self-sustained acoustic oscillations in a Rijke tube system. He proposed a criterion which states that "*If heat be given to the air at the moment of greatest condensation, or be taken from it at the moment of greatest rarefaction, the vibration is encouraged. On the other hand, if heat be given at the moment of greatest rarefaction, or abstracted at the moment of greatest condensation, the vibration is discouraged* [84]". This means the flame can drive the acoustic field in a combustor only if the heat release rate fluctuations in the flame happens to be in phase with the acoustic oscillations in the system. In contrast, when these fluctuations are out of phase with each other (i.e., phase difference being 180°), the acoustic oscillations are damped in the system. This explanation is famously known as the 'Rayleigh criterion'. Till today, this criterion has been used as a necessary condition for the detection of the onset of thermoacoustic instability in a system.

Since these initial investigations, many numerical, analytical, and experimental studies have been performed till date to understand the physical mechanisms responsible for driving of thermoacoustic oscillations and to devise methodologies that can mitigate these oscillations. For more details on such studies on Rijke tube systems, we recommend the readers to go through a number of review papers available on this topic [10, 30, 31, 69, 78, 83, 86].

3.2 Bifurcation Characteristics of a Deterministic Thermoacoustic System

Recent studies [6, 47, 68] investigate a modern variant of the Rijke tube configuration, consisting of a horizontal duct with an assembly of electrically heated wire gauze, also called a heater (Fig. 3.1b). In this system, we need to establish an air flow inside the duct using a blower or a compressor to develop a convective flow field across the heater. This is in contrast to the original configuration of the vertical Rijke tube, where the convecting mean air flow is inherently dependent on the temperature of the external heat source in the system (Fig. 3.1a). The duct of the horizontal Rijke tube is open at both ends. In some systems, we notice that the end of the duct is connected to a decoupler (see Fig. 3.1b), which is a rectangular closed chamber with a small opening for the inlet air flow. The decoupler isolates the duct from the fluctuations in the inlet air supply. It also ensures that the total pressure at the end of the duct is equal to the ambient pressure and hence the pressure fluctuations at the boundary are negligible. This means, the acoustic pressure is zero and the acoustic velocity is maximum at the open end of the duct. This is referred to as the acoustically open boundary condition.

External electric power is supplied to heat the wire gauze to different temperatures. This, in turn, changes the temperature field in the system. An important

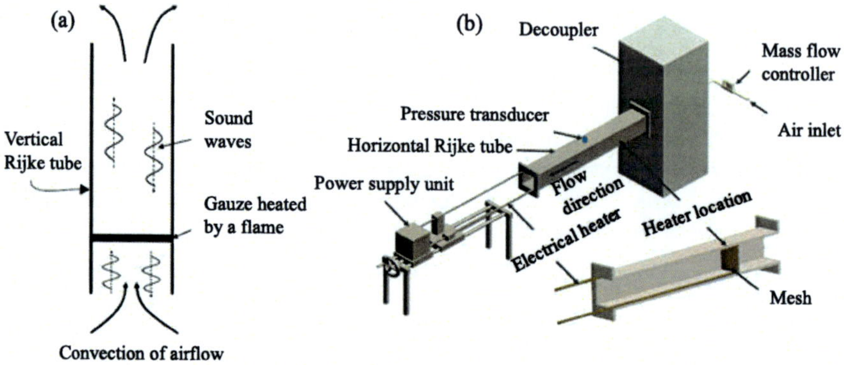

Fig. 3.1 Schematics of (**a**) a vertical Rijke tube system and (**b**) a horizontal Rijke tube system. (**b**) Reproduced with permission from Tandon et al. [98]

advantage of the horizontal Rijke tube system is that it provides us independent controls over various system parameters, i.e., the heater power and mass flow rate of air, which was not possible in Rijke's original system with a vertical duct.

Many researchers have investigated the bifurcation characteristics of a horizontal Rijke tube through experiments [29, 34, 60, 68] and numerical simulations [47, 66, 94, 104] for the variation of different control parameters. These parameters include the heater power, mass flow rate of air, location of the heater in the duct, noise intensity, etc. For a quasi-static variation in the control parameter with the sufficient waiting time, the asymptotic behavior of the system changes from a state of stable fixed point to a stable oscillatory state (limit cycle). Such a qualitative change of the system behavior for a small increment in the control parameter above a critical value is referred to as '*bifurcation*', and the corresponding critical value of the control parameter is called '*bifurcation point*', as we discussed in the previous chapter.

The amplitude spectrum of the acoustic pressure signal observed during the state of limit cycle oscillations (Fig. 3.2a) shows a sharp peak at the frequency corresponding to the fundamental mode of the acoustic field in the system. We also observe a low amplitude peak for a higher harmonic in the spectrum, which primarily occurs due to the inherent nonlinearity present in the system. In Fig. 3.2b, we show the reconstructed phase space of the limit cycle oscillations, which is obtained using the delay embedding theorem proposed by Takens [97]. As discussed in Chap. 2, we can obtain the parameters (embedding dimension and time delay) required for the phase space reconstruction using the methods of false nearest neighbor (or Cao's method) and average mutual information (or autocorrelation), respectively [1, 74]. The reconstructed phase space of limit cycle oscillations shows an elliptic ring-like structure, where the trajectories in the phase space repeat their behavior after a period of the signal.

In Fig. 3.3, we show bifurcation diagrams of the horizontal Rijke tube when the electric power supplied to the wire-gauze (or heater) is varied as a control parameter. When the value of the heater power is below a critical value, we notice

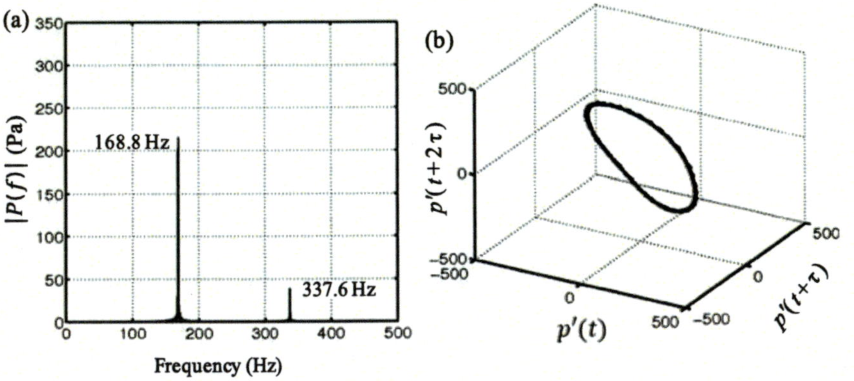

Fig. 3.2 (**a**) The amplitude spectrum and (**b**) the reconstructed phase space of the acoustic pressure fluctuations (p') observed for the state of limit cycle oscillations in a horizontal Rijke tube. Reproduced with permission from Gopalakrishnan and Sujith [33]

a stable steady state in the system. In contrast, as the heater power is increased above a critical value, we observe the occurrence of stable periodic oscillations (limit cycle) in the system. The onset of such limit cycle oscillations from the steady state is referred to as Hopf bifurcation [92]. As mentioned in Chap. 2, Hopf bifurcation is classified into two types: supercritical Hopf and subcritical Hopf. In the horizontal Rijke tube, depending on the value of the air flow rates (i.e., low or high), the system behavior exhibits supercritical Hopf (Fig. 3.3a) or subcritical Hopf (Fig. 3.3b) bifurcation, respectively [29]. Here, we show the bifurcation diagram constructed by plotting the root mean square of the acoustic pressure fluctuations (P_{rms}) as a function of the heater power (K), varied in both forward (increasing K) and reverse (decreasing K) paths.

During supercritical Hopf bifurcation (Fig. 3.3a), we notice that the behavior of the system changes from a state of stable fixed point to a state of low amplitude stable limit cycle oscillations at the Hopf point. Further increase in the heater power leads to a gradual increase in the amplitude of the limit cycle oscillations. On decreasing the heater power in the reverse direction, the behavior of the system is similar to that observed during the forward path. Thus, the system does not exhibit hysteresis in the parameter space of the heater power.

On the other hand, for subcritical Hopf bifurcation (Fig. 3.3b), the behavior of the system suddenly transitions from a state of stable fixed point to a state of large amplitude stable limit cycle oscillations at the Hopf point. Thus, subcritical Hopf bifurcation leads to an abrupt jump in the amplitude of the acoustic pressure oscillations in the system. The amplitude of the limit cycle oscillations continues to increase with further increase in the heater power. With a reverse change in the heater power, we do not observe the transition of the system from the state of stable limit cycle oscillations to stable fixed point at the Hopf point; we need a further reduction in the heater power to a lower value to achieve this transition. The critical value of the heater power at which we observe such a transition in the system is

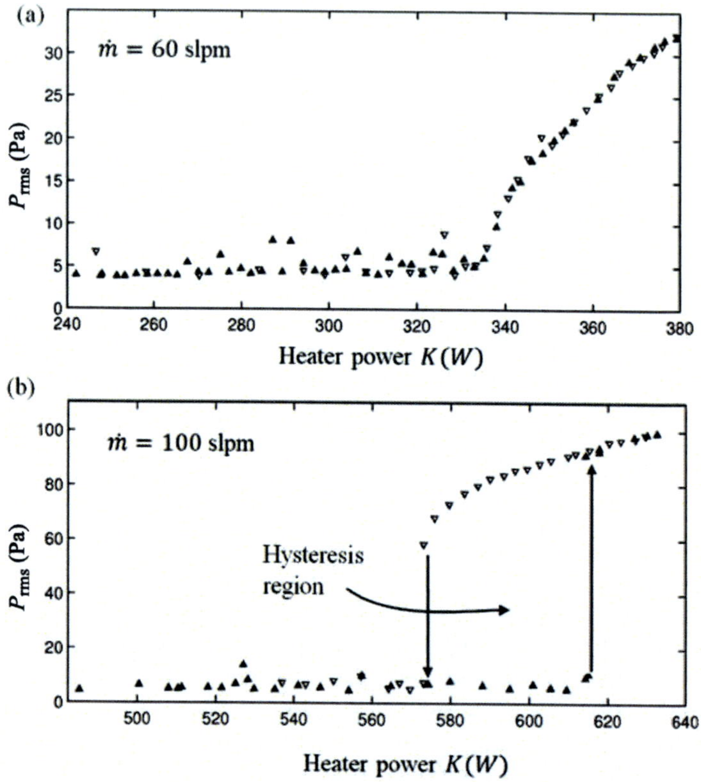

Fig. 3.3 The experimental bifurcation diagrams showing the variation of the root mean square value of acoustic pressure fluctuations (P_{rms}) with the change in the heater power (K) corresponding to (**a**) supercritical Hopf bifurcation and (**b**) subcritical Hopf bifurcation. The symbol filled triangle indicates the forward path and inverted triangle indicates the reverse path. For supercritical Hopf bifurcation, both the forward and the reverse paths follow nearly the same variation. In contrast, for subcritical Hopf bifurcation, these paths show hysteresis region. Adapted with permission from Etikyala and Sujith [29]

called the fold point. Thus, subcritical Hopf bifurcation exhibits hysteresis in the parameter space of the heater power.

In the hysteresis region, the system exhibits bistability, i.e., the dynamics of the system is bounded by two branches: a stable fixed point and a stable limit cycle. In this region, depending on the value of the initial condition or the amplitude of the perturbation, the system dynamics can transition to either of these states due to a phenomenon called '*triggering*', which we will elaborate in Sect. 3.3.3. Further, the boundary of the critical amplitude of the perturbation required for the transition from the steady state to limit cycle oscillations, at each value of the heater power, is referred to as the branch of unstable limit cycle [92]. From Fig. 3.3, we also understand that variation of the secondary control parameter of the system (say, air

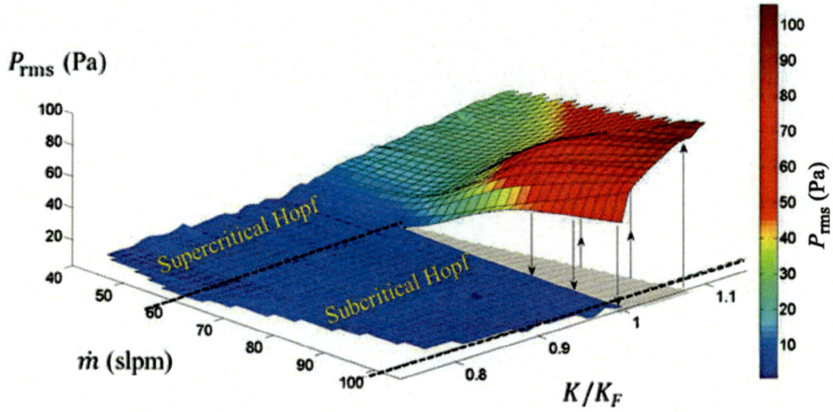

Fig. 3.4 Variation of the root mean square of the acoustic pressure fluctuations (P_{rms}) on a two-parameter bifurcation diagram between the mass flow rate of air ($\dot{m}$) and the normalized heater power (K/K_F). This plot shows the change of criticality of the bifurcation from supercritical Hopf to subcritical Hopf due to a change in K/K_F for different values of $\dot{m}$. The dotted lines separate regions of these bifurcations. Adapted with permission from Etikyala and Sujith [29]

flow rate) can change the criticality of bifurcation of the primary control parameter (say, heater power), indicating that these parameters are inherently coupled with each other in the system.

Figure 3.4 demonstrates the change of such criticality of bifurcation in the Rijke tube, i.e., the change of bifurcation characteristics from supercritical Hopf to subcritical Hopf due to an increase in the mass flow rate of air [29]. We can understand such a change of criticality from a two-parameter bifurcation diagram that shows the variation of the root mean square of the acoustic pressure fluctuations (P_{rms}) as a function of the air flow rate ($\dot{m}$) and the heater power (K). The heater power is normalized by the value of the heater power observed at the fold point (K_F). For lower air flow rates, we notice that the variation in the amplitude of the acoustic pressure fluctuations with the heater power is continuous, indicating the presence of supercritical Hopf bifurcation in the system. In contrast, when the air flow rate is above a critical value, we observe a hysteresis zone in the bifurcation diagram, suggesting a change in the bifurcation from supercritical Hopf to subcritical Hopf in the parameter space of the heater power. With further increase in the air flow rate to a higher value, we notice a corresponding increase in the width of the hysteresis region for the case of subcritical Hopf bifurcation.

Similar to the bifurcation characteristics of a Rijke tube observed for the variation of the air flow rate and the heater power (Fig. 3.4), we next discuss a bifurcation diagram of this system for the variation of the heater location as the control parameter [33]. Figure 3.5 shows the variation of the amplitude of the acoustic pressure fluctuations with a change in the heater location, for two different values of air flow rates. For a lower value of the air flow rate (Fig. 3.5a), when the heater location (x_f) is increased in the forward direction (starting from the inlet of the

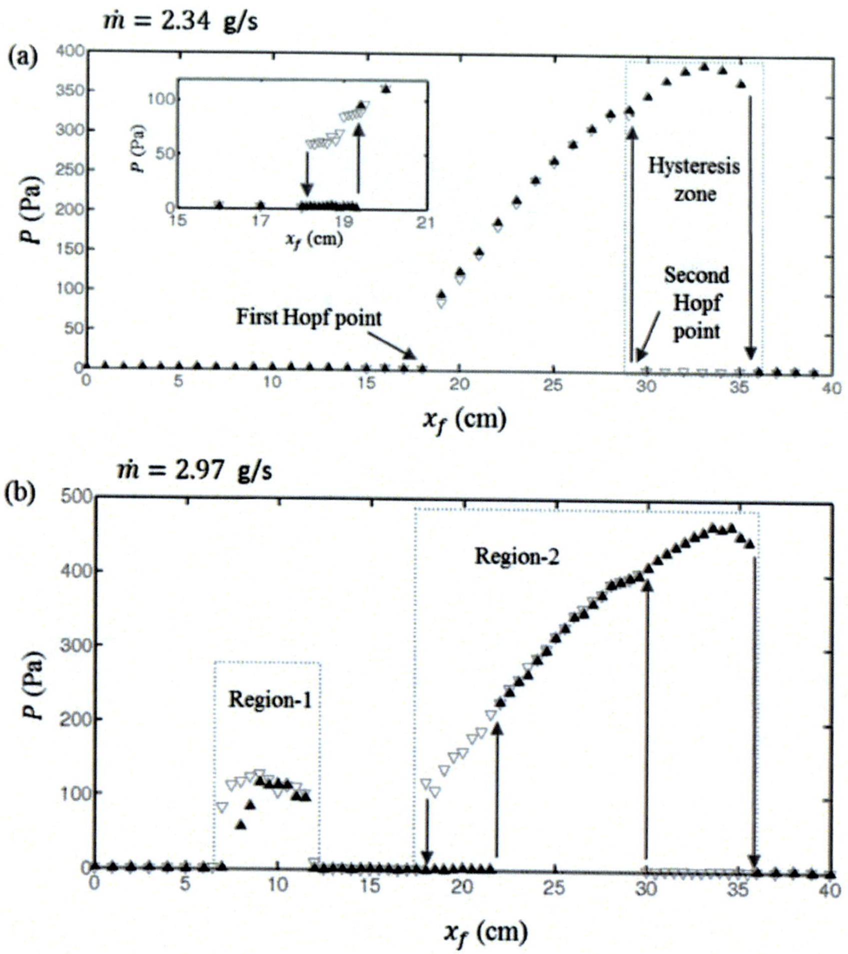

Fig. 3.5 The bifurcation diagrams showing the variation of the amplitude of acoustic pressure fluctuations (P) with the change in the heater location (x_f) in the system, when the mass flow rate of air ($\dot{m}$) is fixed at (**a**) low and (**b**) high values. The inset in (**a**) shows the presence of a hysteresis zone near the first Hopf point. Filled triangle indicates the forward path and inverted triangle indicates the reverse path. We notice that the bifurcation characteristics of the system depend on the value of the underlying air flow rate. The total length of the Rijke tube is 100 cm and x_f is measured from the inlet of the tube. Adapted with permission from Gopalakrishnan and Sujith [33]

duct), we observe the onset of subcritical Hopf bifurcation at $x_f = 19$ cm in the system. At this point, the limit cycle oscillations with a frequency corresponding to the fundamental mode of the duct are generated from the steady state. Here, the total length of the duct is 100 cm. Upon further increase in x_f, the amplitude of limit cycle oscillations shows a continuous increase until it reaches a maximum value at $x_f = 33$ cm; such oscillations then transition to the steady state at $x_f = 36$ cm after a small decrease in their amplitude.

When x_f is decreased in the reverse path, the behavior of the system stays in the steady state till $x_f = 29$ cm and then it transitions to limit cycle oscillations via second Hopf bifurcation. As x_f is further decreased, the system traces the same path in the reverse direction as the forward path and reaches the steady state at $x_f = 18$ cm. We notice hysteresis regions at both the first and the second Hopf bifurcation points. In order to detect the hysteresis region occurring at the first Hopf bifurcation, we need to change the value of x_f in very small steps (see the inset of Fig. 3.5a). In contrast, we clearly observe the wider hysteresis region at the second Hopf point in the bifurcation diagram.

As the air flow rate is increased to a higher value (Fig. 3.5b), we observe a significant change in the bifurcation diagram when compared to the lower air flow rate (Fig. 3.5a). We notice two regions of limit cycle oscillations, occurring through subcritical Hopf bifurcation, in the parameter space of x_f. The variation of the amplitude of acoustic pressure fluctuations in Region-2 is qualitatively similar to that observed in Fig. 3.5a, where limit cycle oscillations occur at the frequency of the fundamental mode of the acoustic field in the system (Fig. 3.2a). In contrast, we observe the occurrence of period-2 limit cycle oscillations in Region-1 of the bifurcation diagram.

To highlight period-2 thermoacoustic oscillations, we show the enlarged portion of Region-1 of Fig 3.5b in Fig. 3.6, where the variation of the amplitude of the acoustic pressure fluctuations is shown as a function of x_f for both the forward (Fig. 3.6a) and the reverse (Fig. 3.6b) paths. In Fig. 3.7, we show the characteristics of the period-2 oscillations. During the state of period-2 oscillations, as the acoustic pressure signal exhibits two local maxima in an oscillation cycle, we notice two distinct points at a particular value of x_f in the corresponding bifurcation diagram (Fig. 3.6). These period-2 oscillations occur due to nonlinear interaction between two linearly unstable modes of the acoustic field in the system (Fig. 3.7a). Hence, the amplitude spectrum of these period-2 oscillations shows two peaks, where the dominant frequency is observed at the second mode (352.8 Hz) and a small peak is observed at the fundamental mode (176.4 Hz) of the system (Fig. 3.7a). The phase portrait of period-2 oscillations shows a two-loop attractor (Fig. 3.7b), where the trajectory repeats its behavior after making two rotations in the phase space.

Thus, from Figs. 3.3, 3.4, 3.5, 3.6, and 3.7, we understand that the bifurcation characteristics of a thermoacoustic system depend on a control parameter that is varied in the system. Having discussed such characteristics of deterministic bifurcations in a thermoacoustic system (horizontal Rijke tube), we next discuss the effect of the presence of noise on the bifurcation characteristics of these systems and also the occurrence of triggering in the system.

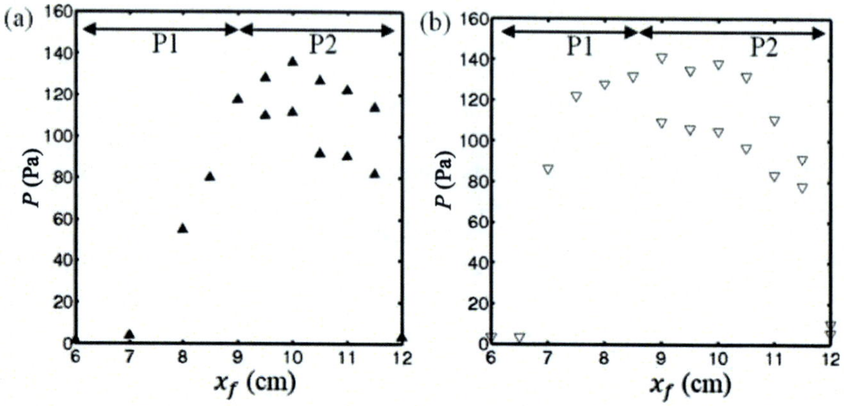

Fig. 3.6 Variation of local positive peaks in the acoustic pressure signal with change in the heater location in the system for (**a**) a forward (increasing x_f) path and (**b**) a reverse (decreasing x_f) path. In these plots, we notice the presence of a single amplitude and two amplitudes at a fixed value of x_f, indicating the existence of period-1 (P1) and period-2 (P2) oscillations in the system, respectively. Adapted with permission from Gopalakrishnan and Sujith [33]

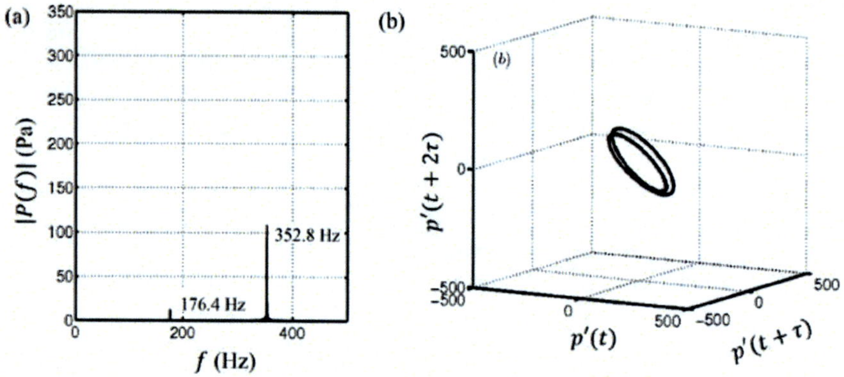

Fig. 3.7 (**a**) The amplitude spectrum and (**b**) the phase portrait of period-2 oscillations observed in the horizontal Rijke tube. Adapted with permission from Gopalakrishnan and Sujith [33]

3.3 Noise-Induced Transition, Triggering, and Stochastic Bifurcation to Limit Cycle

Background fluctuations are inherent in most practical thermoacoustic systems [64]. These are aperiodic fluctuations with a broadband power spectrum. The predictability of such fluctuations in time is very low. The presence of such background fluctuations affects the deterministic characteristics of thermoacoustic oscillations generated in the system by introducing unpredictability in their long-term behavior [63]. Noise also plays a significant role in changing the dynamical transitions (or bifurcation characteristics) of turbulent thermoacoustic systems.

The presence of noise in the system can induce different phenomena, such as triggering (when the system behavior is in the excitable region of operation, see Sect. 3.3.3) and coherence resonance (when the system behavior is in the non-excitable region of operation) [54]. In the case of subcritical Hopf bifurcation, the excitable region is the bistable (hysteresis) region in the parameter space and the non-excitable region (also called the subthreshold region by Kabiraj et al. [52]) is the parameter space prior to the hysteresis region (or prior to the fold point). Triggering due to background noise can lead to the occurrence of thermoacoustic oscillations at a noise strength much lesser than the amplitude level required for crossing the unstable limit cycle branch of a deterministic system [44]. On the other hand, coherence resonance leads to the enhancement of deterministic (or coherent) dynamics for an intermediate noise level in the system [37]. The corresponding noise-induced oscillations can be used to predict the onset of thermoacoustic oscillations in systems with supercritical [58] or subcritical Hopf bifurcations [37, 52, 88]. However, as the type of bifurcation associated with noisy systems is difficult to discern using traditional approaches, we can rely on the method of P-bifurcation [35].

In the following subsections, we discuss the aforementioned aspects of noise-induced phenomena studied in laminar thermoacoustic systems when the external acoustic noise is added during the steady state. We believe that these studies are the initial baby steps towards understanding the complex nonlinear phenomena occurring due to background fluctuations in turbulent thermoacoustic systems, which we will discuss in the rest of the book.

3.3.1 Effect of Noise on Hysteresis (or Bistability) of a Subcritical Hopf Bifurcation

As we discussed before, the presence of noise (or stochastic fluctuations) can make several changes in dynamical properties of a bistable system undergoing subcritical Hopf bifurcation. One of these effects is the change in the range (or width) of the bistability (or hysteresis) observed in the parameter space of the bifurcation diagram. In Figs 3.8 and 3.9, we show the bifurcation characteristics of a horizontal

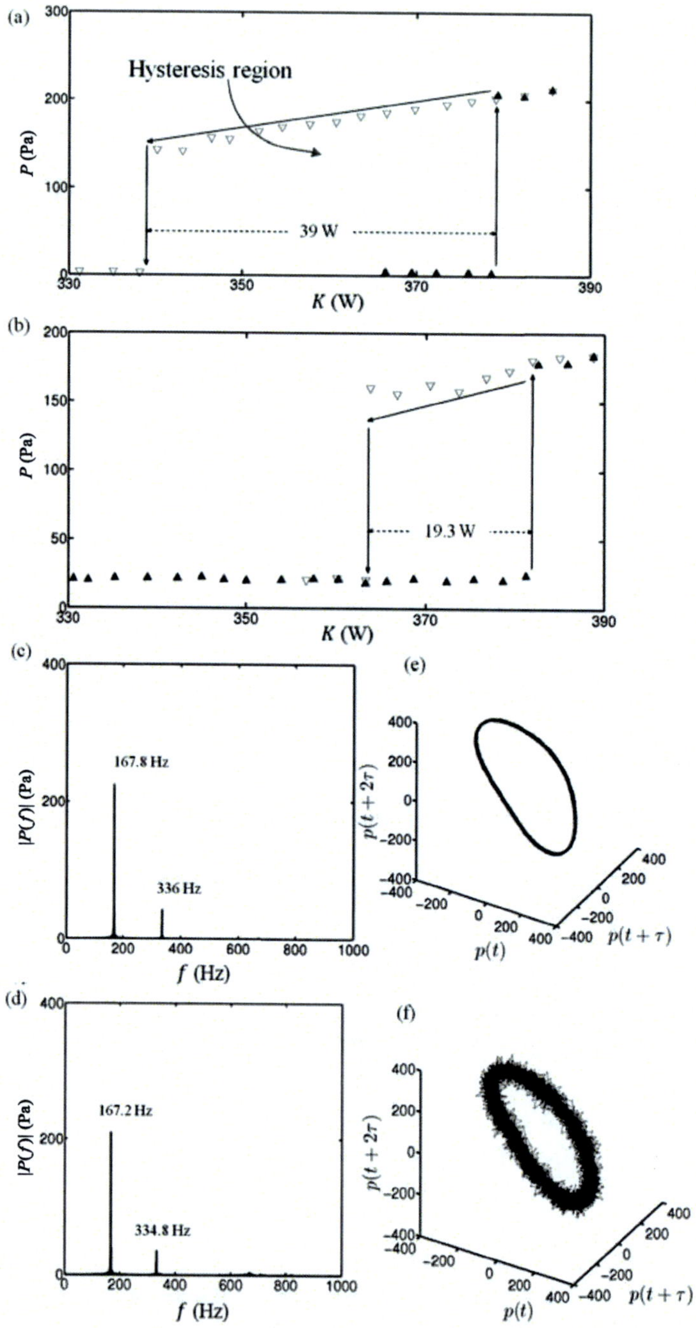

Fig. 3.8 **(a, b)** Plots showing the comparison of subcritical Hopf bifurcations observed in the absence of noise and the presence of noise, respectively, in experiments on a horizontal Rijke tube. **(c, d)** The amplitude spectrum and **(e, f)** the phase portrait of limit cycle oscillations observed for these two cases, respectively. The addition noise in the system reduces the hysteresis width and broadens the thickness of the limit cycle attractor. Adapted with permission from Gopalakrishnan and Sujith [33]

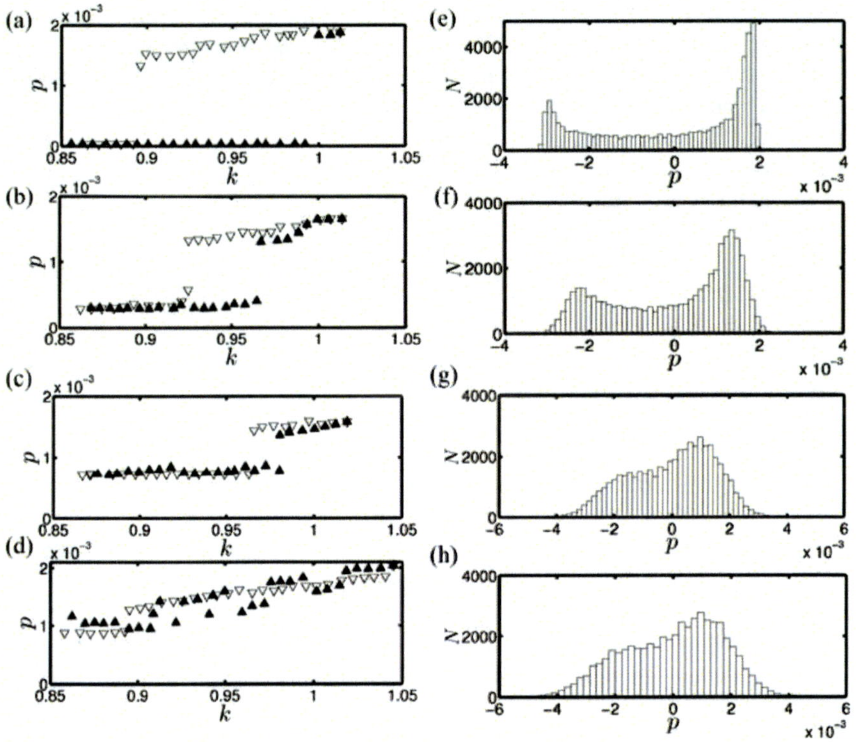

Fig. 3.9 Plots showing the effect of an increase in the noise intensity on **(a)**–**(d)** the bifurcation characteristics of the Rijke tube and **(e)**–**(h)** the distribution (histogram) of the normalized acoustic pressure oscillations (p) observed during the state of thermoacoustic instability. Here, k is the heater power normalized by its value observed at the Hopf point in the absence of noise, and N is the number of data points in the pressure time series. The non-dimensional noise intensity is increased as 0, 0.15, 0.35, and 0.50, respectively. Increasing the noise intensity suppresses the hysteresis width and amplifies the magnitude of background fluctuations in **(a–d)**, and changes the characteristic of the histogram from bimodal to nearly unimodal distribution in **(e–h)**. Adapted with permission from Gopalakrishnan and Sujith [34]

Rijke tube in the presence of external acoustic noise generated through loudspeakers [34]. To obtain these plots, we need to change the electric power supplied to the heater for a fixed noise intensity, such that the behavior of the system transitions from a steady state to limit cycle oscillations and vice versa.

In Fig. 3.8, we show the comparison of the characteristics of subcritical Hopf bifurcation when the underlying flow field in the system is steady, i.e., noise-free laminar (Fig. 3.8a) and when it is contaminated with external stochastic fluctuations (Fig. 3.8b). In the presence of laminar flow field, the system dynamics shows a clean subcritical Hopf bifurcation with a sufficiently larger hysteresis zone, in the parameter space of the heater power (Fig. 3.8a). During the bifurcation, limit cycle oscillations occur at a frequency of 167.8 Hz in the system (Fig. 3.8c). The reconstructed phase space of such limit cycle oscillations shows a thin closed-loop orbit (Fig. 3.8e). In contrast, we notice that the presence of external acoustic noise decreases the width of the bistable region (Fig. 3.8b). However, the addition of noise does not significantly change the frequency of limit cycle oscillations in the system (compare Fig. 3.8c and d). The reconstructed phase space of the noisy limit cycle oscillations shows a thickened close-ring. The thickness of the attractor is a result of the randomness in the local paths of the trajectory while maintaining the global closed-loop orbit of limit cycle attractor in the phase space.

In Fig. 3.9, we present the effect of increase in the noise intensity on the suppression of the hysteresis zone of a subcritical Hopf bifurcation. In this figure, column I compares the changes in bifurcation diagrams of the system and column II indicates the modifications observed in the histogram of the acoustic pressure fluctuations during the state of limit cycle oscillations (i.e., thermoacoustic instability).

In the absence of external noise, we observe sharp jumps in the amplitude of the acoustic pressure fluctuations at both the Hopf point and the fold point of the bifurcation diagram. We also notice the presence of a distinct hysteresis zone (Fig. 3.9a). The histogram of the acoustic pressure fluctuations shows a bimodal distribution with two sharp peaks at the extreme values, i.e., near the maximum amplitude of the signal (Fig. 3.9e). With the addition of low amplitude acoustic noise in the system, we notice a decrease in the width of the hysteresis region and also an increase in the amplitude of background fluctuations in the steady state region of the bifurcation diagram (Fig. 3.9b). The corresponding histogram of the acoustic pressure signal displays a decrease in the heights of the peaks and also a movement of these peaks closer to the mean value (Fig. 3.9f).

Further increase of noise intensity to a higher value makes the jump in the acoustic pressure amplitude at bifurcation points less abrupt. This behavior occurs due to the shifting of the base amplitude of the steady state region to a larger value and shrinking of the hysteresis region (Fig. 3.9c). When the noise intensity is very high (Fig. 3.9d), we observe a complete loss of the hysteresis region. As a result, the transition to thermoacoustic instability for a subcritical Hopf bifurcation appears to be more continuous without showing the existence of distinct Hopf and fold points in the bifurcation diagram. During the presence of high intensity noise, the distribution of the acoustic pressure signal transitions to a unimodal distribution, where the peak in the histogram is located near the mean (Fig. 3.9g, h) rather than at extremes values, as observed in the absence of externally added acoustic noise in the system (Fig. 3.9e). From these results, we understand that the width of the bistable region exhibits a nearly linear decrease with an increase in the intensity of stochastic fluctuations in the system [34].

According to Gopalakrishnan and Sujith [34], the flickering behavior of the acoustic pressure signals makes the transition from the steady state to limit cycle oscillations continuous in the presence of high intensity noise. During flickering, the system dynamics oscillates between a stable oscillatory state and a stable non-oscillatory state [23, 89, 100, 101]. There have been several other studies that reported the loss of hysteresis region in the presence of noise in other fields of science [46, 87, 106]. This change in bifurcation characteristics of the system dynamics from an abrupt jump to a continuous transition, in the presence of high amplitude background fluctuations, makes it difficult to identify the criticality exhibited by the bifurcation in practical thermoacoustic systems working under highly turbulent environments.

In Fig. 3.9a-d, we also notice that the amplitude of the acoustic pressure fluctuations in the steady state region of the bifurcation diagram grows with an increase in the noise intensity. This increase in the amplitude of pressure oscillations in the subthreshold region (i.e., the region before the bistable zone) possibly results from an increase in the coherent nature of the system dynamics [52]. The increase in coherence can be recognized from the sharp peaks in the spectrum of the pressure signal, which is otherwise broadband in the absence of external stochastic fluctuations in the system. The phenomenon of increase in coherence (or determinism) of the system dynamics in the subthreshold region due to noise perturbations is referred to as 'coherence resonance' [81]. For systems exhibiting coherence resonance, the growth (resonance) of background fluctuations is not monotonic with the noise intensity; it peaks for an intermediate level of the noise intensity. Further details on the study of coherence resonance in a thermoacoustic system can be found in Kabiraj et al. [52] and Gupta et al. [37], where they observed similar behavior of increase in the amplitude of acoustic pressure oscillations for an intermediate intensity of stochastic fluctuations in the subthreshold region of a laminar premixed flame combustor.

3.3.2 *Stochastic (or P) Bifurcation*

In the previous sections, we understood that in systems with a noisy environment, the accurate detection of the bifurcation point of thermoacoustic instability is quite challenging. The usual method to detect the onset of bifurcation in a deterministic system is to plot the variation of a quantitative measure of the signal as a function of the control parameter (as seen in Fig. 3.3). However, such representation of the system behavior is not useful in the presence of stochasticity in the underlying flow field. In the presence of high noise intensity, the onset of thermoacoustic instability displays a continuous transition across the bifurcation point, as witnessed in Fig. 3.9. For such stochastic systems, we can detect the dynamical transitions through the variation of a probability density function (PDF) of a signal with a change in the control parameter [70]. We can consider a qualitative change in the PDF when the control parameter crosses a critical value as an indicator for the occurrence of a phenomenological bifurcation (or P-bifurcation) in a stochastic system. The occurrence of such P-bifurcation has been widely studied in both models [7, 24, 106] and experiments [9, 91].

In Fig. 3.10, we present the characteristics of P-bifurcation observed in a horizontal Rijke tube [35]. Such bifurcation analysis is studied by examining the effect of increasing noise intensity on the distribution of the amplitude of acoustic pressure as the system undergoes subcritical Hopf bifurcation. Figure 3.10 shows

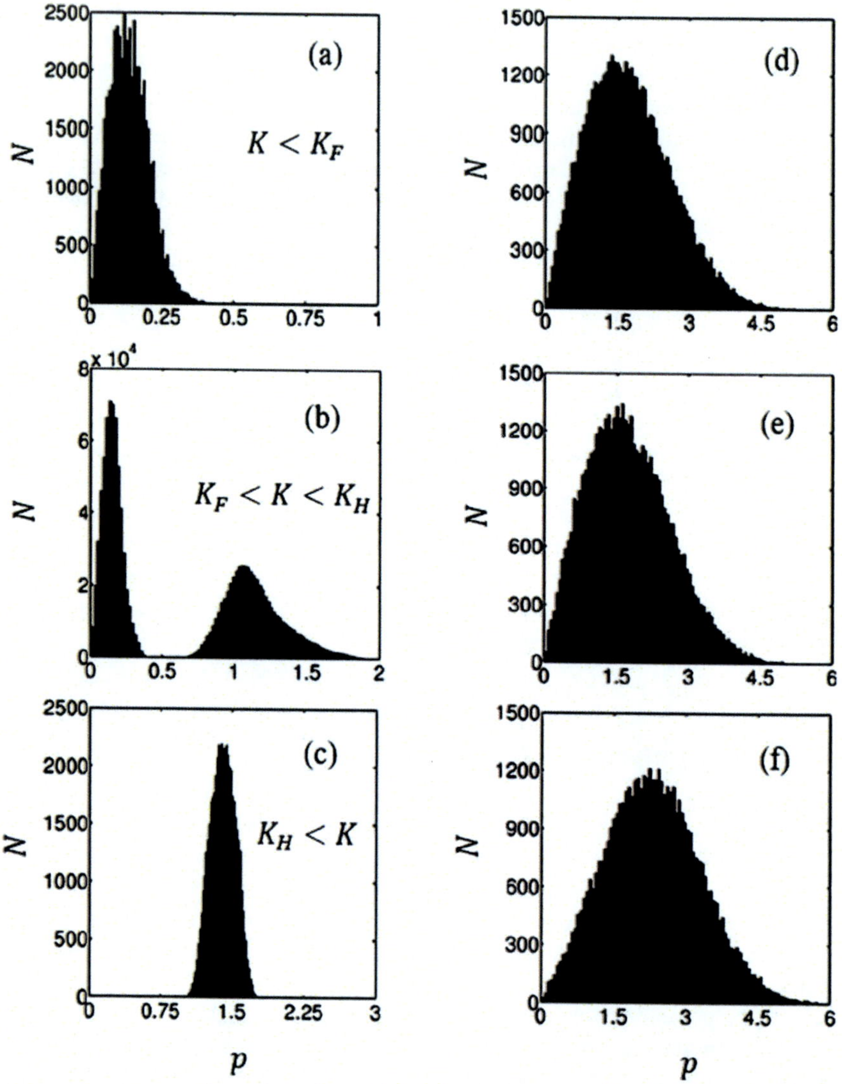

Fig. 3.10 The left and the right columns show the distribution of the positive amplitude of acoustic pressure signal in the presence of low (0.2) and high (0.5) noise intensities, respectively, for (**a**, **d**) $K < K_F$, (**b**, **e**) $K_F < K < K_H$, and (**c**, **f**) $K_H < K$. Here, K is the control parameter and, K_F and K_H correspond to the values of the parameter at the fold point and the Hopf point, respectively. Depending on the intensity of noise, the P-bifurcation characteristics of the system show a significant change. Adapted with permission from Gopalakrishnan et al. [35]

the distribution of the positive amplitude of the acoustic pressure fluctuations for the three different regions of subcritical Hopf bifurcation, which include $K < K_F$ (sub-threshold region), $K_F < K < K_H$ (bistable region), and $K_H < K$ (limit cycle region after the Hopf point). Here, K_F and K_H are the values of the heater power corresponding to the fold and Hopf points, respectively, in the bifurcation diagram of a deterministic system.

In Fig. 3.10a–c, we notice that, when noise level in the system is low, the amplitude distribution of the pressure signal changes from unimodal to bimodal and finally to unimodal, respectively, for the aforementioned three regions of the bifurcation diagram. The peak of the distribution in the first case is near the amplitude level of stable operation, and the peak in the last case is near the amplitude of limit cycle oscillations. The distribution in the bistable region shows two peaks corresponding to these extreme fluctuations observed in the system. This change in the distribution of the pressure amplitude at critical values of K corresponds to the occurrence of P-bifurcation in the system. However, when the noise level in the system is very high (Fig. 3.10d–f), we observe that the amplitude distribution of the signal always remains unimodal during the transition from steady state to limit cycle oscillations. This behaviour of the amplitude distribution indicates the suppression of hysteresis region and, in turn, the absence of P-bifurcation in the system for the condition of very high intensity of noise.

3.3.3 Triggering in Thermoacoustic Systems

In practical thermoacoustic systems, as the amplitude of background turbulent flow field is highly fluctuating, we notice that the onset of thermoacoustic oscillations sometimes occurs much prior to the boundary of unstable operation in the system. The phenomenon of the unexpected switching of the system behavior from stable (steady state) to unstable (limit cycle) operation in the regime of the stable operation (linearly steady state), due to finite amplitude disturbances, is referred to as 'triggering' in thermoacoustic parlance [18, 19, 27] or 'noise-induced transition' [41] and 'noise-induced switching' [45] in the literature of nonlinear dynamics. We come across many studies in the past that are dedicated to understand the phenomenon of triggering in thermoacoustic systems [2, 14, 44, 48, 56, 62, 68, 77, 95, 100, 101, 108, 109]. Triggering has mostly been observed in the solid and liquid rocket motors [3, 11, 99]. Zinn and Lieuwen [109] remarked in the context of gas turbine engines that "*although large-amplitude disturbances are generally required to initiate unstable oscillations in nonlinearly unstable systems, a system may be nonlinearly unstable at low-amplitude disturbances that are of the order of the background noise level*". In such systems, perturbations of magnitude nearly the same as the background fluctuations were observed to trigger the system behavior from stable operation to large amplitude self-sustained thermoacoustic oscillations.

Triggering is an important feature of a nonlinear system. When such systems operate in a linearly stable regime, their behaviour suddenly switches to another stable state when excited with perturbations of amplitude greater than a particular threshold [47]. If the system is operating in the regime with a single stable state, linear stability analysis suggests that all perturbations will eventually die down to that steady state (Fig. 3.11a). However, if the system is operating in the bistable or multistable regime consisting of two or more stable states, then a sufficiently large amplitude perturbation could cause the system to approach the basin of attraction of another stable state. Depending on the stability margin of the system, the perturbation amplitude required to change its stability will be high or low, as shown in Fig. 3.11b, c, respectively.

3.3.3.1 What Causes Triggering?

In thermoacoustic studies, triggering occurs due to a combination of nonlinear and non-normal effects [96]. Nonlinearity in the system arises from different sources, such as nonlinear gas dynamics (which become significant when the acoustic velocity perturbations increases to a value comparable with the speed of sound), heat release rate fluctuations (which become significant when the acoustic velocity perturbations are of the same order as the mean flow velocity), and nonlinear damping [64].

According to Ananthkrishnan et al. [2], for the occurrence of triggering due to nonlinearity, the system should undergo a subcritical Hopf bifurcation or a supercritical Hopf bifurcation followed by a fold bifurcation. In the bistable region of such Hopf bifurcations, the stability analysis shows the possibility of two solutions for the governing equations of the system. One of the solutions is a stable fixed point and the other is the stable limit cycle. For sufficiently large amplitude of perturbations above the critical value, the system behavior switches from stable fixed point to self-sustained limit cycle oscillations, or vice versa, due to the phenomenon of triggering. Combustion community refers to this region of system operation as 'linearly stable but nonlinearly unstable' although both the solutions are dynamically stable [109]. For systems undergoing triggering phenomena, the minimum amplitude of perturbations required to change the stability of a system stays in the regime of the stable limit cycle attractor observed in the bistable regime, as discussed in Sect. 3.2.

On the other hand, non-normality is a linear phenomenon that accounts for the growth of transient fluctuations, before they decay for certain perturbations (initial states), in the regime of linearly stable steady state [6]. For non-normal systems, the eigenvectors are always non-orthogonal.

In Fig. 3.12, we present the difference between normal and non-normal behaviors of a system. In this figure, Φ_1 and Φ_2 are two collinear vectors and F is a resultant vector obtained from the difference between the vectors Φ_1 and Φ_2. For a normal system, the vectors Φ_1 and Φ_2 are orthogonal (Fig. 3.12a). As the time progresses, the resultant vector F decreases monotonically if the individual vectors (Φ_1 and Φ_2) decrease in their magnitudes. In contrast, for a non-normal system, the vectors Φ_1 and Φ_2 are non-orthogonal (Fig. 3.12b). Let us assume that during the time evolution

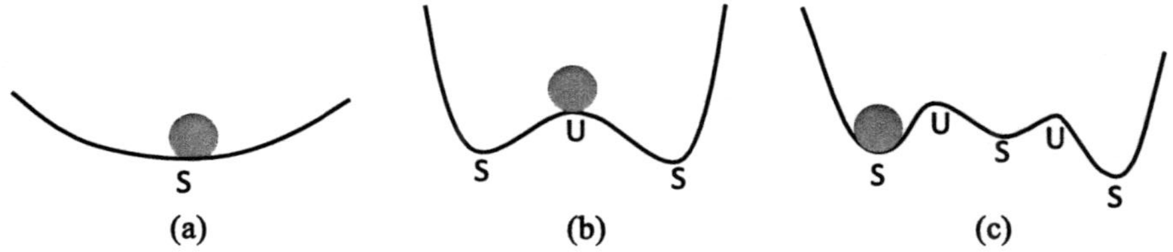

Fig. 3.11 Schematic showing the different aspects of the linear stability of the system. In (**a**), the system is in the influence of a single stable state. While in (**b**) and (**c**), the system is in the influence of multiple stable states. A finite amplitude perturbation can change the stability of such bistable systems. Stable states are indicated by S and unstable states by U

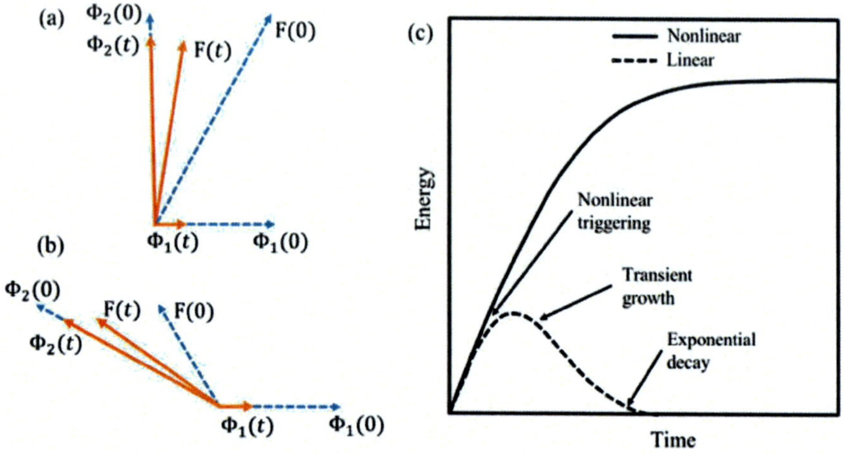

Fig. 3.12 Schematic showing the difference between (**a**) normal and (**b**) non-normal nature of collinear vectors Φ_1 and Φ_2 and their resultant F at different instances of time. For a normal system, the vectors are orthogonal, where the magnitude of F decreases monotonically due to a decrease in the magnitudes of Φ_1 and Φ_2. In contrast, for a non-normal system, the vectors are non-orthogonal, where a decrease in the magnitude of Φ_1 and Φ_2 in time results in a transient growth in the magnitude of F, before it decays to zero. (**c**) Schematic showing the linear and nonlinear evolution of the acoustic energy. Adapted with permission from Balasubramanian and Sujith [6]

of a system, the magnitude of the vector Φ_1 decreases by a larger percentage than that of the vector Φ_2 at each time interval. Then, we observe that, as time progresses, the magnitude of the resultant vector F shows an increase before it ultimately decays to zero, even though the magnitude of individual vectors is decreasing in time. This transient growth of oscillations persists for a sufficiently long time, before exhibiting an eventual exponential decay (Fig. 3.12c).

For a non-normal system, Juniper [47] found that there exists most dangerous initial states whose energy is lower than the energy of unstable limit cycle oscillations. During the transient growth of these initial states, oscillations can cause the system to cross over the unstable limit cycle attractor and transition to the stable limit cycle attractor. In short, non-normality leads to transient growth, even when the individual modes decay exponentially. In a thermoacoustic system that undergoes subcritical Hopf bifurcation, states that have low energy can initially exploit non-normal linear transient growth around the unstable periodic solution before growing from there to the stable periodic solution due to nonlinear effects. This can explain the triggering caused by small-amplitude disturbances, or that order of background noise level [75]. Mariappan et al. [67] provided experimental evidence for the non-normal nature of thermoacoustic system. For detailed studies on non-normality in thermoacoustic systems, the readers may refer to [5, 6, 47, 57, 65, 67, 93, 96, 107, 108]. In the following subsections, we will briefly discuss the experimental studies on triggering phenomenon in thermoacoustic systems.

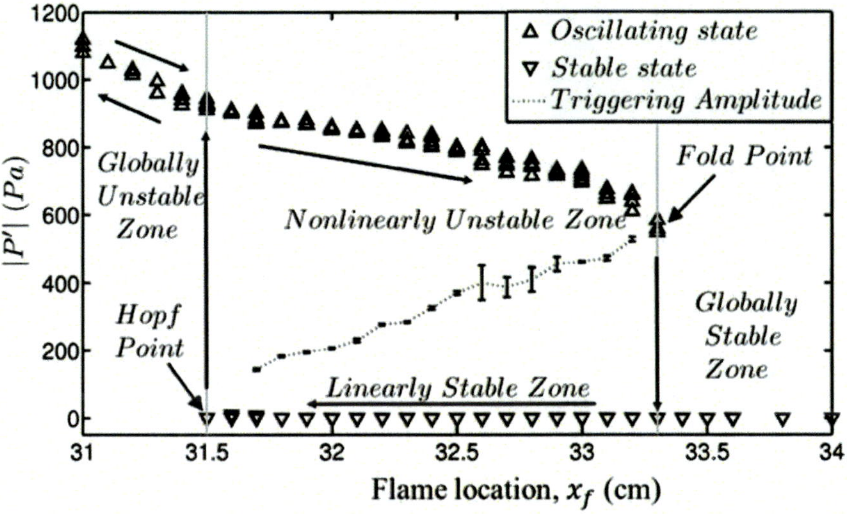

Fig. 3.13 Bifurcation diagram of a non-premixed ducted conical flame combustor undergoing a subcritical Hopf bifurcation, where the amplitude of pressure oscillations (P') is plotted as a function of the flame location (x_f). Arrows indicate the direction of the variation of the control parameter. Adapted with permission from Jagadeesan and Sujith [44]

3.3.3.2 Triggering Due to Deterministic Perturbations

An experimental method often used to study triggering in a thermoacoustic system involves the introduction of perturbations to the system in its bistable region of operation for a short duration, using external sinusoidal (deterministic) fluctuations. The frequency of these external perturbations is fixed at the frequency of self-excited thermoacoustic instability observed in such systems. The amplitude of external perturbations is progressively increased until the system behavior transitions from the state of stable fixed point to stable limit cycle oscillations.

In Fig. 3.13, we show the bifurcation diagram of subcritical Hopf bifurcation observed in a non-premixed ducted conical flame burner [44]. In this plot, the variation of the amplitude of the acoustic pressure fluctuations P is shown as a function of the flame location (x_f) in the combustor. The hysteresis region in the bifurcation diagram consists of a lower branch corresponding to stable operation (stable fixed point) and an upper branch corresponding to unstable operation (stable limit cycle oscillations) of the combustor. During the triggering experiments, the amplitude of external harmonic perturbations is increased in small steps at a particular value of x_f in the bistable region.

In Fig. 3.14a, we show the time series of the acoustic pressure fluctuations for the conditions where the amplitude of harmonic perturbations is lower and higher than a particular threshold. When the amplitude of the harmonic perturbations is less than the threshold, the acoustic pressure fluctuations initially shows a momentary growth

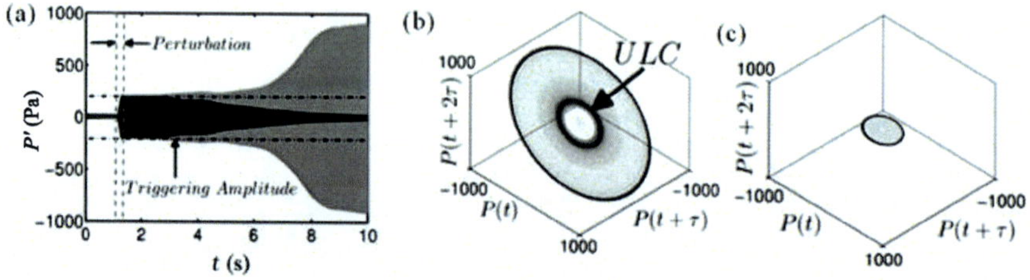

Fig. 3.14 Plots showing the triggering phenomenon observed in the bistable region of a non-premixed ducted conical flame burner. (**a**) The time series of acoustic pressure in gray shows that the system evolves to limit cycle oscillations when the amplitude of the disturbance is sufficiently high to cross the threshold of the triggering amplitude. The time series in black indicates the decay of pressure oscillations to a steady state when the amplitude of the disturbance is less than the threshold. The corresponding phase portraits show the evolution of the trajectory to (**b**) the limit cycle attractor and (**c**) the steady state attractor. *ULC* in (**b**) denotes an unstable limit cycle. Adapted with permission from Jagadeesan and Sujith [44]

in the amplitude due to forcing and eventually decays to the steady state after a transient time is over. Conversely, if the perturbation amplitude is greater than the threshold, the pressure oscillations grow from the steady state and saturate to limit cycle oscillations.

The minimum threshold amplitude of external perturbations at which such switching between the dynamical states takes place in the system is referred to as the 'triggering amplitude'. In the bifurcation diagram (Fig. 3.13), the boundary of the triggering amplitude is indicated by a dotted curve in the bistable zone. This branch also corresponds to a basin boundary of the unstable limit cycle (*ULC*) in the system. For the system behavior to transition from the stable steady state to the stable limit cycle oscillations, the perturbation amplitude has to cross the boundary of the unstable limit cycle. From Fig. 3.13, we notice that the value of the triggering amplitude decreases as X_f approaches the Hopf point in the bistable region. This also means that small amplitude perturbations are sufficient to trigger the system behavior from steady state to limit cycle oscillations in the vicinity of the Hopf bifurcation point.

In Fig. 3.14b, c, we present the reconstructed phase portraits of the acoustic pressure fluctuations observed for the conditions when the perturbation amplitude is just above and below the triggering amplitude, respectively. When the harmonic forcing is switched on, the trajectory in the phase space evolves spirally towards the inner black loop from the steady state (fixed point). This inner loop indicates the threshold amplitude of the unstable limit cycle (*ULC*) that needs to be crossed for the system behavior to transition to the outer black loop of the stable limit cycle oscillations. Once the forcing is switched off and if the amplitude of forcing is above the triggering threshold, the trajectory in the phase space spirals towards the stable limit cycle attractor (Fig. 3.14b). In contrast, if the forcing amplitude is below the triggering threshold, the phase space trajectory hovers for some time around the inner loop and then spirals towards the stable fixed point attractor (Fig. 3.14c).

3.3.3.3 Noise-Induced Triggering

Similar to the triggering phenomenon studied using harmonic perturbations, in the case of noise-induced triggering, the system is forced with external noise (stochastic perturbations) in the bistable region of subcritical Hopf bifurcation [44, 101]. In this region, the switching of a thermoacoustic system from the steady state to the limit cycle primarily depends on the strength of the noise level and the proximity of the control parameter to the Hopf point. When the amplitude of noise perturbations is very low such that it does not trigger the state of limit cycle oscillations in the system, the effect of noise shows the formation of spurts (bursts) in between the noisy oscillations of the acoustic pressure signal [44]. These spurts could be the result of a momentary shift of the system stability to an unstable state causing the growth in the amplitude of acoustic pressure oscillations or the result of the non-normal nature of the system.

Figure 3.15 shows the effect of an increase in the noise amplitude on the triggering of acoustic pressure fluctuations in a non-premixed ducted conical flame burner, when the system is operating in the bistable zone [44]. When the amplitude of the noise level is sufficiently high (Fig. 3.15a), we observe the triggering of the underlying noisy fluctuations to the state of limit cycle oscillations. We also notice the occurrence of low amplitude spurts prior to the transition to limit cycle oscillations (Fig. 3.15a). The presence of such spurts leads to the spread of the central black region in the phase portrait (Fig. 3.15b). During triggering, the system initially switches to an intermediate closed loop attractor of the unstable limit cycle and then spirals towards the outer attractor of the stable limit cycle. With further increase in the noise level (Fig. 3.15c), the behavior of the system shows an alternate switching between the states of periodic oscillations and noisy fluctuations. Thus, at a higher level of noise perturbations, we notice the intermittent switching of the behavior of the system between distinct attractors (Fig. 3.15d), which increases when the noise level is even higher (Fig. 3.15e). The phase portraits for the case of higher noise levels contain more cluttered trajectories (Fig. 3.15f), where both attractors of the fixed point and the stable limit cycle appear to approach each other. The increase in noise level breaks the coupling between the acoustic pressure and heat release rate fluctuations. This, in turn, causes a reduction in the overall amplitude of periodic oscillations triggered in the system. Jagadeesan and Sujith [44] showed that the noise intensity required for triggering at a particular x_f is much smaller than the triggering amplitude of harmonic perturbations required in the deterministic system (Fig. 3.13). With this a brief description on triggering in thermoacoustic systems, we end our discussion on different types of bifurcations observed in laminar thermoacoustic systems during the occurrence of thermoacoustic instability from the steady state.

To summarize, in this chapter, we discussed the bifurcation characteristics of laminar thermoacoustic systems in the context of dynamical systems theory for the variation of different control parameters. We showed that, depending upon the operating condition, a thermoacoustic system can exhibit subcritical Hopf bifurcation or supercritical Hopf bifurcation. Furthermore, we highlighted the effect

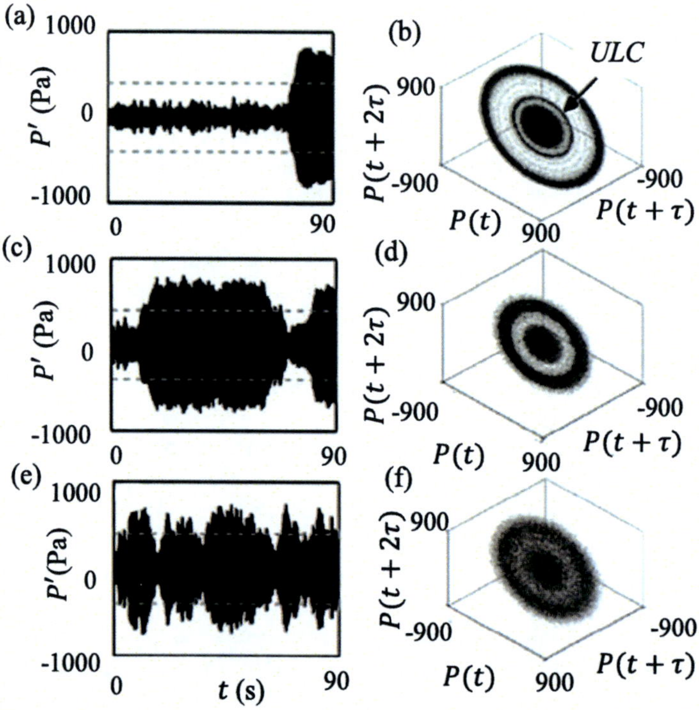

Fig. 3.15 The time series of the acoustic pressure signal (left column) and corresponding phase portraits (right column) showing the effect of increasing noise intensity on the system dynamics, when the noise level is (**a, b**) low (38.8 Pa), (**c, d**) medium (57.2 Pa) and (**e, f**) high (100 Pa) in a non-premixed ducted conical flame combustor. Increasing the noise intensity contaminates the periodicity observed in the system. Adapted with permission from Jagadeesan and Sujith [44]

of external acoustic noise on the dynamics of the acoustic pressure fluctuations in the bistable region of subcritical Hopf bifurcation. We discussed the occurrence of several phenomena of stochastic systems, such as noise-induced transition, P-bifurcation, coherence resonance, and triggering in a thermoacoustic system. Understanding these dynamical behaviors of laminar thermoacoustic systems would help in characterizing the dynamics and bifurcation characteristics of turbulent thermoacoustic systems more appropriately, as discussed in Chap. 6. In the next chapter, we will investigate the occurrence of rich nonlinear behaviors of thermoacoustic instability, beyond the state of limit cycle oscillations, using the framework of dynamical systems theory.

References

1. Abarbanel, H.D.I., Brown, R., Sidorowich, J.J., Tsimring, L.S.: The analysis of observed chaotic data in physical systems. Rev. Mod. Phys. **65**(4), 1331 (1993)
2. Ananthkrishnan, N., Deo, S., Culick, F.E.C.: Reduced-order modeling and dynamics of nonlinear acoustic waves in a combustion chamber. Combust. Sci. Tech. **177**(2), 221–248 (2005)
3. Anderson, W.E., Yang, V.: Liquid Rocket Engine Combustion Instability, vol. 169. Progress in Astronautics and Aeronautics. AIAA, Washington, DC (1995)
4. Balachandran, R., Ayoola, B.O., Kaminski, C.F., Dowling, A.P., Mastorakos, E.: Experimental investigation of the nonlinear response of turbulent premixed flames to imposed inlet velocity oscillations. Combust. Flame **143**(1–2), 37–55 (2005)
5. Balasubramanian, K., Sujith, R.I.: Non-normality and nonlinearity in combustion–acoustic interaction in diffusion flames. J. Fluid Mech. **594**, 29–57 (2008)
6. Balasubramanian, K., Sujith, R.I.: Thermoacoustic instability in a Rijke tube: non-normality and nonlinearity. Phys. Fluids **20**(4), 044103 (2008)
7. Bashkirtseva, I., Ryazanova, T., Ryashko, L.: Stochastic bifurcations caused by multiplicative noise in systems with hard excitement of auto-oscillations. Phys. Rev. E **92**(4), 042908 (2015)
8. Bellows, B.D., Neumeier, Y., Lieuwen, T.: Forced response of a swirling, premixed flame to flow disturbances. J. Propuls. Power **22**(5), 1075–1084 (2006)
9. Billings, L., Schwartz, I.B., Morgan, D.S., Bollt, E.M., Meucci, R., Allaria, E.: Stochastic bifurcation in a driven laser system: experiment and theory. Phys. Rev. E **70**(2), 026220 (2004)
10. Bisio, G., Rubatto, G.: Sondhauss and Rijke oscillations—thermodynamic analysis, possible applications and analogies. Energy **24**(2), 117–131 (1999)
11. Blomshield, F.S., Mathes, H.B., Crump, J.E., Beiter, C.A., Beckstead, M.W.: Nonlinear stability testing of full-scale tactical motors. J. Propuls. Power **13**(3), 356–366 (1997)
12. Bloxsidge, G.J., Dowling, A.P., Langhorne, P.J.: Reheat buzz: an acoustically coupled combustion instability. Part 2. Theory. J. Fluid Mech. **193**, 445–473 (1988)
13. Boudy, F., Durox, D., Schuller, T., Candel, S.: Analysis of limit cycles sustained by two modes in the flame describing function framework. C.R. Mecanique **341**(1–2), 181–190 (2013)
14. Burnley, V.S., Culick, F.E.C.: Influence of random excitations on acoustic instabilities in combustion chambers. AIAA J. **38**(8), 1403–1410 (2000)
15. Candel, S.: Combustion dynamics and control: progress and challenges. Proc. Combust. Inst. **29**(1), 1–28 (2002)
16. Chakravarthy, S.R., Shreenivasan, O.J., Boehm, B., Dreizler, A., Janicka, J.: Experimental characterization of onset of acoustic instability in a nonpremixed half-dump combustor. J. Acoust. Soc. Am. **122**(1), 120–127 (2007)
17. Cheung, W.S., Sims, G., Copplestone, R.W., Tilston, J.R., Wilson, C.W., Stow, S.R., Dowling, A.P.: Measurement and analysis of flame transfer function in a sector combustor under high pressure conditions. In: Proceedings of the ASME Turbo Expo 2003, GT2003-38219, pp. 187–194 (2003)
18. Crocco, L.: Research on combustion instability in liquid propellant rockets. In: Symposium (International) on Combustion, vol. 12, pp. 85–99. Elsevier (1969)
19. Crocco, L., Cheng, S.I.: Theory of combustion instability in liquid propellant rocket motors. Tech. rep., Agardograph no. 8, Butterworths Science Publication (1956)
20. Crocco, L., Mitchell, C.E.: Nonlinear periodic oscillations in rocket motors with distributed combustion. Combust. Sci. Tech. **1**(2), 146–169 (1969)
21. Culick, F.E.C.: Nonlinear behavior of acoustic waves in combustion chambers—II. Acta Astronaut. **3**(9–10), 735–757 (1976)
22. Culick, F.E.C.: Unsteady motions in combustion chambers for propulsion systems. Tech. rep., AGARDograph, NATO/RTO-AG-AVT-039 (2006)

23. Dakos, V., Carpenter, S.R., Brock, W.A., Ellison, A.M., Guttal, V., Ives, A.R., Kefi, S., Livina, V., Seekell, D.A., van Nes, E.H., et al.: Methods for detecting early warnings of critical transitions in time series illustrated using simulated ecological data. PLoS One **7**(7), e41010 (2012)
24. Djeundam, S.R.D., Yamapi, R., Kofane, T.C., Aziz-Alaoui, M.A.: Deterministic and stochastic bifurcations in the Hindmarsh-Rose neuronal model. Chaos **23**, 033125 (2013)
25. Dowling, A.P.: Nonlinear self-excited oscillations of a ducted flame. J. Fluid Mech. **346**, 271–290 (1997)
26. Dowling, A.P.: A kinematic model of a ducted flame. J. Fluid Mech. **394**, 51–72 (1999)
27. Dranovsky, M.L.: Combustion Instabilities in Liquid Rocket Engines: Testing and Development Practices in Russia. Progress in Astronautics and Aeronautics. AIAA, Washington D.C. (2007)
28. Durox, D., Schuller, T., Noiray, N., Candel, S.: Experimental analysis of nonlinear flame transfer functions for different flame geometries. Proc. Combust. Inst. **32**(1), 1391–1398 (2009)
29. Etikyala, S., Sujith, R.I.: Change of criticality in a prototypical thermoacoustic system. Chaos **27**(2), 023106 (2017)
30. Feldman Jr, K.T.: Review of the literature on Rijke thermoacoustic phenomena. J. Sound Vib. **7**(1), 83–89 (1968)
31. Feldman Jr, K.T.: Review of the literature on Sondhauss thermoacoustic phenomena. J. Sound Vib. **7**(1), 71–82 (1968)
32. Gelb, A., Vander Velde, W.E.: Multiple-input Describing Functions and Nonlinear System Design. McGraw-Hill, New York (1968)
33. Gopalakrishnan, E.A., Sujith, R.I.: Influence of system parameters on the hysteresis characteristics of a horizontal Rijke tube. J. Spray Combust. Dyn. **6**(3), 293–316 (2014)
34. Gopalakrishnan, E.A., Sujith, R.I.: Effect of external noise on the hysteresis characteristics of a thermoacoustic system. J. Fluid Mech. **776**, 334–353 (2015)
35. Gopalakrishnan, E.A., Tony, J., Sreelekha, E., Sujith, R.I.: Stochastic bifurcations in a prototypical thermoacoustic system. Phys. Rev. E **94**(2), 022203 (2016)
36. Guan, Y., Gupta, V., Li, L.K.B.: Intermittency route to self-excited chaotic thermoacoustic oscillations. J. Fluid Mech. **894**, R3 (2020)
37. Gupta, V., Saurabh, A., Paschereit, C.O., Kabiraj, L.: Numerical results on noise-induced dynamics in the subthreshold regime for thermoacoustic systems. J. Sound Vib. **390**, 55–66 (2017)
38. Hantschk, C.C., Vortmeyer, D.: Numerical simulation of self-excited thermoacoustic instabilities in a Rijke tube. J. Sound Vib. **227**(3), 511–522 (1999)
39. Heckl, M.A., Howe, M.S.: Stability analysis of the Rijke tube with a Green's function approach. J. Sound Vib. **305**(4–5), 672–688 (2007)
40. Higgins, B.: On the sound produced by a current of hydrogen gas passing through a tube. J. Nat. Phil. Chem. Arts **1**(2), 129–131 (1802)
41. Horsthemke, W., Lefever, R.: Noise-induced transitions in physics, chemistry, and biology. In: Noise-Induced Transitions: Theory and Applications in Physics, Chemistry, and Biology, pp. 164–200. Springer, Berlin (1984)
42. Illingworth, S.J., Juniper, M.P.: When will a flame describing function approach to thermoacoustics work well. In: Proceedings of the 19th International Congress on Sound and Vibration, pp. 1–8 (2012)
43. Jahnke, C.C., Culick, F.E.C.: Application of dynamical systems theory to nonlinear combustion instabilities. J. Propuls. Power **10**(4), 508–517 (1994)
44. Jegadeesan, V., Sujith, R.I.: Experimental investigation of noise induced triggering in thermoacoustic systems. Proc. Combust. Inst. **34**(2), 3175–3183 (2013)
45. Joshi, A., Xiao, M.: Controlling Steady-state and Dynamical Properties of Atomic Optical Bistability. World Scientific, Singapore (2012)
46. Juel, A., Darbyshire, A.G., Mullin, T.: The effect of noise on pitchfork and Hopf bifurcations. Proc. R. Soc. A **453**(1967), 2627–2647 (1997)

47. Juniper, M.P.: Triggering in the horizontal Rijke tube: non-normality, transient growth and bypass transition. J. Fluid Mech. **667**, 272–308 (2011)
48. Juniper, M.P.: Triggering in thermoacoustics. J. Spray Combust. Dyn. **4**(3), 217–237 (2012)
49. Juniper, M.P., Sujith, R.I.: Sensitivity and nonlinearity of thermoacoustic oscillations. Annu. Rev. Fluid Mech. **50**, 661–689 (2018)
50. Kabiraj, L., Saurabh, A., Karimi, N., Sailor, A., Mastorakos, E., Dowling, A.P., Paschereit, C.O.: Chaos in an imperfectly premixed model combustor. Chaos **25**(2), 023101 (2015)
51. Kabiraj, L., Saurabh, A., Wahi, P., Sujith, R.I.: Route to chaos for combustion instability in ducted laminar premixed flames. Chaos **22**(2), 023129 (2012)
52. Kabiraj, L., Steinert, R., Saurabh, A., Paschereit, C.O.: Coherence resonance in a thermoacoustic system. Phys. Rev. E **92**(4), 042909 (2015)
53. Kabiraj, L., Sujith, R.I., Wahi, P.: Bifurcations of self-excited ducted laminar premixed flames. J. Eng. Gas Turbines Power **134**(3), 031502 (2012)
54. Kabiraj, L., Vishnoi, N., Saurabh, A.: A review on noise-induced dynamics of thermoacoustic systems. In: Dynamics and Control of Energy Systems, pp. 265–281. Springer, Berlin (2020)
55. Kashinath, K., Waugh, I.C., Juniper, M.P.: Nonlinear self-excited thermoacoustic oscillations of a ducted premixed flame: bifurcations and routes to chaos. J. Fluid Mech. **761**, 399–430 (2014)
56. Kim, K.T., Hochgreb, S.: Measurements of triggering and transient growth in a model lean-premixed gas turbine combustor. Combust. Flame **159**(3), 1215–1227 (2012)
57. Kulkarni, R., Balasubramanian, K., Sujith, R.I.: Non-normality and its consequences in active control of thermoacoustic instabilities. J. Fluid Mech. **670**, 130–149 (2011)
58. Lee, M., Guan, Y., Gupta, V., Li, L.K.B.: Input-output system identification of a thermoacoustic oscillator near a Hopf bifurcation using only fixed-point data. Phys. Rev. E **101**(1), 013102 (2020)
59. Lei, S., Turan, A.: Nonlinear/chaotic behaviour in thermo-acoustic instability. Combust. Theor. Model. **13**(3), 541–557 (2009)
60. Li, X., Zhao, D., Yang, X.: Experimental and theoretical bifurcation study of a nonlinear standing-wave thermoacoustic system. Energy **135**, 553–562 (2017)
61. Lieuwen, T., Neumeier, Y.: Nonlinear pressure-heat release transfer function measurements in a premixed combustor. Proc. Combust. Inst. **29**(1), 99–105 (2002)
62. Lieuwen, T.C.: Experimental investigation of limit-cycle oscillations in an unstable gas turbine combustor. J. Propuls. Power **18**(1), 61–67 (2002)
63. Lieuwen, T.C.: Statistical characteristics of pressure oscillations in a premixed combustor. J. Sound Vib. **260**(1), 3–17 (2003)
64. Lieuwen, T.C., Yang, V.: Combustion Instabilities in Gas Turbine Engines (Operational Experience, Fundamental Mechanisms and Modeling), vol. 210. Progress in Astronautics and Aeronautics. AIAA, Reston (2005)
65. Mariappan, S., Sujith, R.I.: Thermoacoustic instability in solid rocket motor: non-normality and nonlinear instabilities. In: 48th AIAA Aerospace Sciences Meeting Including the New Horizons Forum and Aerospace Exposition, AIAA 2010-1517 (2010)
66. Mariappan, S., Sujith, R.I.: Modelling nonlinear thermoacoustic instability in an electrically heated Rijke tube. J. Fluid Mech. **680**, 511–533 (2011)
67. Mariappan, S., Sujith, R.I., Schmid, P.J.: Experimental investigation of non-normality of thermoacoustic interaction in an electrically heated Rijke tube. J. Spray Combust. Dyn. **7**(4), 315–352 (2015)
68. Matveev, K.I.: Thermoacoustic instabilities in the Rijke tube: experiments and modeling. Ph.D. thesis, California Institute of Technology (2003)
69. McManus, K.R., Poinsot, T., Candel, S.M.: A review of active control of combustion instabilities. Prog. Energy Combust. Sci. **19**(1), 1–29 (1993)
70. Meunier, C., Verga, A.D.: Noise and bifurcations. J. Stat. Phys. **50**(1–2), 345–375 (1988)
71. Moeck, J., Bothien, M., Schimek, S., Lacarelle, A., Paschereit, C.: Subcritical thermoacoustic instabilities in a premixed combustor. In: 14th AIAA/CEAS Aeroacoustics Conference (29th AIAA Aeroacoustics Conference), AIAA 2008-2946 (2008)

72. Nair, V., Thampi, G., Karuppusamy, S., Gopalan, S., Sujith, R.I.: Loss of chaos in combustion noise as a precursor of impending combustion instability. J. Spray Combust. Dyn. **5**(4), 273–290 (2013)
73. Nair, V., Thampi, G., Sujith, R.I.: Intermittency route to thermoacoustic instability in turbulent combustors. J. Fluid Mech. **756**, 470–487 (2014)
74. Nayfeh, A.H., Balachandran, B.: Applied Nonlinear Dynamics: Analytical, Computational, and Experimental Methods. Wiley, New York (2008)
75. Nicoud, F., Benoit, L., Sensiau, C., Poinsot, T.: Acoustic modes in combustors with complex impedances and multidimensional active flames. AIAA J. **45**(2), 426–441 (2007)
76. Noble, A.C., King, G.B., Laurendeau, N.M., Gord, J.R., Roy, S.: Nonlinear thermoacoustic instability dynamics in a Rijke tube. Combust. Sci. Tech. **184**(3), 293–322 (2012)
77. Noiray, N., Durox, D., Schuller, T., Candel, S.: A unified framework for nonlinear combustion instability analysis based on the flame describing function. J. Fluid Mech. **615**, 139–167 (2008)
78. Oran, E.S., Gardner, J.H.: Chemical-acoustic interactions in combustion systems. Prog. Energy Combust. Sci. **11**(4), 253–276 (1985)
79. Paschereit, C.O., Schuermans, B., Polifke, W., Mattson, O.: Measurement of transfer matrices and source terms of premixed flames. J. Eng. Gas Turbines Power **124**(2), 239–247 (2002)
80. Pawar, S.A., Vishnu, R., Vadivukkarasan, M., Panchagnula, M.V., Sujith, R.I.: Intermittency route to combustion instability in a laboratory spray combustor. J. Eng. Gas Turbines Power **138**(4), 041505 (2016)
81. Pikovsky, A.S., Kurths, J.: Coherence resonance in a noise–driven excitable system. Phys. Rev. Lett. **78**(5), 775–778 (1997)
82. Premraj, D., Pawar, S.A., Kabiraj, L., Sujith, R.I.: Strange nonchaos in self-excited singing flames. Europhys. Lett. **128**(5), 54005 (2020)
83. Raun, R.L., Beckstead, M.W., Finlinson, J.C., Brooks, K.P.: A review of Rijke tubes, Rijke burners and related devices. Prog. Energy Combust. Sci. **19**(4), 313–364 (1993)
84. Rayleigh, J.W.S.: The explanation of certain acoustical phenomena. Nature **18**(455), 319–321 (1878)
85. Rijke, P.L.: LXXI. Notice of a new method of causing a vibration of the air contained in a tube open at both ends. Mag. J. Sci. London Edinburgh Dublin Philos. **17**(116), 419–422 (1859)
86. Sarpotdar, S.M., Ananthkrishnan, N., Sharma, S.D.: The Rijke tube—a thermo-acoustic device. Resonance **8**(1), 59–71 (2003)
87. Sastry, S., Hijab, O.: Bifurcation in the presence of small noise. Syst. Control. Lett. **1**(3), 159–167 (1981)
88. Saurabh, A., Kabiraj, L., Steinert, R., Oliver Paschereit, C.: Noise-induced dynamics in the subthreshold region in thermoacoustic systems. J. Eng. Gas Turbines Power **139**(3) (2017)
89. Scheffer, M., Bascompte, J., Brock, W.A., Brovkin, V., Carpenter, S.R., Dakos, V., Held, H., Van Nes, E.H., Rietkerk, M., Sugihara, G.: Early-warning signals for critical transitions. Nature **461**(7260), 53–59 (2009)
90. Sondhauss, C.: Ueber die schallschwingungen der luft in erhitzten glasröhren und in gedeckten pfeifen von ungleicher weite. Ann. Phys. **155**(1), 1–34 (1850)
91. Song, C., Phenix, H., Abedi, V., Scott, M., Ingalls, B.P., Kærn, M., Perkins, T.J.: Estimating the stochastic bifurcation structure of cellular networks. PLoS Comput. Biol. **6**(3), e1000699 (2010)
92. Strogatz, S.H.: Nonlinear Dynamics and Chaos: With Applications to Physics, Biology, Chemistry, and Engineering. CRC Press, Boca Raton (1994)
93. Subramanian, P.: Dynamical systems approach to the investigation of thermoacoustic instabilities. Ph.D. thesis, Indian Institute of Technology Madras (2011)
94. Subramanian, P., Mariappan, S., Sujith, R.I., Wahi, P.: Bifurcation analysis of thermoacoustic instability in a horizontal Rijke tube. J. Spray Combust. Dyn. **2**(4), 325–355 (2010)
95. Subramanian, P., Sujith, R.I., Wahi, P.: Subcritical bifurcation and bistability in thermoacoustic systems. J. Fluid Mech. **715**, 210–238 (2013)

96. Sujith, R.I., Juniper, M.P., Schmid, P.J.: Non-normality and nonlinearity in thermoacoustic instabilities. J. Spray Combust. Dyn. **8**(2), 119–146 (2016)
97. Takens, F.: Detecting strange attractors in turbulence. Lect. Notes Math. **898**(1), 366–381 (1981)
98. Tandon, S., Pawar, S.A., Banerjee, S., Varghese, A.J., Durairaj, P., Sujith, R.I.: Bursting during intermittency route to thermoacoustic instability: effects of slow–fast dynamics. Chaos **30**(10), 103112 (2020)
99. Wang, Z., Liu, P., Jin, B., Ao, W.: Nonlinear characteristics of the triggering combustion instabilities in solid rocket motors. Acta Astronaut. **176**, 371–382 (2020)
100. Waugh, I., Geuß, M., Juniper, M.: Triggering, bypass transition and the effect of noise on a linearly stable thermoacoustic system. Proc. Combust. Inst. **33**(2), 2945–2952 (2011)
101. Waugh, I.C., Juniper, M.P.: Triggering in a thermoacoustic system with stochastic noise. J. Spray Combust. Dyn. **3**(3), 225–241 (2011)
102. Wicker, J.M., Greene, W.D., Kim, S.I., Yang, V.: Triggering of longitudinal combustion instabilities in rocket motors-nonlinear combustion response. J. Propuls. Power **12**(6), 1148–1158 (1996)
103. Wilhite, J.M., Dolan, B.J., Kabiraj, L., Gomez, R.V., Gutmark, E.J., Paschereit, C.O.: Analysis of combustion oscillations in a staged MLDI burner using decomposition methods and recurrence analysis. In: 54th AIAA Aerospace Sciences Meeting, AIAA 2016-1156 (2016)
104. Yang, X., Turan, A., Lei, S.: Thermoacoustic instability in a Rijke tube with a distributed heat source. J. Thermodyn. **2015**, 1–9 (2015)
105. Yang, X., Turan, A., Lei, S.: Bifurcation and nonlinear analysis of a time-delayed thermoacoustic system. Commun. Nonlinear Sci. Numer. Simul. **44**, 229–244 (2017)
106. Zakharova, A., Vadivasova, T., Anishchenko, V., Koseska, A., Kurths, J.: Stochastic bifurcations and coherence–like resonance in a self-sustained bistable noisy oscillator. Phys. Rev. E **81**(1), 011106 (2010)
107. Zhao, D.: Transient growth of flow disturbances in triggering a Rijke tube combustion instability. Combust. Flame **159**(6), 2126–2137 (2012)
108. Zhao, D., Reyhanoglu, M.: Feedback control of acoustic disturbance transient growth in triggering thermoacoustic instability. J. Sound Vib. **333**(16), 3639–3656 (2014)
109. Zinn, B.T., Lieuwen, T.C.: Combustion instabilities: basic concepts. In: Combustion Instabilities in Gas Turbine Engines: Operational Experience, Fundamental Mechanisms, and Modeling, vol. 210, pp. 3–26. AIAA, Reston, VA (2005)

Chapter 4
Thermoacoustic Instability: Beyond Limit Cycle Oscillations

In the previous chapter, we discussed the bifurcation characteristics of a simple thermoacoustic system (horizontal Rijke tube) exhibiting the transition from a steady state to limit cycle oscillations (thermoacoustic instability), as we change the system parameter. The onset of these limit cycle oscillations happens through Hopf bifurcation. However, we note that the dynamics during thermoacoustic instability is not limited to period-1 limit cycle oscillations, but can exhibit different dynamical states due to the simultaneous excitation of several dominant frequencies (or unstable modes) and their multiples in the system. The interplay between such frequencies makes the system dynamics highly complex and difficult to interpret in most cases. In this chapter, we delineate the possible rich dynamical behaviors of self-excited thermoacoustic instability, in addition to period-1 limit cycle oscillations, observed in different configurations of Rijke-type laminar thermoacoustic systems. Appropriate classification of these dynamical states is just as important as the classification of bifurcations in the system. Such classification would further help in providing efficient control strategies required to either evade or mitigate their occurrence in the systems. Since the occurrence of chaos in nonlinear systems is fascinating, we will also shed light on the evidence of different routes to chaos observed in laminar thermoacoustic systems. Finally, we will discuss the flame-acoustic interactions observed during different states of thermoacoustic instability in such systems as they are relatively easy to study and model when compared to that observed in turbulent combustors.

Prior to 2011, there have been only a few studies that reported the evidence of thermoacoustic oscillations with more than one dominant frequency in laminar as well as turbulent thermoacoustic systems [3, 5, 18, 32, 37, 42, 52]. These discrete frequencies were observed at harmonics (commensurate) or non-harmonics (incommensurate) of each other. The coexistence of such multiple unstable modes makes the system dynamics more complex and difficult to reveal its properties just by observing its signal or power spectrum. However, we can achieve a complete

R. I. Sujith, S. A. Pawar, *Thermoacoustic Instability*, Springer Series in Synergetics,
https://doi.org/10.1007/978-3-030-81135-8_4

description for the characteristics of these complex thermoacoustic signals through the application of different tools from dynamical systems theory.

Although there have been a few instances where dynamical systems theory was applied to delineate the characteristics of thermoacoustic oscillations in the past [8, 20, 30, 34, 35, 55, 56], serious effort towards understanding thermoacoustic oscillations through this framework was seen only in the decade after the year 2011. Hence, this period can be considered as the start of a golden period for thermoacoustics, during which the description of thermoacoustic instability as merely limit cycle oscillations was reexamined using the application of dynamical systems theory. Breakthroughs by different research groups around the world [1, 2, 4, 8, 10, 11, 14, 16, 17, 21–24, 26–30, 34, 39, 40, 43–46, 51, 55, 56, 58, 60–63] have redefined the state of thermoacoustic instability and provided a systematic framework to analyze the onset of thermoacoustic instability, which we will discuss in the present chapter.

4.1 Bifurcation of Thermoacoustic Instability Beyond the State of Limit Cycle

Traditionally, thermoacoustic instabilities observed at discrete modes (tonal sound) of the acoustic field of a combustor were classified as limit cycle oscillations. However, we come across several studies that have shown the evidence of more complex signals than tones (i.e. signals with single frequency and its harmonics) during the state of thermoacoustic instability. In such studies, an appropriate classification of different dynamical states becomes possible due to the application of dynamical systems theory. As we discussed in Chap. 2, since the reconstructed phase space of such complex thermoacoustic oscillations shows definite attractors, we can categorize these dynamical states into specific types.

In this section, we will classify all possible dynamical states of thermoacoustic instability observed in laminar thermoacoustic systems using different tools from dynamical systems theory. Towards this purpose, we take the help of an experimental investigation of Kabiraj et al. [22, 23] performed on a ducted premixed laminar flame combustor (Fig. 4.1). In this system, the location of the flame inside the combustor is varied as the control parameter. In this setup, we can independently study three configurations of the flames, i.e., a single conical flame [17, 21], multiple conical flames [22, 23, 38, 51], and a single V-flame [60]. The flame is stabilized over the burner lip and the glass duct enclosing the flame is moved in the upward direction using the traverse mechanism. This, in turn, alters the relative position of the flame inside the duct. The variation of the flame location changes the characteristics of the local acoustic admittance of the flame in the system, which changes the coupling strength between the acoustic and the heat release rate fields [62]. As a result, we observe the excitation of different modes in the system. At every value of the flame

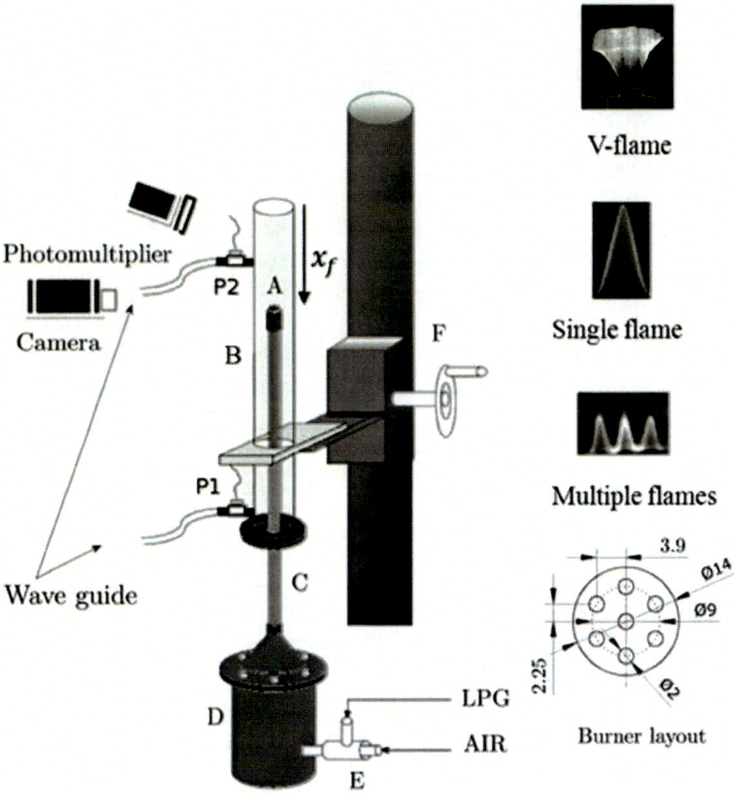

Fig. 4.1 Schematic of the experimental setup of a laminar combustor used to study the bifurcation characteristics of different flame configurations, such as a single conical flame, a single V-flame, and multiple conical flames. During experiments, the equivalence ratio is fixed at the fuel lean condition and the flame location is varied as a control parameter. Parts of the experimental setup are indicated as A: burner tube exit (flame anchoring location), B: open-closed glass duct, C: burner tube, D: decoupler, E: LPG-air premixing chamber, F: traverse, and P1, P2: pressure sensors. The burner layout of the multiple flames configuration is represented on the right side. Adapted with permission from Kabiraj et al. [22]

location (x_f), the dynamic response of the system was measured in terms of the acoustic pressure fluctuations using piezoelectric transducers.

In Fig. 4.2, we show the bifurcation diagram of a ducted laminar multiple flames combustor. The bifurcation diagram illustrates the distribution of local maxima of the acoustic pressure fluctuations in the signal, indicated as dots, as a function of the change in the flame location, x_f [22]. The number of local maxima in a signal indicates a particular dynamical state observed in the system. For example, a period-1 limit cycle exhibits a single point, while a period-2 limit cycle shows two points in such bifurcation diagrams. For aperiodic signals, such as quasiperiodic or chaotic

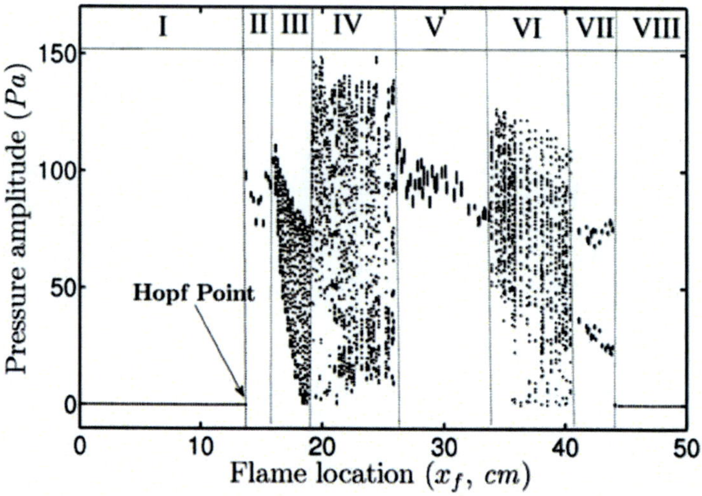

Fig. 4.2 The bifurcation plot of acoustic pressure fluctuations obtained from experiments in a premixed multiple conical flame combustor. As the flame location is varied, dynamical behavior of the system exhibits a sequence of bifurcations causing the occurrence of rich nonlinear states, which are labeled with Roman numerals (I)–(VIII). Regions I and VIII—steady state regions, II—limit cycle, III—quasiperiodic, IV—chaotic, and mode-locked, V—mode-locked, VI—chaotic, and VII—period-2 oscillations. Reproduced with permission from Kabiraj et al. [22]

oscillations, we observe a wide distribution of local maximum values of acoustic pressure fluctuations in the bifurcation diagram.

In region I of Fig. 4.2, we notice the presence of stable steady state in the system behavior, which is observed in the form of a near zero amplitude of the acoustic pressure signal. In region of II of Fig. 4.2, we observe a jump in the amplitude of the pressure signal to higher value. This jump in the pressure amplitude occurs due to the onset of thermoacoustic instability in the system. We can confirm the state of thermoacoustic instability as period-1 limit cycle oscillations from the observation of a single point at a particular x_f in the bifurcation diagram. Thus, during the occurrence of thermoacoustic instability, we observe a qualitative change in the system behavior from a steady state (no oscillations) to finite amplitude single frequency limit cycle oscillations. This transition is also referred to as subcritical Hopf bifurcation of the system (Chap. 2).

With an increase in the flame location, the system behavior stays in the state of limit cycle oscillations for some range of x_f and then transitions to the state of quasiperiodic oscillations (region III in Fig. 4.2), due to the occurrence of a secondary Hopf bifurcation. This transition from limit cycle to quasiperiodic oscillations is also called *Neimark-Sacker bifurcation* or *torus-birth bifurcation* [19]. Since quasiperiodic oscillations are aperiodic in time, we observe distributed values of the pressure amplitude, at a particular value of x_f, in the bifurcation diagram. As the value of x_f is further increased, the system undergoes a *Ruelle–*

Takens–Newhouse bifurcation [6], leading to the occurrence of chaotic oscillations in the system (region IV in Fig. 4.2). Such chaotic oscillations show a larger distribution in the amplitude of the pressure signal in the bifurcation diagram. We also observe small windows of mode-locked (frequency-locked) oscillations in region IV of Fig. 4.2.

A further increase in x_f leads to the transition of the system behavior to a state of mode-locked oscillations (region V in Fig. 4.2), which then transitions to chaotic oscillations (region VI in Fig. 4.2), period-2 oscillations (region VI in Fig. 4.2), and finally to flame blowout (region VII in Fig. 4.2). For mode-locked oscillations and period-2 oscillations, the properties of the signal repeat after a few iterations (for example, period-2 signal repeats after two cycles). Depending on the period of the oscillations, the presence of discrete points are observed for such mode-locked oscillations in the bifurcation diagram. During the regime of flame blowout, we observe that the flame completely extinguishes, resulting in a steady state behaviour of the acoustic pressure signal. Thus, we notice that a change in the control parameter leads to subsequent bifurcations of limit cycle oscillations to more complex dynamical states.

In Fig. 4.4, we discuss a detailed characterization of the aforementioned rich dynamical states of thermoacoustic instability. Toward this purpose, we show the time series, the power spectrum, and the phase portrait of the signals corresponding to these states. The evolution of a signal in the higher-dimensional phase space is presented using the Takens' delay embedding theorem [57]. The topology of the phase space indicates the attractor on which the asymptotic behavior of a given dynamical state is stabilized. Thus, from the shape of an attractor, we can identify the characteristics of a given dynamical state observed in the system (as we discussed in Chap. 2). For an appropriate reconstruction of the phase space, we need optimum values of the embedding dimension and the time delay that can be found from the methods of average mutual information (Fig. 4.3a) and false nearest neighbors (Fig. 4.3b), respectively. These methods are elaborated in detail in Chap. 2. We can choose the optimum time delay as the value of the delay corresponding to the first local minimum of the average mutual information. Furthermore, we can select the minimum embedding dimension as the next value of the dimension at which the percentage of the false nearest neighbors drops to zero.

For the state of limit cycle oscillations, the pressure signal consists of constant amplitude periodic oscillations (Fig. 4.4a-I). The power spectrum of this signal shows a single dominant frequency (f_1) and small peaks at higher harmonics such as $2f_1$ (Fig. 4.4a-II). The reconstructed phase space of limit cycle oscillations shows a closed-loop orbit, where the phase space trajectory repeats its behavior after a period of the oscillation (Fig. 4.4a-III).

In the case of quasiperiodic oscillations, we notice the emergence of another dominant frequency (f_2) at a non-harmonic value of f_1 such that the ratio of these frequencies is incommensurate (Fig. 4.4b-II). As a result, the dynamics of quasiperiodic oscillations never repeats in time, i.e., the dynamics is aperiodic (Fig. 4.4b-I). The reconstructed phase space of such quasiperiodic oscillations

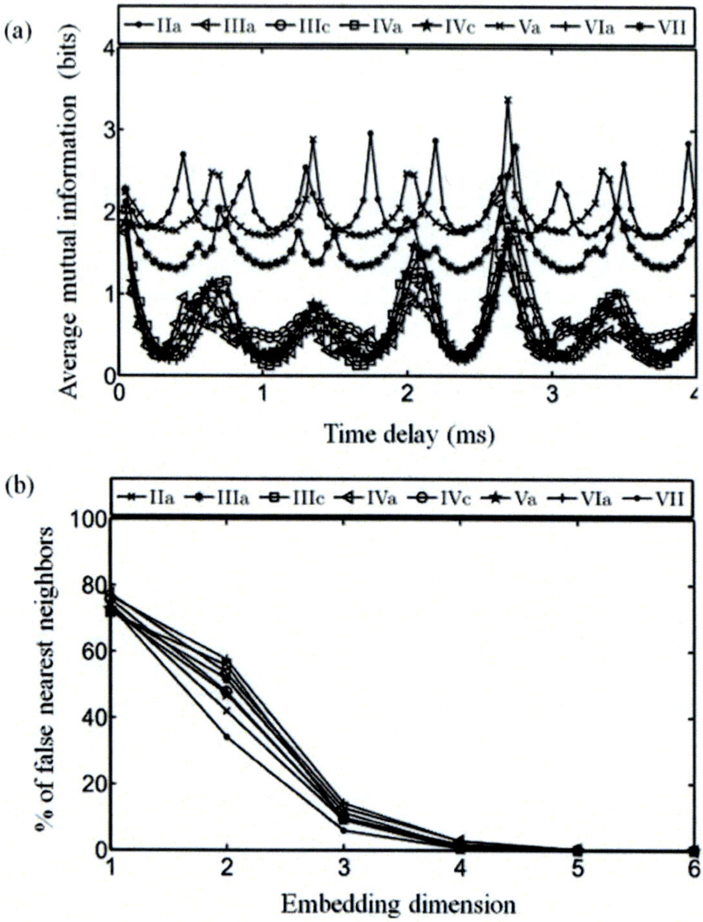

Fig. 4.3 Variation of (**a**) the average mutual information with the time delay and (**b**) the percentage of false nearest neighbors with the embedding dimension for different dynamical regions observed in Fig. 4.2. Reproduced with permission from Kabiraj et al. [22]

demonstrates a 2-torus attractor, wherein the trajectory never repeats its behavior and fills the entire space of a tubular structure (Fig. 4.4b-III).

Contrary to the sharp discrete frequency peaks observed in the power spectrum of limit cycle and quasiperiodic oscillations, the power spectrum of chaotic oscillations (observed in regions IV and VI of Fig. 4.2) are broadband and exhibit multiple low magnitude frequency peaks (see Fig. 4.4c-II and e-II, respectively). The signals of chaotic oscillations are aperiodic in time (Fig. 4.4c-I and e-I). The behavior of the trajectory in the reconstructed phase space for these signals is highly unstable; nevertheless, it is bounded over a strange (fractal) attractor (Fig. 4.4c-III and e-III).

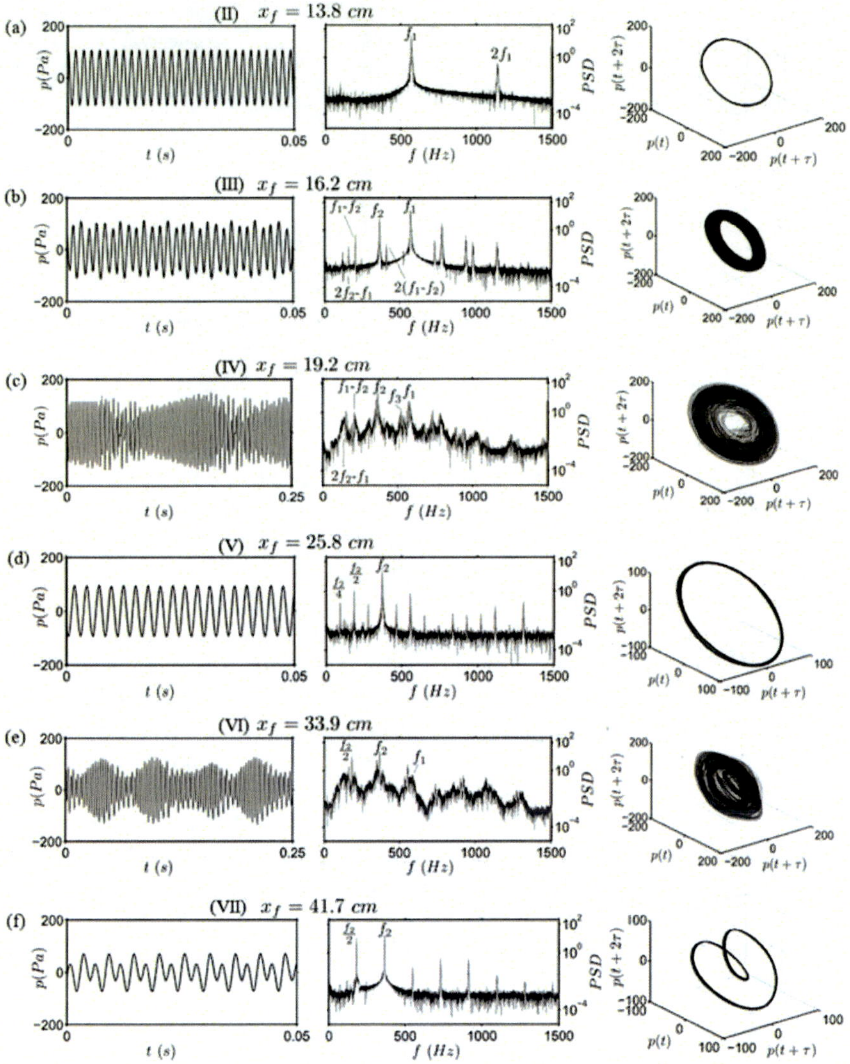

Fig. 4.4 (I)–(III) Time series, power spectrum, and phase portraits, respectively, of the acoustic pressure fluctuations (p) corresponding to the various nonlinear states of thermoacoustic instability, observed in the bifurcation diagram shown in Fig. 4.2. These states are **(a)** limit cycle, **(b)** quasiperiodicity, **(c)** chaos, **(d)** mode-locked, **(e)** chaos, and **(f)** period-2. In all the power spectrums, $f_1 \sim 570$ Hz, $f_2 \sim 364.1$ Hz, $f_3 \sim 524$ Hz. Other dominant frequencies are linear combinations of these frequencies. Flame location (x_f) corresponding to each dynamical state is stated on the top of each plot. Adapted with permission from Kabiraj et al. [22]

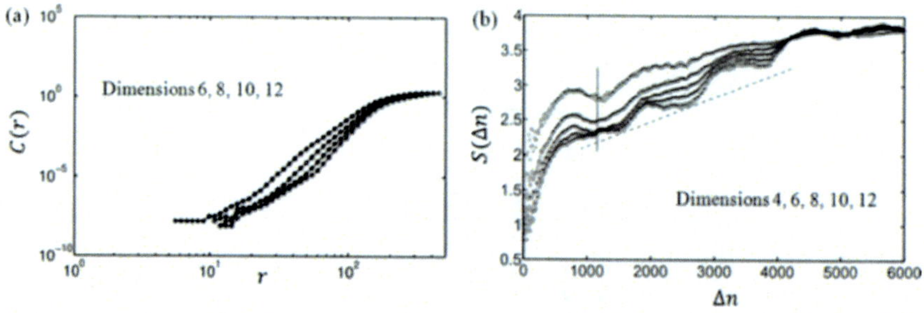

Fig. 4.5 Variation of (**a**) the correlation sum $C(r)$ with the radius r and (**b**) the average separation distance of neighboring phase space trajectories $S(\Delta n)$ with time steps Δn for different embedding dimensions (mentioned in both plots) for chaotic oscillations. Reproduced with permission from Kabiraj et al. [22]

Further, we can confirm the presence of chaos in the system through various measures, such as the Lyapunov exponent and the correlation dimension (as we discussed in Chap. 2), as illustrated in Fig. 4.5. An integer value of correlation dimension indicates the presence of a regular (periodic or quasiperiodic) attractor, while a non-integer value of correlation dimension indicates the presence of a fractal or strange attractor (mostly associated with the chaotic dynamics). Furthermore, a positive value of the maximal Lyapunov exponent indicates the presence of chaotic oscillations and its negative value suggests the presence of periodic oscillations in the system. For chaotic oscillations observed in region IV of Fig. 4.2, the correlation dimension, obtained from the slope of the plot between the correlation sum $C(r)$ and the radius r for different embedding dimensions, shows a value of 5.5 (Fig. 4.5a). The maximal Lyapunov exponent is obtained from the variation of the average separation distance between neighboring phase space trajectories, $S(\Delta n)$, with the increasing time steps, (Δn), for different embedding dimensions [22, 25]. A slope of the linear fit to the plot for different embedding dimensions (4, 6, 8, 10, 12) shows a value of 4.1 ± 1.4 (Fig. 4.5b). A similar analysis performed for chaotic oscillations in region VI of Fig. 4.2 shows the value of the correlation dimension as 4.6 and the value of the maximum Lyapunov as 5.1 ± 0.6 [22]. The non-integer value of the correlation dimension and a positive value of the average Lyapunov exponent together confirm that the dynamical states observed in regions IV and VI of Fig. 4.2 are chaos.

Furthermore, these two chaotic bands in regions IV and VI in Fig. 4.2 are separated by a region of frequency-locked oscillations (see region V in Fig. 4.2). These periodic oscillations repeat their behavior in the phase space after the number of revolutions ascertained from the ratio of locked frequencies. During this state, the power spectrum of the acoustic pressure oscillations shows dominant frequency peaks (f_2, $f_2/2$, and $f_2/4$) that are integer multiples of each other (Fig. 4.4d-II). Further, the phase portrait of such signal exhibits a closed-loop attractor, where the trajectory recurs regularly to its previous location after a few revolutions in the phase

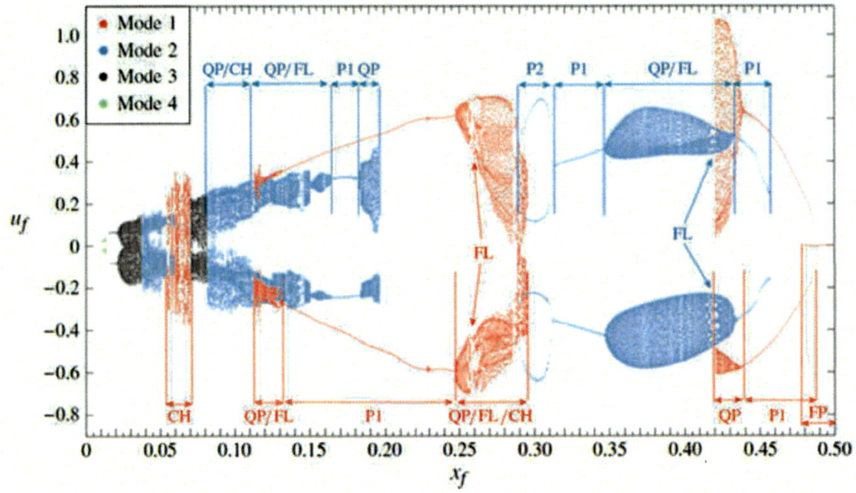

Fig. 4.6 A bifurcation diagram between the acoustic pressure velocity (u_f) and the flame location (x_f), showing the occurrence of several stable dynamical states such as fixed point (FP), period-1 limit cycle (P1), period-2 (P2) limit cycle, frequency-locked or period-k (FL) oscillations, quasiperiodicity (QP), and chaos (CH) in a numerical simulation of a two-dimensional slot-stabilized premixed flame system. The flame location (x_f) is varied in both forward and reverse directions. Colors indicate the excitation of most dominant acoustic modes in the system. The bifurcation diagram shows the abrupt end of some branches due to the presence of unstable solutions. Reproduced with permission from Kashinath et al. [26]

space (Fig. 4.4d-III). A few windows of such mode-locked states are also observed in region IV of Fig. 4.2.

After the chaotic window observed in region VI, the system behavior transitions to a state of regular period-2 oscillations (region VII in Fig. 4.2). For these signals, the oscillations repeat after two cycles and their power spectrum shows two frequency peaks of nearly equal magnitude locked in the ratio of 2 (Fig. 4.4f-II). The phase portrait of this state shows a two-looped attractor (Fig. 4.4f-III).

Similar to the experimental observations of rich dynamical behaviors (Fig. 4.4) in a ducted premixed laminar multiple flame combustor (shown in Fig. 4.2), a numerical study on a two-dimensional slot-stabilized premixed flame system by Kashinath et al. [26] has also shown the occurrence of several complex dynamical states. These states include limit cycle, quasiperiodicity, period-2, frequency-locked, and chaotic oscillations. Figure 4.6 presents a bifurcation diagram corresponding to these states, wherein the variation of the local amplitude (both positive and negative peaks) of acoustic velocity fluctuations (u_f) is plotted as a function of the flame location (x_f) in the duct [26]. They reported the occurrence of these dynamical states through different types of bifurcations, such as supercritical Hopf bifurcation, Neimark–Sacker bifurcation, period-doubling bifurcation, and fold bifurcation.

4.2 Other Dynamical States of Thermoacoustic Instability

In addition to the aforementioned nonlinear states, we also observe a few more distinct dynamical states of thermoacoustic instability in laminar combustors. We will briefly describe these states in the following subsections.

4.2.1 *Strange Nonchaos*

Strange nonchaotic oscillations are a special type of oscillations whose attractor is fractal (similar to that observed for chaotic oscillations), but their dynamics does not possess the property of chaos [13]. This means the phase space trajectories of strange nonchaotic oscillations do not exhibit sensitive dependence on the initial conditions. Thus, such signals display the coexistence of the strange attractor and the regular behavior of the phase space trajectories [13]. We can detect the strangeness of the attractor from the non-integer value of the fractal dimension (or correlation dimension) and the absence of sensitive dependence on the initial conditions from the negative value of the maximal Lyapunov exponent [47]. These oscillations are often observed in quasiperiodically driven nonlinear systems [50]. Furthermore, such oscillations are rarely witnessed in self-excited nonlinear systems. An example of a self-excited system exhibiting strange nonchaos till recent is a pulsating star KIC 5520878 [36]. We can confirm the existence of strange nonchaotic oscillations in a system using different measures [48]. Some of these measures include Lyapunov exponents, fractal dimensions, singular continuous spectrum, phase sensitivity exponent, autocorrelation function, and 0-1 test [9, 47, 50].

Premraj et al. [51] were the first to report the experimental observation of strange nonchaos in the self-excited dynamics of a laminar thermoacoustic system. On the other hand, Guan et al. [15] reported the existence of strange nonchaos in the forced dynamics of a ducted conical flame burner. In Fig. 4.7, we show the dynamic properties of strange nonchaotic oscillations observed in a premixed multiple flames laminar combustor [51]. These strange nonchaotic oscillations are observed in between the transition of the system behavior from quasiperiodic to chaotic oscillations. The phase portrait these of these oscillations shows a strange attractor, which can be observed from the unstable movement of the trajectories in the phase space (Fig. 4.7a). The first return map (or the Poincaré section) of these oscillations shows the presence of multiple wrinkled tori (Fig. 4.7b). These tori possibly occur due to the breakdown of a closed-loop structure observed in the Poincaré section of quasiperiodic oscillations. The power spectrum of these oscillations shows multiple peaks at frequencies that are irrationally related to the dominant frequency of the signal (Fig. 4.7c). For a detailed analysis of strange nonchaos in the experimental data and a model of thermoacoustic systems, readers may refer to [15, 51, 62].

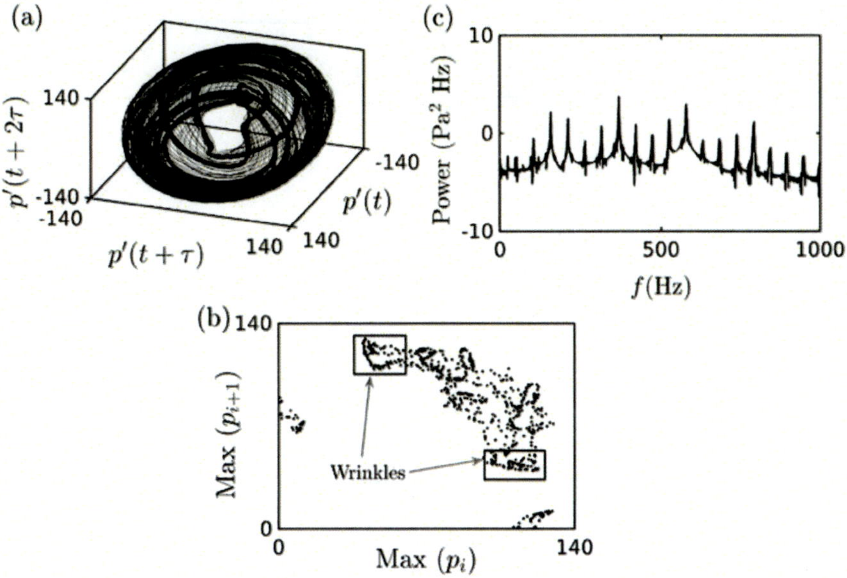

Fig. 4.7 (a)–(c) Phase portrait, Poincaré section, and power spectrum, respectively, of strange nonchaotic oscillations observed from experiments in a premixed multiple flames laminar burner. Adapted with permission from Premraj et al. [51]

4.2.2 *Intermittency*

Intermittency is another form of self-excited oscillations where the system dynamics fluctuates between epochs of low amplitude periodic (laminar or regular) oscillations and bursts of large amplitude chaotic (turbulent or irregular) oscillations at apparently random intervals of time (Fig. 4.8a). Kabiraj and Sujith [21] were the first to report the state of intermittency in thermoacoustic systems. In laminar thermoacoustic systems, the state of intermittency was observed during the route to chaos [17] and also during the route to flame blowout [22].

The power spectrum of an intermittency signal is broadband with a dominant frequency corresponding to periodic oscillations in the signal (Fig. 4.8c). The reconstructed phase space of this signal shows the presence of two dense, concentric disks (Fig. 4.8b). The phase space trajectory revolves in the central disk during the epoch of low amplitude periodic oscillations, and then spirals towards the outer disc formed due to the occurrence of large amplitude chaotic oscillations. At the end of this chaotic burst, the trajectory is re-injected back to the central disk where it continues the periodic rotations until the start of the next burst in the signal. Thus, the trajectory continuously switches back and forth in time between two distant attractors.

We can gain more insights about the dynamics of intermittency from their recurrence plots, as shown in Fig. 4.9. The recurrence plot is a graphical representation

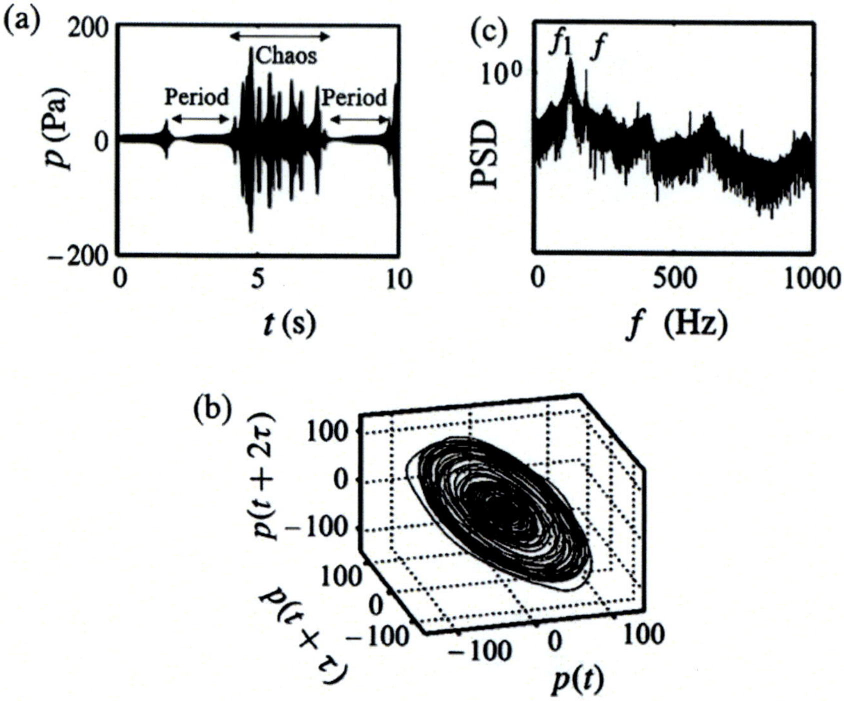

Fig. 4.8 (**a**) The time series, (**b**) phase portrait, and (**c**) power spectrum of the acoustic pressure fluctuations observed from experiments during the occurrence of intermittency prior to flame blowout in a ducted conical laminar flame combustor. Adapted with permission from Kabiraj and Sujith [21]

of the recurrence behavior of the trajectory in the phase space, indicated usually in terms of black and white points (as we discussed in Chap. 2). The recurrence plot of the intermittency signal shows the alternate occurrence of black and white patches. Dense black patches in the recurrence plot indicate the presence of laminar phases (low amplitude oscillations), while white patches illustrate the interruption of these phases by bursts of large amplitude oscillations in the signal [17, 21]. The density of the black points in the recurrence plot becomes sparse during the transition to white regions. A zoomed-in view of the recurrence plot demonstrates the presence of a kite-like elongation on the top right of the black patch. The existence of such structures in the recurrence plot is a characteristic of type-II intermittency [31].

Contrary to the state of intermittency discussed here for laminar thermoacoustic systems, we observe a different type of intermittency prior to the onset of thermoacoustic instability in turbulent combustion systems (refer to Chap. 6). During this state of intermittency, the system behavior alternately switches between epochs of low amplitude chaotic oscillations and bursts of large amplitude periodic oscillations. We note that, intermittency is found to be a very important dynamical

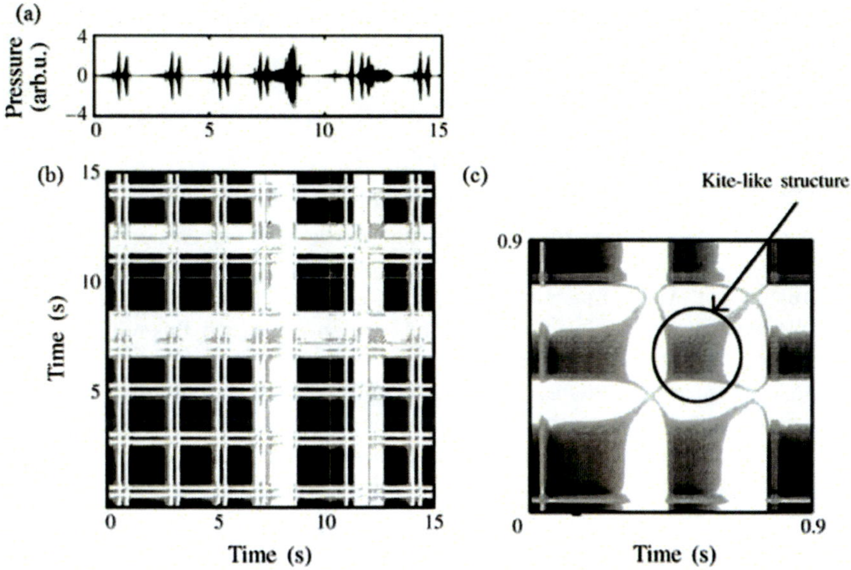

Fig. 4.9 (**a, b**) Recurrence plot for the state of intermittency observed prior to flame blowout. Black regions correspond to the laminar state and white regions correspond to epochs of bursting present in the signal. (**c**) A zoom-in view of the recurrence plot is given on the right to illustrate the presence of kite-like structures, a hallmark of type-II intermittency in the system. Reproduced with permission from Kabiraj and Sujith [21]

state in identifying and understanding the transition to thermoacoustic instability in turbulent systems.

4.3 Routes to Chaos for Thermoacoustic Oscillations

Similar to the observation of periodic oscillations (such as limit cycle oscillations or frequency-locked oscillations), the occurrence of chaos in the self-excited dynamics has also fascinated numerous researchers over the years [19, 33, 54, 59]. Many experimental and numerical studies have reported the existence of chaos in different thermoacoustic systems [8, 10–12, 17, 22, 23, 26, 30, 34, 39, 51, 55, 56, 62]. In this section, we will present a sequence of dynamical transitions (bifurcations) a system undergoes during the occurrence of chaos from periodic oscillations in laminar thermoacoustic systems [17, 22, 26, 34, 55, 56]. Dynamical transitions followed by a system during the occurrence of chaos are referred to as the 'route to chaos' in the literature of nonlinear dynamics [6, 19]. As we discussed in Chap. 2, these routes to chaos are primarily classified into three type: period-doubling route, Ruelle-Takens-Newhouse route, and intermittency route to chaos. In the following subsections, we will discuss the characteristics of these routes to chaos observed in thermoacoustic systems.

4.3.1 Period-Doubling Route to Chaos

The period-doubling route to chaos was first discovered by Feigenbaum [7]. During this route, a change in the control parameter leads to doubling the period of original oscillations at every bifurcation such that after several bifurcations the periodic oscillations eventually transition to chaotic oscillations. Hence, this route is called the period-doubling route to chaos.

In thermoacoustic systems, several numerical studies have shown the presence of a period-doubling route to chaos due to the variation of a bifurcation parameter [8, 26, 34, 55, 56]. However, the experimental evidence of this route is yet to be reported in thermoacoustic systems, to the best of our knowledge. In Figs. 4.10 and 4.11,

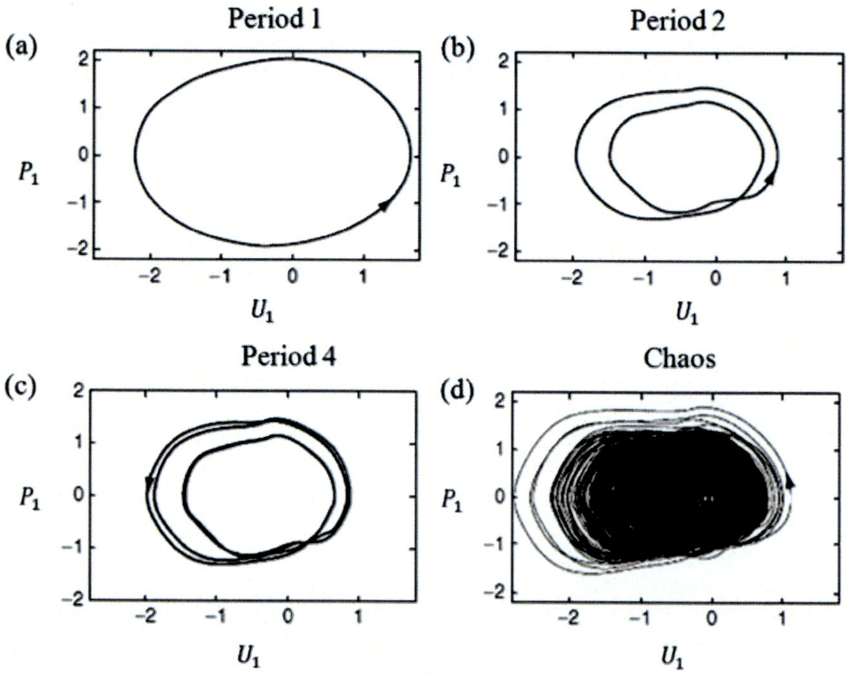

Fig. 4.10 Period-doubling route to chaos observed in a time-lag model of a horizontal Rijke tube consisting of a wire-mesh heater as a compact heat source. **(a)–(d)** Phase portraits between the acoustic pressure (P_1) and acoustic velocity (U_1) variables corresponding to different dynamical states such as limit cycle, period-2, period-4, and chaos, respectively, observed for the variation of non-dimensional heater power as the control parameter. Adapted with permission from Subramanian et al. [56]

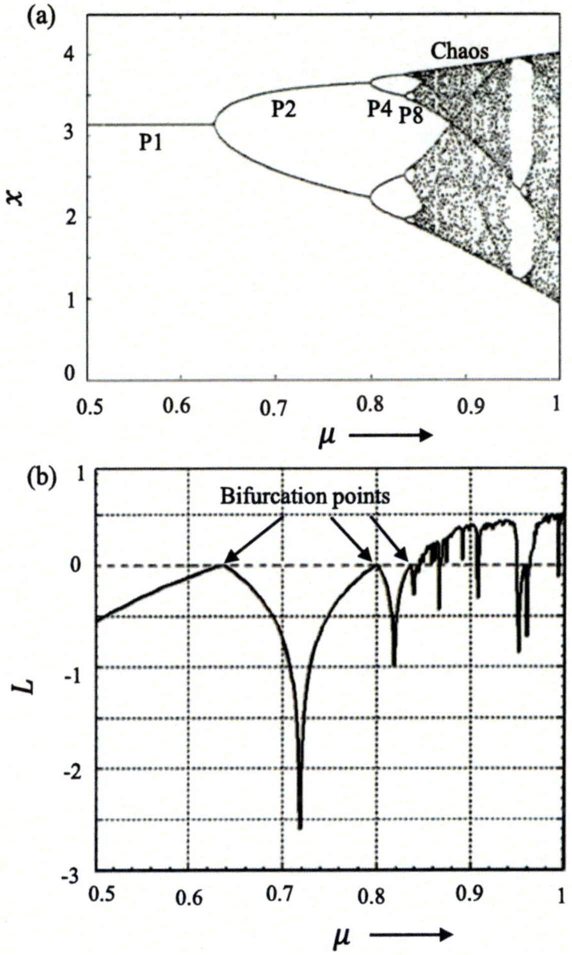

Fig. 4.11 (**a**) Bifurcation diagram showing the variation of the amplitude of acoustic pressure fluctuations (x) with the system parameter (μ) for a time-lag model of a combustion system. It shows a cascade of period-doubling bifurcation, such as period-1 (P1), period-2 (P2), period-4 (P4), period-8 (P8), etc., leading to the occurrence of windows of intermittency and chaos in the system. (**b**) Variation of the Lyapunov exponent (L) with the control parameter (μ), which helps in detecting bifurcation points in the system. Adapted with permission from Lei and Turan [35]

we show the period-doubling routes to chaos observed in reduced-order time-lag models of thermoacoustic systems.

We observe that the change in the control parameter initially leads to the transition of a stable fixed point to limit cycle oscillations (Fig. 4.10a) via Hopf bifurcation. With a further increase in the control parameter above a critical value (i.e., second Hopf point), we observe the occurrence of period-2 oscillations (Fig. 4.10b) due to the emergence of a second frequency at the subharmonic of the first frequency. Further increase in the control parameter causes the addition of two more frequencies at subharmonic of the previous two frequencies; thus, it leads to the occurrence of period-4 oscillations (Fig. 4.10c). In this manner, every bifurcation leads to the doubling the frequency of oscillations and, thus, this period-doubling cascade finally leads to the occurrence of chaotic oscillations in the system (Fig. 4.10d). We can understand this period-doubling bifurcation from the phase portraits of the signals (see Fig. 4.10), where we notice the addition of integer loops in the powers of 2 after every bifurcation (Fig. 4.10a–c). During the occurrence of chaotic oscillations, the phase portrait appears cluttered and fills the entire space due to the highly unstable motions of the trajectory (Fig. 4.10d).

In Fig. 4.11, we show the bifurcation diagram of period-doubling bifurcation observed in a time-lag model of a combustion system studied by Lei and Turan [35]. The variation of the pressure amplitude (x) with the control parameter (μ), which controls the energy/mass addition in the system, shows the bifurcation of period-1 (indicated as a single line) oscillations to period-2 oscillations (indicated as two lines). These period-2 oscillations transition to period-4 oscillations, which then transition to period-8 oscillations, and it continues until chaos is observed in the system dynamics. We also notice the presence of widows of intermittency for some ranges of control parameters.

Bifurcations leading to period-doubling route to chaos in the model can be easily detected through the variation of a nonlinear measure (for example, an average Lyapunov exponent, L) with the control parameter. As we mentioned in Chap. 2, the Lyapunov exponent is a measure of chaos, which shows a positive value during the state of chaotic oscillations. For periodic oscillations, this measure shows a negative value. The approach of the Lyapunov exponent to zero indicates the onset of bifurcation in the system. Thus, period-doubling bifurcations in the system can be easily detected through the crossing of the Lyapunov exponent at zero in the bifurcation diagram.

4.3.2 *Ruelle-Takens-Newhouse Route to Chaos*

Ruelle-Takens-Newhouse route to chaos was discovered by Ruelle and Takens [53] and Newhouse et al. [41] independently and, hence, it is named from their names as the '*Ruelle-Takens-Newhouse route*'. During this route to chaos, the transition from periodic (regular) oscillations to chaotic (irregular) oscillations can occur through

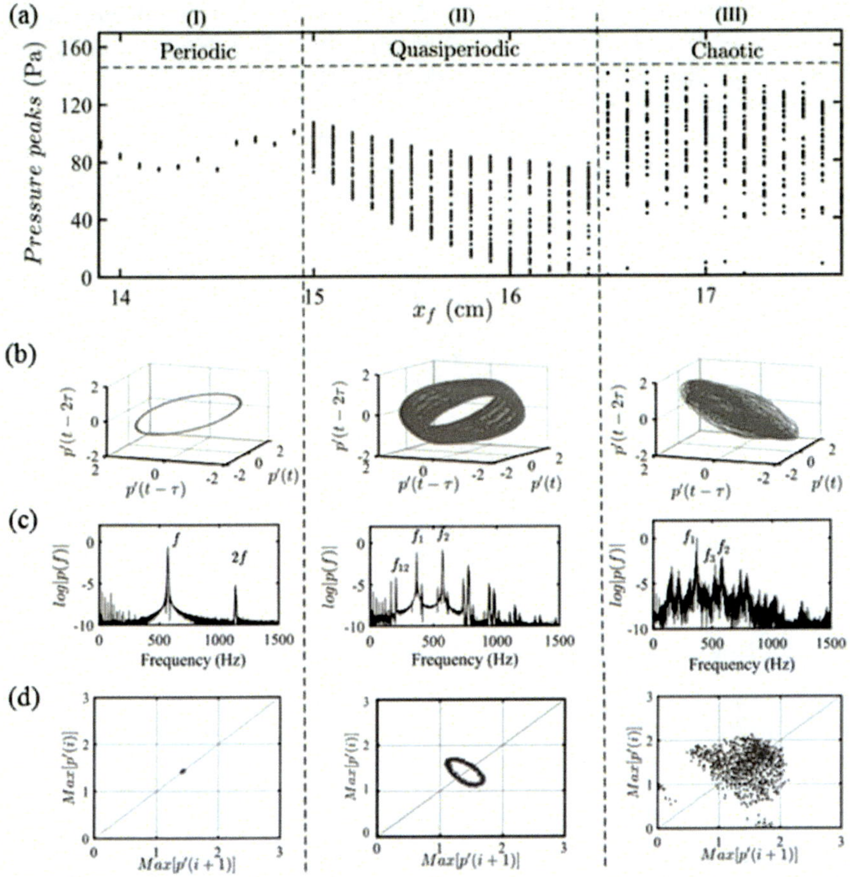

Fig. 4.12 Dynamical properties of quasiperiodicity route to chaos, observed from experiments in a premixed laminar multiple flame burner, are shown in terms of (**a**) bifurcation diagram, (**b**) phase portrait, (**c**) power spectral density, and (**d**) Poincaré section. The dynamical states observed during this route are (I) period-1 limit cycle, (II) quasiperiodicity, and (III) chaos. Adapted with permission from Mondal et al. [38]

three Hopf bifurcations, where every bifurcation leads to the emergence of one non-harmonic frequency with respect to the dominant frequency in the system.

Kabiraj et al. [22] showed the experimental evidence of quasiperiodicity route to chaos in a thermoacoustic system. In Fig. 4.12a, we show a bifurcation diagram of the Ruelle-Takens-Newhouse route observed in the acoustic pressure response of a premixed multiple flame laminar combustor, for the variation of the flame location as a control parameter [22, 38]. The variation of the flame location first leads to the transition of the system behavior from a single-looped steady state to self-sustained limit cycle oscillations (Fig. 4.12b-I) via subcritical Hopf bifurcation. Thus, there is an emergence of a single dominant frequency in the system dynamics (Fig. 4.12c-I).

Further increase in the flame location leads to the excitation of second frequency at a non-integer multiple of the first frequency (Fig. 4.12c-II), as the system undergoes a secondary Hopf bifurcation or Neimark-Sacker bifurcation (Fig. 4.12a). The presence of two incommensurate frequencies (Fig. 4.12c-II) confirms the transition of the system behavior from a single-looped limit cycle attractor (Fig. 4.12b-I) to a 2-torus quasiperiodic attractor (Fig. 4.12b-II). Finally, when the control parameter is increased above a critical value, we notice the occurrence of a third frequency incommensurate with the other two frequencies in the system (Fig. 4.12c-III). As a result, we notice the breaking of a 2-torus attractor of quasiperiodic oscillations (also called torus-death bifurcation) into a strange attractor of chaotic oscillations (Fig. 4.12b-III). During the onset of chaos, the emergence of the third frequency is associated with the simultaneous occurrence of broadband noise in the spectrum (Fig. 4.12c-III). This route can also be referred to as the quasiperiodicity route to chaos [19].

A Poincaré section of this route shows the transition from a fixed point observed for the limit cycle (Fig. 4.12d-I) to a closed-loop structure observed for quasiperiodic oscillations (Fig. 4.12d-II), and then to cluttered points occurring for chaotic oscillations (Fig. 4.12d-III).

4.3.3 Intermittency Route to Chaos

In the intermittency route to chaos, as the name suggests, the transition of regular (periodic) oscillations to irregular (chaotic) oscillations occurs via intermittency. During the state of intermittency, bursts of chaotic oscillations emerge at irregular time intervals amidst periodic oscillations. The frequent occurrence of such bursts is observed as the control parameter approaches the point of the onset of chaos (as we discussed in Chap. 2). This intermittency route to chaos is also called the '*Pomeau-Manneville scenario*' [49].

Guan et al. [17] were the first to report the experimental observation of the intermittency route to chaos in thermoacoustic systems. In Fig. 4.13, we show the dynamic characteristics of the intermittency route to chaos observed in a premixed conical flame laminar combustor. The bifurcation diagram corresponding to this route is shown in Fig. 4.13a. In this plot, we present the variation of the local maxima of the acoustic pressure signal (p') as a function of the flame location normalized with duct length ($\overline{x}$). As the value of the flame location is increased, we observe that the system behavior first transitions from a stable fixed point to large amplitude limit cycle oscillations via Hopf bifurcation. Further increase in the flame location leads to the occurrence of quasiperiodic oscillations from limit cycle oscillations via a secondary Hopf bifurcation. We notice a region of intermittency prior to the transition of quasiperiodic oscillations to chaotic oscillations, as the flame location is further increased. During the state of intermittency, bursts of large amplitude chaotic oscillations occur in between epochs of medium amplitude quasiperiodic fluctuations. Finally, the system behavior transitions to the state of

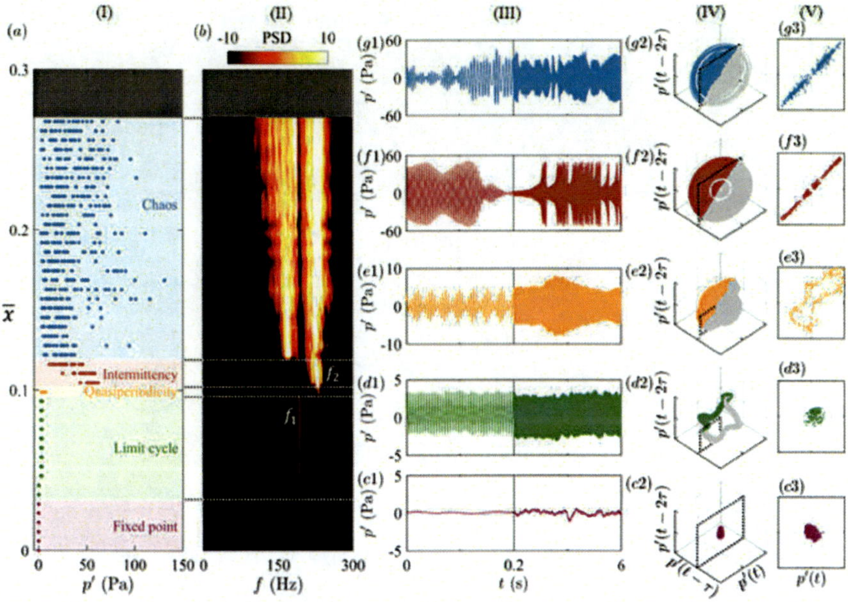

Fig. 4.13 Dynamical properties of intermittency route to chaos observed from experiments in a laminar premixed single conical flame burner for the variation of the normalized flame location ($\overline{x}$) are shown in terms of (I) bifurcation diagram, (II) power spectral density, (III) time series, (IV) phase portrait, and (V) Poincaré section. During this route, the dynamics of the system changes from (c1) stable fixed point, (d1) period-1 limit cycle, (e1) quasiperiodicity, (f1) intermittency, and (g1) chaos. Adapted with permission from Guan et al. [17]

fully chaotic oscillations. We can observe the dynamical properties of these states from the power spectrum, time series, phase portrait, and Poincaré section, as shown in Fig. 4.13(II)–(V), respectively.

4.4 Nonlinear Behavior of Flame-Acoustic Interactions

Having discussed the rich dynamical behaviors exhibited by the acoustic pressure fluctuations in the previous sections, we now present the nonlinear behavior of the flame dynamics occurring during these dynamical states. In Figs. 4.14, 4.15, 4.16, 4.17, 4.18, 4.19, 4.20, we discuss a qualitative description of the flame-acoustic interaction observed for different dynamical states in variety of laminar premixed combustors.

Figure 4.14(I) shows the instantaneous flame images captured over a period of limit cycle oscillations in a ducted premixed conical flame burner. We notice uniform axisymmetric modulations in the flame about the mean flame shape, where a wrinkle (a small projection normal to the flame surface) formed at the base of

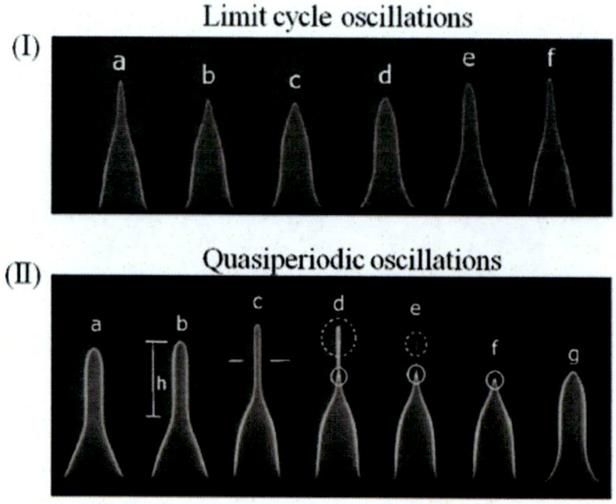

Fig. 4.14 A Sequence of instantaneous high-speed images of a ducted single laminar premixed flame for the states of limit cycle oscillations (top row, I) and the quasiperiodic oscillations (bottom row, II). Alphabets represent the time of acquisition of images. In (II), the arrows and circles mark the evolution of pinch-off during the state of quasiperiodic oscillations. The height of the flame tip is marked with the alphabet h. Adapted with permission from Kabiraj and Sujith [21]

the flame propagates towards its tip. This periodic modulation of the flame shape changes the corresponding heat release rate in the flame that, in turn, drives the acoustic fluctuations in the system. Since the signal is periodic, the flame regains its previous shape after every oscillation cycle (as seen from the similar flame shapes observed at the instants a and f, in Fig. 4.14-I).

On the other hand, the nonlinear effects of acoustic oscillations are more predominant in the instantaneous flame images observed during the occurrence of quasiperiodic oscillations (see Fig. 4.14-II). As these signals are aperiodic in time, we notice that the flame shape never regains its previous shape at any time instant (as the flame images at the instants a and g are not the same, in Fig. 4.14-II). The flame-acoustic interaction leads to the occurrence of several distinct events in addition to the modulation of the flame surface about the mean shape, which are flame elongation, neck formation, pinch-off, and cusping (as seen from flame images at the instants b to f, in Fig. 4.14-II).

In Fig. 4.15, we show a sequence of flame images acquired during the state of intermittency observed prior to flame blowout using a normal video camera (top row) and a high-speed camera (bottom row). During the epoch of low amplitude periodic oscillations of intermittency, we observe axisymmetric oscillations in the flame that is anchored at the burner lip (Fig. 4.15a,b). During the transition to bursts of chaotic oscillations, we notice a gradual increase in the amplitude of oscillations in the flame surface (Fig. 4.15c). However, when the oscillation amplitude is above a certain amplitude, the burner in unable to hold the oscillatory flame and causes the flame to lift-off from the burner. Once the flame is detached, it oscillates at

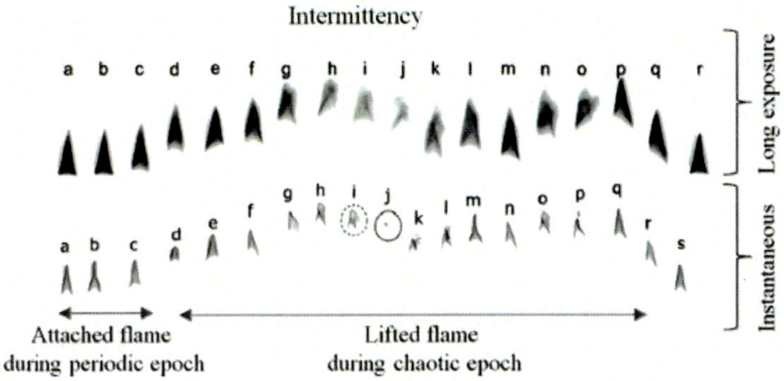

Fig. 4.15 Long exposure (top row) and instantaneous (bottom row) flame images illustrating flame dynamics during the intermittent bursting observed prior to flame blowout in a ducted single laminar flame burner. Every burst is associated with a flame lift-off, oscillations in the lifted flame, and flame reattachment. These events are given from left to right in the order of their occurrence in the flame. Extreme stretching (i, k, l), folding (h), and local extinction (j) states can be seen and are marked with circles. Adapted with permission from Kabiraj and Sujith [21]

a particular distance away from the burner, for a duration corresponding to the epoch of the burst of chaotic oscillations. The flame modulations now no longer remain axisymmetric and the flame surface oscillations are mostly controlled by the underlying hydrodynamics of the circular jet effusing from the burner exit. Far from the burner exit, the jet flow transitions to turbulence. Hence, during the state of the flame lift-off, the flame behaves more like turbulent and not laminar. The wrinkling and stretching in the flame surface, due to the hydrodynamic fluctuations, results in the local heat release rate oscillations in the system. At the end of the epoch of the burst, the flame attaches back to the burner and re-establishes the flame-acoustic coupling in the system.

Similar to a single conical flame burner, the existence of multiple conical premixed flames in the system can also exhibit the complex flame-acoustic interaction. Such a complexity has been observed in the form of variety of dynamical states such as limit cycle, period-n, quasiperiodicity, frequency-locking, strange nonchaos, and chaos [22, 51]. Next, we present the nonlinear interaction of multiple conical flames with the acoustic field in a ducted premixed combustor for the states of limit cycle, frequency-locking, quasiperiodicity, and chaos.

Figure 4.16(I) shows the time series of the acoustic pressure fluctuations and the corresponding instantaneous flame images acquired at different phases of an oscillation cycle (Figs. 4.16-Ia–f), during the state of limit cycle oscillations in a premixed multiple flame combustor. Due to the presence of a single frequency oscillations, the dynamics of this state repeats after every period of oscillation. This can be seen from the similarity in the flame shapes observed for Figs. 4.16(I-d) and (I-h), and Figs. 4.16(I-e) and (I-g). All flames behave in a similar manner during an oscillation cycle. The shape of all flames undergoes periodic modulations, resulting in periodic heat release rate in the system.

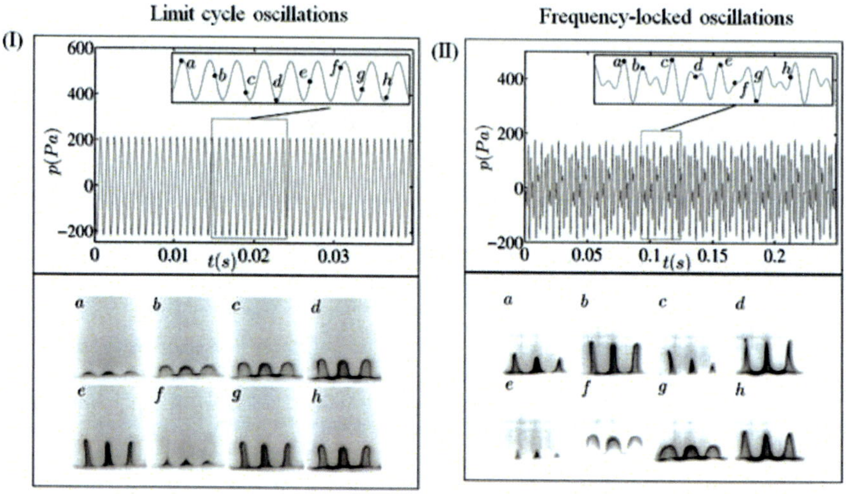

Fig. 4.16 Comparison of the dynamics of multiple flames corresponding to different instances of the acoustic pressure fluctuations for the states of (I) limit cycle and (II) frequency-locked oscillations in a premixed multiple flame burner. Adapted with permission from Kabiraj et al. [23]

During the state of frequency-locked oscillations, two dominant frequencies of the oscillations are rationally related to each other, and the system behavior repeats after a period corresponding to the value of this frequency ratio. In Fig. 4.16(II), the time duration between the point a and the local maxima near point h corresponds to a period of the oscillation [23]. Thus, the period of frequency-locked oscillations is longer than that observed for limit cycle oscillations. As a result, although flame images at points a, b, c, and e correspond to local maxima of the pressure signal, we notice that their behavior does not match each other. Moreover, at some other instances of oscillation cycle (Fig. 4.16-IIc), we notice a difference in the response of all individual flames to the acoustic pressure fluctuations in the system, where these flames exhibit different lengths.

As we know, quasiperiodic oscillations consist of two dominant frequencies that are non-integer multiples of each other. Therefore, the dynamics of acoustic pressure oscillations during this state is aperiodic and, hence, the signal never repeats its behavior in a finite data length. From Fig. 4.17(I), we notice that, although the flame images corresponding to points a, b, d, f, and g are at local maxima of the pressure signal, the flame shapes are significantly different from each other. A similar observation can be made from the flame images corresponding to the local minima of the pressure signal (Figs. 4.17-IIc and IIh). The flame image corresponding to point e in Fig. 4.17(II) shows elongated flames at the point where the pressure is rising to a maximum value. Thus, we observe a loss of the correlation between the pressure and the flame dynamics during this state.

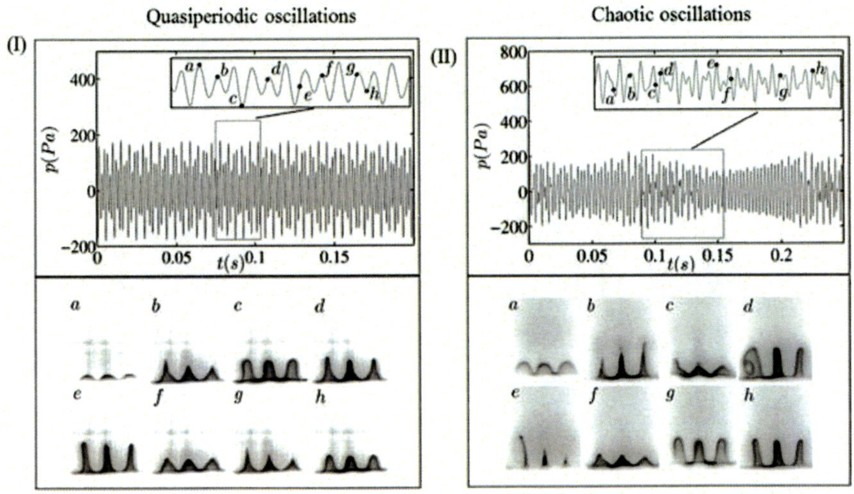

Fig. 4.17 Comparison of the dynamics of multiple flames corresponding to different instances of the acoustic pressure fluctuations for the states of (I) quasiperiodicity and (II) chaotic oscillations in a premixed multiple flame burner. Adapted with permission from Kabiraj et al. [23]

For the state of chaotic oscillations, as the signal contains broadband frequencies, we notice that modulations in the individual flame shape are not regular and mostly asymmetric (Fig. 4.17-II). The flame observed during this state shows complex patterns. The flame shape modulations exhibit different features, which include rolling (image *d*), lifting-off (image *g*), and elongation (images *b* and *e*) of the flame surface. We observe that these patterns do not repeat in the consecutive cycles of chaotic oscillations. Thus, the behavior of multiple flames is highly nonlinear and significantly different for different dynamical states of the acoustic pressure oscillations observed in the system.

The nonlinear interaction of the flame with the inherent acoustic field is more evident in a system of premixed laminar V-flame burner. Figures 4.18, 4.19, 4.20 show the dynamical features of a V-flame observed during the states of period-3, frequency-locked, and quasiperiodic oscillations, respectively.

During the occurrence of period-3 oscillations, we observe that the phase space trajectory repeats its behavior after 3 cycles (Fig. 4.18-III). The amplitude spectrum of this state shows three contributing frequency peaks located at f_2, $f_2/3$, and $2f_2/3$ (Fig. 4.18-II). The instantaneous flame images show two dominating patterns in the flame shape modulation, where pattern one (Fig. 4.18-IV) repeats twice and pattern two (Fig. 4.18-V) occurs once in an oscillation cycle. In the first pattern (Fig. 4.18-IV), we observe the formation of a wrinkle at the base of the flame, which then propagates upwards along its surface. After a certain distance, the wrinkle causes the roll-up of the flame surface. This leads to the growth of the flame surface. When the top part of the flame interacts with its rolled-up portion, we observe a mutual annihilation of the flame in the system. This leads to the separation of islands

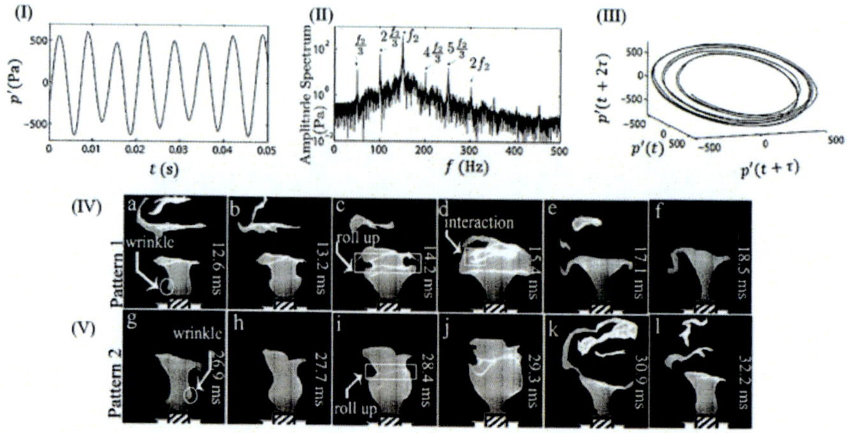

Fig. 4.18 (I)–(III) The time series, amplitude spectrum, and phase portrait of the acoustic pressure fluctuations observed for the state of period-3 oscillations in a ducted premixed laminar V-flame combustor. (IV), (V) The instantaneous images corresponding to different patterns of the flame modulation observed during this state. Adapted with permission from Vishnu et al. [60]

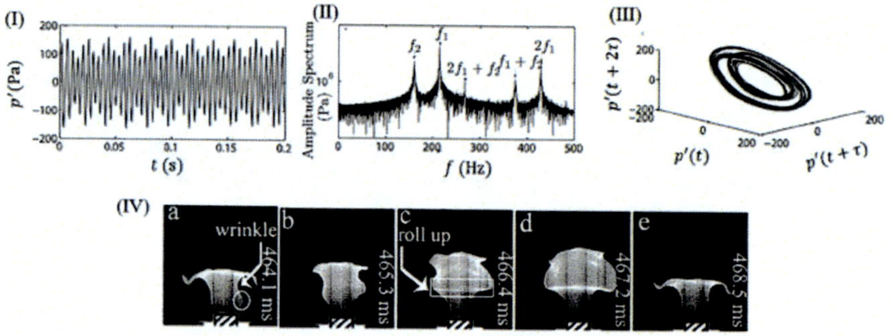

Fig. 4.19 (I)–(III) The time series, amplitude spectrum, and phase portrait of the acoustic pressure fluctuations observed for the state of frequency-locked oscillations in a ducted premixed laminar V-flame combustor. (IV) The instantaneous images corresponding to a single pattern of flame modulation are observed during this state. Adapted with permission from Vishnu et al. [60]

of flame elements, resulting in a reduction of the flame surface area. During the elongation process, the flame base becomes thin. In the second pattern (Fig. 4.18-V), we notice similar processes of flame shape modulations such as a wrinkle formation, propagation, roll up, mutual interaction, annihilation, and reduction of the flame surface area. However, a qualitative difference in the first and the second patterns is that the thinning of the flame base is not observed at the end of the second pattern as compared to that observed for the first pattern.

In Fig. 4.19, we show the dynamics of the acoustic pressure fluctuations and the corresponding flame images during the state of frequency-locked oscillations. The amplitude spectrum of this state shows the locking of frequencies f_1 and f_2, related

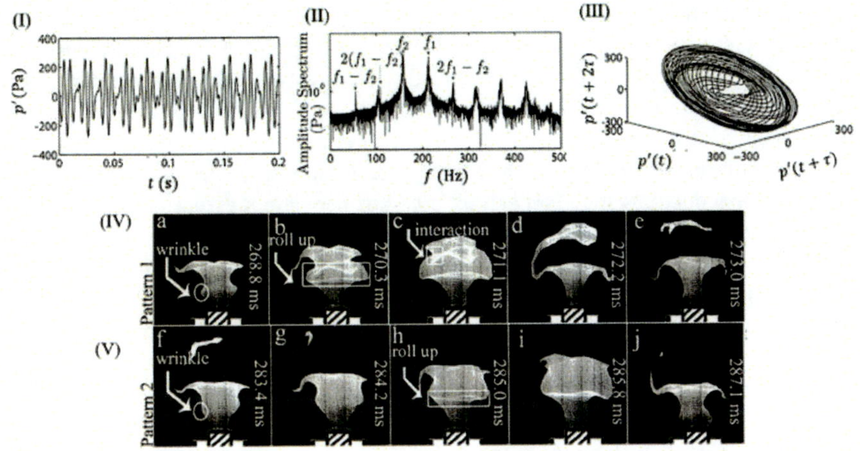

Fig. 4.20 (I)–(III) The time series, amplitude spectrum, and phase portrait of the acoustic pressure fluctuations observed for the state of quasiperiodic oscillations in a ducted premixed laminar V-flame combustor. (IV), (V) The instantaneous images corresponding to different patterns of flame modulation observed during this state. Adapted with permission from Vishnu et al. [60]

in the rational ratio of 4:3 (Fig. 4.19-II). Since these are periodic oscillations, the patterns of the flame oscillation repeat after the cycles corresponding to the ratio of the locked frequencies (Fig. 4.19-III). During this state of oscillations, we notice the repetition of a single pattern of the flame shape modulation (Fig. 4.19-IVa-e). In this pattern, a small wrinkle is formed at the flame base, which while propagating upward along the flame causes the roll-up of the flame surface. This flame roll up is followed by an increase in the flame surface area. The pattern ends with the separation of the top portion of the flame and a reduction in its surface area. We observe a repetition of this pattern after four cycles of the oscillation.

During the state of quasiperiodic oscillations (Fig. 4.20), the amplitude spectrum (Fig. 4.20-II) shows two dominant frequencies (f_1 and f_2) locked at an incommensurate frequency ratio. The phase space trajectory of these oscillations revolves around a torus surface (Fig. 4.20-III). We observe two patterns of the flame surface modulation, where the first pattern (Fig. 4.20-IV) repeats after three cycles which are then followed by a second pattern (Fig. 4.20-V) that occurs once. Both these patterns are observed to repeat on average after four cycles. The process of flame surface modulation, i.e., a formation of a wrinkle, its propagation, roll-up of the flame surface, its mutual interaction with top flame portion, separation of the flame element, and reduction of the flame surface area (as discussed in Fig. 4.18) are observed to repeat in both the patterns. A key difference between these patterns is that a reduction in the surface area at the end of the pattern is nearly uniform and the flame base is broad in the second pattern, as compared to that observed during the first pattern. Each of these patterns of the flame modulation repeats at a frequency of f_1 and together they occur at a low frequency of $f_1 - f_2$.

Thus, all aforementioned observations corresponding to different dynamical states of thermoacoustic instability, in different configurations of laminar burners, affirm that the flame response to the acoustic field involves rich nonlinear dynamics and exhibit different responses depending on the dynamical properties of the underlying acoustic pressure oscillations in the system.

In summary, in this chapter, we demonstrated that even a simpler configuration of a thermoacoustic system can exhibit complex acoustic pressure oscillations and flame shape modulations. By using different tools from dynamical systems theory, we showed that the dynamics of thermoacoustic instability is not limited to limit cycle oscillations alone, but manifests several other complex forms. A secondary Hopf bifurcation of limit cycle oscillations, due to a change in the control parameter, results in the occurrence of different dynamical states, such as quasiperiodicity, period-2, period-3, mode-locked (frequency-locked), chaos, strange nonchaos, and intermittency in a combustor. We also presented the existence of different routes to chaos observed in laminar thermoacoustic systems. Finally, we discussed the nonlinear interaction of the flame with the acoustic field observed during different dynamical states of thermoacoustic instability. In the next chapter, we will discuss the importance of complex systems approach to unravel complexities associated with the dynamics of turbulent thermoacoustic systems.

References

1. Boudy, F., Durox, D., Schuller, T., Candel, S.: Nonlinear flame describing function analysis of galloping limit cycles featuring chaotic states in premixed combustors. In: Proceedings of the ASME Turbo Expo 2012: Turbine Technical Conference and Exposition, GT2012-68998, pp. 713–724. American Society of Mechanical Engineers (2012)
2. Chiocchini, S., Pagliaroli, T., Camussi, R., Giacomazzi, E.: Chaotic and linear statistics analysis in thermoacoustic instability detection. J. Propul. Power, 34(1), 15–26 (2018)
3. Culick, F.E.C.: Some recent results for nonlinear acoustics in combustion chambers. AIAA J. **32**(1), 146–169 (1994)
4. Domen, S., Gotoda, H., Kuriyama, T., Okuno, Y., Tachibana, S.: Detection and prevention of blowout in a lean premixed gas-turbine model combustor using the concept of dynamical system theory. Proc. Combust. Inst. **35**(3), 3245–3253 (2015)
5. Dunstan, W.J., Bitmead, R.R., Savaresi, S.M.: Fitting nonlinear low-order models for combustion instability control. Control Eng. Pract. **9**(12), 1301–1317 (2001)
6. Eckmann, J.P.: Roads to turbulence in dissipative dynamical systems. Rev. Mod. Phys. **53**(4), 643 (1981)
7. Feigenbaum, M.J.: The universal metric properties of nonlinear transformations. J. Stat. Phys. **21**(6), 669–706 (1979)
8. Fichera, A., Losenno, C., Pagano, A.: Experimental analysis of thermo-acoustic combustion instability. Appl. Energy **70**(2), 179–191 (2001)
9. Gopal, R., Venkatesan, A., Lakshmanan, M.: Applicability of 0-1 test for strange nonchaotic attractors. Chaos **23**(2), 023123 (2013)
10. Gotoda, H., Nikimoto, H., Miyano, T., Tachibana, S.: Dynamic properties of combustion instability in a lean premixed gas-turbine combustor. Chaos **21**(1), 013124 (2011)
11. Gotoda, H., Shinoda, Y., Kobayashi, M., Okuno, Y., Tachibana, S.: Detection and control of combustion instability based on the concept of dynamical system theory. Phys. Rev. E **89**(2), 022910 (2014)

12. Gotoda, H., Okuno, Y., Hayashi, K., Tachibana, S.: Characterization of degeneration process in combustion instability based on dynamical systems theory. Phys. Rev. E **92**(5), 052906 (2015)
13. Grebogi, C., Ott, E., Pelikan, S., Yorke, J.A.: Strange attractors that are not chaotic. Physica D **13**(1-2), 261–268 (1984)
14. Guan, Y., Liu, P., Jin, B., Gupta, V., Li, L.K.B.: Nonlinear time-series analysis of thermoacoustic oscillations in a solid rocket motor. Exp. Therm. Fluid Sci. **98**, 217–226 (2018)
15. Guan, Y., Murugesan, M., Li, L.K.B.: Strange nonchaotic and chaotic attractors in a self-excited thermoacoustic oscillator subjected to external periodic forcing. Chaos **28**(9), 093109 (2018)
16. Guan, Y., Li, L.K.B., Ahn, B., Kim, K.T.: Chaos, synchronization, and desynchronization in a liquid-fueled diffusion-flame combustor with an intrinsic hydrodynamic mode. Chaos **29**(5), 053124 (2019)
17. Guan, Y., Gupta, V., Li, L.K.B.: Intermittency route to self-excited chaotic thermoacoustic oscillations. J. Fluid Mech. **894**, R3 (2020)
18. Gutmark, E., Wilson, K., Parr, T., Schadow, K.: Feedback control of multi-mode combustion instability. In: 30th Aerospace Sciences Meeting and Exhibit, AIAA 1992-778 (1992)
19. Hilborn, R.C.: Chaos and Nonlinear Dynamics: An Introduction for Scientists and Engineers. Oxford University Press (2000)
20. Jahnke, C.C., Culick, F.E.C.: Application of dynamical systems theory to nonlinear combustion instabilities. J. Propuls. Power **10**(4), 508–517 (1994)
21. Kabiraj, L., Sujith, R.I.: Nonlinear self-excited thermoacoustic oscillations: intermittency and flame blowout. J. Fluid Mech. **713**, 376–397 (2012)
22. Kabiraj, L., Saurabh, A., Wahi, P., Sujith, R.I.: Route to chaos for combustion instability in ducted laminar premixed flames. Chaos **22**(2), 023129 (2012)
23. Kabiraj, L., Sujith, R.I., Wahi, P.: Bifurcations of self-excited ducted laminar premixed flames. J. Eng. Gas Turbines Power **134**(3), 031502 (2012)
24. Kabiraj, L., Saurabh, A., Karimi, N., Sailor, A., Mastorakos, E., Dowling, A.P., Paschereit, C.O.: Chaos in an imperfectly premixed model combustor. Chaos **25**(2), 023101 (2015)
25. Kantz, H., Schreiber, T.: Nonlinear Time Series Analysis, vol. 7. Cambridge University Press (2004)
26. Kashinath, K., Waugh, I.C., Juniper, M.P.: Nonlinear self-excited thermoacoustic oscillations of a ducted premixed flame: bifurcations and routes to chaos. J. Fluid Mech. **761**, 399–430 (2014)
27. Kasthuri, P., Pavithran, I., Pawar, S.A., Sujith, R.I., Gejji, R., Anderson, W.: Dynamical systems approach to study thermoacoustic transitions in a liquid rocket combustor. Chaos **29**(10), 103115 (2019)
28. Kasthuri, P., Unni, V.R., Sujith, R.I.: Bursting and mixed mode oscillations during the transition to limit cycle oscillations in a matrix burner. Chaos **29**(4), 043117 (2019)
29. Kasthuri, P., Pavithran, I., Krishnan, A., Pawar, S.A., Sujith, R.I., Gejji, R., Anderson, W., Marwan, N., Kurths, J.: Recurrence analysis of slow–fast systems. Chaos **30**(6), 063152 (2020)
30. Keanini, R., Yu, K., Daily, J.: Evidence of a strange attractor in ramjet combustion. In: 27th Aerospace Sciences Meeting, AIAA 1989-624 (1989)
31. Klimaszewska, K., Żebrowski, J.J.: Detection of the type of intermittency using characteristic patterns in recurrence plots. Phys. Rev. E **80**(2), 026214 (2009)
32. Lamraoui, A., Richecoeur, F., Ducruix, S., Schuller, T.: Experimental analysis of simultaneous non-harmonically related unstable modes in a swirled combustor. In: Proceedings of the ASME 2011 Turbo Expo: Turbine Technical Conference and Exposition, GT2011-46701, pp. 1289–1299 (2011)
33. Layek, G.: An Introduction to Dynamical Systems and Chaos. Springer (2015)
34. Lei, S., Turan, A.: Nonlinear/chaotic behaviour in thermo-acoustic instability. Combust. Theor. Model. **13**(3), 541–557 (2009)
35. Lei, S., Turan, A.: Chaotic modelling and control of combustion instabilities due to vaporization. Int. J. Heat Mass Transfer **53**(21-22), 4482–4494 (2010)

36. Lindner, J.F., Kohar, V., Kia, B., Hippke, M., Learned, J.G., Ditto, W.L.: Strange nonchaotic stars. Phys. Rev. Lett. **114**(5), 054101 (2015)
37. Moeck, J.P., Paschereit, C.O.: Nonlinear interactions of multiple linearly unstable thermoacoustic modes. J. Spray Combust. Dyn. **4**(1), 1–27 (2012)
38. Mondal, S., Pawar, S.A., Sujith, R.I.: Synchronous behaviour of two interacting oscillatory systems undergoing quasiperiodic route to chaos. Chaos **27**(10), 103119 (2017)
39. Nair, V., Thampi, G., Karuppusamy, S., Gopalan, S., Sujith, R.I.: Loss of chaos in combustion noise as a precursor of impending combustion instability. J. Spray Combust. Dyn. **5**(4), 273–290 (2013)
40. Nair, V., Thampi, G., Sujith, R.I.: Intermittency route to thermoacoustic instability in turbulent combustors. J. Fluid Mech. **756**, 470–487 (2014)
41. Newhouse, S., Ruelle, D., Takens, F.: Occurrence of strange AxiomA attractors near quasi periodic flows on T^m, $m \geqq 3$. Commun. Math. Phys. **64**(1), 35–40 (1978)
42. Noiray, N., Durox, D., Schuller, T., Candel, S.: A unified framework for nonlinear combustion instability analysis based on the flame describing function. J. Fluid Mech. **615**, 139–167 (2008)
43. Okuno, Y., Small, M., Gotoda, H.: Dynamics of self-excited thermoacoustic instability in a combustion system: Pseudo-periodic and high-dimensional nature. Chaos **25**(4), 043107 (2015)
44. Pawar, S.A., Vishnu, R., Vadivukkarasan, M., Panchagnula, M.V., Sujith, R.I.: Intermittency route to combustion instability in a laboratory spray combustor. J. Eng. Gas Turbines Power **138**(4), 041505 (2016)
45. Pawar, S.A., Seshadri, A., Unni, V.R., Sujith, R.I.: Thermoacoustic instability as mutual synchronization between the acoustic field of the confinement and turbulent reactive flow. J. Fluid Mech. **827**, 664–693 (2017)
46. Pawar, S.A., Raghunathan, M., Reeja, K.V., Midhun, P.R., Sujith, R.I.: Effect of preheating of the reactants on the transition to thermoacoustic instability in a bluff-body stabilized dump combustor. Proc. Combust. Inst. **38**(4), 6193–6201 (2021)
47. Pikovsky, A.S., Feudel, U.: Characterizing strange nonchaotic attractors. Chaos **5**(1), 253–260 (1995)
48. Pikovsky, A.S., Feudel, U., Kuznetsov, S.P.: Strange Nonchaotic Attractors: Dynamics Between Order and Chaos in Quasiperiodically Forced Systems, vol. 56. World Scientific (2006)
49. Pomeau, Y., Manneville, P.: Intermittent transition to turbulence in dissipative dynamical systems. Commun. Math. Phys. **74**(2), 189–197 (1980)
50. Prasad, A., Negi, S.S., Ramaswamy, R.: Strange nonchaotic attractors. Int. J. Bifurc. Chaos **11**(02), 291–309 (2001)
51. Premraj, D., Pawar, S.A., Kabiraj, L., Sujith, R.I.: Strange nonchaos in self-excited singing flames. Europhys. Lett. **128**(5), 54005 (2020)
52. Riley, A., Park, S., Dowling, A., Evesque, S., Annaswamy, A.: Advanced closed-loop control on an atmospheric gaseous lean-premixed combustor. J. Eng. Gas Turbines Power **126**(4), 708–716 (2004)
53. Ruelle, D., Takens, F.: On the nature of turbulence. Les rencontres physiciens-mathématiciens de Strasbourg-RCP25 **12**, 1–44 (1971)
54. Schuster, H.G., Just, W.: Deterministic Chaos: An Introduction. Wiley (2006)
55. Sterling, J.D.: Nonlinear analysis and modelling of combustion instabilities in a laboratory combustor. Combust. Sci. Tech. **89**(1-4), 167–179 (1993)
56. Subramanian, P., Mariappan, S., Sujith, R.I., Wahi, P.: Bifurcation analysis of thermoacoustic instability in a horizontal Rijke tube. J. Spray Combust. Dyn. **2**(4), 325–355 (2010)
57. Takens, F.: Detecting strange attractors in turbulence. Lect. Notes Math. **898**(1), 366–381 (1981)
58. Tandon, S., Pawar, S.A., Banerjee, S., Varghese, A.J., Durairaj, P., Sujith, R.I.: Bursting during intermittency route to thermoacoustic instability: Effects of slow–fast dynamics. Chaos **30**(10), 103112 (2020)
59. Thompson, J.M.T., Stewart, H.B.: Nonlinear Dynamics and Chaos. Wiley (2002)

60. Vishnu, R., Sujith, R.I., Aghalayam, P.: Role of flame dynamics on the bifurcation characteristics of a ducted V-flame. Combust. Sci. Tech. **187**(6), 894–905 (2015)
61. Weng, F., Makeximu, Li, D., Zhu, M.: Low frequency beating instability in a Rijke burner and its active control. J. Low Freq. Noise Vibr. Act. Control **34**(2), 137–152 (2015)
62. Weng, Y., Unni, V.R., Sujith, R.I., Saha, A.: Synchronization framework for modeling transition to thermoacoustic instability in laminar combustors. Nonlinear Dyn. **100**, 3295–3306 (2020)
63. Wilhite, J.M., Dolan, B.J., Kabiraj, L., Gomez, R.V., Gutmark, E.J., Paschereit, C.O.: Analysis of combustion oscillations in a staged MLDI burner using decomposition methods and recurrence analysis. In: 54th AIAA Aerospace Sciences Meeting, AIAA 2016-1156 (2016)

Chapter 5
Thermoacoustic Instability Is Self-Organization in a Complex System

In the previous chapter, we showed that even a simple, laminar thermoacoustic system can give rise to complex behavior arising from chaotic behavior. Industrial combustors are turbulent. Turbulence is characterized by a wide range of scales and a large number of degrees of freedom, and thus makes the thermoacoustic system significantly more complex. Additional complexities arise from chemical kinetics, molecular mixing and interaction with acoustic waves. These complex nonlinear interactions give rise to a rich variety of dynamics with the possibility of chaotic fluctuations on the one hand and ordered periodic oscillations on the other. Further, even combustors with different flame holding mechanisms (e.g., swirl-stabilized or bluff-body-stabilized) exhibit similar behavior. These similarities in systems that have fundamentally different flow physics baffle traditional reductionist thinking. In this chapter, we study thermoacoustic instability in a turbulent combustor using complex system theory. We will introduce the reader to the basic ideas of complex systems theory such as self-organization and emergence. We formulate the onset of thermoacoustic instability as order emerging from chaos, and describe the tools from complex systems theory that can be used to study this transition. Further, we discuss the advantages of using complex systems theory to study this transition and also describe mitigation strategies derived from it.

In a reductionist approach, we split a system into its fundamental constituents and analyze them in detail. Complex systems approach on the other hand focuses on the interaction between the constituent elements [8]. We define a complex system as a system with "*multiple interacting components whose behavior cannot be simply inferred from the behavior of the components*" [63, 65, 82]. Instead of splitting things apart and studying the components in great detail, complex systems approach focuses on what new phenomena can emerge from the interaction of a set of relatively simple components. We list below the key ingredients of complex systems, as suggested by Johnson [27],

R. I. Sujith, S. A. Pawar, *Thermoacoustic Instability*, Springer Series in Synergetics,
https://doi.org/10.1007/978-3-030-81135-8_5

1. Complex systems consist of many interacting elements.
2. The behavior of these elements is affected by feedback.
3. The system is open.

These ingredients are certainly present in a turbulent thermoacoustic system, which indeed qualifies it as a complex system. The following are some of the typical characteristics of complex systems, which are indeed seen in turbulent thermoacoustic systems [27].

- **The system appears to be "alive":** During thermoacoustic instability, the whole thermoacoustic system displays a high level of coherence. Often engineers and technicians talk as though the thermoacoustic system were a living object, often saying things like "the combustor is breathing".
- **The system exhibits emergent behavior:** Work in the 1980s have shown that during the occurrence of thermoacoustic instability, large-scale vortices (coherent structures) emerge in the turbulent flow during the occurrence of thermoacoustic instability [55, 61].
- **Self-organization and emergence occur without any "invisible hand":** During the occurrence of thermoacoustic instability, large amplitude acoustic fluctuations and the large-scale vortices (coherent structures) emerge without any external forcing,
- **The system shows a mix of ordered and disordered behavior:** When the flow field is turbulent, the onset of self-sustained periodic oscillations (thermoacoustic instability) is always presaged by unsteady, intermittent bursts of high amplitude periodic oscillations (order) that appear in an apparently random manner amidst regions of aperiodic low-amplitude fluctuations (disorder) [49]. As Johnson [27] puts it, "*complex systems seem to be able to move between order and disorder on their own accord*". The spatiotemporal field of complex systems show pockets of order amidst pockets of disorder. In a turbulent thermoacoustic system, a spatiotemporal analysis of the flow field during intermittency shows that order emerges from disorder through a chimera state [41].

5.1 Examples of Complex Systems

Complexity is a characteristic of real physical systems. A turbulent flow, the weather and climate system in our planet, our universe, are all examples for complex physical systems. Complex systems are not limited to physical systems alone. Our day-to-day dynamical systems, for example, traffic, economic systems such as the stock market, our language, or biological systems are complex systems, because a great number of interacting elements are involved. A flock of birds, a school of fish, and a herd of cattle can all behave as a complex system—the group behaves as a superorganism with a collective intelligence that helps their wellbeing. Physiological systems are also complex systems; transitions in these systems such as heart attack, epileptic seizures, collapsed immune system, and migraine are emergent phenomena in these complex systems, that arise from sudden spontaneous and unexpected collective action within the body. The emergence of consciousness in living systems is a beautiful example that illustrates the central theme of complex systems theory—that "*the whole is more than its parts*". Nonlinearity and feedback are the important attributes that contribute to the emergence that is a defining characteristic of complex systems.

5.2 Nonlinearity: The Reductionist's Nightmare

How does a small difference in initial condition translate to a huge change, in chaotic systems? The key attribute that contributes to this extreme sensitivity to initial condition is nonlinearity.

A linear relationship between two quantities can be represented on a graph using a straight line. If I pay 3 dollars for a gallon of milk, I will get 2 gallons if I pay 6 dollars, and 5 gallons for 15 dollars. This, in a linear world, we can understand a linear system by understanding its parts individually, and then summing them up. Linear systems are easy to think about, linear equations are always solvable, and solutions of linear problems can be superposed. They are ideally suitable for text books. Every reductionist dreams of living in a linear world.

In contrast to this "peaceful" scenario, in a nonlinear relationship, the effect is not proportional to the cause. The relationship between cause and effect cannot be represented by straight line; instead, it can be represented only by a "nonlinear" curve—which is not a straight line, but a wiggle or a curve. We love to think that if a small change produces a small response, a change twice as big will produce twice as much change. However, in a nonlinear world, you could get a response that is ten times or one tenth or no response at all. In its extreme, nonlinearity can produce bifurcations—a bifurcation occurs when a smooth change in the parameter value of the system causes a sudden qualitative change in its behavior [68].

Compared to linear systems, nonlinear systems are incredibly difficult to deal with. Nonlinear equations cannot be solved in general; their solutions cannot be superposed. They create a plethora of rich behaviors that never occur in linear systems. Nonlinearity is thus a reductionist's nightmare.

5.3 Emergence

No amount of examination of the individual water molecules and their properties can explain the characteristic of wetness of water. That is, the wetness of water cannot be simply obtained by summing up the wetness of the constituent molecules. Wetness is an emergent property which emerges from the interaction between the molecules. While studying traffic flow, a detailed knowledge of the specifications of the cars, their color, shape, and their engine are useless when trying to predict when and where traffic jams will arise. Likewise, the bubbles and crashes that we witness in the stock markets cannot be simply described by summing up the actions of individual traders; they are emergent phenomena that occur due to the nonlinear interactions between the actions of the traders. In a sense, we have to look beyond the players and focus on the rules of the game. From a linear and reductionist perspective, the emergent phenomena are often surprising.

We need to emphasize that the emergent phenomena can arise "without the need for an invisible hand". For our thermoacoustic system—we do not need to force it with an actuator such as loudspeaker to get the standing wave and large coherent structures. At the right operating parameters, thermoacoustic instability onsets, along with the emergence of standing wave and large coherent structures. In fact, the whole point is to understand how this self-organization happens, predict it and control it, and forcing a system with actuator defeats this very purpose.

5.4 Pattern Formation

In a complex system, complex and often beautiful patterns can evolve from a simple set of rules. The science of pattern formation deals with the visible orderly outcomes of self-organization and the common principles behind similar patterns in nature. In a complex nonlinear system such as thermoacoustic system, the interaction creates patterns, and the patterns, in turn, affect the nature of interaction. In dynamical systems, patterns can appear both in time and in space. Patterns emerge in spatially extended, dissipative systems, when they are driven far from equilibrium [64].

The process by which small-scale interactions occurs within a subsystem itself, resulting in the emergence of order at the macro-scales is the pattern forming process. The emergence of large-scale vortices through the collective interaction of small-scale vortices points towards the phenomenon of pattern formation in the hydrodynamic subsystem [20]. Likewise, the emergence of large amplitude coherent standing waves in the acoustic system is due to the pattern formation in the acoustic subsystem. This pattern formation should be distinguished from the interactions across the subsystems, which are referred to as the coupling. Once the patterns form through small scale interactions, the coupling between the subsystems is more ordered. In the absence of this pattern formation, the coupling between the subsystems is disordered. A bidirectional process occurs through which the coupling among the subsystems leads to pattern formation and the patterns, in turn, influence the coupling between the subsystems. The notion that these events occur simultaneously suggests that thermoacoustic instability is a self-organization phenomenon, occurring in the turbulent thermoacoustic system.

5.5 Order Emerging from Chaos

We are amazed when we see bridges made by ants, fireflies flashing in synchrony, and collective animal behavior including bird flocks or school of fish. These are all examples of the phenomena of self-organization [67]. Life itself is an example of self-organization [37, 39].

We expect the entropy of a system to increase, taking it towards disorder. This is the essence of the second law of thermodynamics. Scientists often say that our universe itself is progressing towards disorder, and eventually a thermal death [56, 81]. In this context, this entropy defying self-organization; i.e., order emerging from disorder, may at first seem baffling. Most fundamental theories in physics apply to closed systems, and so is the second law of thermodynamics. Systems that show self-organization are nonlinear open systems;[1] they have matter and/or energy flowing into and out of the system [58]. To decrease entropy, work must be done.

Ilya Prigogne and his co-workers developed the thermodynamic approach to explain the phenomena of self-organization [50]. Prigogine was awarded the Nobel Prize in 1977 for "for his contributions to non-equilibrium thermodynamics, particularly the theory of dissipative structures" [70]. Dissipative structures are spatial structures, which come into being in non-equilibrium dissipative systems. In a thermodynamically non-equilibrium system, the birth of new structures is determined by energy dissipation. Energy dissipation transforms a fraction of the energy of ordered motion into disordered motion, and eventually to heat. It is indeed counter intuitive that dissipation can lead to the origin of structures. Recently, Di Vita [17] has shown that Rayleigh's results are a consequence of Prigogine's results related to open systems.

Turbulent systems have a large number of degrees of freedom, on account of the large number of scales. At the onset of oscillatory instabilities, such as thermoacoustic or aeroacoustic instability in turbulent flows, a huge number of degrees of freedom of macro-systems sharply reduce to a small number of variables (modes) [46].

Next, we will use the language of self-organization to describe the onset of thermoacoustic instability in turbulent flows.

5.6 Onset of Thermoacoustic Instability in Turbulent Combustors

Let us consider the onset of thermoacoustic instability in a simple backward facing combustor shown in Fig. 5.1. During the stable operation, small-scale vortices are

[1]There is no violation of the second law of thermodynamics. Entropy of the system plus the universe will indeed increase.

formed at the dump plane. These vortices are not that coherent, nor are they shed very periodically. Further, they do not have any sort of collective interaction, as illustrated in Fig. 5.2a. These vortices carry the flame and convect downstream to impinge on the bluff-body located at a distance L_c from the backward facing step of a step height d. This impingement leads to intense fine-scale turbulent mixing and the generation of instantaneous release of heat in the system. As the vortices in this turbulent flow are not of consistent size and strength, the sound generated has a broadband characteristic and is not tonal; it has multifractal characteristics [46].

However, as the system behaviour approaches the state of thermoacoustic instability, the vortex shedding frequency comes closer to the acoustic frequency; as a result, the hydrodynamic flow field and the acoustic field start to get synchronized. A phenomenon of collective interaction occurs wherein small-scale vortices interact with each other, resulting in self-organization and order at the larger scales and the formation of large-scale coherent structures [20, 55, 83], as illustrated in the schematic in Fig. 5.2b. This coherence leads to a consistency in the impingement times of vortices on the bluff-body, and it happens in phase with the acoustic pressure fluctuations inside the combustion chamber, leading to their amplification.

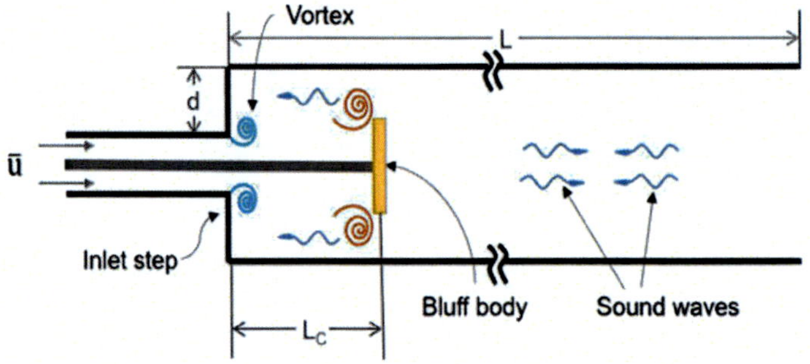

Fig. 5.1 Schematic of flow, flame and acoustic interaction happening in a bluff-body stabilized backward-facing step combustor. Reproduced with permission from Pawar et al. [53]

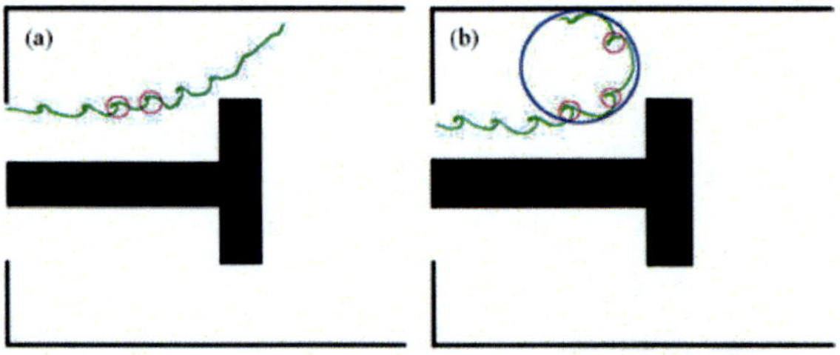

Fig. 5.2 Schematic showing the phenomenon of collective interaction of small-scale vortices (shown by purple circles) into the formation of large-scale vortices (shown by blue circle) along the shear layer in a dump combustor. (**a**) During stable operation, such collective interaction is not observed. In contrast, (**b**) during thermoacoustic instability, we observe self-organization of vortical structures in the system. Reproduced with permission from George et al. [20]

Thus, the positive feedback loop between the hydrodynamics and the acoustic field in the confinement leads to the establishment of self-sustained periodic oscillations.

Many characteristic signatures of complex systems can be seen in this description. We can see that a thermoacoustic system during stable operation has multifractal characteristics. Further, we see synchronization and collective interaction of the elements in the system. The emergence of large coherent structures and the standing wave pattern are both classic examples of emergence and pattern formation in complex systems. We also see a change in the topology of the vortical structures.

We will next look into these attributes of complex systems namely, (1) fractal nature and (2) collective interaction. We will also examine how we can use complex network theory to study the complex topologies and the complex nature of the interactions in the complex system.

5.7 Fractals and Multifractals

Turbulent flows are highly nonlinear and have large degrees of freedom. Turbulence is characterized by fluctuations which have been often described as stochastic [18, 26]. However, Navier-Stokes equations that predict turbulence are deterministic in nature [9, 40, 40].

A system with chaotic dynamics can generate signals that appear noisy. A turbulent combustor in stable operation is never really silent. It generates low amplitude aperiodic pressure fluctuations (known in the thermoacoustics community as combustion noise), which has been shown to be have chaotic nature [48, 71].

Chaotic time series often has a self-similar nature, i.e., if we zoom in on the signal, we will see patterns similar to what we see in the original zoomed out version. Signals with such self-similar structures fill non-integer dimensions, and are called fractals. The concept of fractals is very useful to describe complex systems that have complex dynamics involving fluctuations that span multiple orders of magnitude. Fractal fluctuations appear "noise-like"; however, they do not satisfy the statistics of random variables. The central moments of a random variable are bounded in time; in contrast, the central moments of a fractal signal diverge in time, at least over a limited range of scales. In other words, the signal is non-stationary and there is a variation in the underlying distribution of data values. In a spatiotemporal system, such as thermoacoustic system, the measured fluctuations represent both space and time, which makes the signal non-stationary. For example, the acoustic pressure measurements at a given instant at a given location depend not just on the values at the previous instants, but also on the values of acoustic pressure at other locations in the flow field as well.

Thus, classical measures such as mean or variance are not very useful to describe non-stationary signals, since these measures vary with time over which they are computed. Instead, we characterize non-stationary signals by examining how the moments depend on the time interval over which they are evaluated. This is done by examining the Hurst exponent, which characterizes how the standard deviation of the signal varies with the time duration over which it is calculated. The Hurst exponent measures the amount of correlation or the memory in a time. The Hurst exponent and the fractal dimension D of a time series are related as $D = 2 - H$. If a fractal time series $p(t)$ has a Hurst exponent H, then $p(ct)/c^H$ is also a fractal with the same statistics [78, 79].

Fluctuations in complex systems exhibit a complex scaling behavior that cannot be accounted by a single fractal dimension. Describing such fluctuations involves multiple generalized Hurst exponents. This results in interwoven subsets of varying fractal dimension (varying Hurst exponents), or in other words a 'multifractal' behaviour [52]. For a multifractal signal, fluctuations that have different amplitudes follow different scaling rules.

Even during stable operation, a combustor is never silent. It produces low amplitude aperiodic oscillations, which are generally referred to as "combustion noise". These pressure fluctuations during the occurrence of combustion noise display multifractality. Turbulent reacting flows have multiple spatial/temporal scales that contribute to its dynamics, and it is only natural that the sound produced by such flows have multifractal characteristics. On the other hand, thermoacoustic instability is characterized by a finite number of discrete temporal and spatial scales (one if there is only a single mode or a few scales if there are multiple modes present). As a consequence, the multifractality is lost at the onset of thermoacoustic instability. In fact, there is a gradual loss of multifractality as we approach the onset through an intermittency route; therefore, fractal measures can be used as early warning measures (precursors) that quantify the proximity to an impending thermoacoustic instability, well before any rise in amplitude occurs.

The transition of the thermoacoustic system from a state characterized by a number of scales to one dominated by a few discrete time scales is related to the formation of large-scale coherent structures in the flow field. Next, we examine how such large coherent structures can emerge in a complex system.

5.8 Collective Interaction in Complex Systems

Synchronization is the matching of the rhythm of oscillators upon coupling. Huygens discovered this phenomenon in the seventeenth century, while observing the anti-phase synchronization of pendulums of clocks hanging from a common support, which he referred to as "odd sympathy" [54, 59]. Synchronization is indeed a ubiquitous phenomenon, observed in a wide variety of disciplines including physics [38, 67], chemistry [62], biology [21], various types of engineering systems [25, 60] and ecosystems [10]. Initial studies on synchronization were performed on periodic oscillators [11]. Recent studies show that synchronization occurs for chaotic oscillators as well [12] which is fascinating and surprising, given their sensitive dependence on the initial conditions. Coupling strength, which indicates the strength of the mutual interaction between the coupled oscillators, is a crucial parameter that determines if the oscillators synchronize. Synchronization is usually studied by examining the locking of phase (or frequency) of the signals from the oscillators [54].

For a combustor having turbulent reacting flow, both the the acoustic field and the heat release rate in the combustor may be considered as self-sustained aperiodic oscillators. The presence of continuous perturbations from the inherent turbulent hydrodynamic flow makes these oscillators which are otherwise damped, self-sustained oscillators. Thus, a thermoacoustic system with an underlying turbulent flow field can be viewed as two coupled subsystems—the duct acoustic field and the turbulent fluctuations present in the reacting flow system. These subsystems synchronize at the onset of thermoacoustic instability. The bidirectional coupling between these oscillators results in the adjustment of their rhythms to a common frequency at a particular value of the control parameter. In this view point, during the stable operation, we observe chaotic oscillations in both the acoustic field and the heat release rate fluctuations which are desynchronized. During intermittency, these two signals are synchronized intermittently. During the state of low amplitude thermoacoustic oscillations, they are phase synchronized; while, during the state of large amplitude thermoacoustic oscillations, they exhibit generalized synchronization [53].

Thermoacoustic systems are spatially-extended systems. The reactive flow field can be viewed as a collection of local heat release rate oscillators. During stable operation, these oscillators are incoherent and, during thermoacoustic instability, they are all synchronized with the acoustic field. During intermittency, we see a chimera state, where coherent oscillations co-exist with incoherent ones [41].

The framework of synchronization theory provides avenues for mitigation of thermoacoustic instability. Open-loop forcing has been successfully used to control thermoacoustic instabilities through a phenomenon of asynchronous quenching. Recently, there has been much interest in the phenomena of amplitude death, where two coupled oscillators can reach the same steady state, with the right amount of coupling between them. This concept appears promising to mitigate thermoacoustic instabilities.

In summary, we can view thermoacoustic instability as the synchronization of a network of oscillators. This leads us to the concept of complex networks.

5.9 Complex Networks

The last two decades witnessed an explosive development of the interdisciplinary subject of complex networks which aims at developing a set of unifying principles governing networks in nature, society, and technology. This started with the seminal work on small-world networks by Watts and Strogatz [77] and the work on scale-free networks by Barabási and Albert [5]. Complex networks have pristine mathematical properties and can model relevant phenomena not only in physical systems, but also in biological or technological systems such as airline network, World Wide Web, electrical power grids, etc. Network thinking is the epitome of complex systems approach—we focus on relationships between entities rather than on the details of the entities themselves.

A network is defined by a collection of nodes, connected by links. Nodes are the individual entities in the network, e.g., neurons, websites, people, oscillators, airports. Links are the connections between them, e.g., synapses, web hyperlinks, social relationships, couplings, flights.

Watts and Strogatz [77] proposed that the random addition of a few links reduced the average distance between the nodes drastically. They called such networks that are highly clustered, yet have small characteristic path lengths as small-world networks. However, the networks of Watts and Strogatz did not have the characteristics of many real-world networks. Real-world networks often have a small number of nodes with a large number of connections and a large number of nodes with a small number of connections, and no particular characteristic scale.

In 1999, Barabási and Albert [5] proposed the concept of scale-free networks, which characterize most of the real-world networks such as the internet, World Wide Web, and scientific collaboration networks that have a heavy-tailed distribution of connectivity with no characteristic scale. Airline networks [16, 76], power grids [3, 14, 51], World Wide Web [1, 7], and even biological systems [2, 6] can be characterized as scale-free networks. We know that when the thermoacoustic system transitions from stable operation (occurrence of combustion noise) to unstable operation (occurrence of thermoacoustic oscillations), the amplitude spectrum of the acoustic pressure transitions from one having a broad peaks to one having sharp peaks corresponding to frequencies that are often close to the acoustic modes [13, 46]. Nevertheless, there is indeed a vagueness associated with the definition of a broad peak and a shallow peak. To arrive at a better definition, the pattern emerging during this transition from chaos to order via intermittency needs to be identified and formalized, by formalizing the process of pattern discovery [8]. Complex networks can be used to analyze the various dynamical regimes of turbulent combustors during the transition from combustion noise to thermoacoustic instability via the intermittency route.

Many different types of complex networks can be constructed. Popular examples include visibility graphs and horizontal visibility graphs [66], and recurrence networks [84]. Murugesan and Sujith [42] used visibility graphs and reported scale-free behavior during the state of combustion noise. They also identified that this scale-free behavior disappears during the transition to thermoacoustic instability in a turbulent combustor. Other studies using visibility graphs and also recurrence networks have confirmed this finding [22, 23]. Network measures such as clustering coefficient and characteristic path length can be used to detect precursors to obtain early warning for the onset of thermoacoustic instability and flame blowout [43, 44].

Spatial [73] and spatiotemporal networks [35] of the reaction field of thermoacoustic systems can also be constructed. Unni et al. [73] constructed a correlation-based network from the time series of velocity fluctuations in a turbulent reactive flow field of a bluff-body stabilized dump combustor. The analysis is performed during the transition to thermoacoustic instability via intermittency. They determined critical regions in the flow field during the different dynamical regimes using various network centrality measures. These sensitive locations in the flow field, known as hubs, are critical for the onset of thermoacoustic instability. Disrupting such hubs by tweaking (altering mildly) the flow field or the injector configuration, we can passively control thermoacoustic instability [34, 36, 74]. Spatiotemporal networks of the reaction field can throw light on the processes behind the onset of thermoacoustic instability [35].

5.10 Why Should We Use Complex Systems Approach to Study Thermoacoustic Instability in Turbulent Combustors?

Traditionally, it is believed that when the acoustic driving provided by the flame is balanced by the losses in the system, one attains limit cycle oscillations. Recent experimental studies on thermoacoustic instability in confined systems with ducted laminar flames showed that limit cycle is just one of the several possible end states of the system. They showed that undergoing further bifurcations, the thermoacoustic system can attain states such as quasiperiodicity, period-2, intermittency, frequency-locked, strange non-chaos, and chaos [24, 28–31, 57]. Similar states were observed in numerical simulations a thermoacoustic system comprising a ducted laminar premixed flame by Kashinath et al. [32].

Turbulence introduces a lot more complexity associated with the large number of degrees of freedom [72]. Being a feedback system with a variety of nonlinear phenomenon including molecular mixing, turbulent transport involving a wide range of scales, chemical kinetics, and acoustic wave interaction, etc., thermoacoustic systems offer the possibility of chaotic fluctuations on the one hand to ordered periodic oscillations on the other hand.

We list below a number of phenomena that we observe in turbulent thermoacoustic systems that cannot be described by the traditional thermoacoustics.

1. Traditionally, thermoacoustic instability is modeled as a transition from a fixed point (stable operation) to limit cycle (unstable operation). Recent studies suggest that the stable regime of a combustor with turbulent flow does not correspond to a fixed point. A turbulent combustor is never "silent". Low-amplitude aperiodic pressure fluctuations occur during the stable operation of the combustor. These fluctuations are deterministic and have chaotic behavior [71]. In contrast, thermoacoustic instability is characterized by large amplitude periodic fluctuations; these periodic oscillations represent order. The transition from combustion noise to periodic oscillations happens through an intermediate state of intermittency (Fig. 5.3); a state comprising bursts of large-amplitude periodic oscillations amidst epochs of small-amplitude aperiodic fluctuations, i.e., we see "pockets of order" amidst disorder [49].
2. Nair and Sujith [46] showed that combustion noise is characterized by scale invariance and a multifractal signature. Multifractality of combustion noise

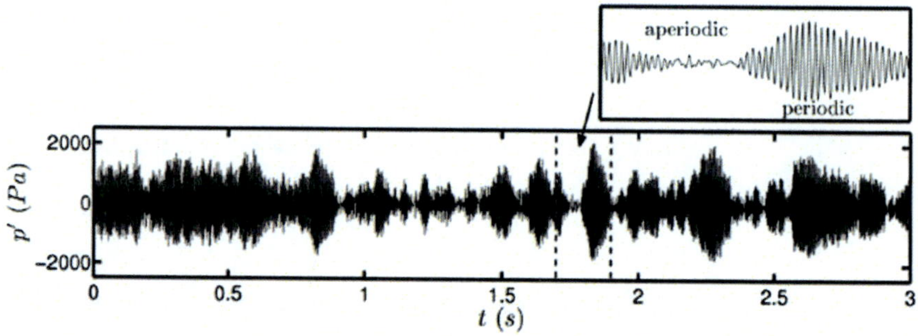

Fig. 5.3 The time series of the acoustic pressure fluctuations observed during the occurrence of intermittency in a bluff-body-stabilized dump combustor, where bursts of large amplitude periodic oscillations occur between epochs of low amplitude aperiodic fluctuations (see inset). Reproduced with permission from Nair et al. [45]

confirms the lack of a single dominant spatial or temporal scale in thermoacoustic systems and reflects the complex nature of the dynamics involved, which arises due to the nonlinear interaction between flow, flame and duct acoustics. In contrast, thermoacoustic instability is characterized by a few dominant length and time scales. As a consequence, multifractality is lost at the onset of thermoacoustic instability.

3. During the stable operation of the combustor, small-scale vortices perturb the flame front causing aperiodic dynamics. During periodic epochs of intermittency, periodic and coherent spatiotemporal patterns emerge as a result of self-organization of the reactive flow field, where small-scale vortices collectively interact leading to the emergence of large-scale vortices. These large coherent structures, upon impingement on the bluff-body and combustor walls, cause localized regions of high heat release rate leading to the emergence of ordered spatiotemporal patterns in the flow field during intermittency [20]. Order emerges from disorder through a chimera state where pockets of order and disorder coexist [41]. During the occurrence of thermoacoustic instability, small-scale vortices collectively interact, leading to sustained periodic emergence of large coherent structures. In these coherent structures, we observe acoustic power production of substantial magnitude that leads to self-sustained periodic acoustic oscillations [20].
4. We see two types of limit cycles in the thermoacoustic system—a "noisy" low amplitude limit cycle that corresponds to the phase synchronization,[2] and a large amplitude "clean" limit cycle that corresponds to generalized synchronization,[2] between the acoustic pressure and heat release rate fluctuations [53].
5. The intermittency route to the onset of thermoacoustic instability and the loss of multifractality which we observe in turbulent thermoacoustic systems are seen

[2]These will be explained in Chap. 8.

in other oscillatory instabilities occurring in turbulent flow such as aeroacoustic [47], aeroelastic [75], and compressor surge [33]. In spite of the many differences between the physics behind the instabilities exhibited by these different systems, they often display similar dynamical behaviour. The traditional reductionist approach which attempts to analyze a complex system in terms of its constituent elements, hits its limit in explaining these similarities in fundamentally different physical systems [4].

Many of these phenomena are well-known to the engineers and technicians on the field. During the occurrence of thermoacoustic instability, engineers and technicians often say that the combustor is breathing. They often describe the combustor as an entity that seems to have a life of its own. Further, they often use phrases such as the "rms is fluctuating", "feeling a pounding on the chest", etc. before the onset of thermoacoustic instability. These indicate that they indeed are very aware of the intermittency route to the onset of thermoacoustic instability in turbulent combustors, in an intuitive sense. Nevertheless, the textbooks and till recently most research papers hardly acknowledge this. Pretending that something does not exist even if it is hard to quantify leads to inadequate understanding, and hence an inability to give the best response to the situation, and also faulty models.

5.11 Practical Applications

Complex systems theory helps us to arrive at strategies to cope with thermoacoustic instability. There are three strategies for managing thermoacoustic instability—(1) evading the instability by obtaining precursors to the onset of instability, (2) smart passive control, and (3) strategies based on synchronization such as amplitude death and open-loop forcing.

1. **Precursors**
   Complex systems approach to thermoacoustic instability has enabled us to obtain early warning measures (precursors) to the onset of thermoacoustic instability based on

   a. Intermittency statistics
   b. Fractal characteristics of the fluctuations
   c. Change in the complex network topology

2. **Smart passive control by using complex network theory**
   Complex network theory enables us to find the critical points in the system that are most vulnerable to disturbances. By disturbing the flow in this region, we attempt to disrupt the instability.
3. **Controls based on synchronization theory**

   a. Open-loop control using the phenomenon of asynchronous quenching: The external periodic oscillations (in flow, fuel, or acoustic fields) are imposed at a frequency far from the natural frequency to quench thermoacoustic oscillations.

b. Quenching oscillations using amplitude death phenomenon: Two thermoacoustic systems in the state of unstable operation are coupled through a tube of specific dimensions (length and diameter) to simultaneously quench thermoacoustic oscillations in both the systems.

Chu [15] has shown that disturbances in a turbulent flow field can be split into acoustic, vorticity, and entropy disturbances. Since 1600s physics has embraced reductionist thinking, which has indeed provided rich dividends in terms of progress in science as epitomized by the success in computing the planetary trajectories using Newton's laws. Hence, following a reductionist approach, we need to study the three subsystems, i.e., acoustics, combustion and flow, in as much detail as possible.

Therefore, a researcher who has a primary disposition towards acoustics will see the problem as an acoustic field that is driven by combustion. Likewise, a combustion expert will think of a thermoacoustic system as a combustion system that is disturbed by an acoustic disturbance, and a fluid mechanics expert will focus on the reacting flow field hydrodynamics, which happens to be influenced by combustion and acoustics. In a lighter vein, many studies in thermoacoustics resemble the story of four blind men studying an elephant [80]. What is often forgotten is that these disturbances are decoupled only in the case of uniform base flow [19], which is certainly not the case in thermoacoustic systems [69]. This coupling between the "subsystems" is strongly nonlinear and can lead to emergent phenomena such as coherent structures, order emerging from chaos, pockets of order amongst pockets of disorder, etc. This emergence seen in the self-organization of the thermoacoustic system indeed resonates with the anti-reductionist slogan "the whole is more than the sum of the parts". Thus, straightjacketing thermoacoustic instability into any single discipline, without emphasizing the interactions is like missing the forest for the trees.

There are no separate subsystems. Our world[3] is a continuum, where we keep drawing boundaries around subsystems depending on what we propose to investigate or discuss. These boundaries or compartments that we create can produce problems when we forget that we have artificially created them, and that they can interact in a strongly nonlinear fashion.

The right boundary rarely coincides with the boundary of an academic discipline. Ideally, we need the mental agility to find the appropriate boundary; however, sadly we are hardly that agile. We develop a strong attachment to the boundaries that our minds happen to be accustomed to. It is indeed liberating to realize that the boundaries are our own creation, and that they depend on the discussion, problem or purpose that we have at the time of creation of these boundaries. It is indeed a challenge to stay detached enough to drop those boundaries that we drew for the last problem, and to find the most appropriate boundary for the current problem that we

[3]By world, we mean the world that we inhabit and look at, with our mind.

are dealing with. Moreover, keep in mind that the interactions between subsystems, the delays, and the nonlinearities are the defining characteristics of a thermoacoustic system.

In spite of what we have majored in, we must follow a system to wherever it leads, and in this case, we have meandered into the realm of complex systems theory, which transcends the disciplines of engineering, physics, and mathematics. As you will see in the subsequent chapters, this is a very rewarding journey.

References

1. Albert, R., Jeong, H., Barabási, A.L.: Diameter of the world-wide web. Nature **401**, 130–131 (1999)
2. Alm, E., Arkin, A.P.: Biological networks. Curr. Opin. Struct. Biol. **13**(2), 193–202 (2003)
3. Arianos, S., Bompard, E., Carbone, A., Xue, F.: Power grid vulnerability: A complex network approach. Chaos **19**(1), 013119 (2009)
4. Barabási, A.L.: The network takeover. Nat. Phys. **8**(1), 14–16 (2012)
5. Barabási, A.L., Albert, R.: Emergence of scaling in random networks. Science **286**(5439), 509–512 (1999)
6. Barabasi, A.L., Oltvai, Z.N.: Network biology: understanding the cell's functional organization. Nat. Rev. Genet. **5**(2), 101–113 (2004)
7. Barabási, A.L., Albert, R., Jeong, H.: Mean-field theory for scale-free random networks. Physica A **272**(1-2), 173–187 (1999)
8. Barabási, A.L., Gulbahce, N., Loscalzo, J.: Network medicine: a network-based approach to human disease. Nat. Rev. Genet. **12**(1), 56–68 (2011)
9. Bergé, P., Pomeau, Y., Vidal, C.: Order Within Chaos: Towards A Deterministic Approach to Turbulence. Wiley, New York: Hermann, Paris (1986)
10. Blasius, B., Huppert, A., Stone, L.: Complex dynamics and phase synchronization in spatially extended ecological systems. Nature **399**(6734), 354–359 (1999)
11. Blekhman, I., Landa, P.S., Rosenblum, M.G.: Synchronization and chaotization in interacting dynamical systems. Appl. Mech. Rev. **48**, 733–752 (1995)
12. Boccaletti, S., Kurths, J., Osipov, G., Valladares, D.L., Zhou, C.S.: The synchronization of chaotic systems. Phys. Rep. **366**(1), 1–101 (2002)
13. Chakravarthy, S.R., Shreenivasan, O.J., Boehm, B., Dreizler, A., Janicka, J.: Experimental characterization of onset of acoustic instability in a nonpremixed half-dump combustor. J. Acoust. Soc. Am. **122**(1), 120–127 (2007)
14. Chen, G., Dong, Z.Y., Hill, D.J., Zhang, G.H., Hua, K.Q.: Attack structural vulnerability of power grids: A hybrid approach based on complex networks. Physica A **389**(3), 595–603 (2010)
15. Chu, B.T.: On the energy transfer to small disturbances in fluid flow (Part I). Acta Mech. **1**(3), 215–234 (1965)
16. Conway, S.: Scale-free networks and commercial air carrier transportation in the United States. In: 24th Congress of the International Council of the Aeronautical Sciences, vol. 29 (2004)
17. Di Vita, A.: Maximum or minimum entropy production? How to select a necessary criterion of stability for a dissipative fluid or plasma. Phys. Rev. E **81**(4), 041137 (2010)
18. Dmitrenko, A.V.: The stochastic theory of the turbulence. In: IOP Conf. Ser.: Mater. Sci. Eng., vol. 468, p. 012021. IOP Publishing (2018)
19. George, J., Sujith, R.I.: Emergence of acoustic waves from vorticity fluctuations: Impact of non-normality. Phys. Rev. E **80**(4), 046321 (2009)
20. George, N.B., Unni, V.R., Raghunathan, M., Sujith, R.I.: Pattern formation during transition from combustion noise to thermoacoustic instability via intermittency. J. Fluid Mech. **849**, 615–644 (2018)

21. Glass, L.: Synchronization and rhythmic processes in physiology. Nature **410**(6825), 277–284 (2001)
22. Godavarthi, V., Unni, V.R., Gopalakrishnan, E.A., Sujith, R.I.: Recurrence networks to study dynamical transitions in a turbulent combustor. Chaos **27**(6), 063113 (2017)
23. Gotoda, H., Hayashi, K., Tsujimoto, R., Domen, S., Tachibana, S.: Dynamical properties of combustion instability in a laboratory-scale gas-turbine model combustor. J. Eng. Gas Turbines Power **139**(4), 041509 (2017)
24. Guan, Y., Gupta, V., Li, L.K.B.: Intermittency route to self-excited chaotic thermoacoustic oscillations. J. Fluid Mech. **894**, R3 (2020)
25. Heagy, J.F., Carroll, T.L., Pecora, L.M.: Synchronous chaos in coupled oscillator systems. Phys. Rev. E **50**(3), 1874–1885 (1994)
26. Humphrey, L.J., Acharya, V.S., Shin, D.H., Lieuwen, T.C.: Modeling the response of turbulent flames to harmonic forcing. Combust. Sci. Tech. **189**(2), 187–212 (2017)
27. Johnson, N.: Simply Complexity: A Clear Guide to Complexity Theory. Simon and Schuster (2009)
28. Kabiraj, L., Sujith, R.I.: Nonlinear self-excited thermoacoustic oscillations: intermittency and flame blowout. J. Fluid Mech. **713**, 376–397 (2012)
29. Kabiraj, L., Saurabh, A., Wahi, P., Sujith, R.I.: Route to chaos for combustion instability in ducted laminar premixed flames. Chaos **22**(2), 023129 (2012)
30. Kabiraj, L., Sujith, R.I., Wahi, P.: Bifurcations of self-excited ducted laminar premixed flames. J. Eng. Gas Turbines Power **134**(3), 031502 (2012)
31. Kabiraj, L., Sujith, R.I., Wahi, P.: Investigating the dynamics of combustion-driven oscillations leading to lean blowout. Fluid Dyn. Res. **44**(3), 031408 (2012)
32. Kashinath, K., Waugh, I.C., Juniper, M.P.: Nonlinear self-excited thermoacoustic oscillations of a ducted premixed flame: bifurcations and routes to chaos. J. Fluid Mech. **761**, 399–430 (2014)
33. Kerres, B., Nair, V., Cronhjort, A., Mihaescu, M.: Analysis of the turbocharger compressor surge margin using a Hurst-exponent-based criterion. SAE Int. J. Engines **9**(3), 1795–1806 (2016)
34. Krishnan, A., Manikandan, R., Midhun, P.R., Reeja, K.V., Unni, V.R., Sujith, R.I., Marwan, N., Kurths, J.: Mitigation of oscillatory instability in turbulent reactive flows: A novel approach using complex networks. Europhys. Lett. **128**(1), 14003 (2019)
35. Krishnan, A., Sujith, R.I., Marwan, N., Kurths, J.: On the emergence of large clusters of acoustic power sources at the onset of thermoacoustic instability in a turbulent combustor. J. Fluid Mech. **874**, 455–482 (2019)
36. Krishnan, A., Sujith, R.I., Marwan, N., Kurths, J.: Suppression of thermoacoustic instability by targeting the hubs of the turbulent networks in a bluff body stabilized combustor. J. Fluid Mech. **916**, A20 (2021)
37. Kurakin, A.: The self-organizing fractal theory as a universal discovery method: the phenomenon of life. Theor. Biol. Med. Modell. **8**, 1–66 (2011)
38. Kuramoto, Y.: Chemical Oscillations, Waves, and Turbulence, vol. 19. Springer Science & Business Media (1984)
39. Luisi, P.L., Ferri, F., Stano, P.: Approaches to semi-synthetic minimal cells: a review. Naturwissenschaften **93**(1), 1–13 (2006)
40. McDonough, J.M.: Introductory lectures on turbulence: physics, mathematics and modeling. In: Mechanical Engineering Textbook Gallery. University of Kentucky (2007)
41. Mondal, S., Unni, V.R., Sujith, R.I.: Onset of thermoacoustic instability in turbulent combustors: an emergence of synchronized periodicity through formation of chimera-like states. J. Fluid Mech. **811**, 659–681 (2017)
42. Murugesan, M., Sujith, R.I.: Combustion noise is scale-free: transition from scale-free to order at the onset of thermoacoustic instability. J. Fluid Mech. **772**, 225–245 (2015)
43. Murugesan, M., Sujith, R.I.: Detecting the onset of an impending thermoacoustic instability using complex networks. J. Propuls. Power **32**(3), 707–712 (2016)

44. Murugesan, M., Zhu, Y., Li, L.K.B.: Complex network analysis of forced synchronization in a hydrodynamically self-excited jet. Int. J. Heat Fluid Flow **76**, 14–25 (2019)
45. Nair, V., Sujith, R.I.: Identifying homoclinic orbits in the dynamics of intermittent signals through recurrence quantification. Chaos **23**(3), 033136 (2013)
46. Nair, V., Sujith, R.I.: Multifractality in combustion noise: predicting an impending combustion instability. J. Fluid Mech. **747**, 635–655 (2014)
47. Nair, V., Sujith, R.I.: Precursors to self-sustained oscillations in aeroacoustic systems. Int. J. Aeroacoust. **15**(3), 312–323 (2016)
48. Nair, V., Thampi, G., Karuppusamy, S., Gopalan, S., Sujith, R.I.: Loss of chaos in combustion noise as a precursor of impending combustion instability. J. Spray Combust. Dyn. **5**(4), 273–290 (2013)
49. Nair, V., Thampi, G., Sujith, R.I.: Intermittency route to thermoacoustic instability in turbulent combustors. J. Fluid Mech. **756**, 470–487 (2014)
50. Nicolis, G., Prigogine, I.: Self-Organization in Nonequilibrium Systems. Wiley-Interscience, New York (1977)
51. Pagani, G.A., Aiello, M.: Power grid complex network evolutions for the smart grid. Physica A **396**, 248–266 (2014)
52. Parisi, G., Frisch, U.: A multifractal model of intermittency. In: Turbulence and Predictability in Geophysical Fluid Dynamics and Climate Dynamics, pp. 84–88. North-Holland, Amsterdam (1985)
53. Pawar, S.A., Seshadri, A., Unni, V.R., Sujith, R.I.: Thermoacoustic instability as mutual synchronization between the acoustic field of the confinement and turbulent reactive flow. J. Fluid Mech. **827**, 664–693 (2017)
54. Pikovsky, A., Rosenblum, M., Kurths, J.: Synchronization: A Universal Concept in Nonlinear Sciences, vol. 12. Cambridge University Press (2003)
55. Poinsot, T.J., Trouve, A.C., Veynante, D.P., Candel, S.M., Esposito, E.J.: Vortex-driven acoustically coupled combustion instabilities. J. Fluid Mech. **177**, 265–292 (1987)
56. Prasad, R.: Classical and Quantum Thermal Physics. Cambridge University Press (2016)
57. Premraj, D., Pawar, S.A., Kabiraj, L., Sujith, R.I.: Strange nonchaos in self-excited singing flames. Europhys. Lett. **128**(5), 54005 (2020)
58. Prigogine, I., Stengers, I.: Order Out of Chaos: Man's New Dialogue with Nature. Verso Books (1984)
59. Ramirez, J.P., Olvera, L.A., Nijmeijer, H., Alvarez, J.: The sympathy of two pendulum clocks: beyond Huygens' observations. Sci. Rep. **6**, 23580 (2016)
60. Roy, R., Thornburg Jr, K.S.: Experimental synchronization of chaotic lasers. Phys. Rev. Lett. **72**(13), 2009–2012 (1994)
61. Schadow, K.C., Gutmark, E., Parr, T.P., Parr, D.M., Wilson, K.J., Crump, J.E.: Large-scale coherent structures as drivers of combustion instability. Combust. Sci. Tech. **64**(4-6), 167–186 (1989)
62. Schreiber, I., Marek, M.: Strange attractors in coupled reaction-diffusion cells. Physica D **5**(2-3), 258–272 (1982)
63. Smith, K., Brighton, H., Kirby, S.: Complex systems in language evolution: the cultural emergence of compositional structure. Adv. Complex Syst. **6**(04), 537–558 (2003)
64. Staliunas, K., Sanchez-Morcillo, V.J.: Transverse Patterns in Nonlinear Optical Resonators, vol. 183. Springer Science & Business Media (2003)
65. Steinmetz, R., Wehrle, K.: Peer-to-peer Systems and Applications, vol. 3485. Springer (2005)
66. Stephen, M., Gu, C., Yang, H.: Visibility graph based time series analysis. PLoS One **10**(11), e0143015 (2015)
67. Strogatz, S.: Sync: The Emerging Science of Spontaneous Order. Penguin UK (2004)
68. Strogatz, S.H.: Nonlinear Dynamics and Chaos: With Applications to Physics, Biology, Chemistry, and Engineering. CRC Press (1994)
69. Sujith, R.I., Juniper, M.P., Schmid, P.J.: Non-normality and nonlinearity in thermoacoustic instabilities. J. Spray Combust. Dyn. **8**(2), 119–146 (2016)

70. The_Nobel_Prize: The Nobel Prize in Chemistry 1977, https://www.nobelprize.org/prizes/chemistry/1977/summary
71. Tony, J., Gopalakrishnan, E.A., Sreelekha, E., Sujith, R.I.: Detecting deterministic nature of pressure measurements from a turbulent combustor. Phys. Rev. E **92**(6), 062902 (2015)
72. Tran, C.V.: The number of degrees of freedom of three-dimensional Navier–Stokes turbulence. Phys. Fluids **21**(12), 125103 (2009)
73. Unni, V.R., Krishnan, A., Manikandan, R., George, N.B., Sujith, R.I., Marwan, N., Kurths, J.: On the emergence of critical regions at the onset of thermoacoustic instability in a turbulent combustor. Chaos **28**(6), 063125 (2018)
74. Unni, V.R., Sujith, R.I., Krishnan, A., Marwan, N., Kurths, J.: System and method for optimizing passive control of oscillatory instabilities in turbulent flows. US Patent US10895382B2, Priority date: February 27 (2018)
75. Venkatramani, J., Nair, V., Sujith, R.I., Gupta, S., Sarkar, S.: Precursors to flutter instability by an intermittency route: a model free approach. J. Fluids Struct. **61**, 376–391 (2016)
76. Verma, T., Araújo, N.A., Herrmann, H.J.: Revealing the structure of the world airline network. Sci. Rep. **4**(1), 1–6 (2014)
77. Watts, D.J., Strogatz, S.H.: Collective dynamics of 'small-world' networks. Nature **393**(6684), 440–442 (1998)
78. West, B.J.: Fractal physiology and the fractional calculus: a perspective. Front. Physiol. **1**, 12 (2010)
79. West, B., Bologna, M., Grigolini, P.: Physics of Fractal Operators. Springer Science & Business Media (2003)
80. Wikipedia: Blind men and an elephant. https://en.wikipedia.org/wiki/Blind_men_and_an_elephant
81. Wikipedia: Heat death of the universe. https://en.wikipedia.org/wiki/Heat_death_of_the_universe
82. Wikipedia: New england complex systems institute. https://en.wikipedia.org/wiki/New_England_Complex_Systems_Institute
83. Yu, H.K., Trouvé, A., Daily, J.W.: Low-frequency pressure oscillations in a model ramjet combustor. J. Fluid Mech. **232**, 47–72 (1991)
84. Zou, Y., Donner, R.V., Marwan, N., Donges, J.F., Kurths, J.: Complex network approaches to nonlinear time series analysis. Phys. Rep. **787**, 1–97 (2019)

Chapter 6
Intermittency—A State That Precedes Thermoacoustic Instability and Blowout in Turbulent Combustors

Let us recall from Chaps. 3 and 4 that during the onset of thermoacoustic instability in laminar systems, the dynamics of the acoustic pressure fluctuations transitions from a state of fixed point to limit cycle oscillations. We observed that this transition in the acoustic pressure fluctuations is direct without the presence of an intermediate state. Additionally, the transition is accompanied by a sudden jump in the pressure amplitude for subcritical Hopf bifurcation or a gradual increase in the pressure amplitude for supercritical Hopf bifurcation. In the present chapter, we will focus on understanding the dynamical changes happening in the acoustic pressure fluctuations during the occurrence of thermoacoustic instability in turbulent thermoacoustic systems. We will first systematically characterize the properties of acoustic pressure fluctuations during the states of stable operation and unstable operation of a turbulent combustor. Subsequently, we will discuss the characteristics of the dynamical transition from stable to unstable operation, and also the properties of the state of intermittency. We will also shed light on the route to the occurrence of another undesired dynamic phenomenon called lean flame blowout. Finally, we will present a universal scaling observed in acoustic pressure fluctuations during the onset of thermoacoustic instability in different turbulent combustors.

We know that, in practical combustion systems, the underlying flow field is turbulent. The presence of turbulence increases the degrees of freedom in the interaction of subsystems, such as the flow, the flame and the acoustic fields of a combustor. As we discussed in the previous chapter, the increase in the degrees of freedom further makes the system dynamics highly complex during the onset of thermoacoustic instability. Therefore, for an engine operator working on industrial combustors, understanding these dynamical transitions in the acoustic pressure fluctuations is necessary. This understanding, in turn, will help in protecting the system and its components from the undesired consequences of thermoacoustic instabilities.

R. I. Sujith, S. A. Pawar, *Thermoacoustic Instability*, Springer Series in Synergetics,
https://doi.org/10.1007/978-3-030-81135-8_6

6.1 Classification of Sound Waves Generated by Turbulent Flame in a Combustor

In most practical combustion systems possessing turbulent flow field, such as propulsion engines (e.g., gas turbines, ramjet, and rockets), industrial power generating units, boilers, and furnaces, the sound (acoustic waves) generated by unsteady combustion is one of the prime concerns. Traditionally, acoustic waves emitting by turbulent flames in such systems are classified into two types: combustion noise (stable operation) and combustion-driven oscillations (unstable operation) [25, 104].

Combustion noise is also referred to as combustion roar in the literature [104]. During this state, we note that the reaction field of a combustor is never silent, due to the presence of turbulence in the underlying flow field. As a result, the flow variables of such systems always show low amplitude irregular fluctuations in the temporal dynamics and a broadband of frequencies in the power spectrum.

On the other hand, combustion-driven oscillations, also referred to as combustion instability or thermoacoustic instability [7, 16, 68], are characterized by large-amplitude, self-sustained oscillations in the acoustic field of the combustor. These oscillations occur primarily due to the positive feedback between the subsystems-the acoustic field, the hydrodynamic field, and the heat release rate fluctuations in the flame. The spectrum of such oscillations displays discrete sharp frequency peaks (tones), amplified around the natural acoustic modes of the system. In the rest of the chapter, we will refer to combustion-driven oscillations as thermoacoustic instability.

We come across several studies in the past that focused on the individual characterization of the states of combustion noise [4, 8, 12, 23, 25, 39, 106, 116, 118, 122, 123] and combustion-driven oscillations [6, 7, 17, 96, 98, 125]. In these studies, the focus was mostly restricted to understanding the dynamical features of these states, physical reasons behind their occurrence, and the control of such oscillations. However, we notice that only a very limited number of investigations have been devoted to study the transition of a combustion system from the state of combustion noise to thermoacoustic oscillations, as the control parameter is systematically varied [43, 125]. During the onset of thermoacoustic instability, aperiodic (disordered) fluctuations disappear and periodic (ordered) oscillations emerge [43, 125]. In the following sections, we first characterize the dynamic properties of the states of combustion noise and thermoacoustic instability, and then discuss the transition from the former state to the latter one.

6.2 What Is Combustion Noise?

In simple words, combustion noise can be defined as low amplitude broadband fluctuations generated by the unsteady combustion processes happening in a turbulent environment [4, 12, 39, 71, 123]. We note that these fluctuations are different from the flow noise that occurs due to the passing of the high velocity unburnt reacting mixture in confined regions of the combustor. The interaction of flow turbulence with chemical reactions happening in the flame causes fluctuations in the local heat release rate. These heat release rate fluctuations further produce local density fluctuations; thus, they generate velocity fluctuations in the combustor.

Combustion noise is generated by both unconfined (i.e., in the free space) [123] and confined (i.e., inside an enclosure) turbulent flames [41]. There is a significant difference in the characteristics of sound generated by these types of flames. Open flames act as a monopole source of sound, which radiates the sound generated in all directions. In contrast, the sound produced by confined flames has preferred direction due to the reflection from combustor boundaries and waveguide effects [41]. The amplitude spectrum of open flames exhibits broadband frequencies, where the excitation of frequencies is dependent on the oscillations in the reaction volume of the combustor. On the other hand, in the case of confined flames, the amplitude spectrum shows broadband frequencies amplified around the natural modes of the combustor. Here, we will discuss only the characteristics of combustion noise generated by confined turbulent flames. For more information on the descriptions of combustion noise due to open flames, the readers may refer to [4, 12, 59, 61, 106, 118, 122, 123, 131].

Depending on the source of generation of acoustic fluctuations during the state of stable operation, combustion noise is primarily classified into two types: direct noise and indirect noise [8, 25, 27]. Direct noise is a result of the local heat release rate fluctuations occurring due to the volumetric expansion and contraction taking place in the reaction field of the combustor [4, 12, 24, 123]. In contrast, indirect noise is a result of the acceleration of entropy waves (generated from hot spots or flame packets) [71] and also the acceleration of vorticity fluctuations (generated from turbulent fluctuations) [55] through non-uniform flow passages (such as turbine blades or convergent section of a nozzle) of an engine. Thus, direct noise is produced by the flame, which then travels towards the combustor exhaust. In contrast, indirect noise is generated away from the flame, due to the propagation of entropy and vorticity fluctuations (that are generated in the combustion zone) through region of non-uniform flows. In this chapter, we will only focus on characterizing the properties of direct combustion noise generated from the unsteady heat release processes occurring in turbulent flame inside a combustor. For more information on indirect noise, the readers may take a look at the reviews by Candel et al. [8] and Dowling [25].

Combustion noise is a complex phenomenon. Here, complexities primarily arise due to the interaction of several processes (such as combustion, turbulence and acoustics) happening at different timescales in the combustor and their interaction with the system boundaries. However, while modelling combustion noise, the effect of sound waves generated from unsteady combustion as well as the influence of the reflected acoustic fluctuations from boundaries on the heat release rate fluctuations are usually not considered. Sometimes, the effects of background turbulent fluctuations on the generation of acoustic waves are either averaged out or neglected [5, 14, 67]. Thus, the system dynamics during the state of combustion noise is modelled as an acoustic problem by decoupling the effects of underlying hydrodynamic fluctuations. However, we found that such formulation which decouples the flow fluctuations and the acoustic fluctuations is unrealistic, as these fluctuations are inherently coupled in practical turbulent combustors.

Furthermore, the dynamics of the acoustic pressure fluctuations observed during the state of combustion noise are usually assumed to be noisy (random) fluctuations [13, 41]. Sometimes, such noisy fluctuations in the acoustic field are considered as a steady state contaminated by the background stochastic fluctuations arising from the underlying flow turbulence [64, 67, 88]. These effects of turbulent fluctuations on the flame dynamics are modelled by incorporating ad-hoc noise terms in the acoustic equations [5, 14, 88]. In these studies, the stochasticity (or randomness) in fluctuations is usually inferred from the observation of Gaussian distribution of the amplitude of the pressure signal [64]. However, in reality, the sources of combustion noise are deterministic as they are derived from different combustion and fluid mechanic processes, that include flame roll-up, vortex coalescence or impingement, fluid dilatation etc. [15]. Therefore, it is important to understand the actual properties of the state of combustion noise to accurately predict the occurrence of thermoacoustic instabilities, and also to improve the models that capture dynamic transitions of the combustor behavior.

In the following subsections, we systematically characterize the properties of acoustic pressure fluctuations observed during the state of combustion noise in laboratory scale turbulent combustors. We present several tests available in dynamical systems theory to detect the deterministic (chaotic) or random properties existing in the pressure signal during this state [34, 48, 129, 135].

6.2.1 The Phase Space Reconstruction

In Fig. 6.1a, we show the time series of the acoustic pressure fluctuations observed during the state of combustion noise in a turbulent bluff-body-stabilized combustor with a backward-facing step (dump plane) at the inlet. A visual inspection of this signal indicates the presence of low amplitude aperiodicity in the pressure fluctuations. Its amplitude spectrum appears to be broadband (Fig. 6.1b), exhibiting multiple frequency peaks amplified around the fundamental mode of the combustor. Such aperiodicity in the temporal dynamics of the system is a result of spatial asynchrony (or incoherence) observed between the coupling of local heat release rate fluctuations and acoustic pressure fluctuations of a combustor, which we will discuss in detail in Chap. 8.

In Fig. 6.1e, we present the projection of the acoustic pressure fluctuations into a higher-dimensional phase space. The phase space is reconstructed using the Takens' delay embedding theorem [127]. In order to construct the phase space, we first need to obtain optimum values of both time delay (τ) and embedding dimension (D) [1]. We can obtain the optimum value of time delay from the first local minima of the average mutual information, I. For the acoustic pressure signal shown in Fig. 6.1a, the first local minimum value of I is observed at $\tau_{opt} = 1.1$ ms (Fig. 6.1c).

To obtain the minimum embedding dimension for the phase space reconstruction, we can use a method developed by Cao [9]. Figure 6.1d shows the variation of

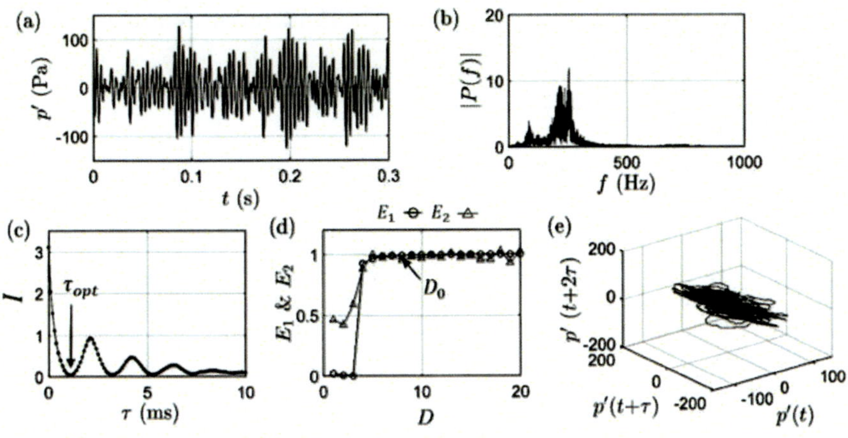

Fig. 6.1 (**a**) Time series of the acoustic pressure oscillations (p'), obtained from a turbulent bluff-body stabilized dump combustor during the state of combustion noise, shows low amplitude chaotic fluctuations. (**b**) The amplitude spectrum shows a broadband of frequencies amplified around the fundamental mode of the combustor. (**c**) The variation of average mutual information I with increasing values of time delay τ, and (**d**) the variation of measures E_1 and E_2 of the Cao's test with increasing values of the dimension, D. (**e**) The reconstructed phase space shows cluttered behavior of the trajectories. The data is described in Nair et al. [86]

measures E_1 and E_2 obtained from the acoustic pressure signal with an increasing value of the attractor dimension (D). As we discussed in Chap. 2, these measures quantify the existence of false nearest neighbors to a state point in the phase space. We notice that the value of E_1 saturates around unity when D is increased above a critical value, which happens when false nearest neighbors of all state points in the phase space are completely resolved. For the acoustic pressure fluctuations shown in Fig. 6.1a, we notice the saturation of the value of E_1 approximately after $D = 7$. Therefore, we can consider $D_0 = 8$ as the embedding dimension required for the phase space reconstruction.

Furthermore, from the variation of the measure E_2 with an increase in D, we can detect the presence of determinism in the acoustic pressure signal [9]. In Fig. 6.1d, we observe that E_2 shows a value lesser than unity for lower values of D and approaches E_1 for higher values of D. This behavior of E_2 indicates the existence of determinism in the dynamics of acoustic pressure fluctuations observed during the state of combustion noise [9]. In contrast, the variation of E_2 remains close to unity for all D for random fluctuations. Thus, the variation of E_1 and E_2 suggests the possibility of moderately high-dimensional deterministic oscillations in the acoustic pressure signal during the state of combustion noise [86].

The reconstructed phase space of the acoustic pressure signal shows a strange attractor (Fig. 6.1e), where we observe that the behavior of the phase space trajectory fills the entire phase space. This is contrary to the behavior of regular attractors, such as limit cycle or quasiperiodic attractor, which shows a definite shape in the phase space (refer to Chap. 2). In order to quantitatively confirm the presence of chaos, we take the help of '0-1 test for chaos' developed by Gottwald and Melbourne [34], as discussed next.

6.2.2 0-1 Test for Chaos

As we discussed in Chap. 2, there are several methods in the literature of dynamical systems theory to detect chaos in a signal [47]. The usual methods employed for the quantitative confirmation of chaos include the calculation of the maximum Lyapunov exponent and the correlation dimension. The observation of characteristics of the power spectrum and the evolution of the phase space trajectory provide qualitative information on the presence of chaos in a signal. For a chaotic signal, the maximum Lyapunov exponent shows a positive value and the correlation dimension exhibits a non-integer number. The power spectrum of a chaotic signal demonstrates a continuous distribution of frequencies (i.e., a broadband spectrum), while the phase space trajectory shows an evolution around a strange attractor.

However, we note that the approach of the Lyapunov exponent is reliable for dynamical systems whose equations are expressed in the form of integer-order differential equations. Therefore, for experimental data or systems with fractional-order differential equations, the detection of chaos through the Lyapunov exponent is not appropriate. Furthermore, the methods of Lyapunov exponent and correlation dimension require the reconstruction of the phase space, that itself requires prior knowledge and optimization of several parameters such as embedding dimension and delay time. Due to the aforementioned limitations of these methods, Gottwald and Melbourne [34] proposed a method called the '0-1 test for chaos' as an alternative to detect chaos in a dynamical system.

For the application of this test, we need a single time series data. We can apply this method to any dynamical system including continuous or discrete systems, integer or fractional-order dynamical systems, or even experimental data. The 0-1 test displays a value near 0 for regular (periodic or quasiperiodic) oscillations and a value near 1 for chaotic oscillations. Thus, we can get a definite conclusion on the existence of chaos in the system dynamics. In the following discussion, we provide a summary of the algorithm of the 0-1 test for chaos [35].

In the 0-1 test for chaos, we first construct two translational variables with sinusoidal basis functions from a single time series of a system variable. From the time evolution of these translational variables in a two-dimensional plane, we then calculate the mean displacement of the trajectory for increasing time duration and examine the growth rate of this displacement.

Let us consider $\varphi(j)$ to be a time series from a dynamical system, where $j = 1, 2, \ldots, N$, and N is the total number of samples in the data. The translational variables derived from $\varphi(j)$ are given as follows,

$$p_c(n) = \sum_{j=1}^{n} \varphi(j)\cos(jc) \quad \text{and} \quad q_c(n) = \sum_{j=1}^{n} \varphi(j)\sin(jc) \tag{6.1}$$

where $n = 1, 2, \ldots\ldots, N$.

The value of c is usually chosen from the interval $(0, \pi)$ [35]. Nair et al. [86] used the value of $c \in (\pi/5, 4\pi/5)$ to detect chaos in the dynamics of combustion noise of a turbulent thermoacoustic system, while Premraj et al. [101] used the value of $c = (\sqrt{5}+1)/2$ for the detection of strange nonchaos in the self-excited dynamics of a laminar premixed burner.

The trajectory obtained for a sequence of n in the (p, q)-plane is referred to as an auxiliary trajectory [35]. The behavior of the auxiliary trajectory helps in distinguishing different dynamical states of a system. For regular (periodic and quasiperiodic) signals, the asymptotic behavior of the auxiliary trajectory is bounded. In contrast, for chaotic signals, the behavior of the trajectory is unbounded (drifting or diffusive). Therefore, to quantify the bounded or diffusive nature of the trajectory, we calculate the mean square displacement, $M(n)$, of the trajectory for different values of n. The value of $M(n)$ is obtained for several values of c as follows,

$$M_c(n) = \frac{1}{N}\sum_{j=1}^{N}\left([p_c(j+n) - p_c(j)]^2 + [q_c(j+n) - q_c(j)]^2\right) \tag{6.2}$$

To get reliable results, we should choose the value of n such that $n \le n_t$ where $n_t \ll N$. The value of $n_t = N/10$ is usually used to obtain good results [35].

For a chaotic signal, the behavior of the trajectory in the (p, q)-plane exhibits a Brownian motion and the variation of $M(n)$ scales linearly with n [35]. On the other hand, for regular (periodic or quasiperiodic) signals, the trajectory is bounded along a circular path in the (p, q)-plane and the variation of $M_c(n)$ remains nearly constant with n [35].

Further, we can modify the mean square displacement of the trajectory to account for better convergence effects as follows [34],

$$D_c(n) = M_c(n) - V_{osc}(c, n) \tag{6.3}$$

where

$$V_{osc}(c, n) = (E\varphi)^2 \frac{1 - \cos nc}{1 - \cos c} \quad \text{and} \quad E\varphi = \lim_{N\to\infty} \frac{1}{N}\sum_{j=1}^{N} \varphi(j) \tag{6.4}$$

In order to quantify the regular or chaotic behavior of a signal, we can calculate the asymptotic growth rate K_c of the change in the modified mean square displacement $D_c(n)$ of the auxiliary trajectory. There are two methods to compute K_c, which include a regression method and a correlation method [34].

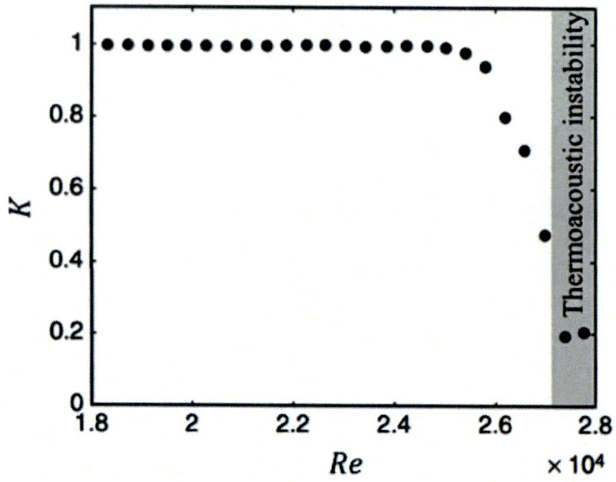

Fig. 6.2 Variation of a measure K of the 0-1 test for chaos with an increase in Reynolds number (Re), obtained for acoustic pressure fluctuations during the onset of thermoacoustic instability in a bluff-body stabilized dump combustor. We notice that the value of K stays near 1 during the state of combustion noise and approaches 0 during the occurrence of thermoacoustic instability. Adapted with permission from Nair et al. [86]

In the regression method, we can calculate the asymptotic growth rate of $D_c(n)$ with increasing values of n on a log-log plot as follows,

$$K_c = \lim_{n\to\infty} \frac{\log D_c(n)}{\log n} \tag{6.5}$$

We can obtain K_c from a straight line fit to the plot between $\log D_c(n)$ and $\log n$.

In the correlation method, we assume vectors $\xi = (1, 2, \ldots, n_t)$ and $\Delta = (D_c(1), D_c(2), \ldots, D_c(n_t))$. Then we can obtain K_c as follows,

$$K_c = \text{corr}(\xi, \Delta) = \frac{\text{cov}(\xi, \Delta)}{\sqrt{\text{var}(\xi)\text{var}(\Delta)}} \in [-1, 1] \tag{6.6}$$

where cov is the covariance of two signals and var is the variance of a signal.

In order to avoid spurious outliers or resonances, we can calculate K_c for various values of c and then obtain median value of K_c (say, K). Thus, we can use the value of the growth rate (K) as an indicator of chaos. The value of K approaches near 1 for a chaotic signal and it is near 0 for a regular (periodic or quasiperiodic) signal.

In Fig. 6.2, we show the variation of K, calculated from the acoustic pressure signal with a change in the Reynolds number Re (a control parameter), during the transition of the system behavior from combustion noise to thermoacoustic instability in a turbulent bluff-body-stabilized dump combustor [86]. During the state of combustion noise (at lower values of Re), we notice that the value of K stays near 1 for a large range of Re. As the system transitions to periodic oscillations, its value displays a gradual fall and eventually approaches 0 at the onset of thermoacoustic instability. From this plot, we can confirm that the dynamics of the acoustic pressure oscillations during the occurrence of combustion noise has the property of chaos.

Furthermore, since the value of K is bounded between 0 and 1, we can use its variation with the change in a control parameter (for example, Reynolds number, Re in Fig. 6.2) as an early warning measure to detect the occurrence of thermoacoustic instability in a system. Using a suitable choice of threshold for the output of the 0-1 test (say, $K \sim 0.9$ where the system behavior shows an indication of loss of chaos [86]), we can make appropriate changes in the operating condition of the system and, thus, prevent the occurrence of thermoacoustic instability. Such precursor technology to evade thermoacoustic instability is independent of the changes in fuel composition, flame stabilization, or combustor geometry. We will describe the details of such precursor technologies in Chap. 12.

Similar to the method of 0-1 test of chaos, we can also use several other methods to detect the presence of chaos or determinism in thermoacoustic systems. These methods include Kaplan-Glass test [86], surrogate test [30, 31, 130], and the calculation of measures such as translational error [32], permutation entropy [31], permutation spectrum [3, 57, 130], and correlation dimension [37, 45, 101]. We encourage readers to look into these references for more information.

6.3 What Is Thermoacoustic Instability?

Thermoacoustic instability (also known as combustion-driven oscillations or combustion instability) refers to the generation of large amplitude, self-sustained periodic oscillations (i.e., loud tonal sound) in a confined reacting flow system. These oscillations occur primarily due to the positive feedback between the subsystems (i.e., flow, flame, and acoustic field) of a combustor. The frequency of thermoacoustic instability is often observed at the unstable natural modes of the combustor. The presence of periodic acoustic waves modulates the flame shape at regular time intervals, leading to periodic heat release rate in the system. During the state of thermoacoustic instability, we observe an in-phase relation between the acoustic pressure fluctuations and the heat release rate fluctuations of the system. Unlike combustion noise, thermoacoustic instabilities are detrimental for the safe operation of the engines and also for the life of the system components [16, 68, 103].

In a turbulent thermoacoustic system, the onset of thermoacoustic instability leads to the periodic shedding of large-scale vortical structures from a flame holding device [10, 29, 40, 51, 62, 70, 97, 108, 112, 119]. These shed vortices accompany cold reactant mixtures and hot products with them. The breaking of such structures results in the periodic heat release rate, which then drive acoustic pressure oscillations in the system. During the state of combustion noise, the dominant frequency of acoustic pressure oscillations is observed to be different from that of the vortex shedding frequency. In contrast, during the onset of thermoacoustic instability, both the acoustic pressure and the vortex shedding frequencies adjust themselves to a common frequency corresponding to the unstable mode of the combustor, known as the phenomenon of *vortex-acoustic lock-in* [10, 92, 117]. Next, we will discuss the transition of the system dynamics from the state of combustion to thermoacoustic instability in turbulent combustors.

6.4 Transition from Combustion Noise to Thermoacoustic Instability in Turbulent Combustors

As we discussed in Sect. 6.2, the dynamics of a turbulent combustion system during the state of combustion noise is chaotic. When the control parameter (for example, Reynolds number of the flow or equivalence ratio) of a system is systematically varied, the transition of these chaotic oscillations to periodic oscillations takes place near the onset of thermoacoustic instability. The accurate prediction of system parameters corresponding to such dynamical transitions and also the values of frequency and amplitude of thermoacoustic instability have been a concern for researchers over the years.

Furthermore, in order to understand the onset of thermoacoustic instability accurately, we need to know the dynamical changes that a system undergoes during the transition from combustion noise (stable operation) to thermoacoustic instability (unstable operation). From such characterization of dynamical states observed prior to thermoacoustic instability, we can devise appropriate strategies that would either forewarn the onset of these instabilities or mitigate these oscillations if they have already established. In the following subsections, we will summarize the recent advances on the understanding of the onset of thermoacoustic instability in turbulent combustors.

6.4.1 Reformulating the Onset of Thermoacoustic Instability as a Loss of Chaos

Traditionally, the onset of thermoacoustic instability in turbulent combustors was described as a direct change of the system behavior from the state of stable operation to unstable operation [64, 97, 119]. During stable operation, the temporal dynamics of the acoustic pressure shows low amplitude aperiodic fluctuations. The power spectrum of these oscillations shows a wide distribution (broadband) of frequencies. In contrast, during unstable operation, the system behavior exhibits limit cycle oscillations with a wide variation in their cycle-to-cycle amplitudes [64, 65, 88]. Such oscillations are observed at discrete frequencies corresponding to natural modes of the combustor.

From the perspective of dynamical systems theory, Lieuwen [64] showed that during the transition from stable to unstable operation, a turbulent system undergoes a Hopf bifurcation. During this bifurcation, a noisy fixed point loses its stability under the influence of strong background disturbances existing in the turbulent field. This, in turn, leads to the emergence of distorted elliptic shaped limit cycle attractor in the phase space. Further, depending on the operating conditions, the system dynamics can exhibit two types of Hopf bifurcations, namely subcritical Hopf or supercritical Hopf [64].

Sometimes, the onset of thermoacoustic instability in turbulent systems has been inferred from the change in the probability density function of the acoustic pressure fluctuations. The probability density function exhibits a transition from nearly Gaussian distribution peaked around the mean amplitude to asymmetric bimodal distribution displaying two peaks around the positive and negative amplitudes of limit cycle oscillations, respectively [64, 88].

Contrary to such traditional views on the onset of thermoacoustic instability, Nair et al. [86, 87] systematically examined the dynamics of stable operation and also the transition from stable to unstable operation using different approaches from dynamical system theory. Nair et al. [86] and Tony et al. [130] showed that the dynamics of acoustic pressure fluctuations observed during stable operation (combustion noise) indeed consists of high-dimensional deterministic chaos and not a noisy fixed point (as discussed in Sect. 6.2). During the onset of unstable operation (thermoacoustic instability), large amplitude ordered (periodic) oscillations gradually emerge from these low amplitude chaotic oscillations. Thus, we can view this onset of thermoacoustic instability as a loss of chaos in the system dynamics. Further, Nair et al. [87] discovered that this gradual loss of chaotic oscillations during the onset of thermoacoustic instability happens via a state of intermittent oscillations.

Note that the transition of chaos to thermoacoustic instability is associated with a change of spectral properties of acoustic pressure fluctuations from a broadband of frequencies to a few discrete peaks. Pavithran et al. [89] referred to this behavior of re-distribution of energy in the spectrum during at the onset of thermoacoustic instability as *spectral condensation*. Such a condensation of energy distributed over a wide range of frequencies to a few frequencies during the occurrence of order (periodicity) is analogous to the condensation phenomenon observed in classical and quantum systems [19, 53]. Furthermore, they found an inverse power law for the variation of spectral measures with the power of the dominant peak during spectral condensation, which we elaborate in Chap. 13.

6.4.2 Intermittency Route to Thermoacoustic Instability

Although the presence of intermittent oscillations prior to thermoacoustic instability has been accepted as a distinct dynamical state only recently in the thermoacoustic literature [87], we can find verbal description of similar dynamical behaviors in previous studies as well [14]. Such dynamical states were indicated as bursting [14] or beating [64] in different turbulent thermoacoustic systems. Clavin et al. [14] clearly described the presence of similar dynamical states in their article as, *"The pressure–time traces observed in combustion chambers of liquid-propellant rocket engines exhibit acoustic oscillations, which sometimes have an erratic evolution of the amplitude* ([2], *especially* [26]). *This random character is more pronounced in the vicinity of critical conditions for the occurrence of high-frequency combustion instabilities and may exhibit itself there in the form of random pressure bursts of strong amplitude.*"

Using the approach of dynamical systems theory, Nair et al. [82, 87] systematically characterized intermittent oscillations observed prior to thermoacoustic instability in turbulent thermoacoustic systems. They showed that such intermittent oscillations consist of bursts of large amplitude periodic oscillations amidst epochs of low amplitude chaotic oscillations and referred to this state as 'intermittency' [99]. As the system dynamics approaches the state of thermoacoustic instability, they noticed an increase in the duration of bursts of periodic oscillations in the intermittency signal. This behavior further suggests that the occurrence of thermoacoustic instability in turbulent combustors from the state of combustion noise is not abrupt, but happens gradually.

Thus, Nair et al. [87] replaced the traditional paradigm for the onset of thermoacoustic instability in turbulent combustors,

$$\text{Stable operation} \rightarrow \text{Unstable operation}$$

with a new paradigm,

$$\text{Combustion noise} \rightarrow \text{Intermittency} \rightarrow \text{Thermoacoustic instability.}$$

We refer to this dynamical transition of the system behavior as the 'intermittency route to thermoacoustic instability' [80]. In recent times, we observe several studies showing the existence of intermittency prior to full-blown thermoacoustic instability in different combustion systems [18, 21, 22, 28, 32, 33, 36, 49, 54, 58, 60, 73, 78, 92–94, 100, 109, 110, 114, 126, 132, 137]. Further, we also observe the presence of intermittency prior to oscillatory instabilities in general systems, such as aeroacoustic [85], aeroelastic [134], and engine compressor [52], which we will describe in more detail in Chap. 13.

In Fig. 6.3, we show a sequence of events in the acoustic pressure fluctuations during the intermittency route to thermoacoustic instability in a turbulent bluff-body stabilized dump combustor [77, 87]. At low values of the flow Reynolds number (Re), we observe a state of combustion noise, where the acoustic pressure signal displays low amplitude chaotic oscillations (Fig. 6.3a). With an increase in the Reynolds number, we notice the occurrence of bursts of large amplitude periodic oscillations amidst epochs of the low amplitude chaotic oscillations at apparently random intervals of time. This state is referred to as intermittency. In the vicinity of the onset of intermittency, we notice a few epochs of bursts of periodic oscillations in the pressure signal (Fig. 6.3b). However, as the Reynolds number approaches the point of the onset of thermoacoustic instability, we observe an increase in the duration of such periodic oscillations during bursts (Fig. 6.3c). Once the value of the Reynolds number crosses a critical value, the system behavior shows the occurrence

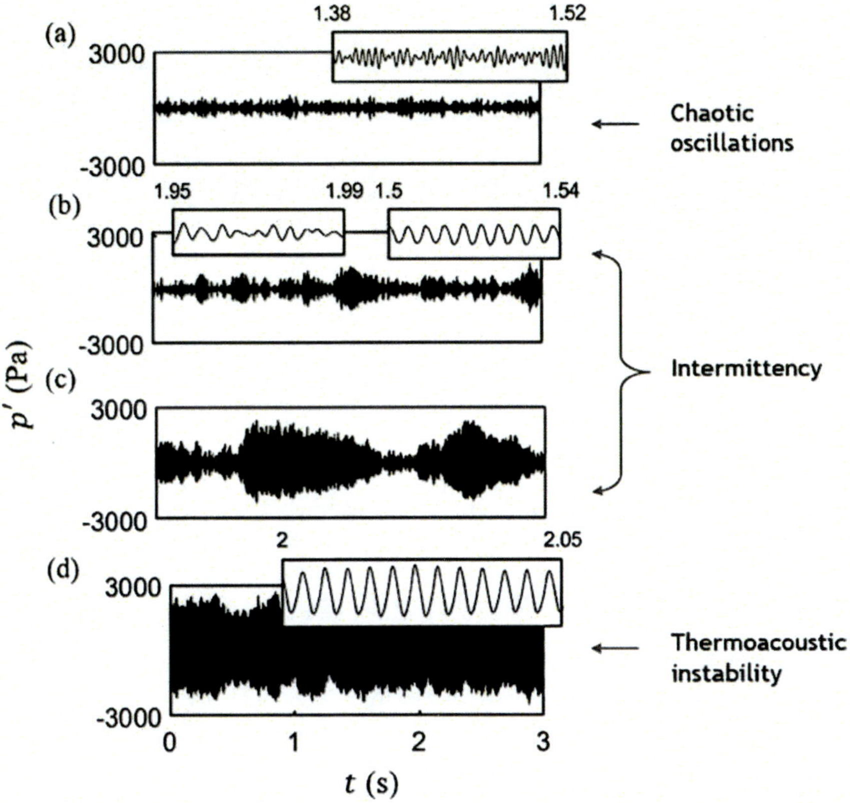

Fig. 6.3 The intermittency route to thermoacoustic instability observed in a turbulent combustor showing the states of (**a**) combustion noise, (**b, c**) intermittency, and (**d**) thermoacoustic instability. The data are presented in Nair et al. [87]

of self-sustained limit cycle oscillations (thermoacoustic instability) in the acoustic pressure signal (Fig. 6.3d). Next, we will characterize the dynamical properties of the intermittency signal observed in turbulent combustors.

6.4.2.1 Characteristics of the Intermittency Signal

As we discussed in Chap. 2, the state of intermittency occurs when the system behavior is under the influence of multiple attractors. This multi-stable behavior of the system causes the phase space trajectory to alternately switch between these distinct attractors. For the state of intermittency observed in a thermoacoustic system, we notice the switching of the system behavior between the attractors of chaotic (observed during combustion noise) and periodic (observed during thermoacoustic instability) oscillations (Fig. 6.4). Due to the presence of an underlying

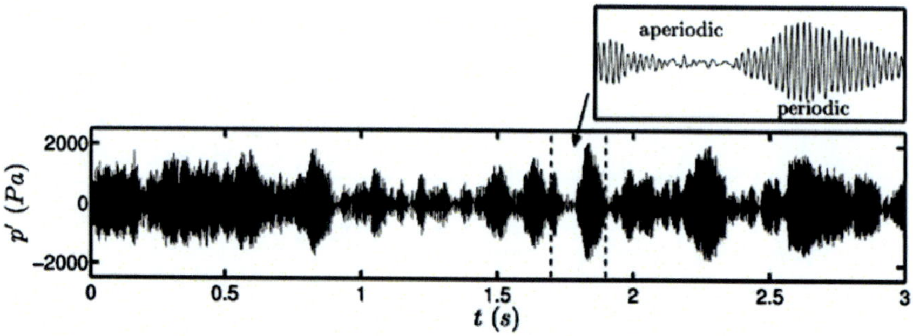

Fig. 6.4 Time series of acoustic pressure fluctuations during the state of intermittency observed prior to thermoacoustic instability in a turbulent combustor. It consists of bursts of high amplitude periodic oscillations amidst low amplitude aperiodic fluctuations, as seen in the inset. Adapted with permission from Nair and Sujith [82]

turbulent flow, the occurrence of bursting events appears apparently randomly in the intermittency signal. The presence of turbulence further modulates the amplitude envelope of such signals. As a result, we do not observe definite patterns for the growth and decay of oscillations in the bursts [128].

To reveal the dynamical features of intermittency, we can project the time series of the acoustic pressure fluctuations (Fig. 6.5a) in the higher-dimensional phase space (Fig. 6.5d) using Takens' delay embedding theorem [127]. We can obtain the optimum values of delay time and embedding dimension required for the phase space reconstruction using appropriate methods mentioned in Chap. 2. Figure 6.5a shows the variation of average mutual information I with increasing values of time lags τ for the intermittency signal. The oscillatory behavior of I is the result of the deterministic characteristics of the signal. We use the value of delay time corresponding to the first local minimum of I, i.e. $\tau_{opt} = 1$ ms, for the calculation of the embedding dimension and the phase space reconstruction.

We can calculate the embedding dimension using Cao's method [9]. In Fig. 6.5b, we show the variation of parameters E_1 and E_2 of this method against increasing value of phase space dimension D. As the value of E_1 saturates after the dimension of 9, we choose $D = D_{0=10}$ as the embedding dimension.

The reconstructed phase space of the intermittency signal is shown in Fig. 6.5d, where we notice the coexistence of two distinct attractors. A chaotic attractor is present at the center and is surrounded by a distorted ring of periodic attractor. Nair et al. [82] showed that these two attractors are connected through a homoclinic orbit [121]. For a homoclinic orbit, the unstable manifold of a hyperbolic equilibrium point joins with its own stable manifold [124]. In Fig 6.5d, the presence of homoclinic orbit leads to transition of the phase space trajectory from a low amplitude chaotic attractor to a large amplitude periodic attractor and back to the chaotic attractor. Thus, the trajectory spirals interchangeably between the central core of the chaotic attractor and the distant ring of the periodic attractor.

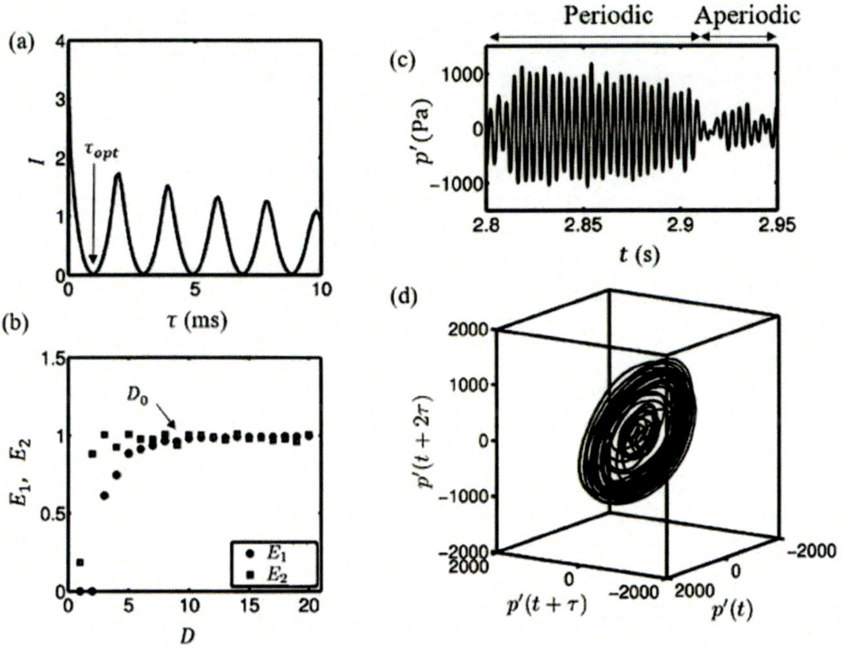

Fig. 6.5 Details of the procedure to construct a phase space from the acoustic pressure signal shown in (**c**) for the state of intermittency. (**a**) The optimum delay (τ_{opt}) is estimated from the first local minimum of average mutual information I. (**b**) The variation of measures E_1 and E_2 with dimension D is used to find the embedding dimension (D_0). (**d**) The reconstructed phase portrait shows the switching of the trajectory between two distant attractors. Adapted with permission from Nair et al. [87]

According to Tandon et al. [128], the intermittent bursting observed prior to thermoacoustic instability is the result of a complex interdependence between the slow and the fast timescales existing in the system. The timescales of oscillations during bursts are fast and usually determined by the frequency of the unstable acoustic modes of the combustor. In contrast, timescales of the amplitude modulation observed during bursts or the occurrence of bursting events in a signal are slow and controlled by the flow (hydrodynamics) timescales present in the system [87]. Here, the hydrodynamic timescales are governed by both the broadband nature of the underlying turbulence and the narrowband dynamics of the unsteady vortex formation processes (i.e., vortex shedding, roll-up, coalescence, or impingement). In a typical thermoacoustic system with underlying turbulent flow field, the acoustic field acts as a fast subsystem and the hydrodynamic field acts as a slow subsystem. In such systems, the intermittent oscillations arise if the acoustic subsystem is modulated by hydrodynamics over a slow timescale [87]. These modulations from the hydrodynamic field further switch the dynamics of the acoustic field back-and-forth across the Hopf point [128].

We note that, while modeling thermoacoustic systems, if the effects of the interdependence of the acoustics and the hydrodynamics are ignored or if the effects of turbulence in the system are averaged out, then the bifurcation of the acoustic pressure fluctuations would just show a direct transition from a fixed point to limit cycle oscillations, i.e., a Hopf bifurcation, giving no room for the occurrence of intermittency in the system. Therefore, turbulent fluctuations need to be included while modeling the system dynamics as they play an important role in the generation of intermittent fluctuations prior to the onset of thermoacoustic instability.

Furthermore, we notice that, similar to the occurrence of intermittency in turbulent thermoacoustic systems, a few studies on laminar or low turbulence thermoacoustic systems have also shown the existence of intermittency or bursting oscillations prior to thermoacoustic instability [50, 91, 136]. In such systems, the interaction between slow scales in the thermal field and fast scales in the acoustic field is found to give rise to bursting dynamics during intermittency [50, 136]. Although we have gained some level of understanding on the physics of the occurrence of intermittency in both laminar and turbulent systems [128], deeper investigations in future are needed to reveal the role of slow-fast interactions in causing intermittent fluctuations in such systems.

6.4.2.2 Bifurcation Analysis of Intermittency Route to Thermoacoustic Instability

As we discussed in Chap. 2, a bifurcation diagram provides a qualitative method to detect dynamical changes that happen in a system due to the variation of a control parameter. Typically, a bifurcation diagram is obtained by plotting the variation of local peaks in the signal with a quasi-static change in the control parameter, as presented in Chap. 4. Note that during the quasi-static variation of the control parameter, we need to provide sufficient waiting time, both to remove the transient dynamics associated with the change in the control parameter and to acquire the experimental data, before the control parameter is changed to the next value. In contrast, during the rate-dependent variation of the control parameter, both the variation of the control parameter and the acquisition of the data from a system are simultaneously performed as a function of time.

For noise-free limit cycle oscillations, as the amplitude of these oscillations remains constant in time, the bifurcation diagram shows a single point. For period-2 oscillations, as the signal repeats its behavior after two oscillation cycles, the corresponding bifurcation diagram shows two distinct points. Since the quasiperiodic or chaotic oscillations are aperiodic in time, we observe a scatter of points in the bifurcation diagram.

Alternatively, we can construct a bifurcation diagram by plotting the variation of a measure of the amplitude of the pressure signal (for example, a global maxima, a median of all local peaks, or the root mean square value of the signal) as a function of the change in a control parameter (see Chap. 3). The bifurcation diagrams obtained using this method help in quantifying the amplitude variation of the signal during

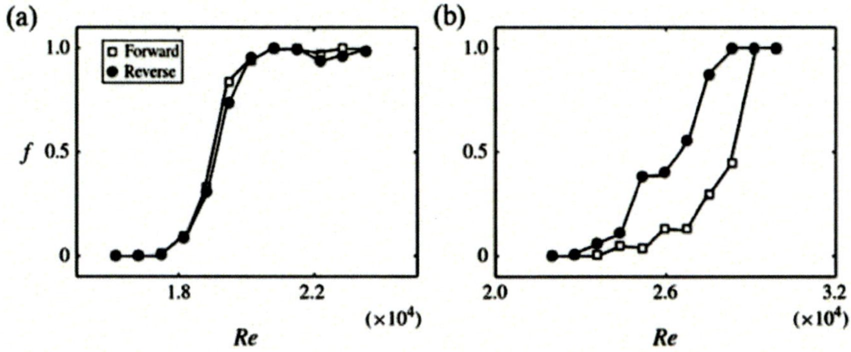

Fig. 6.6 Bifurcation diagrams obtained for (**a**) a swirl-stabilized and (**b**) a bluff-body stabilized turbulent combustors, where the variation of normalized burst count (f) in the acoustic pressure signal is plotted against Reynolds number (Re) during the transition from combustion noise to thermoacoustic instability. We observe that the acoustic subsystem of a swirl-stabilized combustor exhibits supercritical Hopf bifurcation and that of a bluff-body combustor displays subcritical Hopf bifurcation. Adapted with permission from Nair and Sujith [87]

the onset of thermoacoustic instability [91]. However, such bifurcation diagrams do not provide any information about the underlying dynamics exhibited by a system.

In the case of turbulent combustors, we observe a wide variation in the cycle-to-cycle amplitude of the pressure signal during the states of combustion noise and intermittency. Further, the amplitude of the pressure signal is a system dependent property, as it is highly dependent on the operating conditions, and it takes different values for different combustors. In such a situation, we can use a system independent method to compute the bifurcation diagram. In this method, we can construct a bifurcation diagram by plotting the variation of the number of peaks (N) that are above a fixed amplitude threshold in a signal as a function of the control parameter [87]. Here, we can choose the reference threshold as the amplitude that is just above the maximum amplitude of the acoustic pressure oscillations during the state of combustion noise. The value of N is then normalized by the total number of positive peaks in the same signal, i.e., $f = N/N_{tot}$. Thus, the measure f takes a value of zero during the state of combustion noise and a value of 1 during the state of thermoacoustic instability. Therefore, f can serve the purpose of detecting the proximity of thermoacoustic instability in the intermittency regime of the system dynamics.

In Fig. 6.6, we show the variation of the burst count (f) in the acoustic pressure signal with an increase in the value of Reynolds number (Re), for a swirl-stabilized (Fig. 6.6a) and a bluff-body stabilized (Fig. 6.6b) combustor, during the transition from combustion noise to thermoacoustic instability. We notice a smooth change in the values of f for both the forward and the reverse paths of the bifurcation diagram. This smooth variation is the result of the presence of intermittency in the system. During the state of intermittency, bursts occasionally cross the threshold, leading to an increase in the value of f. The shape of the forward and the return paths of f resembles a sigmoid (S-shaped) curve.

We notice a difference in the criticality (type) of bifurcation for both the combustors. For the swirl combustor, both the forward and the reverse paths collapse on top of each other. In contrast, for the bluff-body combustor, we notice the presence of hysteresis in the bifurcation diagram. Thus, these bifurcation diagrams indicate the possibility of supercritical Hopf bifurcation in the acoustic subsystem of the swirl combustor and subcritical Hopf bifurcation in the acoustic subsystem of the bluff-body combustor [87].

Note that these types of bifurcations are not universal to all kinds of such combustors, and they can change from one combustor to another depending on the operating conditions. Further, the variation of different control parameters (for example, equivalence ratio, flame location, or preheat temperature of reactants) could have different effects on the bifurcation plots for the transition from stable to unstable operation in the same thermoacoustic system.

6.4.2.3 Phase Space and Recurrence Analysis of the Intermittency Route to Thermoacoustic Instability

As mentioned previously, the dynamical transitions of a thermoacoustic system during the onset of thermoacoustic instability can be well-understood from the projection of the acoustic pressure signal in the higher-dimensional reconstructed phase space. In Fig. 6.7, we show the reconstructed phase portraits of the acoustic pressure fluctuations observed during different dynamical states, i.e., combustion noise, intermittency, and thermoacoustic instability, in a turbulent combustor. During the state of combustion noise (Fig. 6.7a), the phase space trajectories exhibit cluttered (strange) dynamics; however, the trajectory is bounded around a smaller sized chaotic attractor. For the state of intermittency (Fig. 6.7b), the phase space shows the presence of two distant stable attractors, where the trajectory from the central chaotic attractor first spirals towards a distant limit cycle attractor. After making several periodic revolutions around the limit cycle attractor, the trajectory is re-injected back with a spiral motion towards the central chaotic attractor. During the state of thermoacoustic instability (Fig. 6.7c), the phase space trajectory revolves around the limit cycle attractor and repeats its behavior after a period of the signal. Thus, we notice an emergence of ordered oscillations from the disordered dynamics during the onset of thermoacoustic instability.

Next, we will discuss the use of recurrence plots to characterize the distinguishing features of these aforementioned dynamical states observed during the onset of thermoacoustic instability. As we discussed in Chap. 2, recurrence plots are the graphical representation of the recurrence behavior of a trajectory to its previously visited location in the phase space. In these plots, the recurrences of the phase space trajectory are indicated by black points and non-recurrences are represented through white points. Further, such black points can be arranged into three orientations, which include diagonal lines, vertical/horizontal lines, and isolated points [72].

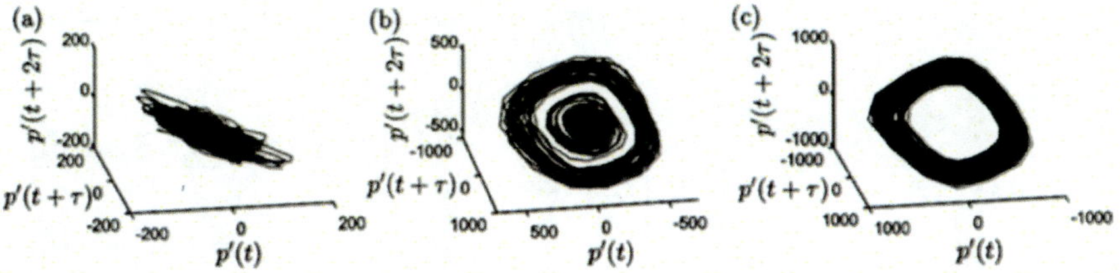

Fig. 6.7 The reconstructed phase space of the acoustic pressure signal for the states of (**a**) combustion noise, (**b**) intermittency, and (**c**) thermoacoustic instability in a turbulent combustor. During the onset of thermoacoustic instability, a chaotic attractor observed for combustion noise transitions to a regular limit cycle attractor. During intermittency, we observe the coexistence of these attractors. The data are presented in Nair et al. [87]

Each of these arrangements of black points indicates specific features of the system dynamics such as determinism (diagonal lines), trapping (vertical/horizontal lines), and randomness (isolated points).

During the state of combustion noise (Fig. 6.8a), the recurrence plot of the acoustic pressure fluctuations shows the existence of grainy structures of black points along with very short diagonal and vertical/horizontal lines. This is an indication of lower levels of determinism existing in aperiodic oscillations of the signal [72]. In the case of intermittency (Fig. 6.8b), we observe the presence of pronounced black and white patches in the recurrence plot. Such black patches occur when the phase space trajectory is trapped in the region of low amplitude chaotic attractor, and white patches occur when the phase space trajectory is in the region of high amplitude limit cycle attractor. During the occurrence of thermoacoustic instability (Fig. 6.8c), the recurrence plot shows the presence of continuous (uninterrupted) diagonal lines separated by a period of the oscillations. The presence of such diagonal lines is an indication of the existence of the highest level of determinism in the signal.

We can derive several measures from the quantification of organization of black points in the recurrence plot (as mentioned in Chap. 2). The variation of these measures with the control parameter during the transition from combustion noise to thermoacoustic instability would help in devising several precursors that can forewarn the onset of thermoacoustic instability [87]. These precursors are discussed elaborately in Chap. 12.

6.5 Intermittency Route to Flame Blowout

In this section, we will briefly discuss the dynamical transition of a turbulent thermoacoustic system post the state of thermoacoustic instability, where the system behavior approaches lean flame blowout [132]. Similar to thermoacoustic instability, flame blowout is an important phenomenon observed in a thermoacoustic

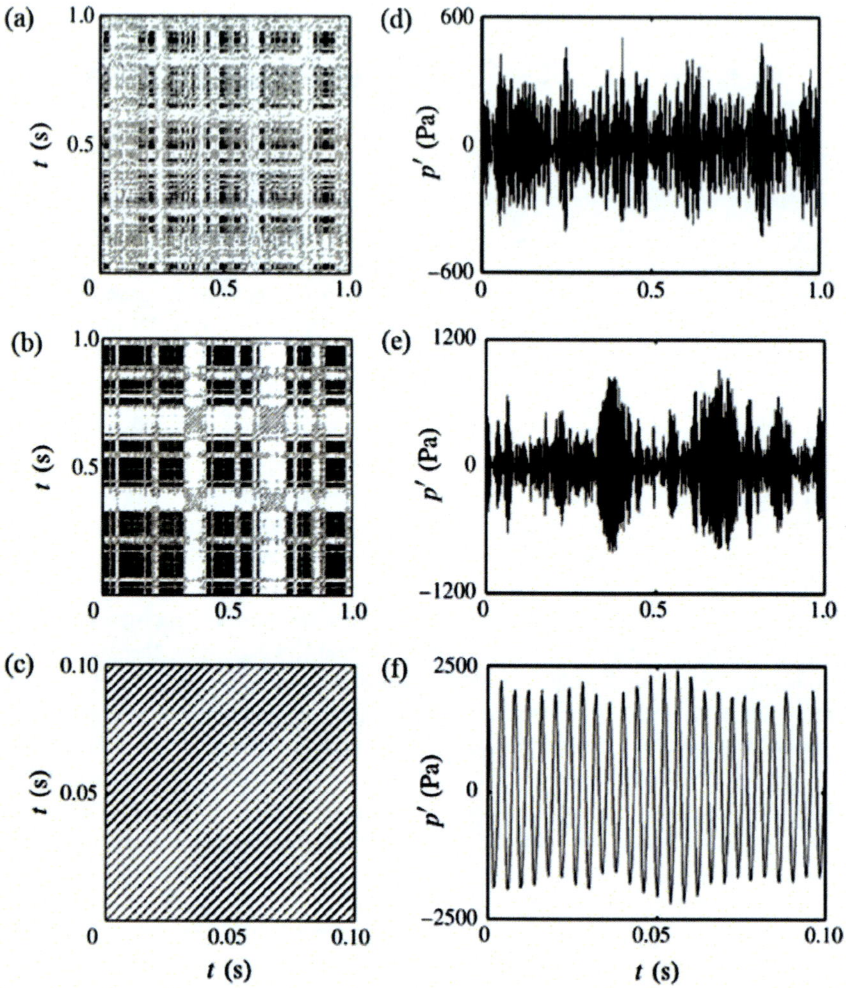

Fig. 6.8 Recurrence plots and the corresponding time series of the acoustic pressure fluctuations for the states of (**a**, **d**) combustion noise, (**b**, **e**) intermittency, and (**c**, **f**) thermoacoustic instability, respectively, in a turbulent combustor. The recurrence plot of combustion noise shows a grainy pattern of black points and that of thermoacoustic instability shows continuous diagonal lines. The recurrence plot of intermittency shows the alternate occurrence of black and white patches, observed for aperiodic and periodic oscillations, respectively. Reproduced with permission from Nair and Sujith [87]

system when a fuel-air mixture is near the lean flammability limit of combustion [30, 74, 107, 138]. Blowout is identified as the complete cessation (extinguishment) of the flame in a combustor. During blowout, the flame becomes more susceptible to high velocity perturbations imposed by the flow [63]. The occurrence of blowout is characterized by an indicator called the Damkohler number [63, 69, 120]. This number accounts for the ratio of the residence (diffusion) timescale of the reactants to the chemical (reaction) timescale of the combustion processes, which is expressed as,

$$Da = \frac{\tau_{\text{residence}}}{\tau_{\text{chemical}}} \tag{6.7}$$

Here, $\tau_{\text{residence}}$ indicates the time for which reactants stay in the combustor volume and τ_{chemical} represents the time required for the reaction to occur.

For lower values of Damkohler number, the residence timescale dominates over the chemical timescale. In such a scenario, the absence of sufficient reaction time leads to the quenching of flame in the combustor. Another factor that has been used as an indicator to blowout is the ratio of the flow speed to the flame speed [95, 105]. This parameter characterizes the collective flamelet extinction event [11], which happens at locations where this ratio exceeds one in the reaction field of the combustor. The interaction of flame fronts with the vortices in the shear layer causes the local stretch rates in the flow to exceed the extinction stretch rates. This results in the local annihilation of flame surfaces. This can alternately be viewed as the local growth of flame holes in the reaction field of the combustor [115].

Traditionally, the onset of flame blowout was considered as a loss of static stability and the occurrence of thermoacoustic instability was presumed to be the loss of dynamic stability [66]. This is because the former case is related to the loss of stability of the reaction field and the latter case is related to the occurrence of large-amplitude periodic oscillations in the system. Contrary to such static features attributed to blowout phenomenon, Kabiraj and Sujith [44] showed that the transition to flame blowout is a dynamically rich event. They showed the existence of intermittency prior to blowout in a premixed laminar burner. During the state of intermittency, the system dynamics alternately switches between regimes of low amplitude periodic oscillations and bursts of large amplitude aperiodic oscillations [44]. Subsequently, several studies have reported the occurrence of intermittency prior to flame blowout in both laminar as well as turbulent combustors [20, 22, 31, 32, 38, 74–76, 79, 81, 84, 111, 132]. Furthermore, Premraj et al. [102] identified the presence of extreme events during the state of intermittency observed prior to flame blowout. Here, extreme events characterize the rare occurrence of very large amplitude bursts above a particular threshold in the system.

In Fig. 6.9, we show the time series of the acoustic pressure fluctuations observed during the transition of a turbulent combustor from the state of combustion noise to blowout via thermoacoustic instability, when the equivalence ratio (control parameter) is decreased from stoichiometric combustion to lean blowout. When the value of the equivalence ratio is near one (Fig. 6.9a), the acoustic pressure

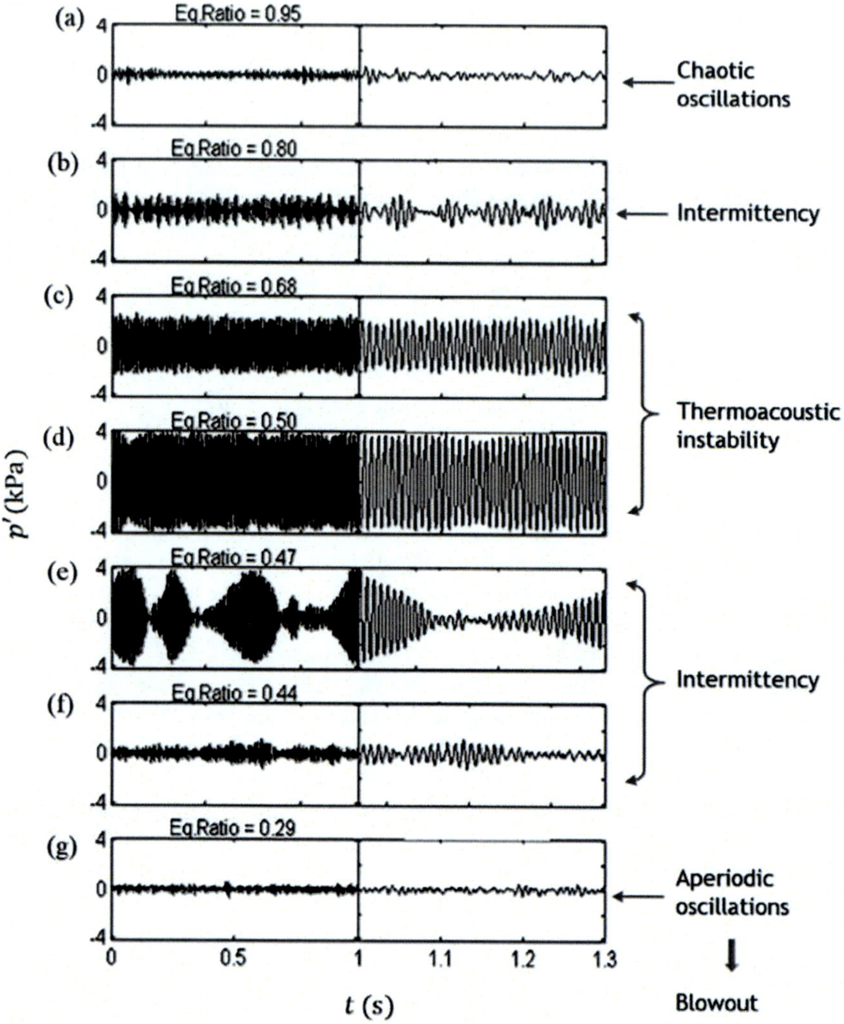

Fig. 6.9 The time series of acoustic pressure oscillations during the transition from the state of combustion noise to flame blowout in a bluff-body-stabilized turbulent combustor. This dynamical route consists of states of (**a**) combustion noise, (**b**) intermittency prior to thermoacoustic instability, (**c**, **d**) low and high amplitude thermoacoustic instability, (**e**, **f**) intermittency post thermoacoustic instability, and (**g**) very low amplitude aperiodic oscillations observed prior to flame blowout. Adapted with permission from Unni and Sujith [132]

fluctuations show low amplitude aperiodic fluctuations called combustion noise. A decrease in the equivalence ratio leads to the occurrence of intermittency (Fig. 6.9b) and then the transition to thermoacoustic instability of low (Fig. 6.9c) and high (Fig. 6.9d) amplitude in the acoustic pressure oscillations. As the equivalence ratio is further decreased, we notice the presence of intermittent oscillations consisting of large-amplitude periodic oscillations amidst epochs of low amplitude aperiodic fluctuations (Fig. 6.9e) post thermoacoustic instability. The amplitude of the periodic oscillations observed in bursts shows a significant reduction as the equivalence ratio is further reduced (Fig. 6.9f). When the value of the equivalence ratio is very low, the system behavior transitions to a state of low amplitude aperiodic fluctuations (Fig. 6.9g), and a subsequent reduction in the equivalence ratio leads to blowout of the flame in the combustor. Therefore, we can view the presence of intermittency in the acoustic pressure signal post thermoacoustic instability as a precursor to blowout event in turbulent combustors.

Thus, the route to lean flame blowout in a turbulent thermoacoustic system traces the following sequence of events [132],

$$\text{Combustion noise} \rightarrow \text{Intermittency} \rightarrow \text{Thermoacoustic instability} \rightarrow$$
$$\text{Intermittency} \rightarrow \text{Low amplitude aperiodic fluctuations} \rightarrow \text{Blowout}$$

Note that this route to flame blowout is not universal, and the dynamical states observed during the transition to blowout depends on the type of system and also the operating conditions of the same system [30–32, 74, 132]. For example, under some operating conditions (e.g., at lower fuel flow rates), Unni et al. [132] showed that the transition to flame blowout can also occur even in the absence of thermoacoustic instability. However, the blowout event is always preceded by the state of intermittency.

6.6 Type of Intermittency En-Route to Thermoacoustic Instability and Its Scaling Laws

From the discussions in the previous sections, we understood that the onset of two dynamically independent events in a turbulent combustor, i.e., thermoacoustic instability and flame blowout, are preceded by the state of intermittency. Thus, intermittency happens to be an important dynamical state in turbulent thermoacoustic systems.

Note that the intermittency route to thermoacoustic instability observed in thermoacoustic systems is qualitatively different from the intermittency route to chaos observed in laminar thermoacoustic systems [38] or in general nonlinear systems [42, 113], as discussed in Chaps. 4 and 2, respectively. Pomeau and Maneville [99] were the first to report the intermittency route to chaos, where they observed the change of the system behavior from the state of limit cycle to chaos via intermittency. During the state of intermittency, bursts of chaotic oscillations occur in between regions of periodic oscillations in an apparently random manner. This is the opposite of what occurs during the transition from chaos to limit cycle oscillations through intermittency in turbulent thermoacoustic systems.

According to Pomeau and Manneville [99], further characterization of the intermittency state is performed by identifying the type of intermittency that a system exhibits during the transition from one stable state to another. As discussed in Chap. 2, intermittency can be classified into three basic types, known as type-I, type-II, and type-III intermittency. Each type of intermittency is associated with three distinct types of bifurcations, namely saddle-node bifurcation, subcritical Hopf bifurcation, and inverse period-doubling bifurcation, respectively [113]. Identifying the type of intermittency from experiments would help in appropriately modeling this phenomenon for a system.

There are several methods to confirm the type of intermittency in a system [113]. One of the easy methods to detect the type of intermittency is the use of recurrence plots [56]. In thermoacoustic systems involving laminar flow field [38, 44] and turbulent flow field [82, 133], the possibility of type-II intermittency has been reported using the method of recurrence plots. In Figs. 6.10a and b, we show the comparison of the recurrence plots observed prior to and post thermoacoustic instability, respectively [133]. For these intermittency signals, the recurrence plots show the alternate occurrence of black and white patches. The magnified portion of such recurrence plots demonstrates a perforated black patch in the lower left corner along with the diagonal lines in the upper right corner. This is a characteristic of type-II intermittency [56]. Thus, intermittent oscillations observed both prior to and post thermoacoustic instability has similar dynamical characteristics of possibly type-II intermittency [20, 133].

In another method, the type of intermittency is confirmed by identifying the scaling behavior of the occurrence of bursting oscillations in the intermittency signals, observed from the point of the onset of intermittency to the onset of only chaotic oscillations in a system [99]. The value of the scaling exponent, obtained from a log-log plot between the mean duration of the laminar phase in a signal and the normalized value of the control parameter, helps in distinguishing different types of intermittency. According to Pomeau and Manneville [99], for type-I intermittency, the scaling exponent is $-1/2$ and, for type-II and type-III intermittency, its value is -1. In order to confirm the type of intermittency using this method of scaling, we need to have longer time series data. This is in contrast with the requirement of a short time series in detecting the type of intermittency using the method of recurrence plots (as discussed before) [56]. By examining the scaling, Pawar et al. [91] reported the existence of type-II intermittency during the transition from combustion noise to thermoacoustic instability in a laboratory scale low turbulence spray combustor.

Contrary of the aforementioned method of scaling, Pavithran et al. [90] proposed a novel method to detect the scaling behavior for the intermittency encountered en route to thermoacoustic instability. In this method, the scaling exponent is obtained by plotting the logarithmic variation of the spectral amplitude (A) corresponding

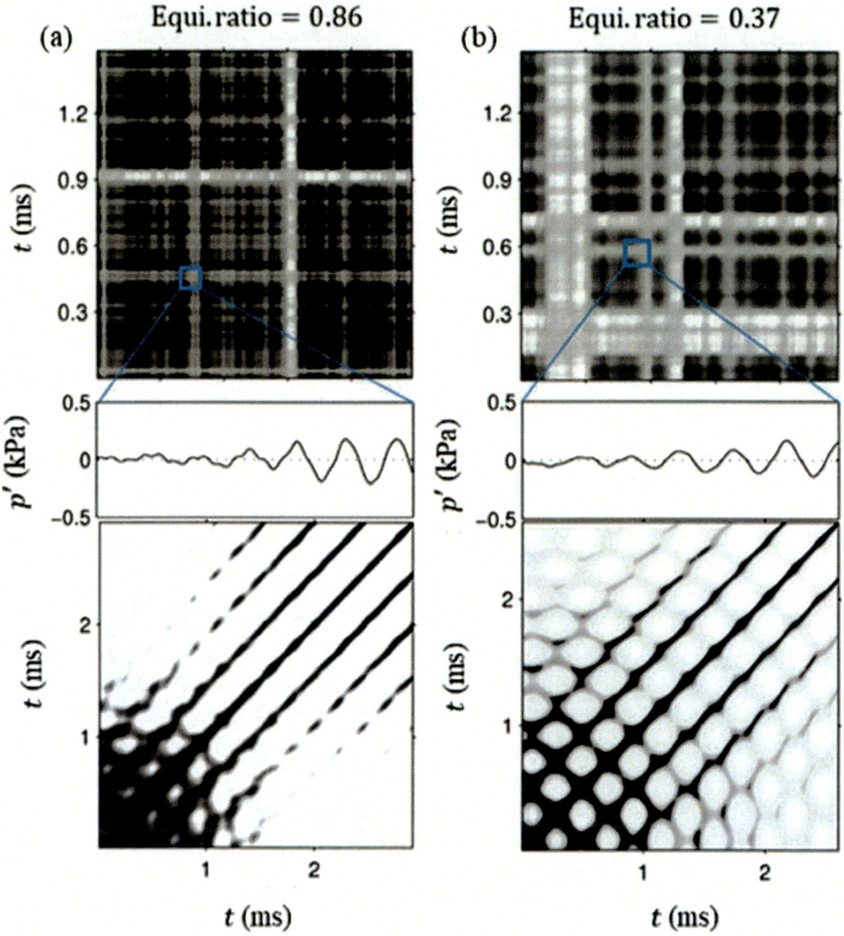

Fig. 6.10 Recurrence plots of intermittent oscillations observed (**a**) prior to thermoacoustic instability and (**b**) prior to lean flame blowout in a turbulent combustor. Both the recurrence plots show characteristics of type-II intermittency. Adapted with permission from Unni and Sujith [133]

to the dominant frequency of a system variable against the Hurst exponent (H). Here, the Hurst exponent (H) measures the fractal behavior of a signal. It is basically a scaling exponent of a plot between the root mean square of the standard deviation of fluctuations in the signal and the increasing size of duration (or time window) of such fluctuations. The value of the Hurst exponent is 0.5 for uncorrelated signals, between 0 and 0.5 for anti-persistent signals, and between 0.5 and 1 for persistent signals [46]. The value of H is observed to be close to 0.25 for chaotic oscillations (during the state of combustion noise) and near 0 for periodic oscillations (thermoacoustic instability) [83]. We will be discussing the details of the Hurst exponent in Chap. 10.

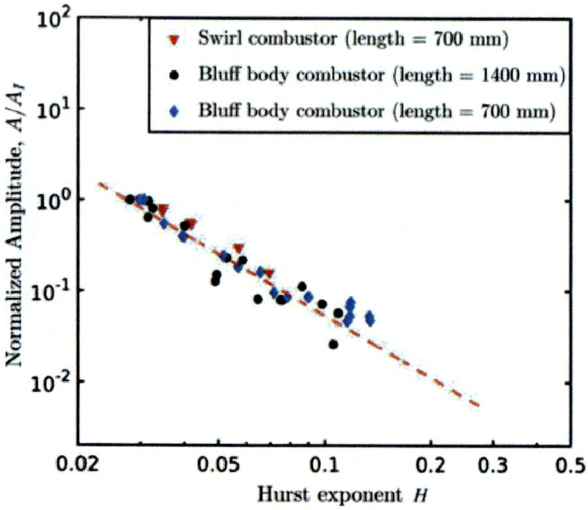

Fig. 6.11 Variation of the normalized spectral amplitude (A/A_I) of the acoustic pressure signal with the corresponding the Hurst exponent (H), during the onset of intermittency to the onset of thermoacoustic instability, for different turbulent combustors. This plot shows an inverse power law with a constant scaling exponent around -2 for all systems. Adapted with permission from Pavithran et al. [90]

In Fig. 6.11, we show the variation of normalized spectral amplitude (A/A_I) of a pressure signal as a function of the Hurst exponent (H) on a log-log plot, for different configurations of turbulent combustors. The calculation is performed for a range of pressure signals observed from the onset of intermittency to the onset thermoacoustic instability in the system. Here, A_I is the spectral amplitude of the dominant frequency corresponding to the state of thermoacoustic instability.

We observe that the variation in normalized spectral amplitude with the Hurst exponent follows a single straight line, regardless of the type of combustor (see Fig. 6.10). This indicates the presence of inverse power-law in the variation of intermittency statistics in these systems. We observe the scaling exponent of the power-law variation is around -2, and this value of the exponent is independent of the dominant frequency (controlled by the length of the combustor) or the physics of the reacting flow (controlled by the type of flame holding mechanism) of the system. This, in turn, suggests that turbulent systems exhibiting intermittency prior to thermoacoustic instability may display a universal scaling of -2 (or inverse square law) in the plot between the spectral amplitude and the Hurst exponent (i.e., $A \propto H^{-2}$). In Chap. 13, we show that such an inverse square power-law is also valid for other turbulent systems, such as aeroacoustic and aeroelastic systems, that exhibit oscillatory instabilities.

According to Pavithran et al. [90], such scaling behavior is unique to turbulent systems exhibiting emergence of order from chaos due to self-organization. However, a similar scaling behavior is not observed for models that display intermittency route to chaos [99]. Therefore, systems that undergo a transition from low amplitude chaotic oscillations to large amplitude periodic oscillations via intermittency might exhibit a different type of intermittency, other than the traditional types of intermittency as mentioned before. Hence, this topic needs further rigorous investigation.

To summarize, in this chapter, we discussed the phenomenon of the onset of thermoacoustic instability (i.e., the transition from stable operation to unstable operation) in confined turbulent combustion systems. We showed that the dynamics of the acoustic pressure fluctuations observed during stable operation is not a noisy fixed point, but indeed consists of low amplitude high-dimensional chaotic oscillations. Further, we presented a novel paradigm describing the onset of thermoacoustic instability in such systems, where the system behavior transitions from the state of combustion noise to thermoacoustic instability via intermittency. Thus, the state of intermittency precedes the occurrence of thermoacoustic instability.

Subsequently, we highlighted the characteristics of the intermittency route to flame blowout, and compared the properties of intermittent oscillations observed prior to thermoacoustic instability and prior to flame blowout. Finally, we showed a universal inverse square law relation between the spectral amplitude of the pressure oscillations and the Hurst exponent, during the intermittency route to thermoacoustic instability for different configurations of turbulent combustors. In the next chapter, we will describe the spatiotemporal interactions taking place between the flame, flow, and acoustic fields during the intermittency route to thermoacoustic instability, in the framework of pattern formation and collective interaction.

References

1. Abarbanel, H.D.I., Brown, R., Sidorowich, J.J., Tsimring, L.S.: The analysis of observed chaotic data in physical systems. Rev. Mod. Phys. **65**(4), 1331 (1993)
2. Anderson, W.E., Yang, V.: Liquid Rocket Engine Combustion Instability, vol. 169. Progress in Astronautics and Aeronautics, AIAA, Washington, DC (1995)
3. Aoki, C., Gotoda, H., Yoshida, S., Tachibana, S.: Dynamic behavior of intermittent combustion oscillations in a model rocket engine combustor. J. Appl. Phys. **127**(22), 224903 (2020)
4. Bragg, S.L.: Combustion noise. J. Inst. Fuel **36**(1), 12–16 (1963)
5. Burnley, V.S., Culick, F.E.C.: Influence of random excitations on acoustic instabilities in combustion chambers. AIAA J. **38**(8), 1403–1410 (2000)
6. Candel, S.: Combustion instabilities coupled by pressure waves and their active control. In: Symposium (International) on Combustion, vol. 24, pp. 1277–1296. Elsevier (1992)
7. Candel, S.: Combustion dynamics and control: Progress and challenges. Proc. Combust. Inst. **29**(1), 1–28 (2002)
8. Candel, S., Durox, D., Ducruix, S., Birbaud, A.L., Noiray, N., Schuller, T.: Flame dynamics and combustion noise: Progress and challenges. Int. J. Aeroacoust. **8**(1), 1–56 (2009)
9. Cao, L.: Practical method for determining the minimum embedding dimension of a scalar time series. Physica D **110**(1-2), 43–50 (1997)
10. Chakravarthy, S.R., Sivakumar, R., Shreenivasan, O.J.: Vortex-acoustic lock-on in bluff-body and backward-facing step combustors. Sadhana **32**(1), 145–154 (2007)
11. Chaudhuri, S., Kostka, S., Renfro, M.W., Cetegen, B.M.: Blowoff dynamics of bluff body stabilized turbulent premixed flames. Combust. Flame **157**(4), 790–802 (2010)
12. Chiu, H.H., Summerfield, M.: Theory of combustion noise. Acta Astronaut. **1**(7-8), 967–984 (1974)
13. Clavin, P., Siggia, E.D.: Turbulent premixed flames and sound generation. Combust. Sci. Tech. **78**(1-3), 147–155 (1991)

14. Clavin, P., Kim, J.S., Williams, F.A.: Turbulence-induced noise effects on high-frequency combustion instabilities. Combust. Sci. Tech. **96**(1-3), 61–84 (1994)
15. Coats, C.M.: Coherent structures in combustion. Prog. Energy Combust. Sci. **22**(5), 427–509 (1996)
16. Culick, F.E.C.: Unsteady motions in combustion chambers for propulsion systems. Tech. rep., AGARDograph, NATO/RTO-AG-AVT-039 (2006)
17. Culick, F.E.C., Yang, V.: Overview of combustion instabilities in liquid-propellant rocket engines. In: Liquid Rocket Engine Combustion Instability, vol. 169, pp. 3–37. Progress in Astronautics and Aeronautics, AIAA Washington, DC (1995)
18. Datta, S., Mondal, S., Mukhopadhyay, A., Sanyal, D., Sen, S.: An investigation of nonlinear dynamics of a thermal pulse combustor. Combust. Theor. Model. **13**(1), 17–38 (2009)
19. Davis, K.B., Mewes, M.O., Andrews, M.R., van Druten, N.J., Durfee, D.S., Kurn, D.M., Ketterle, W.: Bose-Einstein condensation in a gas of sodium atoms. Phys. Rev. Lett. **75**(22), 3969–3973 (1995)
20. De, S., Bhattacharya, A., Mondal, S., Mukhopadhyay, A., Sen, S.: Application of recurrence quantification analysis for early detection of lean blowout in a swirl-stabilized dump combustor. Chaos **30**(4), 043115 (2020)
21. Delage, R., Takayama, Y., Biwa, T.: On–off intermittency in coupled chaotic thermoacoustic oscillations. Chaos **27**(4), 043111 (2017)
22. Domen, S., Gotoda, H., Kuriyama, T., Okuno, Y., Tachibana, S.: Detection and prevention of blowout in a lean premixed gas-turbine model combustor using the concept of dynamical system theory. Proc. Combust. Inst. **35**(3), 3245–3253 (2015)
23. Dowling, A.P.: Flow-acoustic interaction near a flexible wall. J. Fluid Mech. **128**, 181–198 (1983)
24. Dowling, A.P., Ffowcs Williams, J.E.: Sound and Sources of Sound. Horwood (1983)
25. Dowling, A.P., Mahmoudi, Y.: Combustion noise. Proc. Combust. Inst. **35**(1), 65–100 (2015)
26. Dranovsky, M.L., Agarkov, A.F., Denisov, K.P., Zavorohin, I.A., Ivanov, V.N., Pikalov, V.P., Shibanov, A.A.: Injector flame stabilization effects on combustion instability. In: Liquid Rocket Engine Combustion Instability, vol. 169. Progress in Astronautics and Aeronautics, AIAA, Washington, DC (1994)
27. Duran, I., Moreau, S., Nicoud, F., Livebardon, T., Bouty, E., Poinsot, T.: Combustion noise in modern aero-engines. Aerosp. Lab. J. **7**, 1–11 (2014)
28. Ebi, D., Denisov, A., Bonciolini, G., Boujo, E., Noiray, N.: Flame dynamics intermittency in the bistable region near a subcritical Hopf bifurcation. J. Eng. Gas Turbines Power **140**(6) (2018)
29. George, N.B., Unni, V.R., Raghunathan, M., Sujith, R.I.: Pattern formation during transition from combustion noise to thermoacoustic instability via intermittency. J. Fluid Mech. **849**, 615–644 (2018)
30. Gotoda, H., Nikimoto, H., Miyano, T., Tachibana, S.: Dynamic properties of combustion instability in a lean premixed gas-turbine combustor. Chaos **21**(1), 013124 (2011)
31. Gotoda, H., Amano, M., Miyano, T., Ikawa, T., Maki, K., Tachibana, S.: Characterization of complexities in combustion instability in a lean premixed gas-turbine model combustor. Chaos **22**(4), 043128 (2012)
32. Gotoda, H., Shinoda, Y., Kobayashi, M., Okuno, Y., Tachibana, S.: Detection and control of combustion instability based on the concept of dynamical system theory. Phys. Rev. E **89**(2), 022910 (2014)
33. Gotoda, H., Hayashi, K., Tsujimoto, R., Domen, S., Tachibana, S.: Dynamical properties of combustion instability in a laboratory-scale gas-turbine model combustor. J. Eng. Gas Turbines Power **139**(4), 041509 (2017)
34. Gottwald, G.A., Melbourne, I.: A new test for chaos in deterministic systems. Proc. R. Soc. A **460**(2042), 603–611 (2004)
35. Gottwald, G.A., Melbourne, I.: The 0-1 test for chaos: A review. In: Chaos Detection and Predictability, pp. 221–247. Springer (2016)

36. Guan, Y., Liu, P., Jin, B., Gupta, V., Li, L.K.B.: Nonlinear time-series analysis of thermoacoustic oscillations in a solid rocket motor. Exp. Therm. Fluid Sci. **98**, 217–226 (2018)
37. Guan, Y., Li, L.K.B., Ahn, B., Kim, K.T.: Chaos, synchronization, and desynchronization in a liquid-fueled diffusion-flame combustor with an intrinsic hydrodynamic mode. Chaos **29**(5), 053124 (2019)
38. Guan, Y., Gupta, V., Li, L.K.B.: Intermittency route to self-excited chaotic thermoacoustic oscillations. J. Fluid Mech. **894**, R3 (2020)
39. Hassan, H.A.: Scaling of combustion-generated noise. J. Fluid Mech. **66**(3), 445–453 (1974)
40. Hegde, U.G., Reuter, D., Daniel, B.R., Zinn, B.T.: Flame driving of longitudinal instabilities in dump type ramjet combustors. Combust. Sci. Tech. **55**(4-6), 125–138 (1987)
41. Hegde, U.G., Reuter, D., Zinn, B.T.: Sound generation by ducted flames. AIAA J. **26**(5), 532–537 (1988)
42. Hilborn, R.C.: Chaos and Nonlinear Dynamics: An Introduction for Scientists and Engineers. Oxford University Press (2000)
43. Juniper, M.P., Sujith, R.I.: Sensitivity and nonlinearity of thermoacoustic oscillations. Annu. Rev. Fluid Mech. **50**, 661–689 (2018)
44. Kabiraj, L., Sujith, R.I.: Nonlinear self-excited thermoacoustic oscillations: intermittency and flame blowout. J. Fluid Mech. **713**, 376–397 (2012)
45. Kabiraj, L., Saurabh, A., Wahi, P., Sujith, R.I.: Route to chaos for combustion instability in ducted laminar premixed flames. Chaos **22**(2), 023129 (2012)
46. Kantelhardt, J.W., Zschiegner, S.A., Koscielny-Bunde, E., Havlin, S., Bunde, A., Stanley, H.E.: Multifractal detrended fluctuation analysis of nonstationary time series. Physica A **316**(1-4), 87–114 (2002)
47. Kantz, H., Schreiber, T.: Nonlinear Time Series Analysis, vol. 7. Cambridge University Press (2004)
48. Kaplan, D.T., Glass, L.: Direct test for determinism in a time series. Phys. Rev. Lett. **68**(4), 427–430 (1992)
49. Kasthuri, P., Pavithran, I., Pawar, S.A., Sujith, R.I., Gejji, R., Anderson, W.: Dynamical systems approach to study thermoacoustic transitions in a liquid rocket combustor. Chaos **29**(10), 103115 (2019)
50. Kasthuri, P., Unni, V.R., Sujith, R.I.: Bursting and mixed mode oscillations during the transition to limit cycle oscillations in a matrix burner. Chaos **29**(4), 043117 (2019)
51. Ken, H.Y., Trouvé, A., Daily, J.W.: Low-frequency pressure oscillations in a model ramjet combustor. J. Fluid Mech. **232**, 47–72 (1991)
52. Kerres, B., Nair, V., Cronhjort, A., Mihaescu, M.: Analysis of the turbocharger compressor surge margin using a Hurst-exponent-based criterion. SAE Int. J. Engines **9**(3), 1795–1806 (2016)
53. Ketterle, W.: Experimental studies of Bose-Einstein condensation. Phys. Today **52**(12), 30–35 (1999)
54. Kheirkhah, S., Cirtwill, J.D.M., Saini, P., Venkatesan, K., Steinberg, A.M.: Dynamics and mechanisms of pressure, heat release rate, and fuel spray coupling during intermittent thermoacoustic oscillations in a model aeronautical combustor at elevated pressure. Combust. Flame **185**, 319–334 (2017)
55. Kings, N., Bake, F.: Indirect combustion noise: noise generation by accelerated vorticity in a nozzle flow. J. Spray Combust. Dyn. **2**(3), 253–266 (2010)
56. Klimaszewska, K., Żebrowski, J.J.: Detection of the type of intermittency using characteristic patterns in recurrence plots. Phys. Rev. E **80**(2), 026214 (2009)
57. Kobayashi, H., Gotoda, H., Tachibana, S.: Nonlinear determinism in degenerated combustion instability in a gas-turbine model combustor. Physica A **510**, 345–354 (2018)
58. Kobayashi, T., Murayama, S., Hachijo, T., Gotoda, H.: Early detection of thermoacoustic combustion instability using a methodology combining complex networks and machine learning. Phys. Rev. Appl. **11**(6), 064034 (2019)
59. Kotake, S.: On combustion noise related to chemical reactions. J. Sound Vib. **42**(3), 399–410 (1975)

60. Krishnan, A., Sujith, R.I., Marwan, N., Kurths, J.: On the emergence of large clusters of acoustic power sources at the onset of thermoacoustic instability in a turbulent combustor. J. Fluid Mech. **874**, 455–482 (2019)
61. Kumar, R.: Further experimental results on the structure and acoustics of turbulent jet flames. In: 2nd Aeroacoustics Conference, AIAA 1975-523 (1975)
62. Langhorne, P.J.: Reheat buzz: an acoustically coupled combustion instability. Part 1. Experiment. J. Fluid Mech. **193**, 417–443 (1988)
63. Law, C.K.: Combustion Physics. Cambridge University Press (2010)
64. Lieuwen, T.C.: Experimental investigation of limit-cycle oscillations in an unstable gas turbine combustor. J. Propuls. Power **18**(1), 61–67 (2002)
65. Lieuwen, T.C.: Statistical characteristics of pressure oscillations in a premixed combustor. J. Sound Vib. **260**(1), 3–17 (2003)
66. Lieuwen, T.C.: Static and dynamic combustion stability. The Gas Turbine Handbook, pp. 197–203 (2006)
67. Lieuwen, T., Banaszuk, A.: Background noise effects on combustor stability. J. Propuls. Power **21**(1), 25–31 (2005)
68. Lieuwen, T.C., Yang, V.: Combustion Instabilities in Gas Turbine Engines (Operational Experience, Fundamental Mechanisms and Modeling), vol. 210. Progress in Astronautics and Aeronautics, AIAA (2005)
69. Longwell, J.P., Frost, E.E., Weiss, M.A.: Flame stability in bluff body recirculation zones. Ind. Eng. Chem. Res. **45**(8), 1629–1633 (1953)
70. Macquisten, M.A., Dowling, A.P.: Low-frequency combustion oscillations in a model afterburner. Combust. Flame **94**(3), 253–264 (1993)
71. Marble, F.E., Candel, S.M.: Acoustic disturbance from gas non-uniformities convected through a nozzle. J. Sound Vib. **55**(2), 225–243 (1977)
72. Marwan, N., Romano, M.C., Thiel, M., Kurths, J.: Recurrence plots for the analysis of complex systems. Phys. Rep. **438**(5), 237–329 (2007)
73. Mondal, S., Unni, V.R., Sujith, R.I.: Onset of thermoacoustic instability in turbulent combustors: an emergence of synchronized periodicity through formation of chimera-like states. J. Fluid Mech. **811**, 659–681 (2017)
74. Mukhopadhyay, A., Chaudhari, R.R., Paul, T., Sen, S., Ray, A.: Lean blow-out prediction in gas turbine combustors using symbolic time series analysis. J. Propuls. Power **29**(4), 950–960 (2013)
75. Muruganandam, T.: Sensing and dynamics of lean blowout in a swirl dump combustor. Ph.D. thesis (2006)
76. Muruganandam, T.M., Nair, S., Scarborough, D., Neumeier, Y., Jagoda, J., Lieuwen, T., Seitzman, J., Zinn, B.: Active control of lean blowout for turbine engine combustors. J. Propuls. Power **21**(5), 807–814 (2005)
77. Murugesan, M., Sujith, R.I.: Combustion noise is scale-free: transition from scale-free to order at the onset of thermoacoustic instability. J. Fluid Mech. **772**, 225–245 (2015)
78. Murugesan, M., Sujith, R.I.: Physical mechanisms that cause intermittency that presages combustion instability and blowout in a turbulent lifted jet flame combustor. Combust. Sci. Tech. **190**(2), 312–335 (2018)
79. Nair, S.: Acoustic characterization of flame blowout phenomenon. Ph.D. thesis, Georgia Institute of Technology, USA (2006)
80. Nair, V.: Role of intermittent dynamics in the onset of combustion instability. Ph.D. thesis, Indian Institute of Technology Madras, India (2014)
81. Nair, S., Lieuwen, T.: Acoustic detection of blowout in premixed flames. J. Propuls. Power **21**(1), 32–39 (2005)
82. Nair, V., Sujith, R.I.: Identifying homoclinic orbits in the dynamics of intermittent signals through recurrence quantification. Chaos **23**(3), 033136 (2013)
83. Nair, V., Sujith, R.I.: Multifractality in combustion noise: predicting an impending combustion instability. J. Fluid Mech. **747**, 635–655 (2014)

84. Nair, V., Sujith, R.I.: Intermittency as a transition state in combustor dynamics: An explanation for flame dynamics near lean blowout. Combust. Sci. Tech. **187**(11), 1821–1835 (2015)
85. Nair, V., Sujith, R.I.: Precursors to self-sustained oscillations in aeroacoustic systems. Int. J. Aeroacoust. **15**(3), 312–323 (2016)
86. Nair, V., Thampi, G., Karuppusamy, S., Gopalan, S., Sujith, R.I.: Loss of chaos in combustion noise as a precursor of impending combustion instability. J. Spray Combust. Dyn. **5**(4), 273–290 (2013)
87. Nair, V., Thampi, G., Sujith, R.I.: Intermittency route to thermoacoustic instability in turbulent combustors. J. Fluid Mech. **756**, 470–487 (2014)
88. Noiray, N., Schuermans, B.: Deterministic quantities characterizing noise driven Hopf bifurcations in gas turbine combustors. Int. J. Nonlin. Mech. **50**, 152–163 (2013)
89. Pavithran, I., Unni, V.R., Varghese, A.J., Premraj, D., Sujith, R.I., Vijayan, C., Saha, A., Marwan, N., Kurths, J.: Universality in spectral condensation. Sci. Rep. **10**(1), 1–8 (2020)
90. Pavithran, I., Unni, V.R., Varghese, A.J., Sujith, R.I., Saha, A., Marwan, N., Kurths, J.: Universality in the emergence of oscillatory instabilities in turbulent flows. Europhys. Lett. **129**(2), 24004 (2020)
91. Pawar, S.A., Vishnu, R., Vadivukkarasan, M., Panchagnula, M.V., Sujith, R.I.: Intermittency route to combustion instability in a laboratory spray combustor. J. Eng. Gas Turbines Power **138**(4), 041505 (2016)
92. Pawar, S.A., Seshadri, A., Unni, V.R., Sujith, R.I.: Thermoacoustic instability as mutual synchronization between the acoustic field of the confinement and turbulent reactive flow. J. Fluid Mech. **827**, 664–693 (2017)
93. Pawar, S.A., Mondal, S., George, N.B., Sujith, R.I.: Temporal and spatiotemporal analyses of synchronization transition in a swirl-stabilized combustor. AIAA J. **57**(2), 836–847 (2019)
94. Pawar, S.A., Raghunathan, M., Reeja, K.V., Midhun, P.R., Sujith, R.I.: Effect of preheating of the reactants on the transition to thermoacoustic instability in a bluff-body stabilized dump combustor. Proc. Combust. Inst. **38**(4), 6193–6201 (2021)
95. Plee, S.L., Mellor, A.M.: Characteristic time correlation for lean blowoff of bluff-body-stabilized flames. Combust. Flame **35**, 61–80 (1979)
96. Poinsot, T.: Prediction and control of combustion instabilities in real engines. Proc. Combust. Inst. **36**(1), 1–28 (2017)
97. Poinsot, T.J., Trouve, A.C., Veynante, D.P., Candel, S.M., Esposito, E.J.: Vortex-driven acoustically coupled combustion instabilities. J. Fluid Mech. **177**, 265–292 (1987)
98. Polifke, W.: Combustion instabilities. In: Advances in Aeroacoustics and Applications, vol. 5. Von Karman Institute Rhode-St-Genèse, BE (2004)
99. Pomeau, Y., Manneville, P.: Intermittent transition to turbulence in dissipative dynamical systems. Commun. Math. Phys. **74**(2), 189–197 (1980)
100. Premchand, C.P., George, N.B., Raghunathan, M., Unni, V.R., Sujith, R.I., Nair, V.: Lagrangian analysis of intermittent sound sources in the flow-field of a bluff-body stabilized combustor. Phys. Fluids **31**(2), 025115 (2019)
101. Premraj, D., Pawar, S.A., Kabiraj, L., Sujith, R.I.: Strange nonchaos in self-excited singing flames. Europhys. Lett. **128**(5), 54005 (2020)
102. Premraj, D., Suresh, K., Pawar, S.A., Kabiraj, L., Prasad, A., Sujith, R.I.: Dragon-king extreme events as precursors for catastrophic transition. Europhys. Lett. **134**(3), 34006 (2021)
103. Putnam, A.A.: Combustion Driven Oscillations in Industry. Elsevier Publishing (1971)
104. Putnam, A., Faulkner, L.: Overview of combustion noise. J. Energy **7**(6), 458–469 (1983)
105. Radhakrishnan, K., Heywood, J.B., Tabaczynski, R.J.: Premixed turbulent flame blowoff velocity correlation based on coherent structures in turbulent flows. Combust. Flame **42**, 19–33 (1981)
106. Rajaram, R.: Characteristics of sound radiation from turbulent premixed flames. Ph.D. thesis, Georgia Institute of Technology, USA (2007)
107. Rock, N., Emerson, B., Seitzman, J., Lieuwen, T.: Near-lean blowoff dynamics in a liquid fueled combustor. Combust. Flame **212**, 53–66 (2020)
108. Rogers, D.E., Marble, F.E.: A mechanism for high frequency oscillations in ramjet combustors and afterburners. J. Jet Propuls. **26**(1), 456–462 (1956)

109. Roy, A., Singh, S., Nair, A., Chaudhuri, S., Sujith, R.I.: Flame dynamics during intermittency and secondary bifurcation to longitudinal thermoacoustic instability in a swirl-stabilized annular combustor. Proc. Combust. Inst. **38**(4), 6221–6230 (2021)
110. Sampath, R., Chakravarthy, S.R.: Investigation of intermittent oscillations in a premixed dump combustor using time-resolved particle image velocimetry. Combust. Flame **172**, 309–325 (2016)
111. Sarkar, S., Ray, A., Mukhopadhyay, A., Sen, S.: Dynamic data-driven prediction of lean blowout in a swirl-stabilized combustor. J. Spray Combust. Dyn. **7**(3), 209–241 (2015)
112. Schadow, K.C., Gutmark, E.: Combustion instability related to vortex shedding in dump combustors and their passive control. Prog. Energy Combust. Sci. **18**(2), 117–132 (1992)
113. Schuster, H.G., Just, W.: Deterministic Chaos: An Introduction. Wiley (2006)
114. Sen, U., Gangopadhyay, T., Bhattacharya, C., Mukhopadhyay, A., Sen, S.: Dynamic characterization of a ducted inverse diffusion flame using recurrence analysis. Combust. Sci. Tech. **190**(1), 32–56 (2018)
115. Shanbhogue, S.J., Husain, S., Lieuwen, T.: Lean blowoff of bluff body stabilized flames: Scaling and dynamics. Prog. Energy Combust. Sci. **35**(1), 98–120 (2009)
116. Shivashankara, B., Strahle, W., Handley, J.: Combustion noise radiation by open turbulent flames. In: Aeroacoustics Conference, AIAA 1973-1025 (1973)
117. Singh, G., Mariappan, S.: Experimental investigation on the route to vortex-acoustic lock-in phenomenon in bluff body stabilized combustors. Combust. Sci. Tech., 1–29 (2019)
118. Smith, T.J.B., Kilham, J.K.: Noise generation by open turbulent flames. J. Acoust. Soc. Am. **35**(5), 715–724 (1963)
119. Smith, D.A., Zukoski, E.E.: Combustion instability sustained by unsteady vortex combustion. In: 21st Joint Propulsion Conference, AIAA 1985-1248 (1985)
120. Spalding, D.B.: Some Fundamentals of Combustion, vol. 2. Butterworths Scientific Publications (1955)
121. Stone, E., Gorman, M., El-Hamdi, M., Robbins, K.A.: Identification of intermittent ordered patterns as heteroclinic connections. Phys. Rev. Lett. **76**(12), 2061–2064 (1996)
122. Strahle, W.C.: On combustion generated noise. In: 7th Propulsion Joint Specialist Conference, AIAA 1971-735 (1971)
123. Strahle, W.C.: Combustion noise. Prog. Energy Combust. Sci. **4**(3), 157–176 (1978)
124. Strogatz, S.H.: Nonlinear Dynamics and Chaos: With Applications to Physics, Biology, Chemistry, and Engineering. CRC Press (1994)
125. Sujith, R.I., Unni, V.R.: Complex system approach to investigate and mitigate thermoacoustic instability in turbulent combustors. Phys. Fluids **32**(6), 061401 (2020)
126. Taamallah, S., LaBry, Z.A., Shanbhogue, S.J., Ghoniem, A.F.: Thermo-acoustic instabilities in lean premixed swirl-stabilized combustion and their link to acoustically coupled and decoupled flame macrostructures. Proc. Combust. Inst. **35**(3), 3273–3282 (2015)
127. Takens, F.: Detecting strange attractors in turbulence. Lect. Notes Math. **898**(1), 366–381 (1981)
128. Tandon, S., Pawar, S.A., Banerjee, S., Varghese, A.J., Durairaj, P., Sujith, R.I.: Bursting during intermittency route to thermoacoustic instability: Effects of slow–fast dynamics. Chaos **30**(10), 103112 (2020)
129. Theiler, J., Eubank, S., Longtin, A., Galdrikian, B., Farmer, J.D.: Testing for nonlinearity in time series: the method of surrogate data. Physica D **58**(1-4), 77–94 (1992)
130. Tony, J., Gopalakrishnan, E.A., Sreelekha, E., Sujith, R.I.: Detecting deterministic nature of pressure measurements from a turbulent combustor. Phys. Rev. E **92**(6), 062902 (2015)
131. Toong, T.Y.: Instabilities in reacting flows. Acta Astronaut. **1**(3-4), 317–344 (1974)
132. Unni, V.R., Sujith, R.I.: Multifractal characteristics of combustor dynamics close to lean blowout. J. Fluid Mech. **784**, 30–50 (2015)
133. Unni, V.R., Sujith, R.I.: Flame dynamics during intermittency in a turbulent combustor. Proc. Combust. Inst. **36**(3), 3791–3798 (2017)
134. Venkatramani, J., Nair, V., Sujith, R.I., Gupta, S., Sarkar, S.: Precursors to flutter instability by an intermittency route: a model free approach. J. Fluids Struct. **61**, 376–391 (2016)

135. Wayland, R., Bromley, D., Pickett, D., Passamante, A.: Recognizing determinism in a time series. Phys. Rev. Lett. **70**(5), 580–582 (1993)
136. Weng, F., Makeximu, Li, D., Zhu, M.: Low frequency beating instability in a Rijke burner and its active control. J. Low Freq. Noise Vibr. Act. Control **34**(2), 137–152 (2015)
137. Wilhite, J.M., Dolan, B.J., Kabiraj, L., Gomez, R.V., Gutmark, E.J., Paschereit, C.O.: Analysis of combustion oscillations in a staged MLDI burner using decomposition methods and recurrence analysis. In: 54th AIAA Aerospace Sciences Meeting, AIAA 2016-1156 (2016)
138. Zhang, Q., Shanbhogue, S.J., Lieuwen, T.: Dynamics of premixed h2/ch4 flames under near blowoff conditions. J. Eng. Gas Turbines Power **132**(11) (2010)

Chapter 7
Spatiotemporal Dynamics of Flow, Flame, and Acoustic Fields During the Onset of Thermoacoustic Instability

In the previous chapter, we showed that the onset of thermoacoustic instability from the state of combustion noise happens via intermittency in turbulent combustors. The state of combustion noise is indeed complex and exhibits the features of high dimensional chaotic oscillations. During the state of intermittency, bursts of large amplitude periodic oscillations occur amidst epochs of low amplitude chaotic oscillations. Furthermore, the onset of thermoacoustic instability (emergence of order) is associated with the loss of chaos in the temporal field of turbulent combustors. In this chapter, we will discuss the spatiotemporal interactions between the flow, flame, and acoustic subsystems of such combustors, during the transition from combustion noise to thermoacoustic instability via intermittency. We will characterize this transition as a process occurring through pattern formation and collective interaction.

As we mentioned in Chap. 5, thermoacoustic instability is a phenomenon of self-organization. It happens primarily due to a three-way mutual coupling between the acoustic field, the flame dynamics, and the hydrodynamic field of a turbulent combustor (Fig. 7.1a). During the onset of thermoacoustic instability, we observe the periodic emergence of large-scale coherent structures in the reaction field of the combustor [21, 48, 57]. In these coupled subsystems, hydrodynamics (i.e., underlying turbulent fluctuations or vortical structures) affects the flame front, resulting in unsteady heat release rate fluctuations [49, 52, 60]. Such heat release rate fluctuations generate acoustic waves, which then affect the flame as well as the hydrodynamic flow field [9, 19, 49]. The turbulent fluctuations associated with the hydrodynamic field can directly generate acoustic waves in the confinement [18, 54]. The heat transfer from the flame (or the reaction zone) can independently change the hydrodynamic properties (such as flow speed, reactant mixing, or circulation) of the flow, without affecting them through the acoustic field of the combustor [2, 35, 62]. Thus, all these subsystems directly or indirectly affect each other at different time-space scales during all the states of combustor operation, and the interactions between them are highly intricate and more intertwined than what

R. I. Sujith, S. A. Pawar, *Thermoacoustic Instability*, Springer Series in Synergetics,
https://doi.org/10.1007/978-3-030-81135-8_7

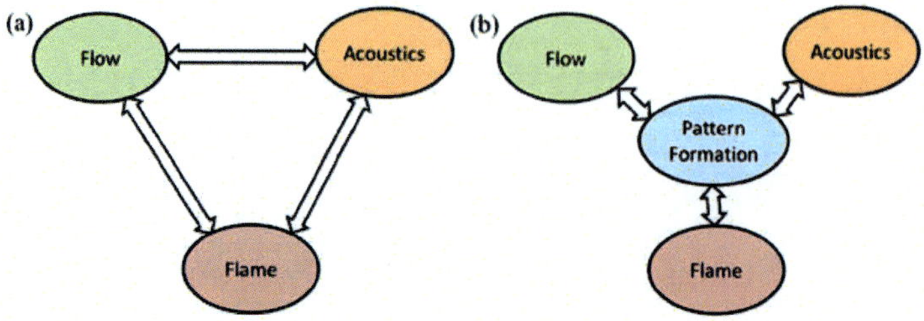

Fig. 7.1 Schematics showing (**a**) the mutual coupling between the flow, the flame and the acoustic subsystems of a combustor, and (**b**) the process of the coupling between these subsystems via pattern formation. The interaction of different subsystems occurs through the emergence of various spatiotemporal patterns. Adapted with permission from George et al. [21]

is described here [38]. We will discuss the dynamical changes happening in all these subsystems of a turbulent combustor during the transition from combustion noise (i.e., disordered fluctuations) to thermoacoustic instability (i.e., self-organized ordered oscillations) in the context of pattern formation [13, 25, 28] and collective interaction [15, 29, 30, 57], see Fig. 7.1b. We have presented brief descriptions of these phenomena at appropriate places in the chapter.

7.1 Pattern Formation

We see patterns everywhere in our day-to-day life (Fig. 7.2). This pattern formation is a universal phenomenon. It is observed in various fields including biology, ecology, astronomy, reaction chemistry, social science, etc. [3, 5, 13, 17, 22, 32, 37, 41, 43, 46, 53, 55, 61, 65]. Familiar examples of patterns that are found in nature are trees, flowers, ripples in sand, Rayleigh-Benard convection, Belousov–Zhabotinsky reaction, seashells, animal fur, human fingertips and body parts, ant foraging, growth of bacterial colonies, galaxies, lightning, etc. [1, 4, 6, 10, 11, 13].

Pattern formation is a non-equilibrium phenomenon that usually happens due to a symmetry-breaking in the system [12]. In simple terms, pattern formation deals with the ability of a particular system to develop self-organized ordered (structured) behavior from the initial disordered (unstructured or homogeneous) behavior throughout the space. Symmetry-breaking leads to a transition from such disordered to ordered dynamics in the spatial or temporal fields of a system.

We observe patterns in various forms, such as periodic, quasiperiodic, or spiral defect chaos [13, 32]. Periodic patterns repeat themselves throughout the space. In contrast, quasiperiodic patterns rarely repeat in either direction. Spiral patterns display regular rotating spiral waves travelling in the outward direction. In contrast, spiral defect chaos is a disordered pattern on a large scale but exhibits rotating spirals and other defects such as patches of strips at the local scale.

Fig. 7.2 Examples of various patterns observed in natural systems. Familiar examples of spiral patterns (from left to right) are (**a**) a nautilus shell, a galaxy, a sunflower, a desert plant, (**b**) a storm formation, a fern bud, an ocean wave, a fingerprint. (**c**) Some examples of fractal patterns are a branched tree, a leaf, a river bed formation, and a cast of the human lungs. Reproduced with permission from Perera and Coppens [47]

During pattern formation, a coherence is developed in the association of different elements of a system (for example, genes, molecules, or particles) due to the positive feedback (mutual coupling) between these elements. The presence of such coherence leads to the occurrence of a self-organized pattern both in the spatial and temporal dynamics of the system. Pattern formation is an emergent property where the global behavior of the system is primarily influenced by the local behavior [42]. Therefore, to control pattern formation, we need to affect the local coupling or the interactions in the system. Furthermore, pattern formation is also related to the adaptation process, where the system adapts its behavior as per its surrounding environment. In such a situation, the parts of the system self-organize in response to changes in the environment [7, 59].

Pattern formation is also seen in both the temporal and spatial dynamics of thermoacoustic systems. We observe a periodic emergence of convecting wrinkles (regular patterns) on laminar flame shapes during the state of thermoacoustic instability [8]. Turbulent flame has spatial structures (patterns) that, in turn, decide the distribution of the local heat release rate in the system [38]. In turbulent combustors, the vortex shedding shows incoherent behavior (a disordered pattern) during the state of stable operation, while the formation of large-scale vortical structures (an ordered pattern) takes place during the state of unstable operation [48]. We observe the formation of standing wave patterns in the spatial domain of the acoustic field during the occurrence of thermoacoustic instability [14]. We need to destroy such standing wave patterns to mitigate thermoacoustic instability, which is often achieved in rocket engines by installing different shaped baffles on the injector plate.

George et al. [21] viewed the spatiotemporal interaction between different subsystems (Fig. 7.1a) of a turbulent combustor during the onset of thermoacoustic instability as pattern formation. Pattern formation acts as a link through which the interaction between these subsystems enhances during the state of thermoacoustic instability (Fig. 7.1b). Here, each subsystem has a direct effect on the spatial or temporal patterns that emerge in the turbulent combustor. During pattern formation, the temporal field of a thermoacoustic system shows periodic (limit cycle) oscillations. These oscillations emerge from the underlying chaotic oscillations observed during the state of combustion noise [45].

7.2 Pattern Formation in Turbulent Thermoacoustic Systems

In the previous chapter, we studied the transition of the temporal dynamics of the acoustic pressure fluctuations from the state of combustion noise to thermoacoustic instability via intermittency. We showed that the acoustic fluctuations observed during the occurrence of combustion noise consist of low amplitude chaotic oscillations and that during thermoacoustic instability exhibit large-amplitude periodic oscillations. During the state of intermittency, the system behavior alternately switches between bursts of large amplitude periodic oscillations and epochs of low amplitude chaotic oscillations. These temporal dynamics of the acoustic pressure fluctuations are the result of nonlinear spatial interactions happening between different subsystems (flow, flame, and acoustics) of a combustor. Therefore, the natural question that arises is: *how do the flow and the flame interact with the global acoustic field during the onset of thermoacoustic instability?* In order to address this question, the primary focus of our subsequent discussions will be on studying the spatial patterns that emerge during the transition of the system behavior from the state of combustion noise (stable operation) to thermoacoustic instability (unstable operation) via intermittency. Before going into the details of the spatiotemporal analysis of a turbulent combustor, we first clarify the difference between two key processes, emergence and self-organization, that occur during the onset of thermoacoustic instability [16].

7.2.1 *Emergence Versus Self-Organization*

Emergence refers to the occurrence of coherent emergents (i.e., properties, behavior, structure, pattern, etc.) at macro-scales in a system due to the complex interactions of its parts at micro-scales [24, 31, 34]. Here, coherence refers to the presence of logical and consistent correlation between parts of the system, where these correlations persist (i.e., maintain identity) for longer duration and also at larger spatial distance in the system. In such a situation, the global dynamical behavior of the system is significantly different from the dynamics of its individual parts. We can alternately describe this behavior as '*the whole is more than the sum of its parts*' [20]. Moreover, the dynamics of a system observed at the macro-level and the micro-level are bidirectionally coupled. As we move in the direction from micro-level to macro-level, we observe an emergent behavior. In contrast, as we move in the direction from macro-level to macro-level, we notice that these emergent structures affect the dynamics at micro-levels. Further, the coherent dynamics formed during emergence is insensitive to external perturbations. This means that failure or removal of a single part of a large system does not affect its global behavior. For example, in a flock of birds or in a car traffic jam, the removal of a single bird or a car does not affect the properties of the emergent phenomena observed during bird flocking or traffic jam.

Self-organization is a dynamic process and adaptive in nature. During self-organization, interacting subsystems acquire and maintain order themselves, in the absence of any external control [16]. Here, organization refers to the arrangement of specific portion of a system to perform a particular task. Self-organization leads to an increase in the order of system behavior. An increase in order implies that the system behavior starts from semi-organized or completely random behavior (i.e., no historical memory). Since self-organization is a dynamic event, changes in the system affect the organized structures and the structures also respond equally to changes, in order to maintain such organized behavior. Self-organization happens only when the system is far from equilibrium [23]. At this condition, the system is very much sensitive to environmental changes; at the same time, it is also dynamic and capable of responding to every change.

According to Wolf and Holvoet [16] "*the essence of emergence is the existence of a global behavior that is novel with respect to the constituent parts of the system. The essence of self-organization is an adaptable behavior that autonomously acquires and maintains an increased order.*" In this section, we study these phenomena of emergence and self-organization using the simultaneous analysis of spatiotemporal information of the flow, flame, and acoustic subsystems of a combustor during the onset of thermoacoustic instability.

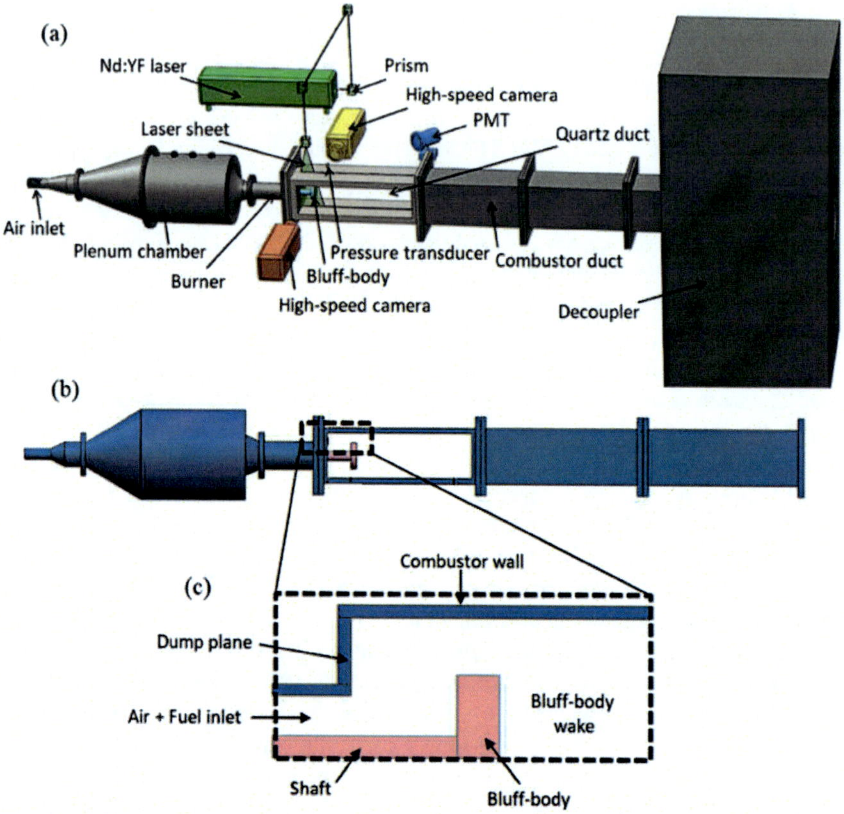

Fig. 7.3 (**a**) Schematic of the experimental arrangement of a turbulent, partially premixed, bluff-body stabilized dump combustor. The acoustic pressure fluctuations are measured using a piezoelectric transducer, the global heat release rate fluctuations in the flame are acquired using a photomultiplier tube (PMT), and the unsteady flame dynamics (OH*/CH* chemiluminescence field) is captured using a high-speed CMOS camera. Particle image velocimetry (PIV) experiments are performed using an Nd:YLF laser that illuminates the seeding particles in the region of a laser sheet. Mie scattering light from the seeding particles is captured using a high-speed camera. (**b**) A two-dimensional view of the experimental setup. (**c**) A zoomed region of the upper half of the bluff-body, which is used for analysing the flame and flow fields of this combustor in the subsequent discussions of the book. Adapted with permission from George et al. [21]

7.2.2 *The Emergence of Patterns During the Onset of Thermoacoustic Instability*

In this section, we discuss the emergence of patterns in the reaction field of a turbulent bluff-body stabilized dump combustor (Fig. 7.3) en-route to thermoacoustic instability [21]. Towards this purpose, the spatiotemporal measurements of the flow field are acquired using particle image velocimetry (PIV) measurements, and that

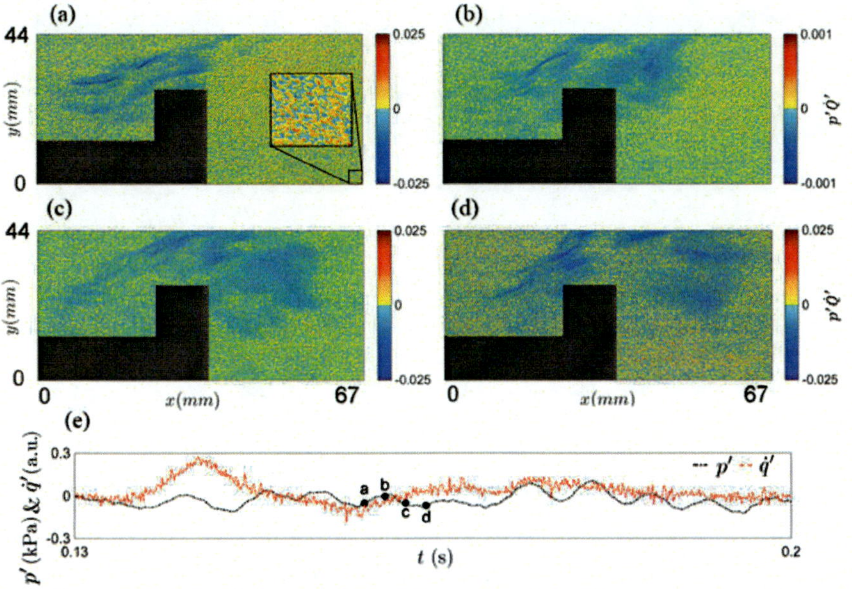

Fig. 7.4 **(a–d)** Instantaneous snapshots of the local acoustic power field $p'\dot{Q}'(x, y, t)$ observed in the upper-half of a bluff-body-stabilized turbulent combustor for the state of combustion noise, corresponding to time instances of acoustic pressure signal (p') and global heat release rate (obtained from photomultiplier tube) signal ($\dot{q}'$) shown in **(e)**. The masked region in all images shows the location of a bluff-body in the combustor. Reproduced with permission from George et al. [21]

of the reaction field (flame) are obtained using CH* chemiluminescence imaging of the flame. Figures 7.4 and 7.5 show instantaneous snapshots of the local acoustic power production $p'\dot{Q}'(x, y, t)$ observed at different time instances of the acoustic pressure signal (p'). These figures correspond to the states of combustion noise and thermoacoustic instability, respectively. Since the length of the reaction zone is much smaller than the acoustic wavelength, we can consider the instantaneous value of the acoustic pressure constant in this zone [39, 44]. Then, we can obtain the instantaneous field of local power production by multiplying the instantaneous pressure fluctuations p' (obtained from a point measurement) with the local heat release rate fluctuations $\dot{Q}'$ observed at each pixel of the flame image.

During the occurrence of combustion noise, we notice the presence of a grainy distribution of the local acoustic power production $p'\dot{Q}'(x, y, t)$ in the reaction field for all instants of time (Fig. 7.4). This grainy appearance of the $p'\dot{Q}'(x, y, t)$ field is a result of the dissimilar value of $p'\dot{Q}'(x, y, t)$ at a particular location than its neighbors when observed at a larger scale. A zoomed view of a small portion of $p'\dot{Q}'(x, y, t)$ field shows the existence of local coherence in $p'\dot{Q}'(x, y, t)$ values with a few immediate neighbors (Fig. 7.4a). George et al. [21] suggested that this behavior of the incoherence at the larger (global) scale and coherence at the smaller

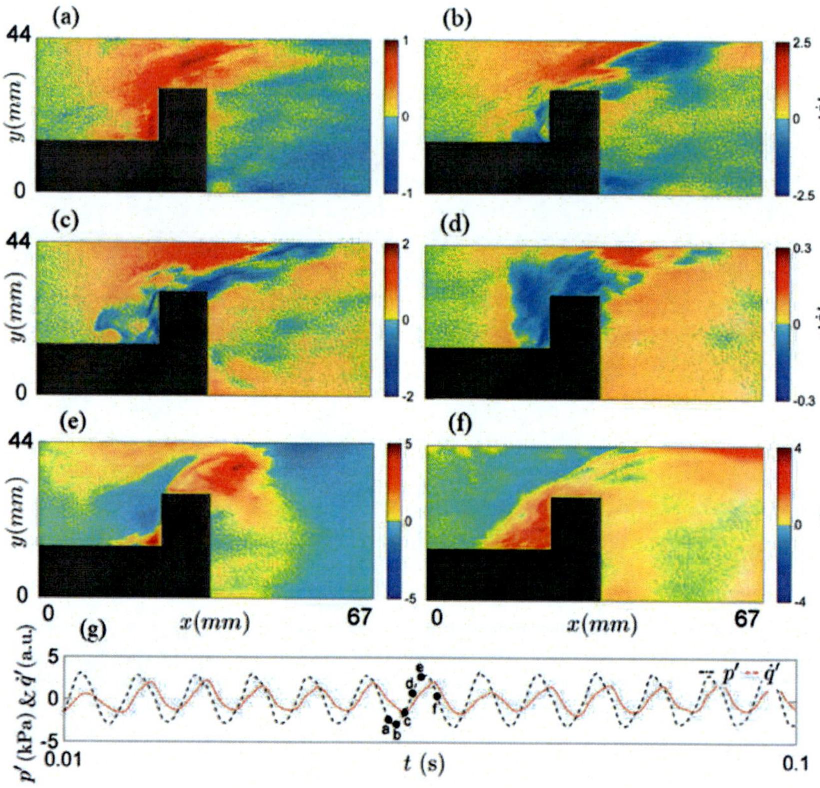

Fig. 7.5 **(a–f)** Instantaneous snapshots of the local acoustic power production field $p'\dot{Q}'(x, y, t)$ observed for the state of thermoacoustic instability, corresponding to time instances of the acoustic pressure signal shown in **(g)**. Reproduced with permission from George et al. [21]

(local) scale is a possibility of the presence of spatiotemporal chaos in the system [40]. Thus, we can refer to such a grainy structure of the spatial $p'\dot{Q}'(x, y, t)$ field observed during the state of combustion noise as a disordered or an incoherent pattern.

Figure 7.5 shows the instantaneous fields of the local acoustic power $p'\dot{Q}'(x, y, t)$ observed over an acoustic cycle during the state of thermoacoustic instability. During this state, we observe the emergence of the large coherent structures of positive $p'\dot{Q}'(x, y, t)$ in the reaction field, as compared to small coherent structures of negative $p'\dot{Q}'(x, y, t)$. A coherent structure (also called a cluster) consist of similar values of $p'\dot{Q}'(x, y, t)$ at different spatial locations in the structure. The presence of the positive coherence at a larger spatial region of the combustor enhances the local coupling between the acoustic pressure and the local heat release rate fluctuations. We observe the periodic formation of such coherent structures in the system. These structures are observed on the top and

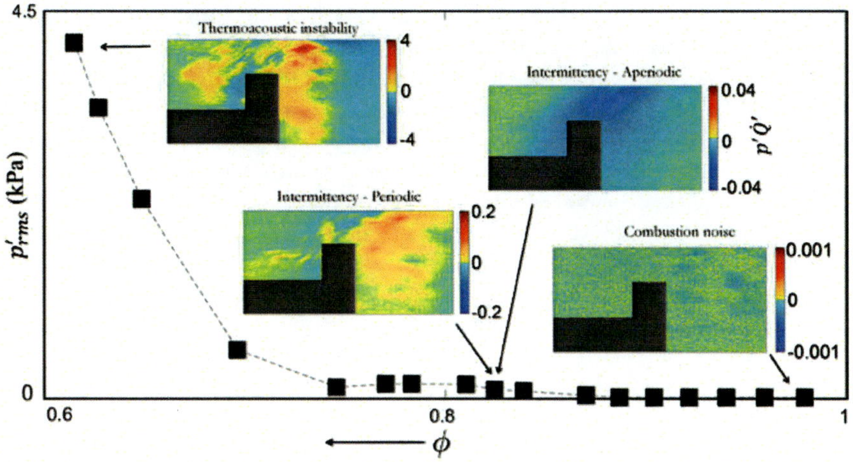

Fig. 7.6 The variation of the root mean square of the acoustic pressure fluctuations ($p'_{\rm rms}$) as a function of the equivalence ratio (ϕ) during the transition from combustion noise to thermoacoustic instability via intermittency. The insets show the instantaneous field of local acoustic power production corresponding to specific values of the equivalence ratio. During the occurrence of periodic oscillations, coherent patterns emerge in the reaction field from the incoherent patterns observed for aperiodic oscillations in the combustor. Reproduced with permission from George et al. [21]

the downstream regions of the bluff-body. We can refer to such regions of positive $p'\dot{Q}'(x, y, t)$ field as ordered or coherent pattern [21].

In Fig. 7.6, we show the variation of a root mean square value of the acoustic pressure fluctuations with decreasing value of equivalence ratio en route to thermoacoustic instability. As the equivalence ratio is reduced, we notice a continuous increase in the amplitude of the acoustic pressure oscillations in the system. In Fig. 7.6, we also show the instantaneous spatial fields of the local acoustic power ($p'\dot{Q}'(x, y, t)$) observed during different dynamical states. These plots show the formation of regions of coherent spatial patterns during the onset of thermoacoustic instability. This happens due to the self-organization of the local heat release rate fluctuations under the influence of the global acoustic field in the system. We notice the emergence of these coherent patterns from the incoherent patterns observed during the state of combustion noise. During the epoch of periodic oscillations of intermittency, one can observe regions of the coherent acoustic power field in the spatial dynamics. In contrast, during the epoch of aperiodic fluctuations of intermittency, the spatial field shows a grainy distribution of acoustic power, where a small region of negative acoustic power is also observed at the top and downstream regions of the bluff-body. Thus, we can observe the alternate occurrence of coherent and incoherent patterns in the reaction field of the turbulent combustor during the state of intermittency. Note that the magnitude of $p'\dot{Q}'(x, y, t)$ in the coherent regions of thermoacoustic instability is significantly higher than that observed during the epochs of periodic oscillations that occur during intermittency.

We note that such a formation of self-organized spatial structured states from initial unstructured states, during the transition from combustion noise to thermoacoustic instability, indicates the existence of pattern formation in a turbulent combustor. We can detect spatial pattern formation through several indicators [36]. Spatial variance [26] is one such measure that has been used for detecting the spatial pattern formation in a thermoacoustic system [21]. We can calculate the spatial variance (σ) of acoustic power field ($p'\dot{Q}'$) as follows,

$$\sigma(t) = \sum_{i=1}^{m}\sum_{j=1}^{n}[p'\dot{Q}'_{ij}(t) - \overline{p'\dot{Q}'_{ij}}(t)]^2 \tag{7.1}$$

where (i, j) is the spatial location in the reaction field ($m \times n$) and $\overline{p'\dot{Q}'_{ij}}(t)$ is the spatial average of acoustic power field at a particular time instant. Strong fluctuations about the mean due to increased recovery time can increase the spatial variance of $p'\dot{Q}'(t)$ field prior to pattern formation in the system [36]. In turbulent combustors, the formation of coherent regions in between incoherent/disordered regions leads to an increase in the spatial variance in the system.

In Fig. 7.7, we show the time series of the acoustic pressure fluctuations (p'), the global heat release rate fluctuations ($\dot{q}'$), and corresponding spatial variance (σ) for the states of combustion noise (Fig. 7.7a,d,g), intermittency (Fig. 7.7b,e,h), and thermoacoustic instability (Fig. 7.7c,f,i), respectively. As the system dynamics transitions from the state of combustion noise to thermoacoustic instability via intermittency, we observe a corresponding increase in the values of spatial variance (σ). We notice maximum values of σ during the state of thermoacoustic instability, indicating the emergence of ordered spatial patterns in the system. Small fluctuations in σ during the state of intermittency suggest the intermittent occurrence of

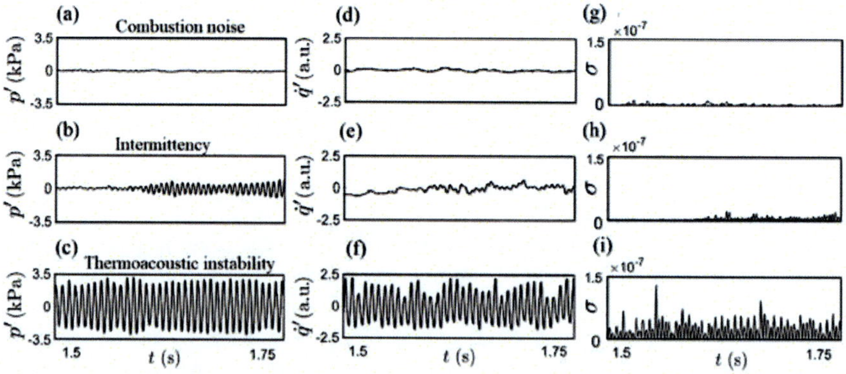

Fig. 7.7 Time series of **(a–c)** the acoustic pressure fluctuations p', **(d–f)** the global heat release rate fluctuations $\dot{q}'$, and **(g–i)** the spatial variance σ of local acoustic power fluctuations $p'\dot{Q}'(x, y, t)$ for the states of combustion noise, intermittency, and thermoacoustic instability, respectively. Reproduced with permission from George et al. [21]

spatial patterns. In contrast, very low values of σ during the state of combustion noise represent the spatial incoherence of local acoustic power in the system.

Having discussed the interactions between the acoustic pressure and the unsteady reaction field during the onset of thermoacoustic instability, we will next attempt to understand the changes in the spatial hydrodynamic field of the turbulent combustor, during the intermittency route to thermoacoustic instability in the context of collective interaction.

7.3 Collective Interaction of Large-Scale Vortices During Thermoacoustic Instability

Collective interaction is an ordered motion that emerges in a self-driven system on a global scale due to the interaction of its parts at the local scale [64]. There are many examples of collective motion in our daily life, which include flocks of birds, herds of animals, crowds, traffic, bacterial colonies, etc. In a thermoacoustic system, the vortex formation observed from the unstable shear layer stabilized over a flame holding device shows collective interaction [38], where the small-scale vortices support the formation of a large-scale vortex in the system. We present the process of collective interaction of vortical structures in the following discussion.

In Figs. 7.8, 7.9, 7.10, we show the hydrodynamic flow field over the bluff-body in a dump combustor (Fig. 7.3c), during the transition from the state of combustion noise to thermoacoustic instability via intermittency. The instantaneous local flow velocity vectors are overlaid on the local vorticity field for specific time instances of the simultaneously acquired acoustic pressure fluctuations in the system. When the flow enters the combustion chamber, it gets deflected towards the combustor walls due to the presence of the bluff-body (Fig. 7.3c). We observe two recirculation zones in the combustor: the first recirculation zone is present in the wake of the dump plane (located at the upper left corner in each image) showing positive vorticity, and the second recirculation zone is present in the wake region of the bluff-body showing negative vorticity.

During the state of combustion noise, we observe the shedding of small scale vortices along the shear layer at random time instances from the tip of the bluff body (indicated by circles Fig. 7.8a–d). This suggests that the vortex shedding is incoherent during the state of stable operation of the combustor. We note that a smaller vortex that can be observed in the system is limited by the resolution of PIV measurements.

In Fig. 7.9, we show the flow dynamics for the state of intermittency. During epochs of periodicity of intermittency (Fig. 7.9e), we notice the periodic shedding of large-scale vortical structures in the wake of the dump plane (Fig. 7.9a–d). These coherent structures convect downstream and break down into small fragments after impingement on the bluff-body and walls of the combustor. This results in an intense fine-scale mixing of the cold reactants and the hot products, leading to a near sudden

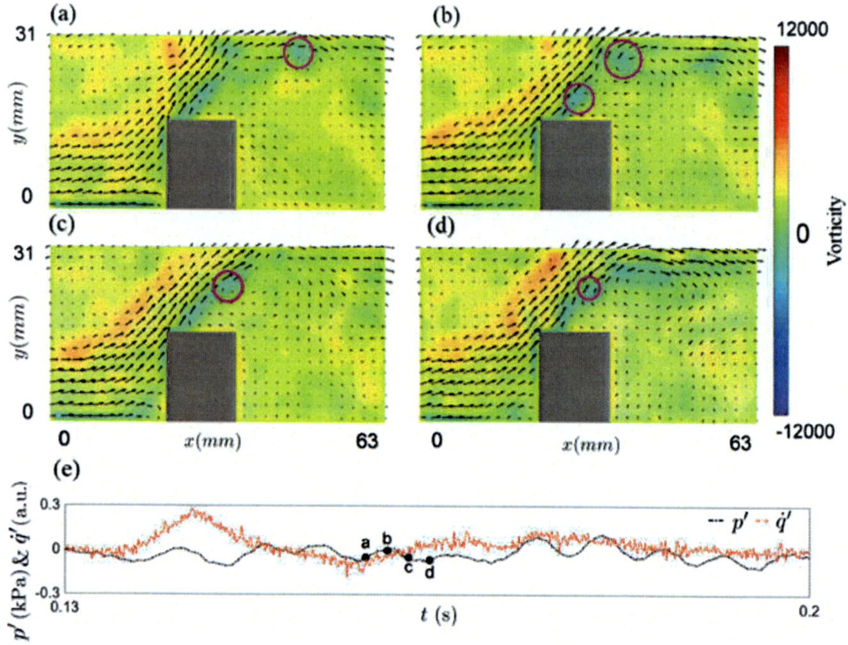

Fig. 7.8 (**a–d**) Instantaneous images of the combustor flow field obtained from PIV measurements for the state of combustion noise, corresponding to time instances of the acoustic pressure signal shown in (**e**). Circles indicate the approximate position and size of vortices generated in the flow field. Adapted with permission from George et al. [21]

release of heat in the system. Subsequently, the flow passes over the bluff-body and forms another large-scale vortex in its wake. This secondary vortex gets diffused in the reaction field as the flow passes through a long duct of the combustor. In contrast, during epochs of aperiodic fluctuations of intermittency (Fig. 7.9j), the flow dynamics (Fig. 7.9f–i) is nearly the same as that observed for the state of combustion noise (Fig. 7.8).

During the state of thermoacoustic instability (Fig. 7.10g), we observe the periodic formation of self-organized large-scale vortices at the dump plane of the combustor (Fig. 7.10d). The frequency of this vortex formation process is equal to the frequency of the acoustic oscillations in the system. The strength of the vorticity of such vortices is larger as compared to that observed during the epochs of periodicity in the intermittency signal. The large-scale vortices formed during this state wrinkle the flame surfaces and produce a coherent field of heat release rate upon impingement on the bluff-body in the combustor.

Alternate approaches that can be used to study the formation of such large-scale coherent structures in the combustor flow field are dynamic mode decomposition (DMD) [58] and backward finite-time Lyapunov exponents (bFTLE) [27]. DMD identifies the fastest growing modes of the system, while bFTLE helps in detecting

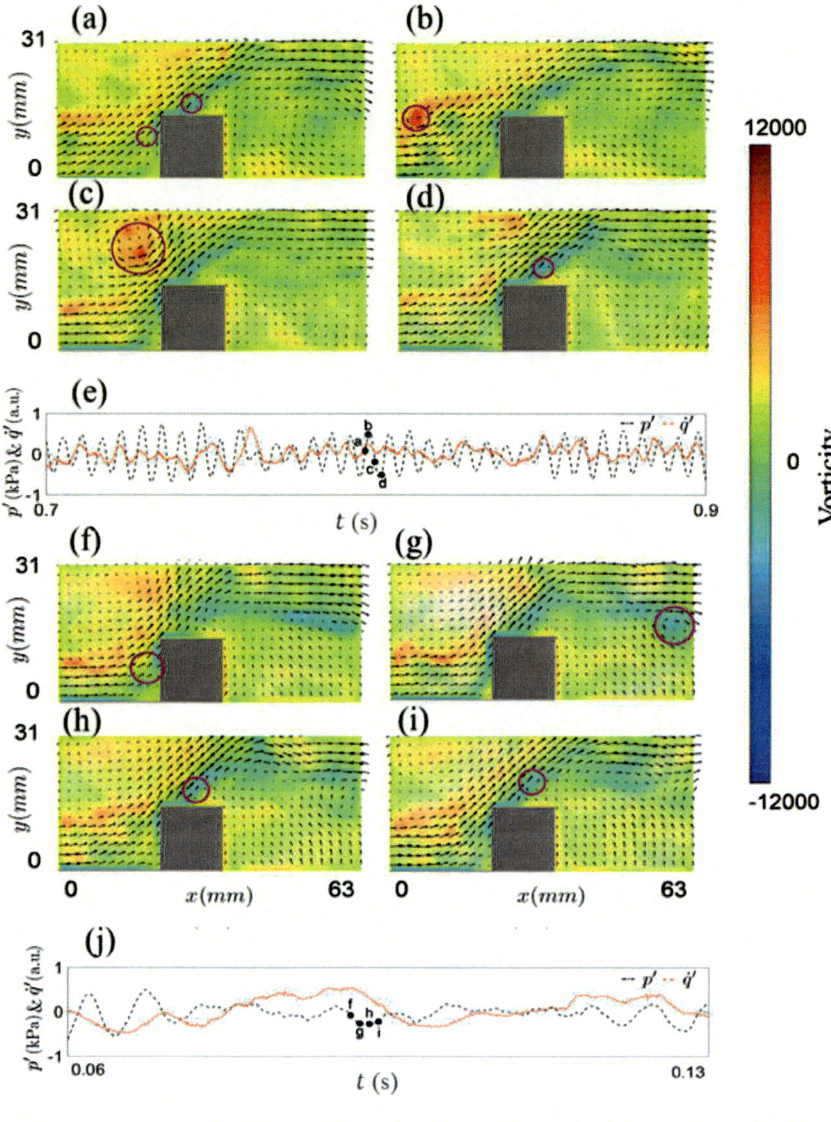

Fig. 7.9 Instantaneous images of the combustor flow field obtained from PIV measurements corresponding to (**a–d**) epochs of periodic fluctuations and (**f–i**) epochs of aperiodic fluctuations of intermittency, corresponding to time instances of the acoustic pressure signal shown in (**e**) and (**j**), respectively. Circles indicate the approximate position and size of vortices generated in the flow field. Adapted with permission from George et al. [21]

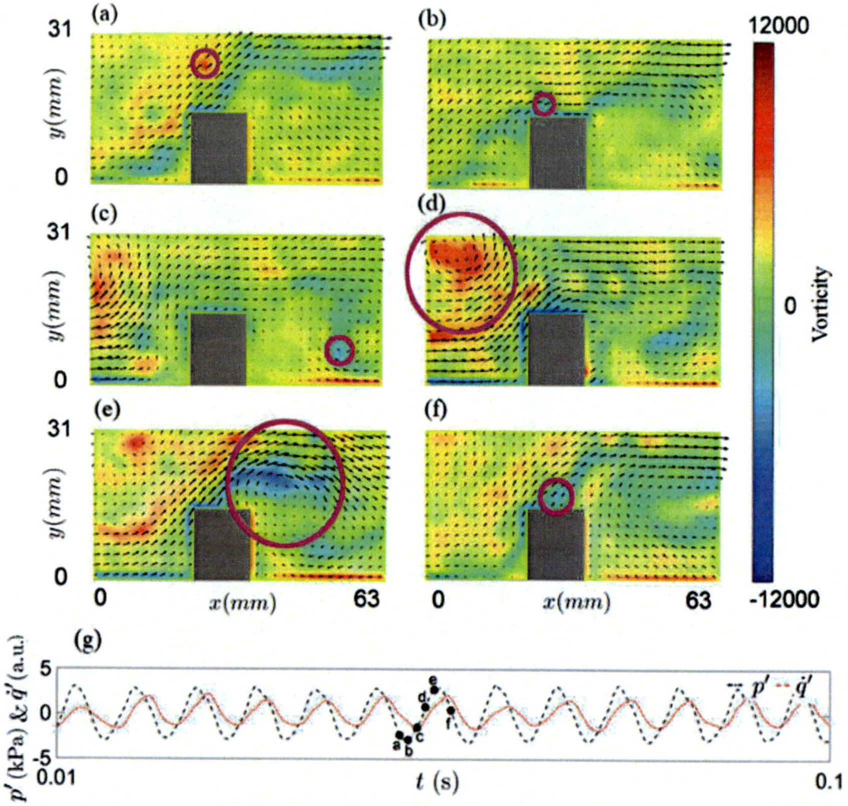

Fig. 7.10 **(a–f)** Instantaneous images of the combustor flow field obtained from PIV measurements for the state of thermoacoustic instability, corresponding to time instances of the acoustic pressure signal shown in **(g)**. Circles indicate the position and the size of vortices generated in the flow field. Adapted with permission from George et al. [21]

the flow structures that are responsible for the sound production in the reaction field of the combustor. The details of these methods for studying the dynamics of a combustor flow field during the onset of thermoacoustic instability can be found in Premchand et al. [50, 51] and Sampath et al. [56].

The existence of the flow-flame interaction under the influence of the acoustic field in a turbulent combustor displays different behaviors of the flame front during the onset of thermoacoustic instability [63]. We can qualitatively study such flame front dynamics from planar Mie scattering images of the reaction field of the combustor, acquired from the upper half of the bluff-body (Fig. 7.11). During these experiments, we need to mix the seeding particles (e.g., olive oil droplets) with the incoming reactants of the combustor. We assume that when these oil droplets pass through the combustion zone, they evaporate and burn. In the presence of a laser sheet, these oil particles existing in the flow (and not in the flame) scatter

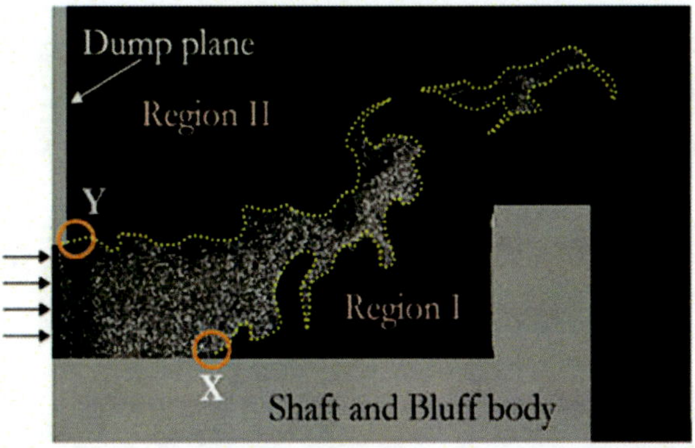

Fig. 7.11 Mie scattering image of the reaction field of a bluff-body combustor, showing the features of the flame front (indicated by dotted yellow contours that envelop the illuminated region of Mie scattered oil droplets) during the state of combustion noise. Arrows on the left side of the image indicate the direction of the reactants entering the combustor. Reproduced with permission from Unni and Sujith [63]

light (called Lorenz-Mie Scattering) [33]. Thus, we can indicate the region of the absence of reaction as the region where oil droplets are present in the combustor, and consider the boundaries of such regions (as indicated by a yellow contour in Fig. 7.11) as the location of the flame fronts [63]. We observe that the flame is stabilized at two points (indicated by X and Y) along the inner shear layer and the outer shear layer (represent regions I and II, respectively). The inner shear layer is formed due to stagnation point flow observed just upstream of the bluff-body and the outer shear layer is formed due to the recirculation zone present downstream of the dump plane. According to Unni and Sujith [63], the stability of these anchor points (X and Y) of the flame determines the dynamical state of the combustor.

In Fig. 7.12, we show such planar Mie scattering images of the combustor reaction field acquired for the states of combustion noise (Fig. 7.12a–e), intermittency (Fig. 7.12f–o), and thermoacoustic instability (Fig. 7.12p–t). During the occurrence of combustion noise (Fig. 7.12u) and epochs of aperiodic oscillations in the intermittency signal (Fig. 7.12w), we notice the formation of small-scale ring vortices (indicated as green circles in each image) at apparently random time intervals at the dump plane (see Figs. 7.12a–e and k–o). These vortices nearly retain their size as they convect along the outer shear, which then perturb the flame tip in an aperiodic manner. The fluctuations in the flame front generate unsteady heat release rate in the system. This generation of unsteady heat release rate, in turn, produces acoustic perturbations. If these acoustic fluctuations are of significant amplitude, they affect the flame front by perturbing the shear layer formed at the dump plane. As the strength of this feedback between the flame and acoustic interaction increases, the system behavior transitions to the state of periodic oscillations [63].

During the epoch of periodic oscillations of intermittency (Fig. 7.12v), we observe a periodic roll-up of the flame front due to the formation of a large-scale

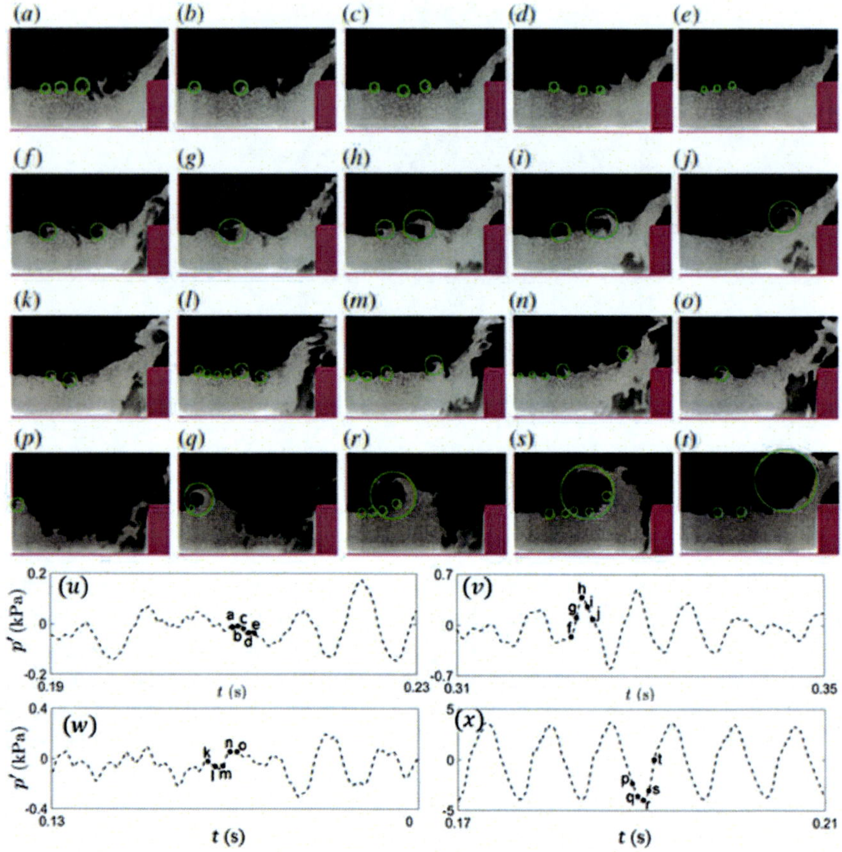

Fig. 7.12 Mie scattering images corresponding to the states of (**a–e**) combustion noise, (**f–j**) epochs of periodic and (**k–o**) aperiodic oscillations of intermittency, and (**p–t**) thermoacoustic instability. (**a–t**) The circles marked in green highlight the vortices in the flow field. The regions marked in purple indicate the bluff-body (at the lower right corner), the central shaft (at the bottom). The dump plane is present at the upper left corner in all images. The time series of pressure fluctuations for the states of (**u**) combustion noise, (**v**) periodic epoch of intermittency, (**w**) aperiodic epoch of intermittency, and (**x**) thermoacoustic instability. Reproduced with permission from George et al. [21]

ring vortex (Fig. 7.12f–j), resulting from the interaction with small-scale vortices developed along the outer shear layer [21]. However, the size of such vortices observed during intermittency is smaller as compared to that observed during the state of thermoacoustic instability (Fig. 7.12r–t). As a result, the roll-up of the flame tip is not that strong during the periodic oscillations of intermittency as compared to that observed during the periodic oscillations of thermoacoustic instability.

During the occurrence of thermoacoustic instability (Fig. 7.12x), a self-organized cascade of small-scale vortices interacting to form a large-scale vortex is observed

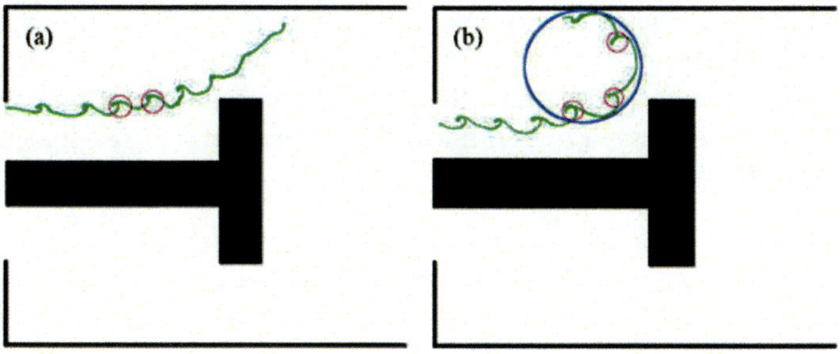

Fig. 7.13 (**a**) Diagrammatic representation of the interaction of small-scale vortices (marked as purple circles) formed along the outer shear layer (green curve) at the dump plane, when collective interaction is absent during epochs of aperiodic oscillations. (**b**) The occurrence of collective interaction leads to the formation of a large-scale vortex (dark blue circle) during epochs of periodic oscillations. Reproduced with permission from George et al. [21]

along the outer shear layer (Fig. 7.12p–t). This interaction of small-scale vortices leading to the formation of a large-scale vortex signifies the existence of collective interaction, resulting in pattern formation in a thermoacoustic system [21]. The collective interaction occurs due to the self-organization of this vortex cascade in every oscillation cycle of the acoustic field. While pattern formation is observed due to the interaction of vortices at micro-scales, it gives rise to self-organized ordered structures at the macro-scales of the system.

Figures 7.13a,b illustrate cartoons of the interaction of small-scale vortices formed along the outer shear layer during epochs of aperiodic and periodic oscillations, respectively. For aperiodic oscillations (Fig. 7.13a), even though small-scale vortices are formed along the unstable outer shear layer, the absence of collective interaction dissipates their strength and, hence, does not lead to the formation of a large-scale structure in the combustor. In contrast, during periodic oscillations (Fig. 7.13b), small-scale vortices formed along the outer shear layer interact with each other strongly to form a large-scale vortex in the system. This vortex then periodically rolls-up the flame front and results in the periodic heat release rate in the system. Such heat release rate fluctuations then periodically feed energy to the acoustic field and establish a positive feedback loop in the combustor. Thus, we observe that a collective interaction of vortices helps to close the feedback loop between the heat release rate oscillations and the acoustic oscillations, resulting in enhanced vortex shedding happening at the acoustic frequency in the system.

Next, we investigate the space-time correlation that arises due to the flow-flame-acoustic interactions observed during the states of combustion noise and thermoacoustic instability. As we discussed before, corrugations are formed along the flame front due to its interaction with different scale vortices formed along the outer shear layer anchored at the dump plane. This flow-flame interaction causes the local heat release rate fluctuations in the system. In contrast, the size of the vortex formed along the shear layer decides the spatial range up to which the correlations are observed in the system. In other words, due to the formation of large-scale structures during the state of thermoacoustic instability, coherence is observed in the local heat release rate fluctuations over a large spatial scale. This results in a stronger spatiotemporal correlation in the local heat release rate fluctuations measured at different locations in the combustor. Thus, we can obtain the space-time correlation, $R(\tau)$, between local heat release rate fluctuations measured at different locations for different time delays as follows [66],

$$R = \frac{\langle \dot{Q}'(x, y, t)\dot{Q}'(x + \Delta x, y, t + \tau)\rangle}{\sqrt{\langle \dot{Q}'^2(x, y, t)\rangle}\sqrt{\langle \dot{Q}'^2(x + \Delta x, y, t + \tau)\rangle}} \tag{7.2}$$

where τ is the time delay, Δx is the distance between two spatial locations (indicated by point A in the inset of Fig. 7.14a), $\langle . \rangle$ is the time average over the entire signal. $\dot{Q}'$ denotes the local heat release rate fluctuations.

Similarly, we can obtain a space-time correlation between the local velocity fluctuations by replacing $\dot{Q}'$ by u' in the above equation. The local heat release rate fluctuations can be obtained from the high-speed OH* chemiluminescence images of the flame, and the local velocity fluctuations can be obtained from PIV measurements of the flow field [21].

In Fig. 7.14, we show the variation of R with τ for $\dot{Q}'$ (Fig. 7.14a,c) and u' (Fig. 7.14b,d) for the states of combustion noise (Fig. 7.14a,b) and thermoacoustic instability(Fig. 7.14c,d). During the occurrence of combustion noise (Fig. 7.14a,b), we observe a very low value of R at all delays for two spatially separated local fluctuations in the combustor field. This further indicates the existence of spatial incoherence in the local heat release rate fluctuations observed during this state. For the state of thermoacoustic instability (Fig. 7.14c,d), a space-time correlation observed between two locations is very high and appears periodic. The time-period of fluctuations in the space-time correlation is the same as that of the acoustic fluctuations in the system. Further, we notice a difference in the delay values at which R shows a maximum value for $\dot{Q}'$ and u' signals. This difference in τ corresponds to the delay in the interaction of these fluctuations in the space. A large value of R observed during the state of thermoacoustic instability is due to the formation of large-scale coherent structures through collective interaction in the system. This large-scale vortex entrains the fresh reactants and hot products over a longer length scale; their mixing and subsequent reaction result in the coherence of the local heat release rate fluctuations over a larger space in the combustor. Further, we note that the spatial coherence in the reaction field of the combustor is not only dependent on the convective nature of the bulk flow, but also depends on the emergence of coherent flow structures in the flow field.

To summarize, in this chapter, we presented the spatiotemporal dynamics of a turbulent combustor during the transition from combustion noise to thermoacoustic instability via intermittency. We showed that the interaction of the flow, the flame, and the acoustic processes during these states results in the formation of different types of patterns in the local acoustic power field of the combustor. This pattern formation acts as a connecting link between the interactions of various subsystems of a turbulent thermoacoustic system. The distribution of the acoustic

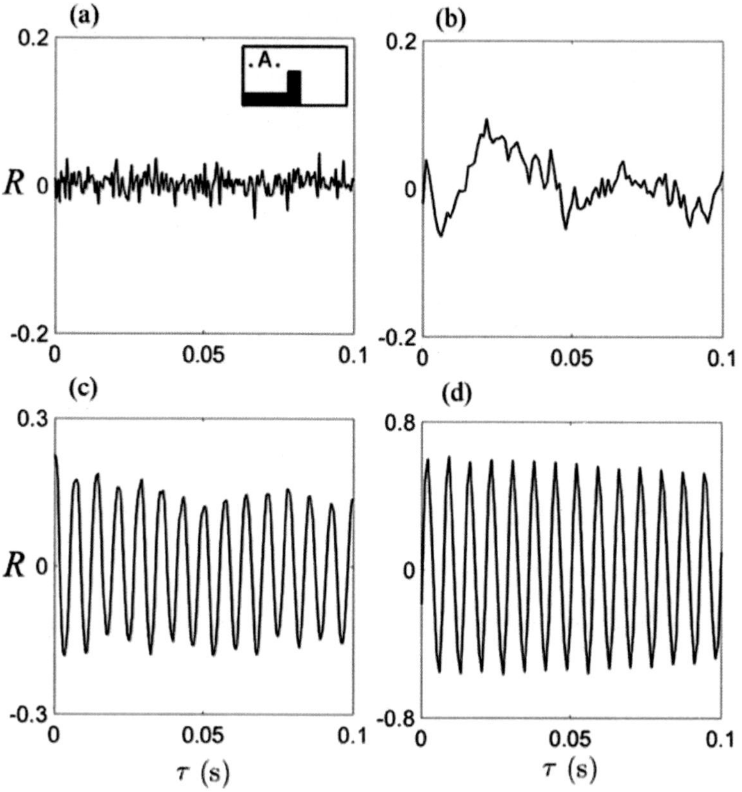

Fig. 7.14 Variation of the space-time correlation, R, as a function of an increase in the spacing between a pair of points τ of (**a**, **c**) the local heat release rate fluctuations and (**b**, **d**) the local velocity fluctuations, corresponding to the states of (**a**, **b**) combustion noise and (**c**, **d**) thermoacoustic instability. Reproduced with permission from George et al. [21]

power field is grainy (incoherent pattern) during the state of combustion noise, while it shows large regions of positive acoustic power (coherent pattern) during the state of thermoacoustic instability. During the state of intermittency, these patterns alternately emerge in the reaction field of the combustor. Furthermore, we investigated the flow-flame interaction and showed that a cascade of small vortices shed along the unstable shear layer exhibits the formation of collective interaction during epochs of periodic oscillations in the system. This collective interaction results in the formation of large-scale vortices in the combustor that, in turn, results in a large-scale roll-up of the flame front. Such collective interaction between small-scale vortices is not observed in the system during the occurrence of aperiodic oscillations.

In the next chapter, we will discuss a quantitative approach based on the framework of synchronization theory to study the coupled interaction between the self-sustained oscillations observed in the acoustic field and the turbulent reacting field during the onset of thermoacoustic instability.

References

1. Ball, P., Borley, N.R.: The Self-made Tapestry: Pattern Formation in Nature, vol. 198. Oxford University Press (1999)
2. Ballal, D.R.: Studies of turbulent flow-flame interaction. AIAA J. **24**(7), 1148–1154 (1986)
3. Baxter, G.W., Behringer, R.P., Fagert, T., Johnson, G.A.: Pattern formation in flowing sand. Phys. Rev. Lett. **62**(24), 2825–2828 (1989)
4. Belik, V., Geisel, T., Brockmann, D.: Natural human mobility patterns and spatial spread of infectious diseases. Phys. Rev. X **1**(1), 011001 (2011)
5. Buka, A., Kramer, L.: Pattern Formation in Liquid Crystals. Springer Science & Business Media (2012)
6. Busse, F.H.: Non-linear properties of thermal convection. Rep. Prog. Phys. **41**(12), 1929–1967 (1978)
7. Camazine, S., Deneubourg, J.L., Franks, N.R., Sneyd, J., Bonabeau, E., Theraula, G.: Self-organization in Biological Systems. Princeton University Press (2003)
8. Candel, S.: Combustion dynamics and control: Progress and challenges. Proc. Combust. Inst. **29**(1), 1–28 (2002)
9. Candel, S., Durox, D., Ducruix, S., Birbaud, A.L., Noiray, N., Schuller, T.: Flame dynamics and combustion noise: Progress and challenges. Int. J. Aeroacoust. **8**(1), 1–56 (2009)
10. Croquette, V.: Convective pattern dynamics at low Prandtl number: Part I. Contemp. Phys. **30**(2), 113–133 (1989)
11. Croquette, V.: Convective pattern dynamics at low Prandtl number: Part II. Contemporary physics **30**(3), 153–171 (1989)
12. Cross, M., Greenside, H.: Pattern Formation and Dynamics in Nonequilibrium Systems. Cambridge University Press (2009)
13. Cross, M.C., Hohenberg, P.C.: Pattern formation outside of equilibrium. Rev. Mod. Phys. **65**(3), 851 (1993)
14. Culick, F.E.C.: Combustion instabilities in propulsion systems. In: Unsteady Combustion, pp. 173–241. Springer (1996)
15. Czirók, A., Vicsek, T.: Collective behavior of interacting self-propelled particles. Physica A **281**(1-4), 17–29 (2000)
16. De Wolf, T., Holvoet, T.: Emergence versus self-organisation: Different concepts but promising when combined. In: International Workshop on Engineering Self-organising Applications, pp. 1–15. Springer (2004)
17. Dolnik, M., Zhabotinsky, A.M., Epstein, I.R.: Resonant suppression of Turing patterns by periodic illumination. Phys. Rev. E **63**(2), 026101 (2001)
18. Dowling, A.P.: Flow-acoustic interaction near a flexible wall. J. Fluid Mech. **128**, 181–198 (1983)
19. Ducruix, S., Schuller, T., Durox, D., Candel, S.: Combustion dynamics and instabilities: Elementary coupling and driving mechanisms. J. Propuls. Power **19**(5), 722–734 (2003)
20. Flood, R.L., Carson, E.R.: Dealing with Complexity: An Introduction to the Theory and Application of Systems Science. Springer Science & Business Media (2013)
21. George, N.B., Unni, V.R., Raghunathan, M., Sujith, R.I.: Pattern formation during transition from combustion noise to thermoacoustic instability via intermittency. J. Fluid Mech. **849**, 615–644 (2018)
22. Gierer, A., Meinhardt, H.: A theory of biological pattern formation. Kybernetik **12**(1), 30–39 (1972)

23. Glansdorff, P., Prigogine, I.: Thermodynamic theory of structure, stability and fluctuations (1971)
24. Goldstein, J.: Emergence as a construct: History and issues. Emergence **1**(1), 49–72 (1999)
25. Gollub, J.P., Langer, J.S.: Pattern formation in nonequilibrium physics. Rev. Mod. Phys. **71**(2), S396 (1999)
26. Guttal, V., Jayaprakash, C.: Spatial variance and spatial skewness: leading indicators of regime shifts in spatial ecological systems. Theor. Ecol. **2**(1), 3–12 (2009)
27. Haller, G.: Lagrangian coherent structures. Annu. Rev. Fluid Mech. **47**, 137–162 (2015)
28. Henderson, R.D.: Nonlinear dynamics and pattern formation in turbulent wake transition. J. Fluid Mech. **352**, 65–112 (1997)
29. Ho, C.M., Huang, L.S.: Subharmonics and vortex merging in mixing layers. J. Fluid Mech. **119**, 443–473 (1982)
30. Ho, C.M., Nosseir, N.S.: Dynamics of an impinging jet. part 1. the feedback phenomenon. J. Fluid Mech. **105**, 119–142 (1981)
31. Holland, J.H.: Emergence: From Chaos to Order. OUP Oxford (2000)
32. Hoyle, R., Hoyle, R.B.: Pattern Formation: An Introduction to Methods. Cambridge University Press (2006)
33. Hulst, H.C., van de Hulst, H.C.: Light Scattering by Small Particles. Courier Corporation (1981)
34. Johnson, S.: Emergence: The Connected Lives of Ants, Brains, Cities, and Software. Simon and Schuster (2002)
35. Kadowaki, S., Hasegawa, T.: Numerical simulation of dynamics of premixed flames: flame instability and vortex–flame interaction. Prog. Energy Combust. Sci. **31**(3), 193–241 (2005)
36. Kéfi, S., Guttal, V., Brock, W.A., Carpenter, S.R., Ellison, A.M., Livina, V.N., Seekell, D.A., Scheffer, M., van Nes, E.H., Dakos, V.: Early warning signals of ecological transitions: methods for spatial patterns. PLoS One **9**(3), e92097 (2014)
37. Lappa, M.: Thermal Convection: Patterns, Evolution and Stability. Wiley (2009)
38. Lieuwen, T.C.: Unsteady Combustor Physics. Cambridge University Press (2012)
39. Lieuwen, T., Zinn, B.T.: Application of multipole expansions to sound generation from ducted unsteady combustion processes. J. Sound Vib. **235**(3), 405–414 (2000)
40. Manneville, P.: Instabilities, Chaos and Turbulence, vol. 1. World Scientific (2010)
41. Meron, E.: Pattern formation in excitable media. Phys. Rep. **218**(1), 1–66 (1992)
42. Moran, P.A.P.: Notes on continuous stochastic phenomena. Biometrika **37**(1/2), 17–23 (1950)
43. Murray, J.: Mathematical Biology II: Spatial Models and Biomedical Applications, vol. 3. Springer (2001)
44. Nair, V.: Role of intermittent dynamics in the onset of combustion instability. Ph.D. thesis, Indian Institute of Technology Madras, India (2014)
45. Nair, V., Thampi, G., Karuppusamy, S., Gopalan, S., Sujith, R.I.: Loss of chaos in combustion noise as a precursor of impending combustion instability. J. Spray Combust. Dyn. **5**(4), 273–290 (2013)
46. Pearson, J.E.: Complex patterns in a simple system. Science **261**(5118), 189–192 (1993)
47. Perera, A.S., Coppens, M.O.: Re-designing materials for biomedical applications: from biomimicry to nature-inspired chemical engineering. Philos. Trans. Royal Soc. A **377**(2138), 20180268 (2019)
48. Poinsot, T.J., Trouve, A.C., Veynante, D.P., Candel, S.M., Esposito, E.J.: Vortex-driven acoustically coupled combustion instabilities. J. Fluid Mech. **177**, 265–292 (1987)
49. Preetham, Santosh, H., Lieuwen, T.: Dynamics of laminar premixed flames forced by harmonic velocity disturbances. J. Propuls. Power **24**(6), 1390–1402 (2008)
50. Premchand, C.P., George, N.B., Raghunathan, M., Unni, V.R., Sujith, R.I., Nair, V.: Lagrangian analysis of intermittent sound sources in the flow-field of a bluff-body stabilized combustor. Phys. Fluids **31**(2), 025115 (2019)
51. Premchand, C.P., George, N.B., Raghunathan, M., Unni, V.R., Sujith, R.I., Nair, V.: Lagrangian analysis of flame dynamics in the flow field of a bluff body-stabilized combustor. J. Eng. Gas Turbines Power **142**(1), 011015 (2020)

52. Renard, P.H., Thevenin, D., Rolon, J.C., Candel, S.: Dynamics of flame/vortex interactions. Prog. Energy Combust. Sci. **26**(3), 225–282 (2000)
53. Rietkerk, M., Van de Koppel, J.: Regular pattern formation in real ecosystems. Trends Ecol. Evol. **23**(3), 169–175 (2008)
54. Rockwell, D., Naudascher, E.: Self-sustained oscillations of impinging free shear layers. Annu. Rev. Fluid Mech. **11**(1), 67–94 (1979)
55. Sachs, T.: Pattern Formation in Plant Tissues, vol. 25. Cambridge University Press (2005)
56. Sampath, R., Mathur, M., Chakravarthy, S.R.: Lagrangian coherent structures during combustion instability in a premixed-flame backward-step combustor. Phys. Rev. E **94**(6), 062209 (2016)
57. Schadow, K.C., Gutmark, E.: Combustion instability related to vortex shedding in dump combustors and their passive control. Prog. Energy Combust. Sci. **18**(2), 117–132 (1992)
58. Schmid, P.J.: Dynamic mode decomposition of numerical and experimental data. J. Fluid Mech. **656**, 5–28 (2010)
59. Schweisguth, F., Corson, F.: Self-organization in pattern formation. Dev. Cell **49**(5), 659–677 (2019)
60. Shin, D.H., Lieuwen, T.: Flame wrinkle destruction processes in harmonically forced, turbulent premixed flames. J. Fluid Mech. **721**, 484–513 (2013)
61. Sivashinsky, G.I.: Instabilities, pattern formation, and turbulence in flames. Annu. Rev. Fluid Mech. **15**(1), 179–199 (1983)
62. Stöhr, M., Boxx, I., Carter, C.D., Meier, W.: Experimental study of vortex-flame interaction in a gas turbine model combustor. Combust. Flame **159**(8), 2636–2649 (2012)
63. Unni, V.R., Sujith, R.I.: Flame dynamics during intermittency in a turbulent combustor. Proc. Combust. Inst. **36**(3), 3791–3798 (2017)
64. Vicsek, T., Zafeiris, A.: Collective motion. Phys. Rep. **517**(3-4), 71–140 (2012)
65. Walgraef, D.: Spatio-temporal Pattern Formation: with Examples from Physics, Chemistry, and Materials science. Springer Science & Business Media (2012)
66. Wallace, J.M.: Space-time correlations in turbulent flow: A review. Theor. App. Mech. Lett. **4**(2), 022003 (2014)

Chapter 8
Synchronization Between the Acoustic Field of the Confinement and the Turbulent Reacting Flow

In the previous chapter, we showed that the phenomenon of thermoacoustic instability is a result of complex interaction of various coupled nonlinear processes such as chamber acoustics, unsteady flame dynamics, and hydrodynamics present in a thermoacoustic system. Among these processes, the feedback mechanism between the acoustic field in the chamber and the heat release rate fluctuations in the flame plays a dominant role in the genesis of such thermoacoustic instability. In this feedback coupling, the heat release rate fluctuations act as a source of driving for the acoustic oscillations in a duct and also receive perturbations that are generated by acoustic oscillations. Therefore, we need to study the mutual coupling between these two processes for achieving a deeper understanding of the onset of thermoacoustic instability. In the present chapter, we will investigate the coupled interaction of the acoustic pressure and heat release rate fluctuations during the onset of thermoacoustic instability in turbulent systems, using the framework of mutual synchronization. Furthermore, we will independently present the characteristics of the forced response of self-excited acoustic pressure fluctuations and hydrodynamic fluctuations to external periodic perturbations, using the framework of forced synchronization. Such investigations may help us understand the flame-acoustic coupling in thermoacoustic systems and devise control strategies required for the mitigation of thermoacoustic instability.

Lord Rayleigh [82] pioneered in putting forth a physical explanation for the occurrence of thermoacoustic instability. According to his criterion for thermoacoustic instability, the heat release rate in the flame adds energy to the acoustic field only if the heat is supplied at the maximum of the acoustic pressure cycle or removed at the minimum of the acoustic pressure cycle. Further, this criterion was modified to incorporate the effect of damping in the system. It was found that the heat release rate fluctuations in the flame can drive the acoustic fluctuations in the system if the phase difference between these fluctuations lies between −90 to +90 degrees and when the acoustic driving overcomes the damping in the system [106]. Mathematically the Rayleigh's criterion is expressed as,

R. I. Sujith, S. A. Pawar, *Thermoacoustic Instability*, Springer Series in Synergetics,
https://doi.org/10.1007/978-3-030-81135-8_8

$$\int_0^T \int_0^V p'\dot{Q}' dv dt > \int_0^T \int_0^V \Phi dv dt \tag{8.1}$$

where T is the period of oscillation, V is the control volume (the combustor volume), Φ is the acoustic dissipation term, p' is the acoustic pressure fluctuation, and $\dot{Q}'$ is the heat release rate fluctuation.

Therefore, for the onset of thermoacoustic instability, the value of the term on the left-hand side (i.e., acoustic driving or also called Rayleigh's integral) has to be greater than the term on the right-hand side (i.e., acoustic dissipation) in (8.1). This inequality is satisfied only when an in phase relationship is established between p' and $\dot{q}'$ fluctuations in the system.

In Chap. 6, we showed that the onset of thermoacoustic instability in turbulent combustors exhibits the transition of the system behavior from a state of low amplitude aperiodic fluctuations (combustion noise) to large amplitude self-sustained periodic oscillations (thermoacoustic instability) via intermittency. During the state of intermittency, the system behavior abruptly switches between epochs of low amplitude aperiodic fluctuations and bursts of large amplitude periodic oscillations. Since the integral of the acoustic driving term in (8.1) is averaged over an oscillation cycle (i.e., a period of the oscillation, T), this calculation is possible only for signals that have a definite time period (specifically, for periodic signals observed during the state of thermoacoustic instability). Therefore, the calculation of Rayleigh's integral for the signals observed during the states of combustion noise and intermittency would be ambiguous due to the absence of definite timescales. In such a situation, we need an alternate framework that can characterize the interaction between the coupled subsystems more accurately during these dynamical states.

To address this situation, Pawar et al. [72] proposed the framework of mutual synchronization of coupled oscillators [79] to study the interaction between the acoustic field and the heat release rate field during the intermittency route to thermoacoustic instability in turbulent combustors. With the help of synchronization theory, we can systematically study the instantaneous interaction of these subsystems. Further, the simultaneous characterization of these subsystems may help us in realizing the underlying phenomenon responsible for the occurrence of different dynamical states in a thermoacoustic system. This framework can also help us in detecting the onset of thermoacoustic instability in a system [26]. In the following subsection, we provide an overview of the synchronization phenomenon of coupled oscillators.

8.1 Basics of Synchronization of Coupled Oscillators

The synchronization phenomenon was discovered in the seventeenth century by Huygens [9, 39, 40, 79, 103] while observing the coupled motions of two pendulum clocks hung on a common wooden beam. At the time of synchronization, he noticed the out-of-phase locking of rhythms of both the pendulum clocks. This 'odd sympathy' of coupled pendulum clocks has become the basis for later studies on understanding the coupled interaction between nonlinear oscillators. The observation of such synchronization phenomenon is ubiquitous in nature. Since the time of Huygens, synchronization phenomenon has been reported in various fields including ecology, biology, geology, and engineering systems. We see many examples of synchronization in our day-to-day life; a few of them include, beating of heart cells, firing of neurons in the brain, flashing of fireflies in unison, chirping of crickets, flocking of birds, swarming of fish schools, and rhythmic applause of an audience [97].

To study the phenomenon of synchronization, we require a minimum of two oscillators. Further, the inherent dynamics of these individual oscillators need to be self-sustained in the uncoupled state [79]. Here, the term self-sustained refers to the fact that the oscillators do not need any external energy (or forcing) to maintain their oscillations. However, the generation and maintenance of these oscillations are solely dependent on the balance between the internal energy sources and the dissipation losses in the system [43]. The mutual interaction of two or more such self-sustained oscillators, at appropriate coupling conditions, engenders the oscillators to adjust their time or frequency scales to a common value, resulting in a state of mutual synchronization.

In Fig. 8.1, we demonstrate the synchronization characteristics of two pendulums hung over a common beam. When the oscillatory pendulums are mounted on separate beams, even if they start at the same initial condition, the relative motion

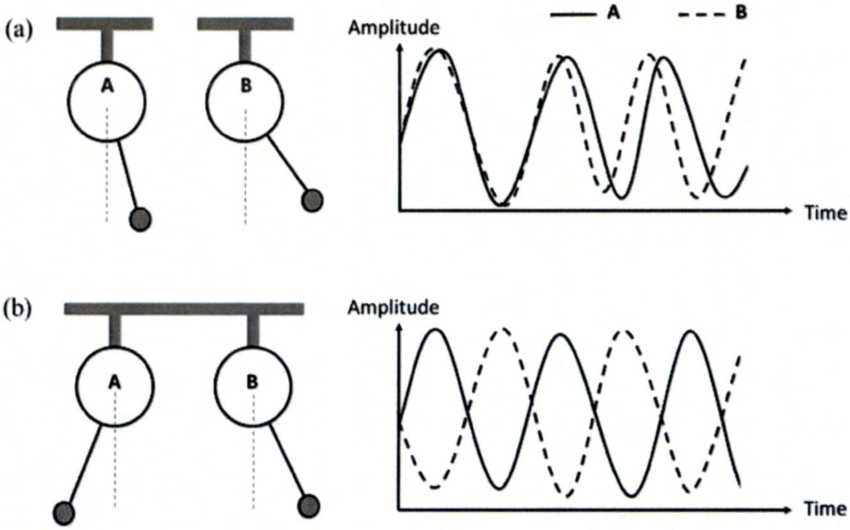

Fig. 8.1 Schematic showing the importance of coupling in synchronization of two oscillatory pendulum clocks. (**a**) When the pendulum clocks are mounted on separate beams, their oscillatory motion is desynchronized, and (**b**) when they are mounted on the same beam, their motions are locked in time (i.e., they exhibit synchronization)

between these pendulums exhibits a drift in time. However, when these pendulums are hung on a common beam, the mutual interaction induced by the beam drives the pendulums to adjust their motions to a common phase and frequency. In the state of synchronization (or phase locking), a small perturbation through an external source may momentarily disturb the synchronized motion of the oscillators. However, upon removal of the effect of such a disturbance, these oscillators return to their synchronization state.

The synchronization phenomenon has been studied for both periodic (limit cycle) and chaotic oscillators [6, 79]. Synchronization of periodic oscillators is easier to achieve as compared to chaotic oscillators [12]. Chaotic oscillators can synchronize, despite their sensitivity to initial conditions, has fascinated many researchers over the years [15, 16, 48, 55, 85, 86, 88].

Depending on the type of interaction existing between the oscillators, we can classify synchronization of oscillators as mutual synchronization and forced synchronization [6]. Mutual synchronization is characterized by the existence of bidirectional coupling between the oscillators, where every oscillator is in the direct (or indirect) influence of every other oscillator through coupling (Fig. 8.2a). On the other hand, forced synchronization indicates the existence of unidirectional coupling between the oscillators, where a forcing oscillator only affects the dynamics of a forced oscillator and does not receive any influence from it in return (Fig. 8.2b). In the state of mutual synchronization, the synchronization frequency of oscillators lies in between their natural frequencies observed during the uncoupled state. In contrast, in the state of forced synchronization, the frequency of the forced (driven) oscillator locks to the frequency of the forcing (driver) oscillator.

Furthermore, depending on the resultant dynamics exhibited by synchronized oscillators, we can classify the state of synchronization into different types such as phase synchronization, lag synchronization, complete synchronization, and generalized synchronization [49, 79]. In the presence of weak coupling, synchronization of oscillators leads to the locking of their instantaneous phases without affecting their amplitudes. Hence, we refer to this state of synchronization as *phase synchronization* [85]. In the presence of strong coupling, we notice a perfect locking between the instantaneous phases as well as amplitudes of the oscillators, such that the resultant dynamics of these oscillators appear identical. We call such a form of synchronization as *complete (identical) synchronization* [25, 76]. In the case of *lag synchronization*, the oscillators show perfect locking of their amplitudes; however, they exhibit a finite phase difference in time [86]. In the case of in-phase synchronization, the oscillators are locked with a zero-degree phase difference. In contrast, in the case of anti-phase synchronization, the oscillators are locked with a phase difference of 180 degrees.

In some cases, synchronization of oscillators leads to the existence of a functional relationship between them. We refer to this state of synchronization as *generalized synchronization* [88]. For example, if $X(t)$ and $Y(t)$ denote the behavior of two non-identical oscillators, then during the state of generalized synchronization, these oscillators exhibit a functional relationship such that $Y(t) = \Phi(X(t))$. This means, the response $Y(t)$ of one oscillator can be uniquely derived from the dynamics

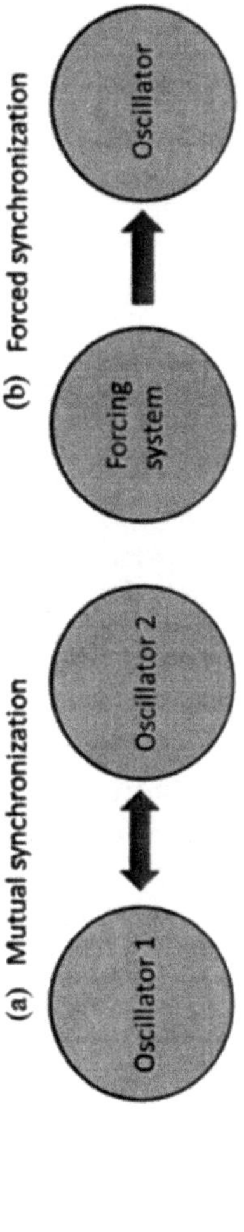

Fig. 8.2 Diagrammatic illustration of the interaction between oscillators that are **(a)** mutually coupled and **(b)** forced. Here, an arrow indicates the direction of the interaction between the oscillators. This suggests that coupled oscillators affect each other during mutual synchronization, while only forcing system affects the dynamics of an oscillator during forced synchronization

$X(t)$ of the other oscillator in the state of synchronization. Sometimes, a state of *intermittent synchronization* occurs prior to the state of complete synchronization, where oscillators exhibit alternate synchronized and desynchronized oscillations at irregular intervals of time [14].

When the number of oscillators in a population is high (greater than two), we observe complex synchronized patterns during the state of synchronization or while transitioning to the state of synchronization [49, 56, 57, 96]. During the occurrence of synchronization, the coupled behavior of all oscillators in a population displays identical frequency and retains a constant phase difference. Under certain coupling conditions, the population of oscillators splits into two or more synchronized groups, also known as clusters. We can refer to this state of coupled oscillators as *clustering* [27]. During the state of clustering, oscillators belonging to a particular cluster exhibit the identical dynamics (i.e., the same phase and amplitude) and such dynamics are different for oscillators in other clusters. We note that the clusters in a population of oscillators are always synchronized with each other (i.e., clusters have a constant non-zero phase difference) and the frequency of oscillators in all the clusters remains the same.

The emergence of a small desynchrony (i.e., a mismatch in frequencies), in a population of otherwise synchronized identical oscillators, divides the population into groups of synchronized and desynchronized oscillators. This state of coexistence of synchrony and desynchrony in the same population of oscillators is referred to as *chimera* [47]. The name chimera is derived from the Greek mythological creature having the body of a lion, the head of a goat, and the tail of a serpent [2]. Tinsley et al. [100] and Hagerstrom et al. [36] reported the experimental evidence of such a chimera state, for the first time, in a system of chemical oscillators and coupled-map lattice, respectively.

After a brief introduction to synchronization phenomenon of coupled oscillators, we next summarize the tools that have been developed to detect the synchrony between them. We can compute synchronization of any two coupled oscillators from the locking of their instantaneous phases (ϕ) or instantaneous frequencies (ω) in time. During the state of synchronization, the instantaneous phase difference ($\Delta\phi$) between such oscillators remains bounded and exhibits a constant value in time. We can refer to this condition as *phase locking* [79]. We can confirm the phase locking of a pair of oscillators using the following condition,

$$\Delta\phi_{i,j}(t) = |\phi_i(t) - \phi_j(t)| = \text{constant} \tag{8.2}$$

where i and j are two oscillators that are coupled. The frequencies of both these oscillators remain the same during the state of phase locking (i.e., $\omega_i = \omega_j$). We can call this condition as *frequency locking*.

During the state of synchronization, the instantaneous phase difference between two oscillators remains constant for noise-free signals and shows small fluctuations about a mean value for noisy signals. When the oscillators are desynchronized, the instantaneous phase difference between them displays a continuous drift (growth or decay) in time.

Over the years, several tools have been developed to detect the synchronization of coupled oscillators [42, 49, 79]. One common method to study synchronization of two oscillators is to calculate the instantaneous properties (i.e., phase, amplitude and frequency) of their signals. These instantaneous properties are usually calculated using an analytic signal approach based on the Hilbert transform [79]. Here, an analytic signal is obtained by generating a complex signal whose real part is the original signal and the imaginary part is its Hilbert transform. Thus, we can express the analytic signal as,

$$\zeta(t) = x(t) + i x_H(t) = A(t)e^{i\phi(t)} \tag{8.3}$$

Here, $x(t)$ is the original signal whose instantaneous properties we need to calculate, and $x_H(t)$ is the Hilbert transform of $x(t)$ which we can obtain as,

$$x_H(t) = \frac{1}{\pi}\text{P.V.}\int_{-\infty}^{\infty}\frac{x(\tau)}{t-\tau}d\tau \tag{8.4}$$

where P.V. is the Cauchy principal value of the integral. Therefore, we can calculate the instantaneous phases of an oscillator as $\phi(t) = \tan^{-1}[x_H(t)/x(t)]$ and the instantaneous amplitudes as $A(t) = \sqrt{x^2(t) + x_H^2(t)}$. The instantaneous angular frequency of an oscillator is obtained as $\omega(t) = d\phi(t)/dt$.

Furthermore, we note that the instantaneous phase defined by the Hilbert transform has a physical meaning only if the phase trajectory in the analytic plane rotates around a unique center of rotation (Fig. 8.3a), i.e., the attractor is *phase coherent* [49]. Here, an analytic plane is obtained by plotting the original signal against its Hilbert transformed signal. Therefore, the calculation of the Hilbert phase is valid only for signals with a narrowband spectrum. For signals with a broadband spectrum, i.e., signals consisting of multiple time scales (Fig. 8.3b), the attractor is *phase non-coherent* [49]. Hence, it is not straightforward to define the phase of such signals. For analysing the synchronization properties of phase non-coherent signals, we can use methods such as trajectory curvature [70], intrinsic mode decomposition [102], method of projection [70], cross wavelet transform [42], correlation between two signals [7, 93], and recurrence based analysis [84]. In this chapter, we will discuss the method based on the recurrence plot in detail, while for the information on other methods, readers may refer to the relevant references presented above.

In some cases, the signal may possess particular marker events (i.e., the occurrence of local maximas, minimas, or zero-crossings) that repeat after a certain time interval. For example, the recurrence of R-peaks in the ECG signal of the human heart indicates the occurrence of local maxima in consecutive cycles of the signal [79]. For such signals, we can use the method of Poincaré section to calculate their instantaneous phases. Here, we consider that the consecutive occurrence of such discrete events (t_k) increases the phase of the signal by 2π radians. Thus, the phase of a signal exhibits a piece-wise linear progression between the marker events, which we can obtain as follows [79],

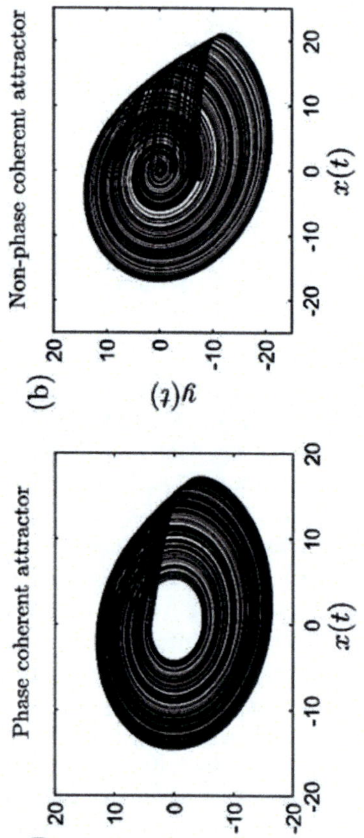

Fig. 8.3 Difference between (**a**) a phase-coherent and (**b**) a phase non-coherent Rösseler attractor: $\dot{x} = -y - z$; $\dot{y} = x + ay$; $\dot{z} = 0.2 + z(x - 10)$, where $a = 0.15$ for the phase coherent and $a = 0.25$ for the non-phase coherent attractor. For the phase-coherent attractor, the trajectory in the analytic plane displays a smeared limit cycle with a single center of rotation. For the phase non-coherent attractor, the trajectory shows cluttered behavior with multiple centers of rotation

$$\phi(t) = 2\pi k + 2\pi \frac{t - t_k}{t_{k+1} - t_k} \tag{8.5}$$

where $t_k \leq t < t_{k+1}$, $k = 1, 2, 3, \ldots, N_k$, and N_k is the total number of peaks in the signal.

8.2 Mutual Synchronization of the Acoustic Field and the Turbulent Reactive Flow Field in Thermoacoustic Systems

From the previous section, we have understood that for implementing the framework of synchronization theory, two criteria need to be satisfied: the minimum number of oscillators should be two and these oscillators should exhibit self-sustained oscillations.

In a turbulent thermoacoustic system, we have two oscillators. The acoustic field in a duct with turbulent flow is a self-sustained oscillator. In the absence of combustion, the acoustic fluctuations generated by the turbulent flow in a duct show low amplitude, self-sustained chaotic oscillations [26]. Further, in the absence of acoustic perturbations, open turbulent flames display broadband fluctuations in the heat release rate [81, 95]. Thus, turbulent flames uncoupled with the duct acoustic field can behave like a self-sustained oscillator with aperiodic fluctuations. In summary, we can consider the acoustic field and the heat release rate field in the combustor as self-sustained chaotic oscillators, under the influence of turbulent flow. Otherwise, both these oscillators are damped oscillators. The presence of continuous turbulent flow fluctuations make them self-sustained oscillators. Therefore, in a turbulent combustor, we study the onset of thermoacoustic instability using the framework of synchronization.

When these oscillators are present in the same combustion system, the presence of confinement engenders weak inherent coupling between them during the occurrence of combustion noise (i.e., stable operation). This scenario is analogous to the human respiratory system, where the synchronization between the cardiovascular and the respiratory rhythms is studied even though these subsystems cannot be considered to be independent, as they are weakly coupled through our body [92]. During the state of thermoacoustic instability (i.e., unstable operation), both the oscillators, the acoustic field and the turbulent reacting field, exhibit strongly coupled in-phase periodic oscillations [72]. Thus, using the application of the synchronization theory, we can investigate and quantify the coupled interaction between the acoustic pressure and heat release rate fluctuations during the emergence of order from chaos in turbulent thermoacoustic systems.

In contrast, for laminar thermoacoustic systems, both the acoustic field and the reacting flow field appears as damped oscillators during the state of stable operation and display a steady state behavior in their dynamics [101]. Due to the absence of self-sustained oscillations, the application of synchronization theory to study the dynamics of stable operation and the transition from stable to unstable operation in such systems would be questionable. Nevertheless, laminar thermoacoustic systems display many self-oscillatory states (e.g., limit cycle, quasiperiodicity, strange nonchaos, and chaos), as we discussed in Chap. 4. Thus, we can successfully apply tools from synchronization theory to analyze the coupled interaction between the acoustic pressure and the heat release rate fluctuations in the flame, during the transition from one state of thermoacoustic oscillations to another, as elaborated in [63, 101].

In the following subsections, we focus on investigating both the temporal and the spatiotemporal interactions between the coupled acoustic field and heat release rate field in turbulent thermoacoustic systems, during the intermittency route to thermoacoustic instability.

8.2.1 *Temporal Dynamics of the Acoustic Pressure Field and the Heat Release Rate Field in a Turbulent Combustor*

The dynamical behavior of a thermoacoustic system, due to the variation of a control parameter, is usually characterized in terms of the acoustic pressure fluctuations and the global (or local) heat release rate fluctuations [20, 98]. The simultaneous characterization of oscillations of both these subsystems may help in studying the instantaneous coupling existing between them. Pawar et al. [72] investigated the coupled interaction of the simultaneously acquired acoustic pressure fluctuations and global heat release rate fluctuations during the intermittency route to thermoacoustic instability in a turbulent bluff-body stabilized dump combustor (refer to Fig. 7.3 in Chap. 7).

In Fig. 8.4, we show the dynamical behavior of a turbulent thermoacoustic system for different operating conditions of the mean velocity of the air flow, $\bar{u}$. With the variation in the flow velocity, we observe that the system behavior changes from a state of combustion noise to thermoacoustic instability via intermittency [72]. During the state of combustion noise, both the acoustic pressure and the heat release rate signals exhibit low amplitude chaotic fluctuations, and these oscillations are unlocked in time (Fig. 8.4a). When the system behavior displays intermittency (Fig. 8.4b), both the signals show locking of their oscillations during the epochs of bursts of periodicity and unlocking of their oscillations during the epochs of low amplitude chaos.

As the system dynamics transitions to the state of thermoacoustic instability, we observe the locking of periodic (limit cycle) oscillations of both the signals in time. Further, we notice two types of thermoacoustic instabilities in the system, which are: 'weakly correlated limit cycle' and 'strongly correlated limit cycle'. During weakly correlated limit cycle oscillations, both signals exhibit a wide variation in their cycle-to-cycle amplitudes; however, their instantaneous phases are locked in time (Fig. 8.4c). In contrast, for strongly correlated limit cycle oscillations, both the signals display a high level of correlation in their instantaneous amplitudes as well as in phases (Fig. 8.4d). Pawar et al. [72, 74] showed that during the state of strongly correlated limit cycle, the instantaneous frequency of the vortex shedding is nearly the same as the natural frequency of the acoustic pressure oscillations. In contrast, the vortex shedding frequency is slightly mismatched with the acoustic frequency during the state of weakly correlated limit cycle. As a result, the amplitude of the acoustic pressure oscillations observed during the state of strongly correlated limit cycle is much larger than that observed for the state of weakly correlated limit cycle. Furthermore, they found that mean phase difference between the acoustic pressure and heat release rate signals during the state of weakly correlated limit cycle is sufficiently large (around 50 deg) and that during the state of strongly correlated limit cycle is near zero deg.

The frequency domain analysis of these coupled acoustic pressure (p') and heat release rate ($\dot{q}'$) signals shows interesting features of their frequency-locking behavior during the onset of thermoacoustic instability (see Fig. 8.5). During the state of combustion noise, we observe different values of the dominant frequencies for the acoustic pressure (p') and the heat release rate ($\dot{q}'$) signals (see green arrows corresponding to f_a and f_h, respectively, in Fig. 8.5b and a shaded region in Fig. 8.5c). The dominant frequency of the global heat release rate fluctuations (f_h) is observed in the lower frequency range and it results primarily from the unsteady hydrodynamic field of the combustor (as discussed in Chap. 7). In contrast, the

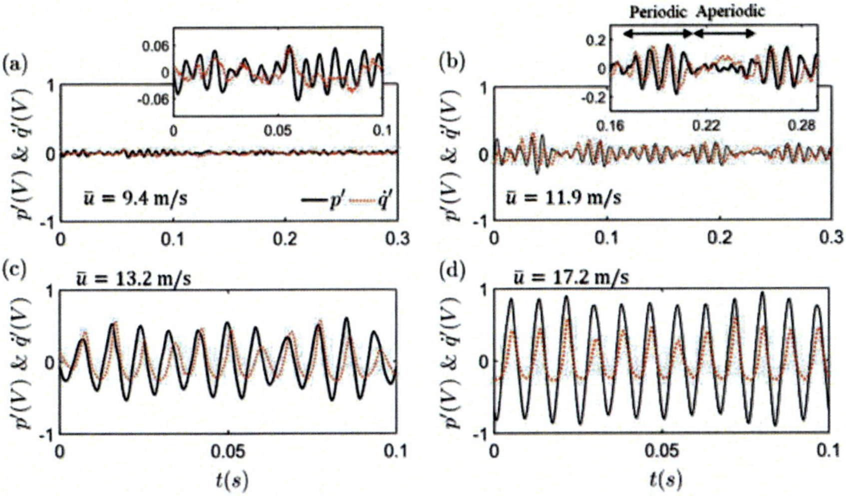

Fig. 8.4 Time series of the acoustic pressure (p' in black) and the heat release rate fluctuations ($\dot{q}'$ in red) observed for different dynamical states such as **(a)** combustion noise, **(b)** intermittency, **(c)** weakly correlated, and **(d)** strongly correlated limit cycle oscillations (thermoacoustic instability) in a turbulent bluff-body stabilized dump combustor, when the flow velocity ($\bar{u}$) is varied as the control parameter. Adapted with permission from Pawar et al. [72]

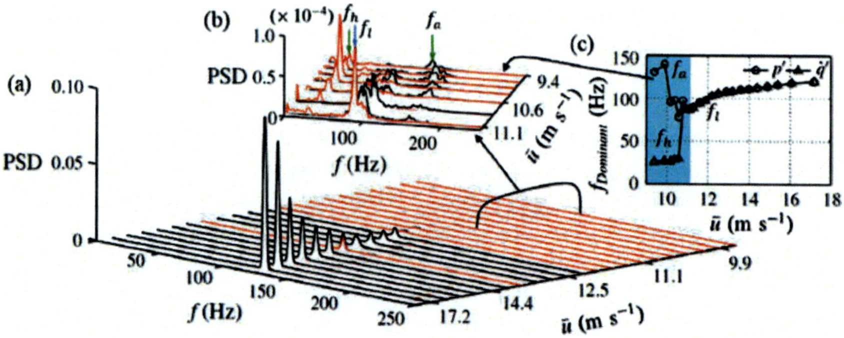

Fig. 8.5 (a, b) Waterfall diagrams showing the variation of power spectral density (PSD) of the acoustic pressure (p' in black) and the heat release rate ($\dot{q}'$ in red) signals for the intermittency route to thermoacoustic instability. (c) The variation of the dominant frequency of these oscillations with the mean velocity of the flow ($\overline{u}$). During the state of combustion noise (the shaded region in c), we notice the presence of dominant frequencies at different values in the spectrum of acoustic pressure (f_a) and heat release rate (f_h) signals. These frequencies then shift to the common dominant frequency (f_l) during the transition from combustion noise to intermittency in a turbulent thermoacoustic system. Reproduced with permission from Pawar et al. [72]

dominant frequency of acoustic pressure oscillations (f_a) is observed to be higher than f_h and it corresponds to the fundamental mode (in the present case, a quarter-wave mode) of the longitudinal acoustic oscillations in the system.

During the state of intermittency, both the acoustic pressure and the heat release rate signals show a common dominant frequency (indicated as f_l with a blue arrow in Fig. 8.5b). Note that the region of intermittency starts immediately after the shaded region in Fig. 8.5c. The value of this common frequency of both the acoustic pressure and heat release rate signals lies between the dominant frequencies of these signals observed during the state of combustion noise (Fig. 8.5c). Therefore, during the epoch of periodic oscillations of intermittency, both the acoustic pressure and heat release rate signals show a common frequency (frequency-locking) at f_l. While during the epoch of aperiodic oscillations of intermittency, these signals display different dominant frequencies at f_a and f_h, respectively. We also note that, as soon as both the acoustic pressure and the heat release rate oscillations exhibit frequency locking (or phase locking) during the state of intermittency, their instantaneous dynamics displays growth in the amplitude as well as the transition from aperiodic to periodic oscillations (Fig. 8.4b).

As the system dynamics approaches thermoacoustic instability, the periodic content in both the acoustic pressure and the heat release rate signals becomes stronger and sustained for a longer duration. This results in an increase in the amplitude of the common frequency peak in the respective spectra of these signals (Fig. 8.5a). Furthermore, the dominant frequency of these oscillations displays a continuous increase with a corresponding increase in the flow velocity (Fig. 8.5c). At the onset of thermoacoustic instability, aperiodic dynamics observed in both the signals completely transitions to periodic dynamics. During this state, both the signals exhibit a single frequency corresponding to the unstable mode of the acoustic field in the combustor, which happens due to the phenomenon called *vortex-acoustic lock-in* [4, 18, 94]. During the occurrence of vortex-acoustic lock-in, the frequency of both the acoustic pressure and heat release rate signals remains nearly constant, although the mean flow velocity is increased to a very high value [74].

Next, we investigate the coupled behavior of the acoustic pressure and the heat release rate signals (Fig. 8.4) using the framework of mutual synchronization, for the intermittency route to thermoacoustic instability.

8.2.2 Temporal Synchronization of the Acoustic Pressure and the Global Heat Release Rate Fluctuations

Here, we discuss the coupled interaction between the acoustic pressure and the global heat release rate signals using a statistical measure of synchronization based on recurrence plots, called the probability of recurrence, $P(\tau)$ [72]. As we discussed in Chap. 2, recurrence indicates the behavior of a phase space trajectory to visit its previous location or the neighborhood after some divergence in the phase space. We can calculate the probability of the trajectory to visit its former location in the phase space, say x_i, after a time lag τ as follows [84],

$$P(\tau) = \frac{1}{N-\tau} \sum_{i=1}^{N-\tau} \Theta(\epsilon_i - \|x_i - x_{i+\tau}\|) \tag{8.6}$$

where $P(\tau)$ is the probability of recurrence (or the generalized autocorrelation function), Θ is the Heaviside function, N is the total number of delayed vectors, ϵ_i is the predefined recurrence threshold centered at x_i, and $\|.\|$ is the Euclidean norm. The value of $P(\tau)$ varies between 0 and 1.

The probability of recurrence computes the correlation between a state vector and its delayed version for increasing values of the delay time (τ). At the zero time lag (i.e., $\tau = 0$), the value of $P(\tau)$ is always unity. For a periodic signal, $P(\tau)$ shows a peak near unity after every period of the oscillations and shows a value near zero at every other instances. On the other hand, for chaotic or aperiodic signals, since the period of the oscillations is not constant in time, the peaks of $P(\tau)$ show values lower than unity and are located at unequal intervals of time.

In order to study synchronization properties of two signals, we need to detect the similarity in the variation of $P(\tau)$ plots of the signals. The coincidence of time instances corresponding to the maximum values of $P(\tau)$ indicates the phase-locking of the signals, while the concurrence of maximum values of $P(\tau)$ at these instances signifies amplitude-locking behavior of these signals. Therefore, the coincidence of only the positions of the peaks of $P(\tau)$ for two signals indicates the existence of phase synchronization between them [84]. Conversely, the locking of both the

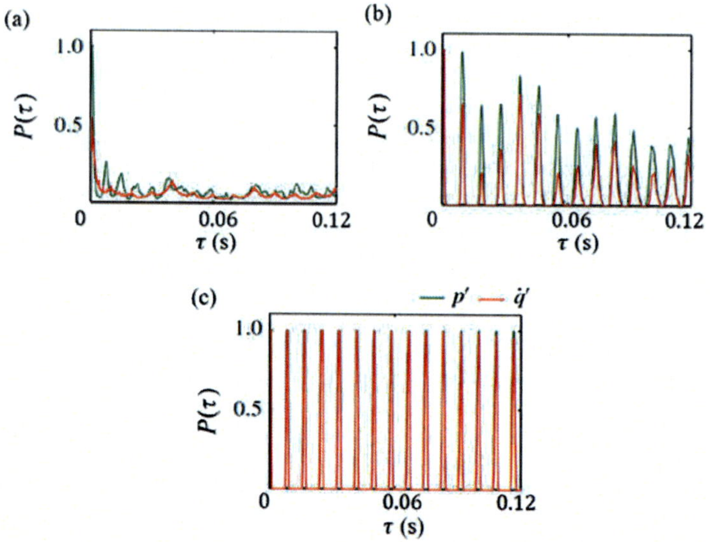

Fig. 8.6 The plots showing the variation of the probability of recurrence $P(\tau)$ as a function of the time lag (τ) for the states of (**a**) combustion noise, (**b**) weakly and (**c**) strongly correlated limit cycle oscillations. (**a**) For the state of combustion noise, the variation of $P(\tau)$ with τ is uncorrelated in time. (**b**) For weakly correlated limit cycle, the variation of $P(\tau)$ with τ shows locking of only positions of the peaks and, (**c**) for strongly correlated limit cycle, it shows the locking of both the positions and the heights of the peaks. Reproduced with permission from Pawar et al. [72]

position and the height of the peaks of $P(\tau)$ of the signals represent the existence of generalized synchronization between them [84].

For the state of combustion noise (Fig. 8.6a), we observe that the variation of $P(\tau)$ with τ does not match for both the acoustic pressure and heat release rate signals, due to the difference in their time scales. Thus, the signals are desynchronized in time. During the states of thermoacoustic instability (Figs. 8.6b,c), we notice the locking of the position of the peaks of $P(\tau)$ for both the signals, indicating the locking of their instantaneous phases. For the case of weakly correlated limit cycle oscillations (Fig. 8.6b), we observe the matching of only the position of the peaks of $P(\tau)$ and not their magnitudes for the acoustic pressure and heat release rate signals. This suggests the existence of phase synchronization between these signals during the state of weakly correlated limit cycle oscillations. On the other hand, for the case of strongly correlated limit cycle oscillations (Fig. 8.6c), both the positions and the heights of the peaks of $P(\tau)$ show a perfect matching for the acoustic pressure and the heat release rate signals. Such a locking behavior of $P(\tau)$ indicates the existence of generalized synchronization between them during the state of strongly correlated limit cycle oscillations.

For the state of intermittency (Fig. 8.7), both the acoustic pressure and the heat release rate signals show the characteristics of phase synchronization during epochs of periodic oscillations (Fig. 8.7a) and desynchronization during epochs of

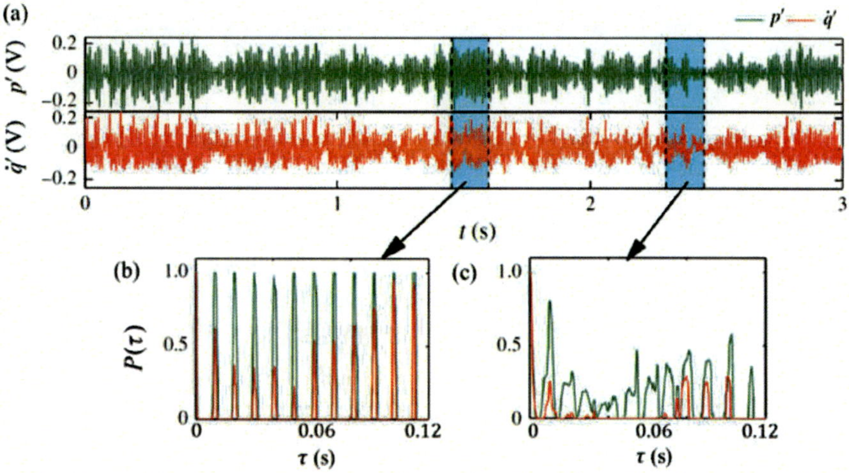

Fig. 8.7 (**a**) Signals of acoustic pressure (p') and heat release rate ($\dot{q}'$) fluctuations during the state of intermittency. (**b, c**) Plots of $P(\tau)$ versus τ show the characteristics of phase synchronization during epochs of periodic oscillations and desynchronization during epochs of aperiodic oscillations of p' and $\dot{q}'$ signals, respectively. Reproduced with permission from Pawar et al. [72]

aperiodic oscillations (Fig. 8.7b). Thus, both the signals demonstrate the features of intermittent phase synchronization during the occurrence of intermittency in the system.

From the above discussion, we conclude that the onset of thermoacoustic instability in turbulent combustors can be viewed as a synchronization phenomenon between the acoustic pressure and the heat release rate signals. In other words, the acoustic field and the reacting turbulent field are synchronized during the occurrence of thermoacoustic instability. During this synchronization transition, these fields change their coupled behavior from the state of desynchronization to generalized synchronization (GS) via intermittent phase synchronization (IPS) and phase synchronization (PS). Note that during the state of desynchronization, both the acoustic pressure and the heat release rate signals are chaotic, while during the state of synchronization, they are periodic. Therefore, we can diagrammatically write this synchronization transition of turbulent thermoacoustic systems as,

$$\text{Desync Chaos} \longrightarrow \text{IPS} \longrightarrow \text{PS} \longrightarrow \text{GS}$$

As we discussed in Sect. 8.1, during the state of generalized synchronization, the dynamics of coupled oscillators exhibits a functional relationship with each other. Similarly, we also observe the presence of a possible functional relation between the acoustic pressure and the heat release rate fluctuations during the state of generalized synchronization in a turbulent combustor [72]. The visual inspection of the experimental signals obtained during this state shows features

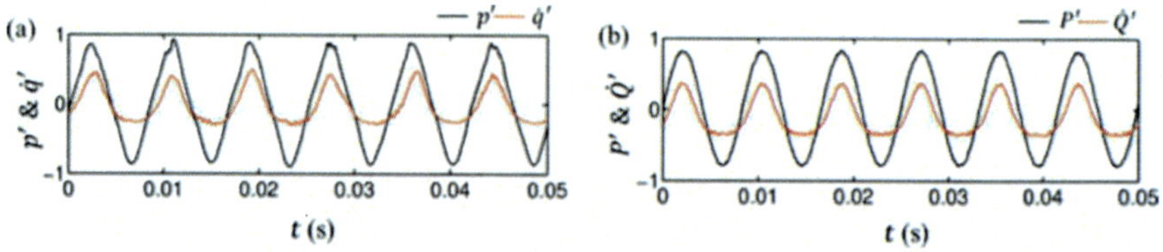

Fig. 8.8 (**a**) The time series of the acoustic pressure (p') and the heat release rate fluctuations ($\dot{q}'$) obtained from experiments in a turbulent combustor during the state of generalized synchronization. (**b**) The time series of these fluctuations (P' and $\dot{Q}'$) constructed using a model. We notice the presence of a functional relationship between the pressure and the heat release rate signals during the state of generalized synchronization. Adapted with permission from Pawar et al. [72]

of sinusoidal oscillations for the acoustic pressure signal and features of slow-fast (spiky) oscillations for the heat release rate signal (Fig. 8.8a). In the spiky heat release rate oscillations, the signal is nearly flat for a longer time (slow epoch), then it peaks for a short duration (fast epoch) and again becomes flat on the completion of an oscillation cycle [46]. Based on these properties of the experimental signals, we can reconstruct the acoustic pressure (p') and the heat release rate ($\dot{q}'$) signals using the following simple expressions such that these signals exhibit a functional relation (i.e., $\dot{Q}'(t) = f[P'(t)]$) as follows [72],

$$P' = [a_1 + n_{p1}(t)] \sin [\omega t + \phi_1 + n_{p2}(t)] \tag{8.7}$$

and,

$$\dot{Q}' = [a_2 + n_{q1}(t)] \sin [\omega t + a_3 \sin (\omega t + \phi_2) + n_{q2}(t)] \tag{8.8}$$

where a_1, a_2, and a_3 are constants, ϕ_1 and ϕ_2 are the initial phases, and n_{p1}, n_{p2}, n_{q1}, and n_{q2} are Gaussian white noise terms added to the amplitude and phase parts of the signals.

The comparison of the signals p' and $\dot{q}'$ obtained from experiments in a turbulent combustor during the state of generalized synchronization (Fig. 8.8a) and the signals of P' and $\dot{Q}'$ obtained from the aforementioned functional forms (Fig. 8.8b) show a qualitative similarity. Thus, the signals of the acoustic pressure and the heat release rate fluctuations exhibit a functional relationship (as shown in (8.7) and (8.8)) during the state of strongly correlated limit cycle oscillations, where the dynamical properties of the heat release rate signal can be derived from the acoustic pressure signal.

Next, we discuss the properties of coupled acoustic pressure and heat release rate fluctuations, for different states of the system dynamics, using the multivariate recurrence plot analysis. There are two ways to perform the simultaneous characterization of a pair of signals: cross recurrence plot and joint recurrence plot [59], as we discussed in Chap. 2.

Using joint recurrence plot (JRP), we analyze the simultaneous recurrence of the phase space trajectories of two signals computed in their individual phase space [83]. Thus, JRP counts the simultaneous occurrence of black points at particular time instances in the recurrence plots of both the signals. As JRP is constructed through the product of recurrence matrices of two signals, the number of data points in these signals need to be the same. We can calculate JRP as follows,

$$JR_{i,j} = \Theta(\epsilon_x - \|x_i - x_j\|)\Theta(\epsilon_y - \|y_i - y_j\|) \tag{8.9}$$

where x_i and y_j are state vectors of length N such that, $x_i \in \mathbb{R}^d,\ y_j \in \mathbb{R}^m;\ i, j = 1, 2, \ldots, N$. Therefore, $JR_{i,j} = 1$ if $\|x_i - x_j\| < \epsilon_x$ and $\|y_i - y_j\| < \epsilon_y$; otherwise, $JR_{i,j} = 0$.

Likewise, using cross recurrence plots (CRP), we can perform the simultaneous characterization of the recurrence properties of two signals [105]. In this method, we first need to embed the time series of both the normalized signals in the same phase space, and then measure the recurrence behavior of a state point (x_i) on the trajectory of one signal with respect to the neighborhood location of the state points (y_i) on the trajectory of another signal. The total number of data points in both the signals need not be the same for the calculation of CRP. As a result, the recurrence matrix of CRP can be asymmetric $[N \times M]$. We can compute the cross recurrence matrix of two signals as follows,

$$CR_{i,j} = \Theta(\epsilon - \|x_i - y_j\|) \tag{8.10}$$

where $x_i, y_j \in \mathbb{R}^d$; $i = 1, 2, \ldots, N$ and $j = 1, 2, \ldots, M$. Therefore, $CR_{i,j} = 1$ if $\|x_i - y_j\| < \epsilon$; otherwise, $CR_{i,j} = 0$. Similar to the recurrence plots, the long diagonal structures in CRP indicate a similar behavior of both the signals in the phase space.

In Fig. 8.9, we present the recurrence behavior of the simultaneously acquired acoustic pressure (p') and heat release rate ($\dot{q}'$) fluctuations, during the four states of synchronization (depicted in Figs. 8.6 and 8.7), in terms of joint recurrence plot (Fig. 8.9a–d) and cross recurrence plot (Fig. 8.9e–h). A black point in the JRP indicates the simultaneous recurrence (or synchronization) of both the signals at a particular time instant in the phase space. During the occurrence of desynchronized aperiodic fluctuations (Fig. 8.9a), we notice the irregular occurrence of black points in JRP. Furthermore, as the simultaneous recurrence of both the acoustic pressure and the heat release rate signals is rare in the phase space, we notice the lower values of the recurrence density of the state points in JRP.

During the state of intermittent phase synchronization (Fig. 8.9b), black patches in JRP indicate the simultaneous occurrence of low amplitude aperiodic fluctuations in both the acoustic pressure and the heat release rate signals, while broken diagonal lines represent the presence of weakly correlated periodic oscillations in these signals. As the amplitudes of both the acoustic pressure and the heat release rate signals do not remain constant from one oscillation cycle to another during the state of phase synchronization, the individual recurrence plots of these

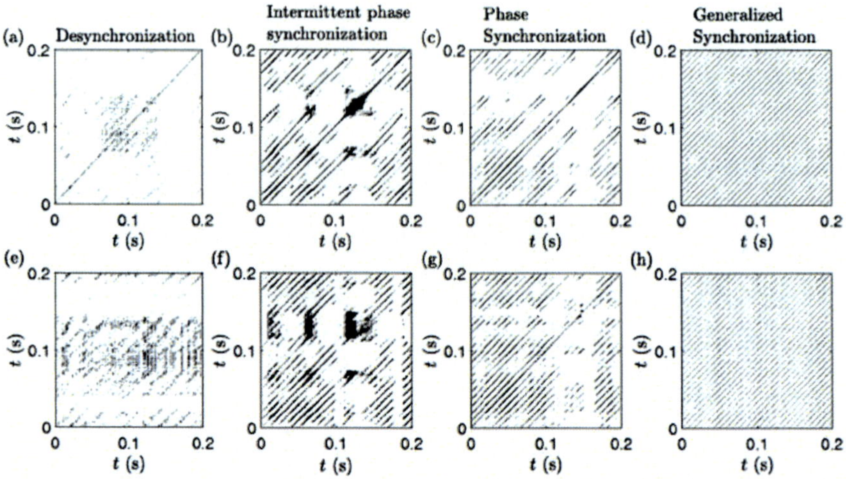

Fig. 8.9 **(a–d)** Joint recurrence plots (JRP) and **(e–h)** cross recurrence plots (CRP) between coupled p' and $\dot{q}'$ signals for different states of synchronization observed in a turbulent bluff-body stabilized dump combustor. Reproduced with permission from Godavarthi et al. [26]

signals are non-identical [74] and their JRP shows discontinuous diagonal lines (Fig. 8.9c). During the state of generalized synchronization, as these signals are highly correlated in both amplitudes and phases the recurrence plots of their individual signals are identical [74] and are similar to JRP (Fig. 8.9d). Hence, for the state of generalized synchronization, all these recurrence plots show only continuous (unbroken) diagonal lines occurring parallel to the main diagonal line.

On the other hand, a black point in the cross recurrence plot (CRP) indicates the coexistence (or closeness) of trajectories of two signals at a particular time instant in the same embedded space. This, in turn, indicates the synchronization characteristic of these signals. As we mentioned before, unlike JRP, CRP can be asymmetric about the main diagonal line.

During the state of desynchronization of oscillations (Fig. 8.9e), as both the acoustic pressure and the heat release rate signals are non-identical, their CRP shows a nearly irregular occurrence of black points. During the state of intermittent phase synchronization (Fig. 8.9f), the parallel running trajectories observed during epochs of periodic oscillations exhibit broken (shorter) diagonal lines in CRP and the trajectories trapped in the region of low amplitude oscillations display black patches in CRP. During the state of phase synchronization (Fig. 8.9g), as both the acoustic pressure and the heat release rate signals have weakly correlated periodic oscillations, the corresponding CRP shows only broken diagonal lines. In contrast, during the state of generalized synchronization (Fig. 8.9h) as these signals consist of strongly correlated periodic oscillations, the corresponding CRP demonstrates continuous diagonal lines. We notice that JRP and CRP of both the acoustic pressure and the heat release rate signals are identical during the state of generalized synchronization, wherein the signals exhibit a functional relationship with each other [26].

Next, we present the synchronization transition of the acoustic pressure and the heat release rate signals during the onset of thermoacoustic instability, using quantitative measures derived from the joint and cross recurrence properties of these signals [26, 84]. We below provide a list of these measures.

i. **Recurrence rate (RR_J):** It is a measure of the density of black points present in the joint recurrence plot.

$$RR_J = \frac{1}{N^2} \sum_{i,j=1}^{N} JR_{i,j} \tag{8.11}$$

where $JR_{i,j}$ is the joint recurrence matrix, obtained from (8.9).

ii. **Determinism (DET_J):** It measures the percentage of black points that form diagonal lines in the joint recurrence plot.

$$DET_J = \frac{\sum_{l=l_{\min}}^{N} lF(l)}{\sum_{l=1}^{N} lF(l)} \tag{8.12}$$

where $F(l)$ is the frequency distribution of the diagonal lines of length l in the joint recurrence plot and $l_{\min} = 2$.

iii. **Correlation of probability of recurrence (CPR):** It measures the linear correlation between the plots of probability of recurrence $P(\tau)$ of two signals, which we can calculate as follows,

$$CPR = \frac{\langle \bar{P}_1(\tau > \tau_c)\bar{P}_2(\tau > \tau_c)\rangle}{\sigma_1\sigma_2} \tag{8.13}$$

where $\bar{P}_1$ and $\bar{P}_2$ are the mean subtracted values of probability of recurrence P_1 and P_2 of two signals, respectively, and σ_1 and σ_2 are the standard deviation of P_1 and P_2, respectively. The value of CPR is calculated only after the lag τ_c, where τ_c corresponds to the value of delay at which the correlation of the signal becomes less than $1/e$ [29]. When the locations of the peaks in the plots of $P(\tau)$ for both the signals are locked, CPR achieves a value close to 1. Thus, CPR helps in detecting the onset of phase synchronization in the system.

iv. **Joint probability of recurrence (JPR):** It is the average probability of joint recurrences in the phase space, which we obtain by the following equation,

$$JPR = \frac{\frac{RR_J}{RR} - RR}{1 - RR} \tag{8.14}$$

where RR_J is the joint recurrence rate of two signals and RR in the recurrence rate of the individual signal, $RR = 1/N^2 \sum_{i,j=1}^{N} \Theta(\epsilon_x - \|x_i - x_j\|)$. During the state of generalized synchronization, the recurrence plots of both the signals become identical [74], leading to $RR_J = RR$, and hence, JPR becomes equal to 1.

In Fig. 8.10, we show the variation of the aforementioned recurrence quantifiers, i.e., determinism (DET_J), recurrence rate (RR_J), correlation of probability of recurrence (CPR), and joint probability of recurrence (JPR), with the change in the mean flow velocity ($\bar{u}$) in the system, during the transition to thermoacoustic instability [26]. Here, DET_J and RR_J are calculated from the joint recurrence plots. We observe that the value of DET_J achieves a near unity during the onset of thermoacoustic oscillations in the system (Fig. 8.10a), i.e., during the states of phase synchronization and, generalized synchronization. However, the value of RR_J shows a plateau during the state of phase synchronization (Fig. 8.10b), as the recurrence properties of both signals remain nearly the same during this state. During the transition to generalized synchronization, the value of RR_J exhibits a gradual increase, due to an increase in the simultaneous recurrence behavior of both the signals in the system.

The measures CPR and JPR (Fig. 8.10c) quantify the similarity in the plots of the probability of recurrence, $P(\tau)$, of both the acoustic pressure and the heat release rate signals (as shown in Figs. 8.6 and 8.7). As the variation of CPR captures the locking of positions of the peaks in the plots of $P(\tau)$ for these signals, it achieves a maximum value closer to unity during the onset of phase synchronization. The value of CPR remains nearly the same during the transition from phase synchronization to generalized synchronization, whereas the value of JPR shows an increase during this transition. Thus, we can use DET_J and CPR to detect the occurrence of phase synchronization and, RR_J and JPR to detect the occurrence of generalized synchronization between the coupled acoustic pressure and heat release rate oscillations in a thermoacoustic system [26].

8.2.3 Spatiotemporal Synchronization of the Turbulent Reacting Flow Field with the Duct Acoustics

In the previous section, we discussed the occurrence of synchronization phenomenon in the temporal dynamics of the acoustic pressure and the global heat release rate fluctuations during the onset of thermoacoustic instability in a turbulent combustor. These temporal oscillations in the system variables are in fact the result of spatial interactions that take place between various subsystems, such as the duct acoustics, the hydrodynamics, and the flame, at different spatial length scales and time scales in the combustor [54].

As we discussed in Chap. 7, the spatial dynamics of the reacting field in a turbulent combustor is incoherent during the state of combustion noise, while the formation of large-scale coherent structures is observed during the state of thermoacoustic instability. In such systems, the vortical structures shed from a flame holding device break down due to the impingement on obstacles in their path

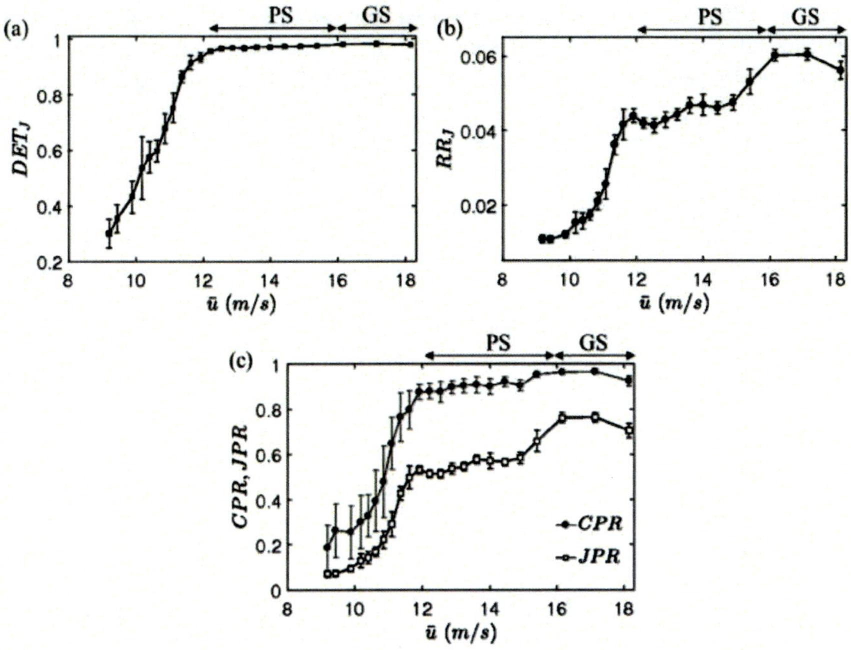

Fig. 8.10 The variation of (**a**, **b**) DET_J and RR_J, calculated from the joint recurrence plots, respectively, and (**c**) CPR and JPR as a function the flow velocity ($\overline{u}$), during the transition from combustion noise to thermoacoustic instability via intermittency. Using these recurrence measures, we can detect the onset of phase synchronization and generalized synchronization states in a turbulent thermoacoustic system. Adapted with permission from Godavarthi et al. [26]

(such as a bluff-body, combustor walls or neighboring vortices), leading to fine-scale turbulence mixing [91]. This intense turbulence mixing of the cold reactants and the hot products causes a sudden release of heat in the system. The acoustic oscillations generated due to this near instantaneous heat release affect the vortex formation process through a feedback mechanism and enforce the shedding of another vortex for stronger feedback. The completion of such a feedback loop during the state of thermoacoustic instability, in turn, helps in maintaining sustained synchronized oscillations in all the subsystems of a combustor. In this section, we will present the emergence of order (observed during thermoacoustic instability) in a disordered reaction field (observed during combustion noise) in the context of mutual synchronization.

Towards this purpose, we need to simultaneously study the dynamics of CH* chemiluminescence field of the flame and the acoustic pressure fluctuations during different dynamical states in a turbulent thermoacoustic system [63]. The CH* chemiluminescence field of the flame is a line-of-sight integrated two-dimensional image of the three-dimensional reaction field. We can consider the heat release fluctuations at each pixel of this image as a local heat release rate oscillator. Thus, we can view the entire image of the reaction field as a network of local heat release rate oscillators. Then, we can study the interaction among such oscillators using the framework of synchronization theory. Since the underlying flow field is highly turbulent and the reaction is a continuous process, the turbulent reactive field exhibits self-sustained chaotic oscillations even in the state of stable operation [63].

Further, we note that the coupling between these heat release rate oscillators can be of different forms [63]. As the flow field is convective, these oscillators are coupled through a non-local delay coupling. Moreover, as the turbulent mixing of the reactants happens through a diffusion process at the molecular level, the existence of diffusive coupling between the oscillators is also possible. In addition to these couplings, the acoustic field imposes a global coupling on the coupled local heat release rate oscillators. Therefore, the characterization of the exact role of individual coupling mechanisms and the corresponding coupling strengths would be difficult for such complex systems involving the turbulent reaction field.

In order to study the synchronization characteristics of the network of local heat release oscillators with the acoustic pressure oscillator, we need to examine the instantaneous phase field between them [63]. The phase field can be obtained from the calculation of the instantaneous phase difference between the dynamics of each heat release rate oscillator and the acoustic oscillator using an analytic signal approach based on Hilbert transform (refer to Sect. 8.1). In Fig. 8.11, we show the spatiotemporal pattern of such instantaneous phase field observed for different dynamical states, such as combustion noise, intermittency, and thermoacoustic instability, of a turbulent combustor.

For the state of combustion noise, we notice a nearly randomly distributed instantaneous phase field, observed as a grainy structure in the reaction field (Fig. 8.11a). This incoherent phase field further indicates the presence of desynchrony between the local heat release rate oscillators and the acoustic pressure oscillator. As a result, these oscillators experience a weak coupling with each other, causing their dynamics to be in the form of low amplitude aperiodic oscillations [63].

During the state of intermittency, as the temporal dynamics fluctuates between low amplitude aperiodic oscillations and the high amplitude periodic oscillations, we observe interesting patterns in the phase field of the reaction zone (Fig. 8.11b,c). During this state, the reaction field shows the coexistence of patches of phase synchrony as well as phase asynchrony at particular instances of time in the system. The phase synchronous region corresponds to the regions where the phase difference between the acoustic pressure (p') and the heat release rate ($\dot{q}'$) fields lie between $-\pi/2$ to $\pi/2$ rad; otherwise, it is considered as a phase asynchronous region. Therefore, during an epoch of low amplitude aperiodicity (Fig. 8.11b), the reaction zone is dominated by phase asynchronous regions of p' and $\dot{q}'$ oscillations. In contrast, during an epoch of bursts of large amplitude periodic oscillations (Fig. 8.11c), the reaction zone is dominated by phase synchronous region of p' and $\dot{q}'$. We also notice small regions of phase asynchrony in the phase synchronous regions (Fig. 8.11c), which possibly arise due to the presence of background turbulence fluctuations in the reacting field.

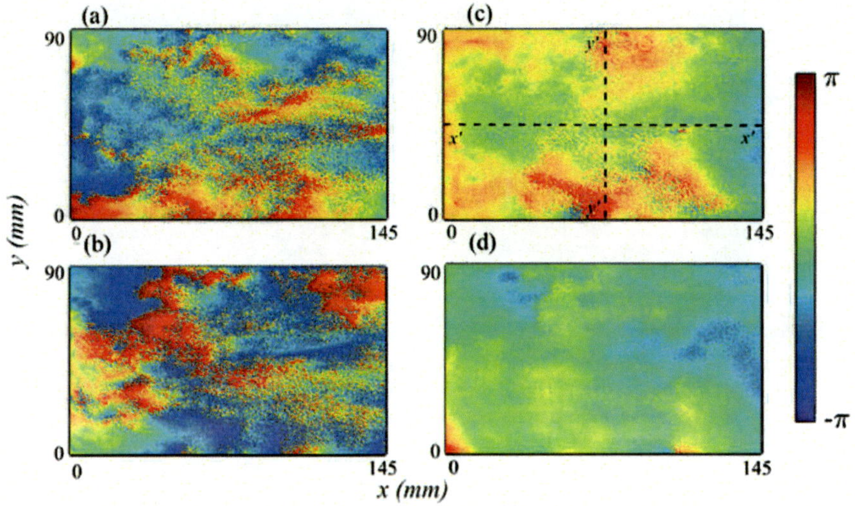

Fig. 8.11 Instantaneous snapshots of the phase field of the reaction zone in a turbulent combustor during the states of (**a**) combustion noise, (**b, c**) aperiodic and periodic epochs of intermittency, respectively, and (**d**) thermoacoustic instability. For aperiodic oscillations, the instantaneous phase is randomly distributed between the interval $-\pi$ to π. For periodic oscillations, the phase distribution is more coherent with the value of the relative phase bounded between $-\pi/2$ to $+\pi/2$. Reproduced with permission from Mondal et al. [63]

For the state of thermoacoustic instability (Fig. 8.11d), we notice that the reaction zone is mostly dominated by phase synchronous oscillations, where the relative phase between p' and $\dot{q}'$ oscillations lies between $-\pi/2$ to $\pi/2$ rad. The presence of such coherent in-phase oscillations in most of the oscillators in the reaction field acts as a source of driving in the system, due to the satisfaction of the Rayleigh's criterion. Thus, synchronization of all such oscillators results in the emergence of order in the local as well as the global dynamics of the system. We note that the extent of the synchronized oscillators existing in the reaction field has a direct impact on the amplitude of periodic oscillations observed in the acoustic pressure signal [63]. We observe smaller regions of synchronized heat release rate oscillators with the acoustic pressure oscillator during the epochs of bursts of periodic oscillations of the intermittency, as compared to that observed during the state of thermoacoustic instability. Hence, the amplitude of the periodic oscillations observed in such bursts is lower than that observed during thermoacoustic instability.

Furthermore, we can depict the synchronous transition of the local heat release rate oscillators in the reacting field through the use of phasor plots, as shown in Fig. 8.12. The phasor plots represent the distribution of arrows of unit length, where the orientation of each arrow indicates the instantaneous phase angle of the local heat release rate oscillator in the reaction field. The regions of the random orientation of arrows indicate the existence of local desynchrony between the

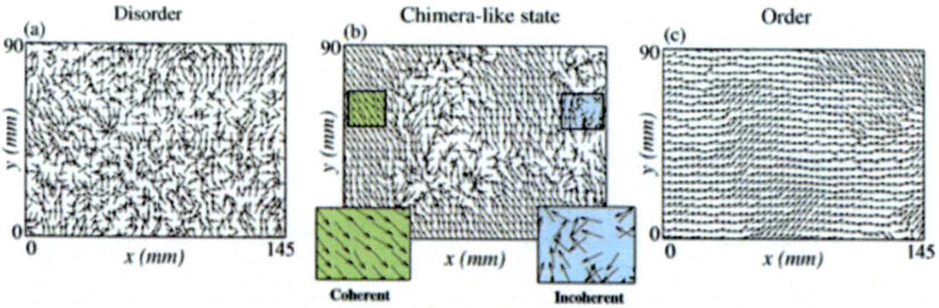

Fig. 8.12 Instantaneous snapshots of the phasor field of the reaction zone for the states of (**a**) combustion noise, (**b**) periodic epoch of intermittency, and (**c**) thermoacoustic instability. For the state of combustion noise, the arrows are randomly distributed. In contrast, for the state of thermoacoustic instability, arrows exhibit coherent alignment in a particular direction. During the occurrence of intermittency, the coexistence of regions of ordered and disordered orientation of arrows is observed, indicating the presence of chimera. Reproduced with permission from Mondal et al. [63]

oscillators. In contrast, the regions of coherent alignment of arrows represent the presence of phase synchrony between the oscillators.

During the state of combustion noise (Fig. 8.12a), we notice the random orientation of phase arrows in the reaction field, signifying the disordered behavior of the oscillators. In contrast, during the state of thermoacoustic instability (Fig. 8.12c), we observe the locally coherent alignment of arrows throughout the reaction field. This indicates the ordered behavior of the oscillators.

During the state of intermittency (Fig. 8.12b), we notice an interesting pattern of the coexistence of phase synchrony and phase asynchrony at a given time instance in the reaction field. In the phase synchronous region, the orientation of arrows is ordered. In contrast, in the phase asynchronous region, such orientation of arrows is disordered. Mondal et al. [63] referred to the coexistence of such synchrony and desynchrony in a population of the local heat release rate oscillators as chimera [47]. Hashimoto et al. [37] reported the presence of chimera-like state in the spatiotemporal dynamics of a model rocket combustor. Furthermore, Mondal et al. [63] observed that the regions of synchrony and desynchrony change their locations with time due to the convective flow field of the combustor. They referred to such a behavior of the time-varying global synchrony of the reaction field as *breathing chimera* of oscillators [3]. Thus, we can describe the transition to thermoacoustic instability in the spatiotemporal dynamics as the emergence of order from disorder through the occurrence of chimera. As the system dynamics approaches a state of synchronization between the acoustic pressure and the heat release rate oscillators, the low amplitude desynchronized aperiodic oscillations of these oscillators change to large amplitude periodic oscillations.

In Fig. 8.13(I), we show a space-time plot for an array of oscillators at a particular horizontal location $(x' - x')$ of the reaction field (as marked a dashed line in Fig. 8.11c). The distribution of the instantaneous phases of this array of oscillators

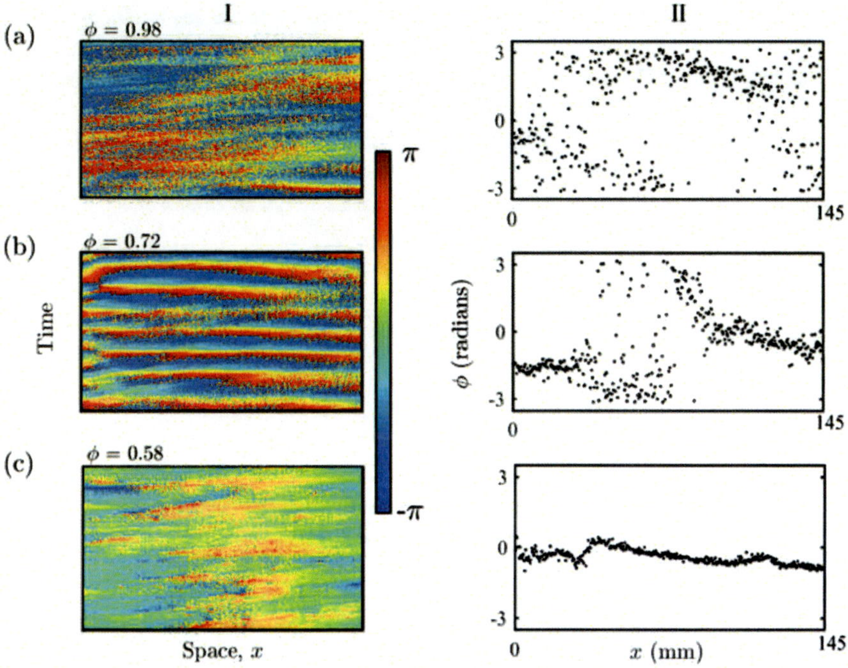

Fig. 8.13 (I) Space-time plots of the instantaneous phase field and (II) distribution of the instantaneous phases for the states of (**a**) combustion noise, (**b**) intermittency, and (**c**) thermoacoustic instability. The relative phase of an array of oscillators is (**a**) nearly randomly distributed for the state of combustion noise, (**c**) nearly bounded for the state of thermoacoustic instability, and (**b**) shows the presence of bounded and unbounded phase distribution for the state of intermittency. Reproduced with permission from Mondal et al. [63]

is shown in Fig. 8.13(II). An apparently random distribution of phases in the space-time plot for the state of combustion noise indicates the desynchronization of the oscillators (Fig. 8.13a). However, during the state of thermoacoustic instability, we observe that this distribution of phases is concentrated around a phase shift of nearly zero degrees, suggesting the phase synchronization of all the oscillators (Fig. 8.13c-II). We can notice the simultaneous presence of both synchronized and desynchronized oscillators at a particular time instant in the reaction field during the state of chimera (Fig. 8.13b-II).

Further, we can quantify the spatial synchronization of all non-locally coupled heat release rate oscillators in a thermoacoustic system using a measure called Kuramoto order parameter R [47]. We can obtain the value of R as follows,

$$R(t) = \frac{1}{K}\left|\sum_{j=1}^{K} e^{i\theta_j(t)}\right| \tag{8.15}$$

where K is the number oscillators in a population, θ_j is the instantaneous phase of the jth oscillator. When the oscillators are globally synchronized at a particular time instant, the value of $R(t)$ approaches 1, while for completely desynchronized oscillators, the value of $R(t)$ lies near 0.

In order to study the synchronization transition of heat release rate oscillators in the system, we can plot the variation of the mean ($\bar{R}$) and the variance (σ^2) of the order parameter with a change in the control parameter. The values of $\bar{R}$ and σ^2 are calculated as follows,

$$\bar{R} = \frac{1}{N}\sum_{t=1}^{N} R(t) \quad \text{and} \quad \sigma^2 = \frac{1}{N}\sum_{t=1}^{N}[R(t) - \bar{R}]^2 \tag{8.16}$$

where N is the number data point in the heat release rate signal.

In Fig. 8.14, we show the change in the time-averaged value of the order parameter $\bar{R}$ (Fig. 8.14a) and the variance of the order parameter σ^2 (Fig. 8.14b) as a function of a decrease in the equivalence ratio, ϕ. During the state of combustion noise ($\phi \sim 1$), all the oscillators exhibit a global desynchrony in time. Hence, the value of $\bar{R}$ is near zero. In contrast, during the state of thermoacoustic instability ($\phi \sim 0.6$), all the oscillators are phase synchronized with each other. As a result, $\bar{R}$ takes a higher value. Note that the value of $\bar{R}$ is lower than 1, which is possibly due to the existence of small regions of phase incoherence in the otherwise phase coherent state of the reaction field (as observed in Fig. 8.11d). Figure 8.14a shows that this transition from combustion noise to thermoacoustic instability happens through a gradual increase in $\bar{R}$.

However, we observe a contradictory behavior in the variation of σ^2 (see Fig. 8.14b). We notice that the value of σ^2 is low for both the states of combustion noise and thermoacoustic instability; however, it displays higher values for the state of intermittency. The large fluctuations observed in the order parameter during the state of intermittency indicate the existence of breathing chimera [3] in the system.

Similar observations of the synchronization transition of the acoustic pressure and the turbulent reacting flow dynamics have been reported by Pawar et al. [74] in a partially premixed swirl-stabilized combustor. In another study, Guan et al. [34] showed that the underlying unstable hydrodynamic mode of the system plays a dominant role in the synchronization characteristics of the acoustic pressure and the heat release rate fluctuations. Their study on a turbulent swirl-stabilized liquid-fueled diffusion flame combustor demonstrated that, in the presence of a strong hydrodynamic mode, these fluctuations are weakly phase synchronized or desynchronized during the state of thermoacoustic instability. In contrast, in the absence of such hydrodynamic mode, these fluctuations show generalized synchronization with each other. Pawar et al. [75] studied the synchronization transition of multiple conical laminar flamelets with the acoustic pressure fluctuations during the onset of thermoacoustic instability in a ducted low-turbulence laboratory spray combustor. They observed the emergence of collective synchronization in multiple flamelets during the onset of periodic oscillations in the system.

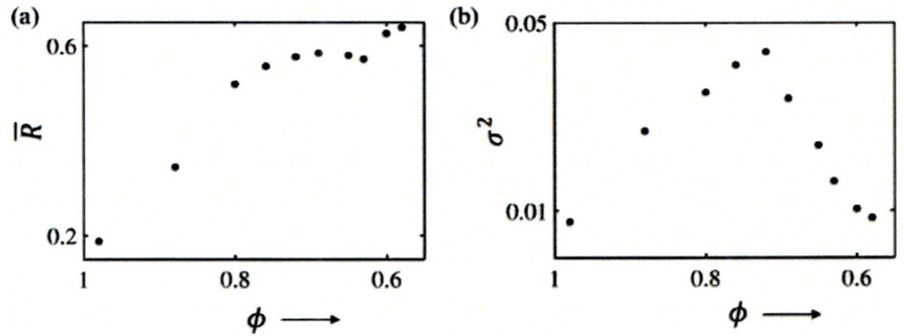

Fig. 8.14 Variation of (**a**) the mean $\bar{R}$ and (**b**) the variance σ^2 of the Kuramoto order parameter with the change in the equivalence ratio (ϕ), during the transition from combustion noise to thermoacoustic instability via intermittency. The synchronization transition of the heat release rate oscillators in the reaction field is observed during the onset of thermoacoustic instability. Reproduced with permission from Mondal et al. [63]

Until now, we discussed the features of mutual synchronization of the self-excited acoustic field and heat release rate field during the onset of thermoacoustic instability in a turbulent combustor. Next, we will focus on the characterization of the forced response of self-excited thermoacoustic oscillations and the unstable hydrodynamic field in the framework of forced synchronization. In this analysis, we will reveal different properties of the forced response of such oscillatory instabilities using various tools from synchronization theory and nonlinear time series analysis.

8.3 Forced Synchronization of Limit Cycle Oscillations in Thermoacoustic Systems

8.3.1 Basics of Forced Synchronization of an Oscillator

As we mentioned in Sect. 8.1, when the self-sustained oscillations with a discrete natural frequency (ω_n) of a system are forced with the external periodic perturbations of amplitude A_f and frequency ω_f, a unidirectional coupling is developed from the forcing system to the forced system [6, 79]. The imposed forcing then dictates the characteristics of the phase, amplitude, and frequency of oscillations that are forced. During the state of forced synchronization, the frequency of forced oscillations completely disappears, and the oscillations are observed only at the forcing frequency. A variation of the values of the forcing amplitude (A_f) and the forcing frequency (ω_f, or the normalized frequency ω_f/ω_n) required for forced synchronization of an oscillator displays a V-shaped plot, commonly known as the Arnold tongue [79], see Fig. 8.15a. This also means that the value of the forcing amplitude required for synchronization of an oscillator exhibits a linear relationship with an increase in the difference between the forcing and natural frequencies (refer to the solid lines of the Arnold tongue in Fig. 8.15a). Thus, the larger the value of the frequency difference, the higher is the value of the forcing amplitude required for forced synchronization of an oscillator.

We can characterize the process of forced synchronization of an oscillator by two routes: the *locking route* and the *suppression route* [6], as shown in Fig. 8.15a. When the forcing frequency is close to the natural frequency of an oscillator, the synchronization transition exhibited by the forced system is perceived to happen via the locking route (refer to route A in Fig. 8.15a). During this route, since the forcing and the natural frequencies are very close to each other (as observed from the amplitude spectrum of the forced signal), the natural frequency exhibits a gradual movement towards the forcing frequency as the forcing amplitude is progressively increased. On the other hand, for larger values of the frequency detuning, the forced system exhibits the synchronization transition via a suppression route (refer to route B in Fig. 8.15a). During this route to forced synchronization, the suppression of the natural frequency of forced oscillations occurs without undergoing any move-

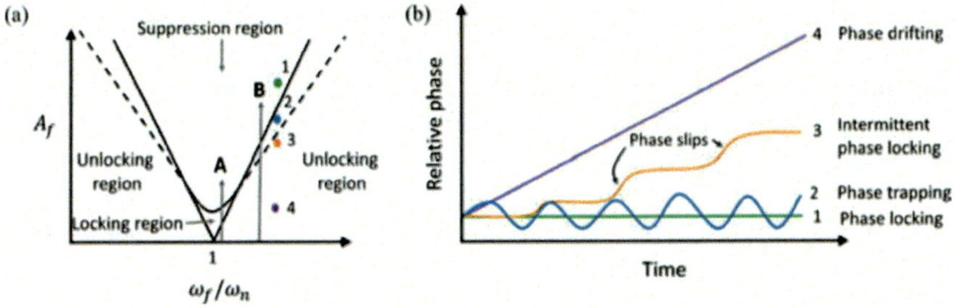

Fig. 8.15 Schematics of (**a**) the Arnold tongue, a plot between the forcing amplitude (A_f) and the normalized frequency ratio (ω_f/ω_n) required for forced synchronization of an oscillator, and (**b**) the temporal variation of the relative phase between the forced oscillator and the forcing system for different states of forced synchronization. In (**a**), paths A and B indicate the locking and suppression routes to forced synchronization. The relative phase plots corresponding to points 1 to 4 in (**a**) and (**b**) indicate the states of phase locking, phase trapping, intermittent phase locking, and phase drifting, respectively

ment towards the forcing frequency, when the forcing amplitude is progressively increased to a higher value.

Different ways have been developed to detect the occurrence of locking or suppression routes to forced synchronization of an oscillator [6]. One way to identify these routes is to analyze the value of the beating frequency (modulation frequency of the amplitude envelope) of the forced signal as a function of the change in the forcing amplitude. When the frequency detuning is a small finite value, the increase in the forcing amplitude leads to a transition from limit cycle oscillations (observed during the unforced state) to quasiperiodic oscillations. Such quasiperiodic oscillations can be identified from modulations in the amplitude envelop of the forced oscillations, where the fast oscillations occur at the natural frequency and the slow oscillations in the amplitude envelope occur at the frequency corresponding to the difference between the forcing and natural frequencies. According to Balanov et al. [6], the value of the beating frequency shows a gradual decrease for the locking route but exhibits a sudden decrease to zero for the suppression route.

Alternatively, we can differentiate the routes to forced synchronization in a system from the properties of the instantaneous phase difference (or the relative phase) between the forcing and the natural oscillations [6]. These instantaneous phase difference plots show different behaviors, such as phase drifting, intermittent phase locking, phase trapping, and phase locking.

During the state of phase drifting, the relative phase between the forcing and natural oscillations remains unbounded and shows almost a monotonic growth in time (see plot 4 in Fig. 8.15b). For the state of phase trapping, the relative phase between these oscillations is trapped within a finite value of the phase difference, wherein the relative phase exhibits oscillations at the beating frequency, i.e., the difference between the forcing and natural frequencies (see plot 2 in Fig. 8.15b). Prior to the transition to the state of phase trapping, the relative phase between the forcing and natural oscillations shows the behavior of intermittent phase locking (see plot 3 in Fig. 8.15b). During this state, epochs of phase-locked oscillations between these signals (observed as plateau regions in plot 3 in Fig. 8.15b) are intermittently separated by short epochs of phase-slips. During the occurrence of phase-slips, the relative phase between the signals shows jumps of integer multiples of 2π rad. In the case of phase locking, the instantaneous phase difference between the forcing and natural oscillations remains constant in time (see plot 1 in Fig. 8.15b). Thus, the route to forced synchronization can happen either through the presence or the absence of a phase trapping state in between the states of phase drifting and phase locking (Figs. 8.15a and b). The occurrence of phase trapping during the route to forced synchronization confirms the existence of the suppression route; otherwise, it indicates the presence of the locking route in the system [6].

Having introduced the basics of forced synchronization of an oscillator, we now move our attention to discuss and understand the forced response of different subsystems (i.e., acoustics, hydrodynamics, and flame) of a thermoacoustic system during the state of self-sustained oscillations (or also called oscillatory instabilities). Several studies in the past have investigated the forced response of the acoustic field during the state of thermoacoustic instability [8, 30–34, 45, 64, 87, 89] and, that of the reacting [44, 52, 73] and the non-reacting [50, 51, 53, 67] hydrodynamic fields during the state of global instability. In these studies, the forced response of natural oscillations of a system is analyzed using the framework of forced synchronization (as discussed in Fig. 8.15). In the following subsections, we separately discuss the characteristics of forced synchronization of each subsystem of a thermoacoustic system.

8.3.2 Forced Response of the Self-Excited Acoustic Field

Over the years, we have come across many studies that are dedicated to understand the response of the self-excited thermoacoustic instabilities to external acoustic forcing, and also to use this methodology for devising active open-loop controls required for mitigation of such instabilities [5, 10, 11, 32, 35, 62, 64, 71, 87, 89]. In the earlier studies on forcing of thermoacoustic instability, the phenomenon of lock-in of the natural frequency of these instabilities with the forcing frequency, at critical values of the forcing parameters (i.e., amplitude and frequency), was referred to as frequency entrainment [11]. However, the rigorous application of the framework of forced synchronization in characterizing the response of thermoacoustic oscillations was started by Balusamy et al. [8] in a turbulent thermoacoustic system and, subsequently followed by several other detailed parametric studies in different laminar thermoacoustic systems [32, 33, 45, 64, 87, 89] and thermoacoustic engines [41, 77, 90, 104]. Apart from improving the understanding on the locking phenomenon of natural thermoacoustic oscillations with forcing oscillations, the synchronization framework has also helped us in devising controls that can mitigate thermoacoustic instability through a phenomenon of asynchronous quenching

[32, 64, 87, 89]. We will elaborate a detailed description of this phenomenon and its relevance for mitigation of thermoacoustic instability in Chap. 12.

In this section, we first present a systematic analysis of forced synchronization of limit cycle oscillations in a laminar thermoacoustic system and then in a turbulent thermoacoustic system. Since it is easy to conduct forcing experiments in great detail and for longer duration in laminar systems as compared to turbulent systems, the results obtained from forced laminar thermoacoustic systems provide a stepping stone to understand the forced response of complex turbulent thermoacoustic systems. We will start with a detailed experimental study of forced synchronization conducted on a simple thermoacoustic system, known as a horizontal Rijke tube [64].

8.3.2.1 Forced Synchronization of Limit Cycle Oscillations in a Horizontal Rijke Tube

As we discussed in Chap. 3, a horizontal Rijke tube is a very simple prototypical model of a practical thermoacoustic system. In this system, an electrically heated wire mesh is used as a concentrated heat source and the mean air flow is restricted to laminar conditions [28, 58, 61]. Depending on the variation of different control parameters associated with this system, i.e., the location of the heat source, the air flow rate, and the heater power, the system behavior can undergo Hopf bifurcation, leading to the onset of limit cycle oscillations from a steady state. Once the stable limit cycle oscillations are established, we can conduct forcing experiments on the system [64]. We can measure the forced acoustic response of the system in terms of the acoustic pressure fluctuations using pressure transducers. The natural frequency of limit cycle oscillations (f_{n0}) corresponds to the unstable mode of the acoustic field in the system. We can use external loudspeakers to provided controlled periodic forcing of particular frequency (f_f) and amplitude (A_f) around the natural frequency (f_{n0}) of limit cycle oscillations.

In Fig. 8.16, we show the forced response of limit cycle oscillations for different conditions of the forcing signal, shown in the parameter plane of the forcing amplitude (A_f) and the normalized forcing frequency (f_f/f_{n0}). We notice that the variation of forcing parameters corresponding to the onset of forced synchronization of limit cycle oscillations (see green lines in Fig. 8.16) follows a nearly V-shaped plot with the origin at $f_f/f_{n0} = 1$. This plot is usually referred to as an 'Arnold tongue' [79]. We can also call this plot as the '1:1 synchronization map', as the forcing frequency is the same as the natural frequency at the origin of the Arnold tongue.

The regions marked across the Arnold tongue correspond to different states of forced synchronization observed in the acoustic response of the system. Region-I corresponds to phase locking, region-II represents phase trapping, and region-III shows the phase drifting behavior of the forced system. We notice the existence of two routes to forced synchronization, occurring during the transition from phase drifting to phase locking at the Arnold tongue. When the forcing frequency (f_f) is near the natural frequency (f_{n0}) of limit cycle oscillations, the transition from

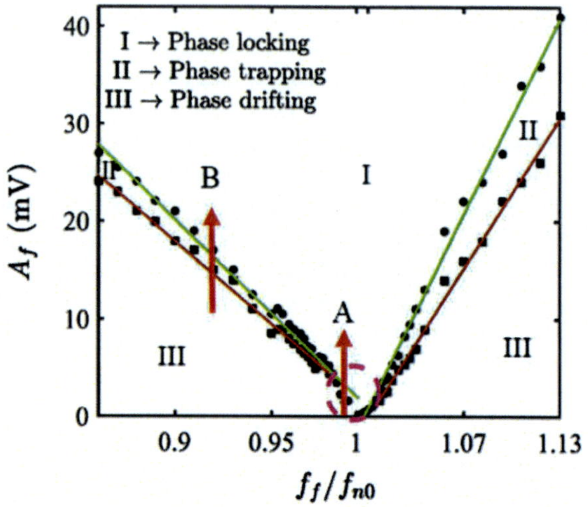

Fig. 8.16 A plot between the normalized forcing frequency (f_f/f_{n0}) and forcing amplitude (A_f) shows the Arnold tongue observed for forced limit cycle oscillations in experiments on a horizontal Rijke tube. The regions of (I) phase locking, (II) phase trapping, and (III) phase drifting are separated by different color lines. Arrows A and B indicate the existence of the locking (inside the pink dotted circle) and the suppression routes to forced synchronization in the system, respectively. f_{n0} is the frequency of limit cycle oscillations in the unforced state. Adapted with permission from Mondal et al. [64]

phase drifting to phase locking happens via the locking route (as shown by arrow A inside a pink dotted circle around $f_f/f_{n0} = 1$ in Fig. 8.16). On the other hand, for larger values of the frequency detuning ($\Delta f = |f_f - f_{n0}|$), such a transition ensues through the intermediate state of phase trapping during the suppression route (as shown by arrow B in Fig. 8.16). Thus, we observe the existence of both the routes to forced synchronization in a thermoacoustic system. This behavior of a thermoacoustic system is similar to Fig. 8.15, discussed for the forced synchronization of limit cycle oscillations in general [6]. In Figs. 8.17 and 8.18, we present the characteristics of different states of forced synchronization and compare the difference between two routes of forced synchronization, respectively.

Figure 8.17 shows the dynamical and the spectral properties of different states of forced synchronization observed in the horizontal Rijke tube when the forcing frequency is varied at a fixed value of the forcing amplitude. We observe that when the forcing frequency is far from the natural frequency, the effect of forcing on limit cycle oscillations is negligible (see the inset in Fig. 8.17a-I). The first return map (or Poincaré section) for this state shows a small cluster of points (Fig. 8.17a-I), indicating the return of the phase space trajectory nearly at the same point after a period of the oscillation. The relative phase between the forcing and the natural oscillations shows the characteristic of phase drifting, which is observed in the form of a monotonic increase in the instantaneous phase difference between these oscillations (see the inset of Fig. 8.17a-II).

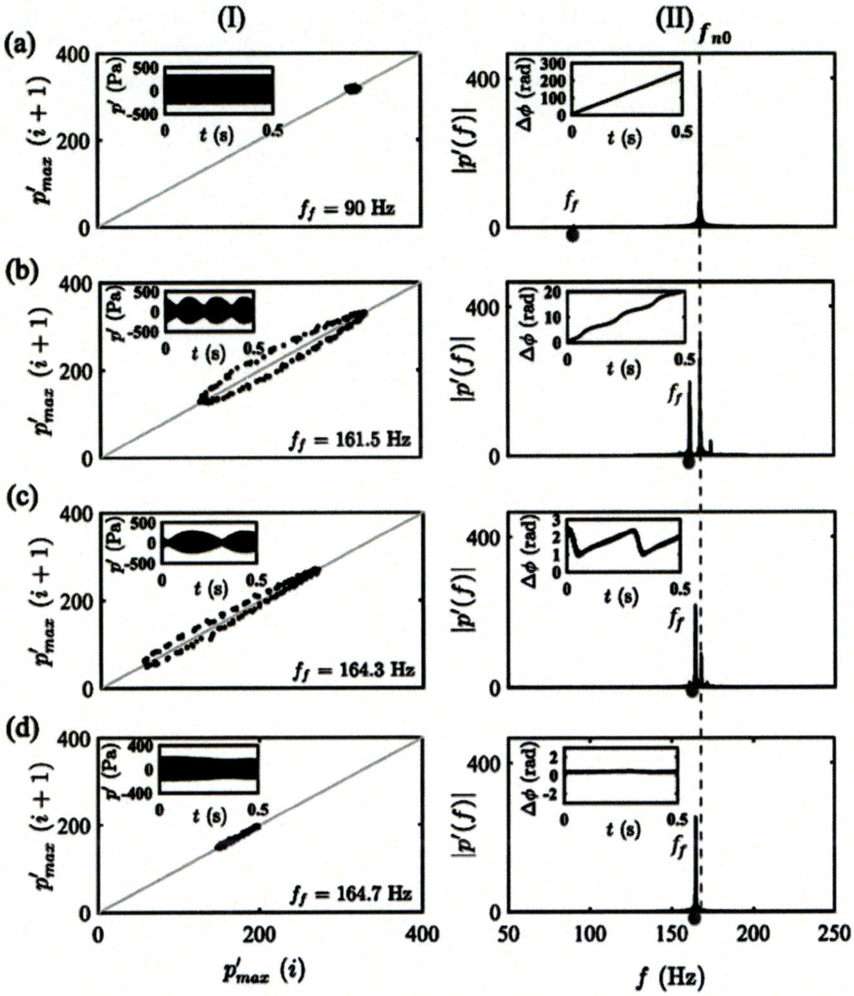

Fig. 8.17 Plots of (I) the first return map and (II) the amplitude spectrum of the forced acoustic pressure (p') signal for different values of forcing frequency when forcing amplitude fixed at $A_f =$ 5 mV. The insets in I and II correspond to the time series of the pressure signal and the relative phase ($\Delta\phi$) between the pressure and the forcing signals, respectively. In (II), a black dot on the x-axis indicates the location of f_f and a vertical dotted line represents the position of f_{n0} at 168.8 Hz in the spectrum. During the onset of forced synchronization, $\Delta\phi$ shows the transition from the state of (**a**) phase drifting to (**d**) phase locking via (**b**) intermittent phase locking and (**c**) phase trapping. Reproduced with permission from Mondal et al. [64]

When the forcing frequency is near the natural frequency, limit cycle oscillations start showing response to forcing, as observed from the modulation in the amplitude envelope of the pressure signal (see the inset of Fig. 8.17b-I). Since the frequency difference between the forcing and the natural oscillations is incommensurate, the resultant dynamics of the pressure signal shows quasiperiodic oscillations. We can confirm this characteristic of the existence of quasiperiodic signal from the presence of a closed-loop circle in the return map (Fig. 8.17b-I), which occurs due to the presence of a 2-torus attractor in the three-dimensional phase space. The relative phase plot of this state shows the characteristics of intermittent phase synchronization (inset of Fig. 8.17b-II), wherein the plot shows a staircase-like structure whose flat regions correspond to epochs of phase locking and short epochs of phase jumps correspond to desynchronization of two signals. These phase jumps, also called phase-slips, always correspond to integer multiples of 2π rad.

When frequencies of forcing and natural oscillations are very close to each other, the relative phase between these signals exhibits the behavior of phase trapping (inset of Fig. 8.17c-II). During this state, the relative phase plot shows periodic oscillations about the value of the mean phase difference, whose frequency equals to the frequency difference between the forcing and the natural oscillations. The return map of this state shows a closed-loop attractor (Fig. 8.17c-I), indicative of quasiperiodic oscillations in the forced response of the system.

When the forcing frequency is very close to the natural frequency, we notice the quenching of the natural frequency and the sole presence of the forcing frequency in the spectrum of pressure fluctuations (Fig. 8.17d-II). This demonstrates the occurrence of frequency-locking (or frequency entrainment or forced synchronization) in the system. During this state, the relative phase between the signals of forcing and forced limit cycle oscillations remains bounded and shows small fluctuations around the constant phase shift (inset of Fig. 8.17d-II). The return map of this state shows a concentrated cluster of state points along the diagonal line (Fig. 8.17d-I), indicative of periodic oscillations occurring at the forcing frequency in the system. Thus, we notice that the acoustic pressure response undergoes a dramatic change during the occurrence of forced synchronization of the limit cycle in the system.

Further, in Fig. 8.18, we explore the difference between the locking and the suppression routes to forced synchronization of limit cycle oscillations in a thermoacoustic system. In this figure, column I shows the pressure dynamics corresponding to the locking route and column II shows the pressure dynamics corresponding to the suppression route. In this analysis, the forcing frequency is kept constant and the forcing amplitude is changed gradually from zero to a high value such that the state of phase locking is attained in the system. For the suppression route, the forcing frequency ($f_f = 155$ Hz) is fixed sufficiently far from the natural frequency ($f_{n0} = 168.8$ Hz). While for the locking route, the forcing frequency ($f_f = 168$ Hz) is fixed very close to the natural frequency.

For the suppression route, we observe the presence of the state of phase trapping (Fig. 8.18c-I) prior to the state of phase locking (Fig. 8.18d-I) in the system. In contrast, the state of phase trapping is absent for the locking route (Fig. 8.18II). Furthermore, during the suppression route, we observe that the natural frequency

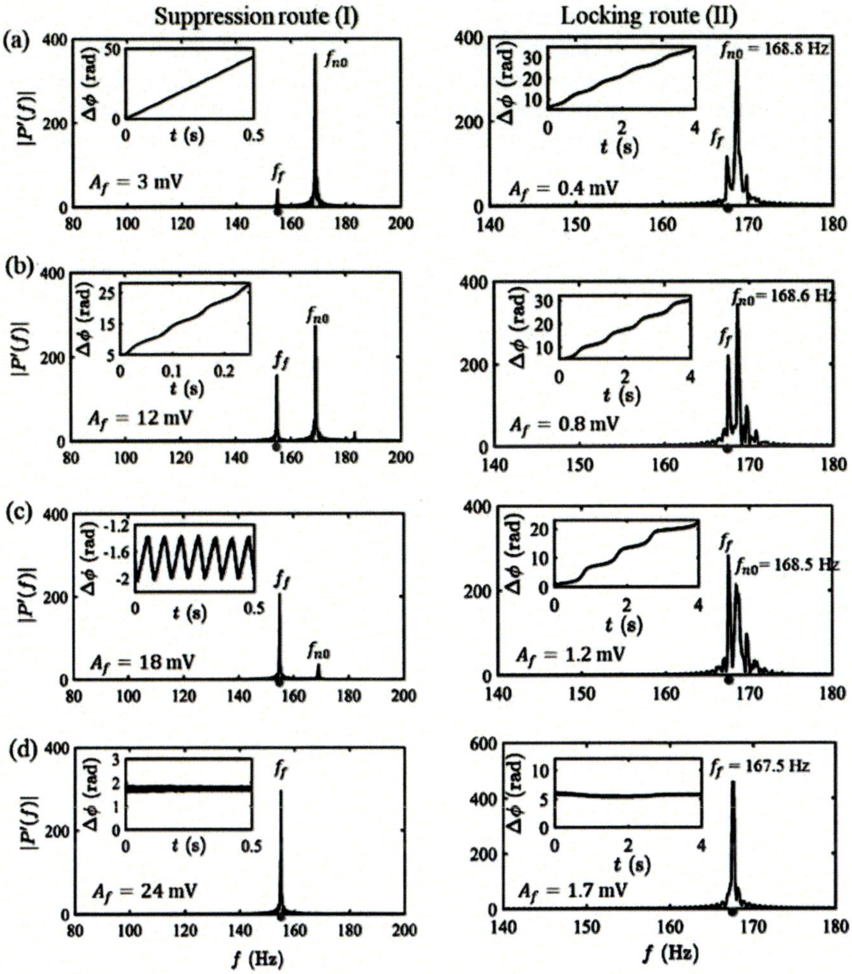

Fig. 8.18 Difference between two routes to forced synchronization: (I) suppression route and (II) locking route in a horizontal Rijke tube. Every plot shows the amplitude spectrum and the temporal variation of the relative phase ($\Delta\phi$) between the forcing and the pressure signals (in the inset) for different values of forcing amplitude when the forcing frequency is fixed at 155 Hz in (I) and at 16.5 Hz in (II). We observe the presence of phase drifting state (c) in the suppression route (I), which is absent in the locking route (II). A black dot on the x-axis of the spectrum indicates the location of f_f. Adapted with permission from Mondal et al. [64]

does not exhibit any movement due to a change in the forcing amplitude and it completely disappears at the onset of forced synchronization in the system (observe the variation of f_{n0} in the spectrums of column I of Fig. 8.18a–d). On the contrary, for the locking route, the natural frequency shows a small shift towards the forcing frequency prior to frequency-locking, due to an increase in the forcing amplitude (compare the spectrums of column II of Fig. 8.18a–d). Further, we note that the forcing amplitude required for forced synchronization of limit cycle oscillations in the suppression route is much higher than that required for the locking route.

Next, we discuss the variation in the amplitude response of the forced limit cycle oscillations when the forcing frequency is varied across the natural frequency at a constant value of the forcing amplitude (Fig. 8.19). This amplitude response of the forced acoustic pressure fluctuations illustrates interesting features of both synchronization and amplification (resonance) when the forcing frequency is close to the natural frequency. We can clearly study these features of forced oscillations from the plot showing the variation of the response amplitude of the acoustic pressure signal as a function of the change in the frequency ratio (f_f/f_{n0}). Here, we can obtain the response amplitude of a signal in terms of the root mean square value (P_{rms}) of the entire signal or from the spectral amplitudes corresponding to forcing (A_f) and the natural (A_n) frequency peaks of these oscillators. When the forcing frequency is far from the natural frequency, the plots of the response amplitudes do not show any significant change. However, when the forcing frequency is near the natural frequency, we notice a significant change in all plots of the response amplitude.

As the frequency ratio approaches unity from the left hand side (Fig. 8.19), we observe a gradual decrease in the amplitude of the natural frequency peak (A_n), while that of the forcing frequency (A_f) displays a corresponding increase. As the amplitude of the natural frequency falls to zero, the amplitude of the forcing frequency shows an abrupt jump to a significantly high value. With further increase in the frequency ratio, the amplitude of the natural frequency stays at zero, and that of the forcing frequency demonstrates a gradual decrease. We refer to this region of the response amplitude plot where A_n is zero as the phase/frequency locking region, similar to that observed inside the Arnold tongue in Fig. 8.16. In this region, the system dynamics is dominated only by the forcing oscillations and not by the natural oscillations. As soon as the frequency ratio crosses the phase-locking boundary of the Arnold tongue, we observe that the amplitude of the natural frequency regains a non-zero value and exhibits a gradual increase with further increase in the frequency ratio. We observe a combined response of the variation of spectral amplitudes of the forcing and natural frequencies in the variation of the normalized amplitude (P_{rms}/P_{rms}^0) of the acoustic pressure signal.

The behavior of suppression of the natural frequency in the phase-locking region happens due to a phenomenon called *synchronous quenching* and the significant growth in the amplitude of the acoustic pressure oscillations occurs due to a phenomenon called *resonance amplification* [1, 68, 69]. Altogether, we can refer to the coexistence of such synchronization and resonance behavior of a forced system in the Arnold tongue as *synchronance* [64]. Furthermore, we notice the asymmetry

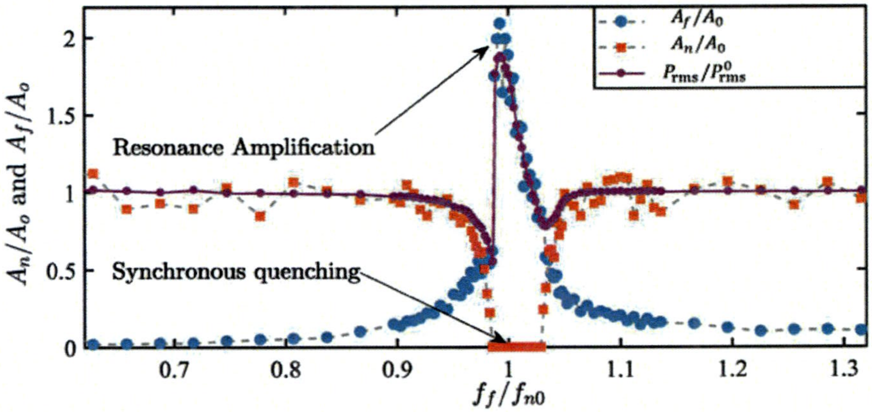

Fig. 8.19 The variation of the response amplitudes of the natural frequency (A_n) and the forcing frequency (A_f) in the amplitude spectrum and the root mean square value of the pressure signal (P_{rms}) as a function of the change in the normalized forcing frequency (f_f/f_{n0}) across the natural frequency (f_{n0}) at a fixed value of the forcing amplitude ($A_f = 5$ mV). P^0_{rms} is the rms value of the pressure signal in the unforced state. We notice the coexistence of synchronous quenching of natural oscillations and resonance amplification of the forcing oscillations during the regime of forced synchronization in the system. Adapted with permission from Mondal et al. [64]

in the amplitude of response of the forced limit cycle oscillations across the natural frequency, which happens due to a nonlinear interaction of the forcing and natural oscillations [45].

8.3.3 Forced Synchronization of the Acoustic Field and the Heat Release Rate Field in Laminar Combustors

In the previous section, we discussed the response of limit cycle oscillations (or thermoacoustic instability) observed in the acoustic field of a Rijke tube system, for different conditions of the forcing amplitude and forcing frequency. Similarly, we can also characterize the simultaneous response of the acoustic pressure and the heat release rate fluctuations in the flame to external acoustic forcing [31, 32, 45, 87]. Since thermoacoustic instability is a result of mutual interaction of these fluctuations, the simultaneous analysis of these signals with the forcing signal reveals several hidden features of forced thermoacoustic systems that cannot be detected from the investigation of the acoustic pressure alone.

For example, in the previous section, we mentioned that the forcing of thermoacoustic instability far from its natural frequency can completely quench these instabilities through the phenomenon of asynchronous quenching [32, 64, 87]. During the occurrence of asynchronous quenching, we observe that the acoustic power production in the system approaches zero due to the breaking of the coupling between the acoustic pressure and the heat release rate fluctuations [32, 64, 87], which we have elaborated this in Chap. 12. Thus, we can achieve the mitigation of thermoacoustic instability through external forcing.

8.3.4 Forced Synchronization of Quasiperiodic and Chaotic Thermoacoustic Oscillations

In Chap. 4, we showed that thermoacoustic instabilities not only occur in the form of period-1 limit cycle oscillations, but also exhibit other self-sustained aperiodic oscillations observed in the form of quasiperiodicity and chaos. For quasiperiodic oscillations, the power spectrum shows of two incommensurate frequencies and the phase space consists of 2-torus attractor. In contrast, chaotic oscillations consist of three or more incommensurate frequencies in the power spectrum and a strange attractor in the phase space (refer to Sect. 2.4 in Chap. 2).

Similar to the investigation of forced synchronization of period-1 limit cycle oscillations (as discussed in the previous sections), Kashinath et al. [45] performed a numerical study on forced synchronization of quasiperiodic and chaotic oscillations in a laminar premixed flame burner. Further, Guan et al. [33] showed a detailed experimental characterization of forced synchronization of quasiperiodic oscillations in a ducted laminar premixed burner. They showed that the complete synchronization of quasiperiodic oscillations occurs through three steps: when the forcing amplitude is increased for a fixed value of the forcing frequency, the system behavior transitions from (i) three-frequency quasiperiodic oscillations (desynchronization) to (ii) two-frequency quasiperiodic oscillations (partial synchronization), and then to (iii) period-1 limit cycle oscillations (complete synchronization). The forced response of quasiperiodic oscillations shows the existence of two Arnold tongues, concentrated around the unstable natural frequencies of these oscillations. In order to achieve the mitigation of such oscillations through asynchronous quenching, we need to force the system around the weaker mode (low amplitude frequency peak) at an amplitude just sufficient for complete synchronization of the oscillations. The forcing of quasiperiodic oscillations near the stronger mode (large amplitude frequency peak) leads to the resonant amplification of oscillations in the system.

On the other hand, forced synchronization of chaotic oscillations is more intricate due to the presence of several frequencies in the spectrum and the unstable movement of the trajectories in the phase space [6]. Such studies are usually performed to control the chaotic dynamics in the system. For the chaos control in thermoacoustic systems, we need force the chaotic oscillations around the dominant frequency. This, in turn, helps us in stabilizing the system dynamics at one of the infinitely looping unstable periodic orbits of the chaotic attractor. Kashinath et al. [45] showed the stabilization of the strange attractor of chaos onto a regular stable period-3 attractor in a laminar thermoacoustic system.

Thus, we note that the control of self-excited quasiperiodic and chaotic oscillations in practical thermoacoustic systems is possible. In this control method, we need to first force these aperiodic oscillations at their dominant frequency to stabilize them into a stable periodic state. Then, we can conduct the usual forced synchronization studies (as discussed in the previous section) to control the properties of such periodic oscillations.

8.3.5 Characteristics of Forced Synchronization of Limit Cycle Oscillations in Turbulent Combustors

Unlike numerous studies on forced synchronization of thermoacoustic instability in laminar systems, there is hardly any literature on the forced synchronization of turbulent systems. Balusamy et al. [8] studied the characteristics of forced synchronization of limit cycle oscillations in a premixed swirl-stabilized turbulent combustor. They observed the existence of quasiperiodicity prior to forced synchronization of limit cycle oscillations and also different states of forced response such as phase locking, phase drifting, and phase trapping in the system. These behaviors are similar to those observed in laminar thermoacoustic systems (as discussed in Sect. 8.3.2.1). In addition to these characteristics of forced synchronization,

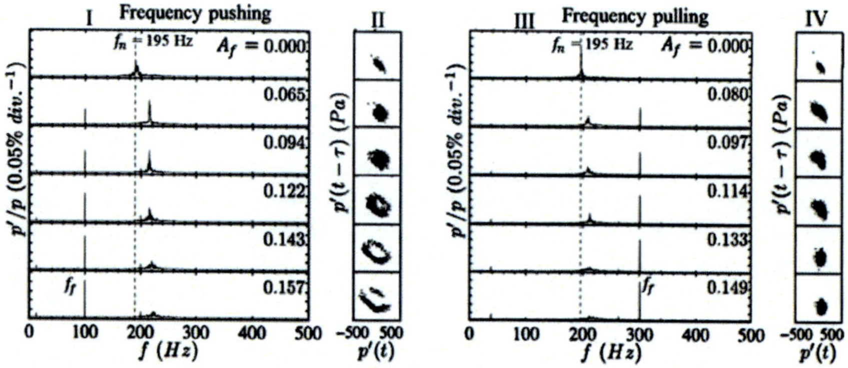

Fig. 8.20 The amplitude spectrum (I, III) and the corresponding Poincaré section (II, IV) of forced limit cycle oscillations in acoustic pressure in a turbulent combustor when the frequency ratio is fixed at $f_f/f_n = 0.51$ (column I) and $f_f/f_n = 1.54$ (column III). The amplitude of forcing ($A_f = u'/u$) is indicated in each plot, where u' is the acoustic velocity and u is the mean flow velocity in the system. The natural frequency of the self-excited oscillations is $f_n = 195$ Hz, indicated by a dotted line. As A_f is gradually increased, we observe the movement of f_n away from f_f during the frequency pushing behavior and towards f_f during the frequency pulling behavior. Adapted with permission from Balusamy et al. [8]

turbulent thermoacoustic systems show two more key features, i.e., frequency pulling and frequency pushing, in their forced response, which we will discuss below.

During the occurrence of forced synchronization of thermoacoustic oscillations in laminar systems, the frequency peak corresponding to natural oscillations completely disappears and the system response exhibits oscillations only at the forcing frequency. In contrast, there are a few studies on turbulent combustors that reported the occurrence of frequency pulling and frequency pushing when thermoacoustic oscillations are forced away from their natural frequency [8, 11]. Frequency pulling indicates the movement of the natural frequency towards the forcing frequency in the spectrum and frequency pushing signifies the movement of the natural frequency away from the forcing frequency, when the forcing amplitude is varied as a control parameter. As a result, the difference between the forcing and the natural frequencies increases for the case of frequency pushing and decreases for the case of frequency pulling, with an increase in the forcing amplitude.

In Fig. 8.20, we show the spectral response of forced thermoacoustic instability in a turbulent combustor for different values of forcing amplitude when the forcing frequency is lower (column I in Fig. 8.20) and higher (column III in Fig. 8.20) than the natural frequency of thermoacoustic instability [8]. The position of the natural frequency ($f_n = 195$ Hz) in the uncoupled state is indicated by a dotted line. Columns II and IV in Fig. 8.20 show two-dimensional phase portraits (Poincaré sections) of the forced acoustic response of the system.

When the forcing frequency is lower than the natural frequency (column I in Fig. 8.20), as the forcing amplitude increases, the pressure dynamics transition from the state of limit cycle oscillations to quasiperiodic oscillations through a torus-birth bifurcation. Furthermore, the frequency corresponding to natural oscillations shows a gradual shift away (pushing) from the frequency peak corresponding to external forcing. The presence of quasiperiodic oscillations is confirmed from the distribution of state points along a closed-loop in the Poincaré plot (column II in Fig. 8.20). On the other hand, when the forcing frequency is greater than the natural frequency (column III in Fig. 8.20), the increase in the forcing amplitude causes the natural frequency to move towards the forcing frequency. This appears as a frequency pulling behavior of the system. Due to the presence of such frequency pushing/pulling behavior in the system, we need a larger value of the forcing amplitude to achieve forced synchronization of thermoacoustic instability, as compared to that required for the system whose natural frequency does not exhibit any movement under the influence of forcing [8].

Having discussed in detail the features of forced synchronization of self-excited acoustic pressure fluctuations during the state of thermoacoustic instability in different laminar and thermoacoustic systems, we next summarize a few studies that characterize the forced response of self-excited hydrodynamic field to external sinusoidal acoustic perturbations.

8.3.6 Forced Synchronization of Self-Excited Oscillations in the Hydrodynamic Field

As we discussed in Chap. 7, thermoacoustic instabilities are developed due to a coupled interaction of multiple subsystems, such as the duct acoustics, the hydrodynamics, and the flame, present in a confined environment of a combustion system. In the absence of a strong positive feedback from the acoustic field (i.e., during the state of stable operation), the flame or the underlying hydrodynamic field (together can be referred to as reacting flow field) can exhibit self-sustained oscillations due to the presence of the inherent nonlinearity associated with these fields [22]. In a thermoacoustic system, the interaction of the acoustic field with the oscillatory (narrowband fluctuations) reacting flow field would be significantly different from that observed with the non-oscillatory (broadband fluctuations) reacting flow field. In order to study the interaction of such unstable reacting flow field and the acoustic field, an alternate method has been used, where the response of the reacting flow field is characterized for different conditions of external forcing amplitudes and frequencies. Next, we examine the characteristics of forced reacting flow field using an example of the preheated reacting wake of a bluff-body stabilized turbulent combustor subjected to different parameters of harmonic forcing [21, 23].

In a turbulent bluff-body combustor, the key flow features that control the dynamics of the bluff-body stabilized flame are the separating free shear layers and the wake [17, 54]. Here, the free shear layer is a region in the flow arising due to the presence of a significant velocity gradient in fluid streams, and the wake is a region

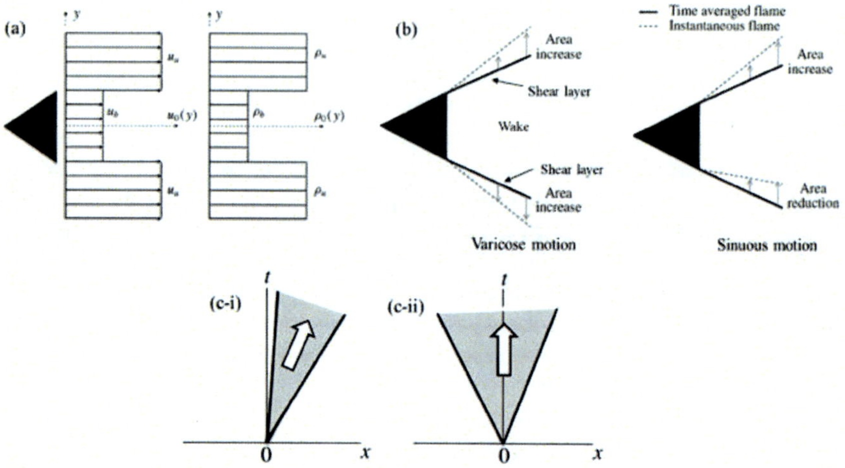

Fig. 8.21 The schematics of (**a**) the velocity and the density profiles of a bluff-body wake flow, (**b**) the flame fluctuations about the averaged value under the influence of varicose and sinuous type of vortex shedding. The schematic of the evolution of a disturbance induced at $(x_0, t_0) = (0, 0)$ in (**c-i**) convectively unstable and (**c-ii**) globally unstable flows. (**a, b**) Adapted with permission from Emerson and Lieuwen [21]. (**c**) Reproduced with permission from Lieuwen [54]

behind the bluff-body consisting of two shear layer and a recirculation zone (i.e., a region of negative velocity), see Fig. 8.21b. Shear layers are also referred to as the mixing layers, as these are formed along the boundary of mixing of the fuel and the oxidizer or the boundary of mixing of the cold reactants and the hot products. As a consequence of the flame residing in the shear layer, this region tends to be at a high temperature having low densities.

At high Reynolds number, the separated shear layers become unstable due to the presence of Kelvin-Helmholtz instability [80]. The roll-up of the shear layers leads to the formation of a sequence of vortices, often referred to as the Von Karman vortex street [78]. The symmetric and the asymmetric roll-up of the shear layers indicate the occurrence of varicose and sinuous modes of vortex shedding, respectively (see Fig. 8.21b). This vortex shedding modulates the flame surface area and, hence, controls the heat release rate fluctuations in the flame. When these heat release rate fluctuations occur in-phase with the acoustic pressure fluctuations, we observe the onset of thermoacoustic instability in the system. This, in turn, suggests that the underlying hydrodynamic field controls the coupling between the acoustic pressure and heat release rate fluctuations and, hence, we need to investigate the properties of such hydrodynamic field.

Furthermore, as we know, the presence of combustion affects the stability of the global hydrodynamic mode of the wake. Due to combustion, the density of the wake flow is lower than the free stream flow (see Fig. 8.21a). The ratio of densities of the burnt products and the fresh reactants across the shear layer plays an important role in deciding the stability of the wake [22, 24]. In the presence of a high density ratio (non-preheated reactants), the wake is convectively unstable (see image corresponding to $\sigma_\rho = 3.1$ in Fig. 8.22a). In contrast, for a low density ratio (preheated reactants), the wake is globally unstable (see image corresponding to $\sigma_\rho = 1.7$ in Fig. 8.22a). Note that to change the density ratio, the density of the reactants is changed by varying their temperature, which can be achieved by compressing the air to a high pressure ratio (similar to the flow entering the combustor through the compressor and/or diffusor) or by partially preheating the air (similar to the hot flow entering the afterburner) [54].

In Fig. 8.22a, we show the amplitude spectrum of the fluctuations in the shear layer of the wake flows consisting of a high density ratio (convectively unstable) and a low density ratio (globally unstable). For convectively unstable wake flows, the fluctuations in the shear layer are broadband. While for globally unstable wake flows, these fluctuations are narrowband.

For globally unstable flows [19, 38], perturbations initiated in the flow field grow exponentially in both (upstream and downstream) directions (Fig. 8.21c-ii) and eventually contaminate the entire flow. As a result, the local flow field exhibits self-sustained oscillations. On the other hand, for convectively unstable flows [19, 38] (another form of flow instability), externally imposed disturbances grow only in the downstream direction and decay at the point they are introduced (Fig. 8.21c-i). The important characteristic of convectively unstable flows is that these flows respond to all external perturbations and, hence, are considered as amplifiers [38]. On the contrary, globally unstable flows act as an oscillator, where they offer resistance to the low amplitude forcing, but show characteristics of the forced synchronization, such as phase or frequency locking, for higher values of the forcing amplitude [13, 60]. Therefore, we can conduct forced synchronization studies only on globally unstable flows and not on convectively unstable flows. Note that the transition from convectively unstable to globally unstable flow fields in a bluff-body combustor does not occur abruptly, but takes place gradually through the state of intermittency where the flow dynamics fluctuates alternately between these flow fields [22, 99].

In Fig. 8.22b, we present the amplitude spectrum of the forced globally unstable wake flows. In the absence of forcing, we notice a single peak in the spectrum corresponding to the vortex shedding frequency (f_n) in the wake (curve 1 in Fig. 8.22b). In the presence of low amplitude forcing, the magnitude of the vortex shedding frequency shows a small decrease; thus, we observe two peaks corresponding to forcing (f_f) and natural (f_n) frequencies in the spectrum (curve 2 in Fig. 8.22b). When the forcing amplitude is sufficiently high, we notice a complete suppression of the frequency corresponding to vortex shedding and the existence of only the frequency corresponding to external forcing (f_f) in the spectrum (curve 3 in Fig. 8.22b). We refer to this state of the forced response of the globally unstable wake flows as frequency-locking or frequency entrainment. During this state, forcing solely controls the dynamics of the wake [23]. With an increase in the value of $|f_f - f_n|$, the value of the forcing amplitude required for frequency-locking in the system also shows a corresponding increase [21]. This behavior is similar to that we described for forced synchronization of self-excited acoustic pressure fluctuations in the previous sections.

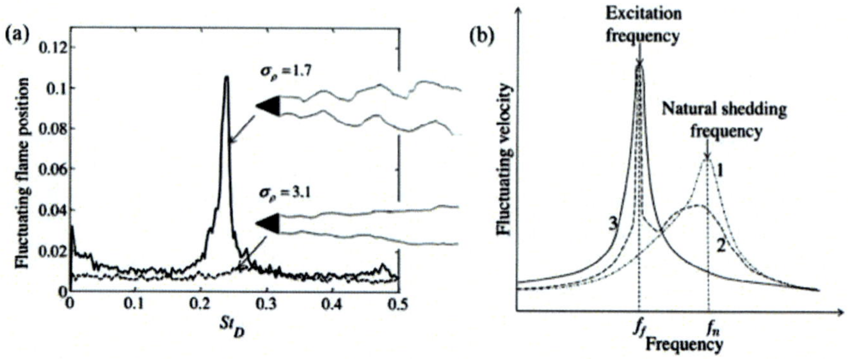

Fig. 8.22 (**a**) The amplitude spectrum of the flame edge fluctuations at a given location along the flame in a bluff-body combustor for two values of the density ratios (σ_ρ). In (**a**), the snapshots of the low density ratio $\sigma_\rho = 1.7$ (globally unstable) and high density ratio $\sigma_\rho = 3.1$ (convectively unstable) wakes are shown. (**b**) The amplitude spectrum of the flame edge fluctuations for different values of the forcing amplitude at a fixed forcing frequency. The spectra corresponding to 1, 2, and 3 indicate the unforced, forced, and synchronized states of oscillations, respectively. Adapted with permission from Lieuwen [54]

Similarly, we can also characterize the forced response of the self-excited dynamics of hydrodynamic jets, which could be non-reacting [44, 51, 53] or reacting [44, 52]. Such investigations on the forced response of the unstable reacting or non-reacting jets help in improving the understanding on the flame-acoustic interactions present in practical thermoacoustic systems.

Low-density hydrodynamic jets, when discharged into a quiescent environment, exhibit a strong region of absolute instability at the base of the jet due to Kelvin-Helmholtz instability [65]. Such Kelvin-Helmholtz instability arises due to the presence of inflection points in the radial velocity profile of the jet and the presence of a steeper density gradient in the opposite direction of the velocity gradient. At a critical value of these gradients in the density and the velocity profiles, the shear layer separating the jet and the ambient air becomes unstable. Such unstable shear layer then demonstrates a self-excited global mode, which show a sharp peak in the spectrum of the velocity fluctuations [66]. In order to study forced synchronization of such jets, we need to force them using external acoustic perturbations of different amplitude and frequency.

In Fig. 8.23, we present the forced response of velocity fluctuations observed in the globally unstable non-reacting helium jet [51, 53]. We notice that when the forcing frequency (f_f) is close to f_n (i.e., $f_f/f_n \sim 1.04$), the relative phase plot shows the characteristics of phase drifting for low values of A_f (Fig. 8.23d). As A_f approaches the synchronization boundary, we observe the property of intermittent phase synchronization. The power spectrum of the velocity fluctuations for these forcing conditions show the presence of multiple frequencies arising due to a nonlinear interaction between f_f and f_n. This interaction between coupled frequencies leads to modulations in the signal of velocity fluctuations (Fig. 8.23e).

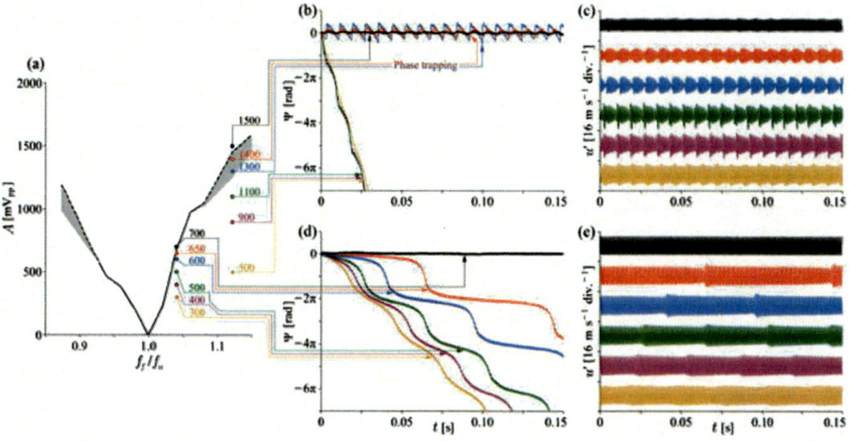

Fig. 8.23 Forced synchronization characteristics of a low-density helium jet, where (**a**) indicates the Arnold tongue, (**b**, **d**) the relative phase (Ψ) between the velocity fluctuations (shown as u' in **c** and **e**, respectively) and the forcing signal. The shaded regions in (**a**) represent the presence of phase trapping state in the system. We observe the presence of a phase trapping state during the suppression route of forced synchronization in (**b**); however, this state is not observed during the locking route of forced synchronization in (**d**). Reproduced with permission from Li et al. [53]

The phase space of these oscillations displays the surface of a stable quasiperiodic attractor [53]. The instantaneous phase difference $\Psi(t)$ between the forcing and velocity fluctuations show the property of intermittent phase locking, where the long constant epochs of $\Psi(t)$ are interrupted by short epochs of phase jumps (or phase slips). At the onset of forced synchronization, the frequency peak corresponding to f_n disappears completely, the jet oscillates only at f_f, and the relative phase $\Psi(t)$ remains constant in time (Fig. 8.23d).

When f_f is far from f_n (i.e., $f_f/f_n \sim 1.12$), we observe the presence of a phase trapping state in between the states of phase drifting and phase locking (Fig. 8.23b,c). During phase trapping, the relative phase $\Psi(t)$ is bounded and oscillates at a frequency equal to $|f_n - f_f|$.

Thus, we observe the presence of both the locking route (for $f_f/f_n \sim 1.04$) and the suppression route (for $f_f/f_n \sim 1.12$) of synchronization in the forced response of the oscillatory helium jet. This behavior is similar to that we discussed for forced synchronization of limit cycle oscillations in a general nonlinear oscillator [6, 79] in Sect. 8.3 and also in the self-excited dynamics of a thermoacoustic system [32, 64] in Sect. 8.3.2. This also means, we can observe the similar characteristics of forced synchronization in any nonlinear physical system that exhibits oscillatory instabilities.

In summary, in this chapter, we investigated both the mutual and the forced synchronization characteristics of a thermoacoustic system. We showed that the mutually coupled acoustic field and the reactive flow dynamics of a turbulent combustor undergo a synchronization transition during the onset of thermoacoustic instability. Several measures derived from the recurrence property of the phase space trajectory of the signal can be used to characterize the coupled interaction of these oscillations in the system.

We also discussed the effect of harmonic forcing to control the phase and the amplitude properties of the self-excited limit cycle, quasiperiodicity, and chaotic oscillations in a system. Further, we delineated manifestations of several underlying phenomena such as synchronance and asynchronous quenching in a forced thermoacoustic system. Finally, we studied the forced response of the oscillatory hydrodynamic field resulting from absolute instability. We note that most of these forced synchronization studies are based on laminar thermoacoustic systems, and only a limited knowledge of forced synchronization of thermoacoustic oscillations in turbulent systems is available. Hence, we need further investigations to reveal the underlying features of the forced response of thermoacoustic oscillations observed in such systems.

We need an accurate understanding of the nature of coupling or synchronization properties of the acoustic pressure and the heat release rate fluctuations to implement appropriate mitigation strategies required for quenching of thermoacoustic instability in a combustion system. We will discuss in detail the control methodologies based on the framework of synchronization theory (such as amplitude death and active open-loop control through asynchronous quenching) for the suppression of thermoacoustic instability in Chap. 12. In the next chapter, we will discuss the phenomenological model developed for studying the synchronization transition of the acoustic field and the heat release rate field in a thermoacoustic system.

References

1. Abel, M., Ahnert, K., Bergweiler, S.: Synchronization of sound sources. Phys. Rev. Lett. **103**(11), 114301 (2009)
2. Abrams, D.M., Strogatz, S.H.: Chimera states for coupled oscillators. Phys. Rev. Lett. **93**(17), 174102 (2004)
3. Abrams, D.M., Mirollo, R., Strogatz, S.H., Wiley, D.A.: Solvable model for chimera states of coupled oscillators. Phys. Rev. Lett. **101**(8), 084103 (2008)
4. Alemela, P.R., Casado, J.R., Kumar, S., Kok, J.: Thermoacoustic analysis of the dynamic pressure inside a model combustor during limit cycle oscillations. J. Spray Combust. Dyn. **5**(1), 25–47 (2013)
5. Annaswamy, A.M., Ghoniem, A.F.: Active control in combustion systems. IEEE Control Syst. Mag. **15**(6), 49–63 (1995)
6. Balanov, A., Janson, N., Postnov, D., Sosnovtseva, O.: Synchronization: From Simple to Complex, vol. 17. Springer (2009)
7. Balasubramanian, K., Sujith, R.I.: Thermoacoustic instability in a Rijke tube: Non-normality and nonlinearity. Phys. Fluids **20**(4), 044103 (2008)
8. Balusamy, S., Li, L.K.B., Han, Z., Juniper, M.P., Hochgreb, S.: Nonlinear dynamics of a self-excited thermoacoustic system subjected to acoustic forcing. Proc. Combust. Inst. **35**(3), 3229–3236 (2015)
9. Bell, A.E.: The horologium oscillatorium of Christian Huygens. Nature **148**(3748), 245–248 (1941)

10. Bellows, B.D., Neumeier, Y., Lieuwen, T.: Forced response of a swirling, premixed flame to flow disturbances. J. Propuls. Power **22**(5), 1075–1084 (2006)
11. Bellows, B.D., Hreiz, A., Lieuwen, T.: Nonlinear interactions between forced and self-excited acoustic oscillations in premixed combustor. J. Propuls. Power **24**(3), 628–630 (2008)
12. Blekhman, I.I.: Synchronization in Science and Technology. ASME Press, New York (1988)
13. Blevins, R.D.: The effect of sound on vortex shedding from cylinders. J. Fluid Mech. **161**, 217–237 (1985)
14. Boccaletti, S., Valladares, D.L.: Characterization of intermittent lag synchronization. Phys. Rev. E **62**(5), 7497–7500 (2000)
15. Boccaletti, S., Kurths, J., Osipov, G., Valladares, D.L., Zhou, C.S.: The synchronization of chaotic systems. Phys. Rep. **366**(1), 1–101 (2002)
16. Boubaker, O., Jafari, S.: Recent Advances in Chaotic Systems and Synchronization: From Theory to Real World Applications. Academic Press (2018)
17. Cardell, G.S.: Flow past a circular cylinder with a permeable wake splitter plate. Ph.D. thesis, California Institute of Technology, USA (1993)
18. Chakravarthy, S.R., Shreenivasan, O.J., Boehm, B., Dreizler, A., Janicka, J.: Experimental characterization of onset of acoustic instability in a nonpremixed half-dump combustor. J. Acoust. Soc. Am. **122**(1), 120–127 (2007)
19. Chomaz, J.M.: Global instabilities in spatially developing flows: non-normality and nonlinearity. Annu. Rev. Fluid Mech. **37**, 357–392 (2005)
20. Ducruix, S., Schuller, T., Durox, D., Candel, S.: Combustion dynamics and instabilities: Elementary coupling and driving mechanisms. J. Propuls. Power **19**(5), 722–734 (2003)
21. Emerson, B., Lieuwen, T.: Dynamics of harmonically excited, reacting bluff body wakes near the global hydrodynamic stability boundary. J. Fluid Mech. **779**, 716–750 (2015)
22. Emerson, B., O'Connor, J., Juniper, M., Lieuwen, T.: Density ratio effects on reacting bluff-body flow field characteristics. J. Fluid Mech. **706**, 219–250 (2012)
23. Emerson, B., O'Connor, J., Noble, D., Lieuwen, T.: Frequency locking and vortex dynamics of an acoustically excited bluff body stabilized flame. In: 50th AIAA Aerospace Sciences Meeting including the New Horizons Forum and Aerospace Exposition, AIAA 2012-451 (2012)
24. Erickson, R.R., Soteriou, M.C.: The influence of reactant temperature on the dynamics of bluff body stabilized premixed flames. Combust. Flame **158**(12), 2441–2457 (2011)
25. Fujisaka, H., Yamada, T.: Stability theory of synchronized motion in coupled-oscillator systems. Prog. Theor. Phys. **69**(1), 32–47 (1983)
26. Godavarthi, V., Pawar, S.A., Unni, V.R., Sujith, R.I., Marwan, N., Kurths, J.: Coupled interaction between unsteady flame dynamics and acoustic field in a turbulent combustor. Chaos **28**(11), 113111 (2018)
27. Golomb, D., Hansel, D., Shraiman, B., Sompolinsky, H.: Clustering in globally coupled phase oscillators. Phys. Rev. A **45**(6), 3516–3530 (1992)
28. Gopalakrishnan, E.A., Sujith, R.I.: Influence of system parameters on the hysteresis characteristics of a horizontal Rijke tube. J. Spray Combust. Dyn. **6**(3), 293–316 (2014)
29. Goswami, B., Ambika, G., Marwan, N., Kurths, J.: On interrelations of recurrences and connectivity trends between stock indices. Physica A **391**(18), 4364–4376 (2012)
30. Guan, Y., Liu, P., Jin, B., Gupta, V., Li, L.K.B.: Nonlinear time-series analysis of thermoacoustic oscillations in a solid rocket motor. Exp. Therm. Fluid Sci. **98**, 217–226 (2018)
31. Guan, Y., Murugesan, M., Li, L.K.B.: Strange nonchaotic and chaotic attractors in a self-excited thermoacoustic oscillator subjected to external periodic forcing. Chaos **28**(9), 093109 (2018)
32. Guan, Y., Gupta, V., Kashinath, K., Li, L.K.B.: Open-loop control of periodic thermoacoustic oscillations: experiments and low-order modelling in a synchronization framework. Proc. Combust. Inst. **37**(4), 5315–5323 (2019)
33. Guan, Y., Gupta, V., Wan, M., Li, L.K.B.: Forced synchronization of quasiperiodic oscillations in a thermoacoustic system. J. Fluid Mech. **879**, 390–421 (2019)

34. Guan, Y., Li, L.K.B., Ahn, B., Kim, K.T.: Chaos, synchronization, and desynchronization in a liquid-fueled diffusion-flame combustor with an intrinsic hydrodynamic mode. Chaos **29**(5), 053124 (2019)
35. Gutmark, E., Parr, T.P., Hanson-Parr, D.M., Schadow, K.C.: On the role of large and small-scale structures in combustion control. Combust. Sci. Tech. **66**(1-3), 107–126 (1989)
36. Hagerstrom, A.M., Murphy, T.E., Roy, R., Hövel, P., Omelchenko, I., Schöll, E.: Experimental observation of chimeras in coupled-map lattices. Nat. Phys. **8**(9), 658–661 (2012)
37. Hashimoto, T., Shibuya, H., Gotoda, H., Ohmichi, Y., Matsuyama, S.: Spatiotemporal dynamics and early detection of thermoacoustic combustion instability in a model rocket combustor. Phys. Rev. E **99**(3), 032208 (2019)
38. Huerre, P., Monkewitz, P.A.: Local and global instabilities in spatially developing flows. Annu. Rev. Fluid Mech. **22**(1), 473–537 (1990)
39. Huygens, C.: Letter to de Sluse, Oeuveres Completes de Christian Huygens (letters; No. 1333 of 24 February 1665, No. 1335 of 26 February 1665, No. 1345 of 6 March 1665), (Société Hollandaise des Sciences, Martinus Nijhoff, La Haye, 1893) (1665)
40. Huygens, C.: Horoloqium Oscilatorium. Apud F. Muquet, Parisiis (1673); English translation: Christiaan Huygens's the pendulum clock or geometrical demonstrations concerning the motion of pendula as applied to clocks (Iowa State University Press, Ames, 1986) (1673)
41. Hyodo, H., Biwa, T.: Phase-locking and suppression states observed in forced synchronization of thermoacoustic oscillator. J. Phys. Soc. Jpn. **87**(3), 034402 (2018)
42. Issartel, J., Bardainne, T., Gaillot, P., Marin, L.: The relevance of the cross-wavelet transform in the analysis of human interaction–a tutorial. Front. Psychol. **5**, 1566 (2015)
43. Jenkins, A.: Self-oscillation. Phys. Rep. **525**(2), 167–222 (2013)
44. Juniper, M.P., Li, L.K.B., Nichols, J.W.: Forcing of self-excited round jet diffusion flames. Proc. Combust. Inst. **32**(1), 1191–1198 (2009)
45. Kashinath, K., Li, L.K.B., Juniper, M.P.: Forced synchronization of periodic and aperiodic thermoacoustic oscillations: lock-in, bifurcations and open-loop control. J. Fluid Mech. **838**, 690–714 (2018)
46. Kasthuri, P., Pavithran, I., Krishnan, A., Pawar, S.A., Sujith, R.I., Gejji, R., Anderson, W., Marwan, N., Kurths, J.: Recurrence analysis of slow–fast systems. Chaos **30**(6), 063152 (2020)
47. Kuramoto, Y., Battogtokh, D.: Coexistence of coherence and incoherence in nonlocally coupled phase oscillators. arXiv preprint cond-mat/0210694 (2002)
48. Lakshmanan, M., Murali, K.: Chaos in Nonlinear Oscillators: Controlling and Synchronization, vol. 13. World Scientific (1996)
49. Lakshmanan, M., Senthilkumar, D.V.: Dynamics of Nonlinear Time-delay Systems. Springer Science & Business Media (2011)
50. Li, L.: Forcing of globally unstable jets and flames. Ph.D. thesis, University of Cambridge, UK (2012)
51. Li, L.K.B., Juniper, M.P.: Lock-in and quasiperiodicity in a forced hydrodynamically self-excited jet. J. Fluid Mech. **726**, 624–655 (2013)
52. Li, L.K.B., Juniper, M.P.: Lock-in and quasiperiodicity in hydrodynamically self-excited flames: Experiments and modelling. Proc. Combust. Inst. **34**(1), 947–954 (2013)
53. Li, L.K.B., Juniper, M.P.: Phase trapping and slipping in a forced hydrodynamically self-excited jet. J. Fluid Mech. **735**, R5 (2013)
54. Lieuwen, T.C.: Unsteady Combustor Physics. Cambridge University Press (2012)
55. Maistrenko, Y.L., Mosekilde, E., Postnov, D.: Chaotic Synchronization: Applications to Living Systems, vol. 42. World Scientific (2002)
56. Manoj, K., Pawar, S.A., Sujith, R.I.: Experimental investigation on the susceptibility of minimal networks to a change in topology and number of oscillators. Phys. Rev. E **103**(2), 022207 (2021)
57. Manrubia, S.C., Mikhailov, A.S.: Emergence of Dynamical Order: Synchronization Phenomena in Complex Systems. World Scientific (2004)

58. Mariappan, S., Sujith, R.I., Schmid, P.J.: Experimental investigation of non-normality of thermoacoustic interaction in an electrically heated Rijke tube. J. Spray Combust. Dyn. **7**(4), 315–352 (2015)
59. Marwan, N., Romano, M.C., Thiel, M., Kurths, J.: Recurrence plots for the analysis of complex systems. Phys. Rep. **438**(5), 237–329 (2007)
60. Masselin, M., Ho, C.M.: Lock-on and instability in a flat plate wake. In: Shear Flow Control Conference, pp. AIAA 1985–571 (1985)
61. Matveev, K.I.: Thermoacoustic instabilities in the Rijke tube: experiments and modeling. Ph.D. thesis, California Institute of Technology, USA (2003)
62. McManus, K.R., Vandsburger, U., Bowman, C.T.: Combustor performance enhancement through direct shear layer excitation. Combust. Flame **82**(1), 75–92 (1990)
63. Mondal, S., Unni, V.R., Sujith, R.I.: Onset of thermoacoustic instability in turbulent combustors: an emergence of synchronized periodicity through formation of chimera-like states. J. Fluid Mech. **811**, 659–681 (2017)
64. Mondal, S., Pawar, S.A., Sujith, R.I.: Forced synchronization and asynchronous quenching of periodic oscillations in a thermoacoustic system. J. Fluid Mech. **864**, 73–96 (2019)
65. Monkewitz, P.A., Sohn, K.D.: Absolute instability in hot jets. AIAA J. **26**(8), 911–916 (1988)
66. Monkewitz, P.A., Lehmann, B., Barsikow, B., Bechert, D.W.: The spreading of self-excited hot jets by side jets. Phys. Fluids A **1**(3), 446–448 (1989)
67. Murugesan, M., Zhu, Y., Li, L.K.B.: Complex network analysis of forced synchronization in a hydrodynamically self-excited jet. Int. J. Heat Fluid Flow **76**, 14–25 (2019)
68. Ohe, K., Takeda, S.: Asynchronous quenching and resonance excitation of ionization waves in positive columns. Contrib. Plasma Phys. **14**(2), 55–65 (1974)
69. Ohsawa, T.: Synchronous quenching due to nonlinear mode coupling in beam-plasma system. J. Phys. Soc. Jpn. **49**(6), 2340–2348 (1980)
70. Osipov, G.V., Hu, B., Zhou, C., Ivanchenko, M.V., Kurths, J.: Three types of transitions to phase synchronization in coupled chaotic oscillators. Phys. Rev. Lett. **91**(2), 024101 (2003)
71. Paschereit, C., Gutmark, E.: Control of thermoacoustic instabilities in a premixed combustor by fuel modulation. In: 37th Aerospace Sciences Meeting and Exhibit, AIAA 1999-711 (1999)
72. Pawar, S.A., Seshadri, A., Unni, V.R., Sujith, R.I.: Thermoacoustic instability as mutual synchronization between the acoustic field of the confinement and turbulent reactive flow. J. Fluid Mech. **827**, 664–693 (2017)
73. Pawar, S.A., Sujith, R.I., Emerson, B., Lieuwen, T.: Characterization of forced response of density stratified reacting wake. Chaos **28**(2), 023108 (2018)
74. Pawar, S.A., Mondal, S., George, N.B., Sujith, R.I.: Temporal and spatiotemporal analyses of synchronization transition in a swirl-stabilized combustor. AIAA J. **57**(2), 836–847 (2019)
75. Pawar, S.A., Panchagnula, M.V., Sujith, R.I.: Phase synchronization and collective interaction of multiple flamelets in a laboratory scale spray combustor. Proc. Combust. Inst. **37**(4), 5121–5128 (2019)
76. Pecora, L.M., Carroll, T.L.: Synchronization in chaotic systems. Phys. Rev. Lett. **64**(8), 821–824 (1990)
77. Penelet, G., Biwa, T.: Synchronization of a thermoacoustic oscillator by an external sound source. Am. J. Phys. **81**(4), 290–297 (2013)
78. Perry, A.E., Chong, M.S., Lim, T.T.: The vortex-shedding process behind two-dimensional bluff bodies. J. Fluid Mech. **116**, 77–90 (1982)
79. Pikovsky, A., Rosenblum, M., Kurths, J.: Synchronization: A Universal Concept in Nonlinear Sciences, vol. 12. Cambridge University Press (2003)
80. Prasad, A., Williamson, C.H.K.: The instability of the shear layer separating from a bluff body. J. Fluid Mech. **333**, 375–402 (1997)
81. Rajaram, R., Lieuwen, T.: Acoustic radiation from turbulent premixed flames. J. Fluid Mech. **637**, 357–385 (2009)
82. Rayleigh, J.W.S.: The explanation of certain acoustical phenomena. Nature **18**(455), 319–321 (1878)

83. Romano, M.C., Thiel, M., Kurths, J., von Bloh, W.: Multivariate recurrence plots. Phys. Lett. A **330**(3-4), 214–223 (2004)
84. Romano, M.C., Thiel, M., Kurths, J., Kiss, I.Z., Hudson, J.: Detection of synchronization for non-phase-coherent and non-stationary data. Europhys. Lett. **71**(3), 466–472 (2005)
85. Rosenblum, M.G., Pikovsky, A.S., Kurths, J.: Phase synchronization of chaotic oscillators. Phys. Rev. Lett. **76**(11), 1804–1807 (1996)
86. Rosenblum, M.G., Pikovsky, A.S., Kurths, J.: From phase to lag synchronization in coupled chaotic oscillators. Phys. Rev. Lett. **78**(22), 4193–4196 (1997)
87. Roy, A., Mondal, S., Pawar, S.A., Sujith, R.I.: On the mechanism of open-loop control of thermoacoustic instability in a laminar premixed combustor. J. Fluid Mech. **884**, A2 (2020)
88. Rulkov, N.F., Sushchik, M.M., Tsimring, L.S., Abarbanel, H.D.I.: Generalized synchronization of chaos in directionally coupled chaotic systems. Phys. Rev. E **51**(2), 980–994 (1995)
89. Sahay, A., Roy, A., Pawar, S.A., Sujith, R.I.: Dynamics of coupled thermoacoustic oscillators under asymmetric forcing. Phys. Rev. Appl. **15**(4), 044011 (2021)
90. Sato, M., Hyodo, H., Biwa, T., Delage, R.: Synchronization of thermoacoustic quasiperiodic oscillation by periodic external force. Chaos **30**(6), 063130 (2020)
91. Schadow, K.C., Gutmark, E.: Combustion instability related to vortex shedding in dump combustors and their passive control. Prog. Energy Combust. Sci. **18**(2), 117–132 (1992)
92. Schäfer, C., Rosenblum, M.G., Kurths, J., Abel, H.H.: Heartbeat synchronized with ventilation. Nature **392**(6673), 239–240 (1998)
93. Sethares, W.A.: Rhythm and Transforms. Springer Science & Business Media (2007)
94. Singh, G., Mariappan, S.: Experimental investigation on the route to vortex-acoustic lock-in phenomenon in bluff body stabilized combustors. Combust. Sci. Tech., 1–29 (2019)
95. Strahle, W.C.: Combustion noise. Prog. Energy Combust. Sci. **4**(3), 157–176 (1978)
96. Strogatz, S.H.: From Kuramoto to Crawford: exploring the onset of synchronization in populations of coupled oscillators. Physica D **143**(1-4), 1–20 (2000)
97. Strogatz, S.: Sync: The Emerging Science of Spontaneous Order. Penguin UK (2004)
98. Sujith, R.I., Unni, V.R.: Complex system approach to investigate and mitigate thermoacoustic instability in turbulent combustors. Phys. Fluids **32**(6), 061401 (2020)
99. Suresha, S., Sujith, R.I., Emerson, B., Lieuwen, T.: Nonlinear dynamics and intermittency in a turbulent reacting wake with density ratio as bifurcation parameter. Phys. Rev. E **94**(4), 042206 (2016)
100. Tinsley, M.R., Nkomo, S., Showalter, K.: Chimera and phase-cluster states in populations of coupled chemical oscillators. Nat. Phys. **8**(9), 662–665 (2012)
101. Weng, Y., Unni, V.R., Sujith, R.I., Saha, A.: Synchronization framework for modeling transition to thermoacoustic instability in laminar combustors. Nonlinear Dyn. **100**, 3295—3306 (2020)
102. Yalçınkaya, T., Lai, Y.C.: Phase characterization of chaos. Phys. Rev. Lett. **79**(20), 3885–3888 (1997)
103. Yoder, J.G.: Christiaan Huygens, book on the pendulum clock (1673). In: Landmark Writings in Western Mathematics 1640-1940, pp. 33–45. Elsevier (2005)
104. Yoshida, T., Yazaki, T., Ueda, Y., Biwa, T.: Forced synchronization of periodic oscillations in a gas column: Where is the power source? J. Phys. Soc. Jpn. **82**(10), 103001 (2013)
105. Zbilut, J.P., Giuliani, A., Webber Jr, C.L.: Detecting deterministic signals in exceptionally noisy environments using cross-recurrence quantification. Phys. Lett. A **246**(1-2), 122–128 (1998)
106. Zinn, B.T.: Pulse combustion: recent applications and research issues. In: Symposium (International) on Combustion, vol. 24, pp. 1297–1305. Elsevier (1992)

Chapter 9
Model for Intermittency Route to Thermoacoustic Instability

In the previous chapters, we discussed dynamical transitions in turbulent combustors during the onset of thermoacoustic instability. We showed that the dynamics of such systems changes from the state of combustion noise to thermoacoustic instability via intermittency. The onset of thermoacoustic instability is thus a process of a gradual emergence of periodic oscillations from chaotic oscillations. Further, we explored the synchronization transition of the coupled acoustic pressure and heat release rate fluctuations during the intermittency route to thermoacoustic instability. We showed that these oscillations are desynchronized during the state of combustion noise, intermittently phase synchronized during state of intermittency, phase synchronized during the state of weakly correlated limit cycle oscillations, and finally exhibit generalized synchronization during the state of strongly correlated limit cycle oscillations. In this chapter, we introduce a reduced-order model developed for capturing the intermittency route to thermoacoustic instability in a bluff-body stabilized dump combustor. Using this model, we also study the coupled behavior of the acoustic pressure and the heat release rate fluctuations en route to thermoacoustic instability.

9.1 Model Description

In Chap. 6, we discussed the transition from stable operation (combustion noise) to unstable operation (thermoacoustic instability) that happens via intermittency. During the state of intermittency, the dynamics of the system alternately switches between bursts of large amplitude periodic oscillations and epochs of low amplitude aperiodic oscillations. Several previous studies have successfully modeled the combustion dynamics observed either during the state of stable operation [7, 28], the state of unstable operation [5, 14, 17, 26] or a direct transition from stable to unstable operation without the intermediate states [12]. However, studies that model

R. I. Sujith, S. A. Pawar, *Thermoacoustic Instability*, Springer Series in Synergetics,
https://doi.org/10.1007/978-3-030-81135-8_9

the transition from stable to unstable operation via intermittency, as the system parameter is varied, are very few [15, 18, 23, 24].

In this section, we discuss recent advances in developing a reduced-order (also called phenomenological or minimal) model for the intermittency route to thermoacoustic instability in turbulent combustors. In such combustors, vortex-acoustic interaction plays an important role in deciding the resultant dynamical state of the combustor [3, 4, 6, 8, 9, 11, 13, 19, 20, 22, 25, 27, 29], as we discussed in Chap. 7. While developing a reduced-order model, we ignore some details of the system that are not very critical and consider only the key features related to the specific phenomenon that govern thermoacoustic instability. These models make the analytical understanding of such phenomena much easier and, in turn, help in deriving new insights about the system behavior [1, 10]. Unlike the case of high-fidelity (CFD or LES) simulations, the results obtained from reduced-order models of a system are computationally very inexpensive.

To reproduce the experimental results pertaining to the intermittency route to thermoacoustic instability, we start with a reduced-order kicked oscillator model developed by Matveev and Culick [13]. This model helps us in studying the dynamics of a turbulent bluff-body combustor with a backward-facing step (dump plane) at the inlet [15, 23]. In this combustor, the formation of a vortex takes place at the dump plane (Fig. 9.1). In the model, we treat the process of the vortex formation as a build-up of the vorticity (or circulation, Γ) in time. As soon as the vorticity crosses a particular threshold (Γ_{cr}), we consider a vortex to be shed from the dump plane. The shed vortex then convects downstream and impinges on the bluff-body. The impingement of the vortex results in an intense fine scale-mixing of the fuel-air mixture, causing a sudden (localized) heat release in the system [29]. We refer to the time instant at which the vortex hits the bluff-body and produces near instantaneous heat release rate in the system as a 'kick'. Therefore, we call this model as a 'kicked oscillator model'. The localized heat release rate generates acoustic oscillations, which then affect the vortex shedding processes (or hydrodynamics) at the dump plane. Thus, a three-way coupling is developed between the combustion, the acoustic, and the hydrodynamic fields in the system. During the occurrence of lock-in of the kicking time and the dominant acoustic timescale, we observe the onset of thermoacoustic instability in the system.

In this model, we retain the key geometrical properties of a turbulent combustor considered in the experimental study (Chaps. 6–8), which include the length of the combustor duct (L), the height of the backward-facing step (d), and the location of the bluff-body from the step (L_c). We consider the geometry of the combustor as acoustically closed at the inlet and acoustically open at the exit (i.e., closed-open duct). Thus, at $x = 0$, we have the acoustic velocity $u' = 0$ and, at $x = L$, we have the acoustic pressure $p' = 0$ [14]. Further, we consider only a one-dimensional acoustic field in the system, i.e., $p'(t, x)$ and $u'(t, x)$. The flow field in the system is turbulent and the mean flow velocity is $\overline{u}$.

The linearized continuity, momentum, and energy equations for the one-dimensional acoustic field in a duct with a low Mach number flow are given below [16]. The detailed derivations of these equations are provided in Appendix A.

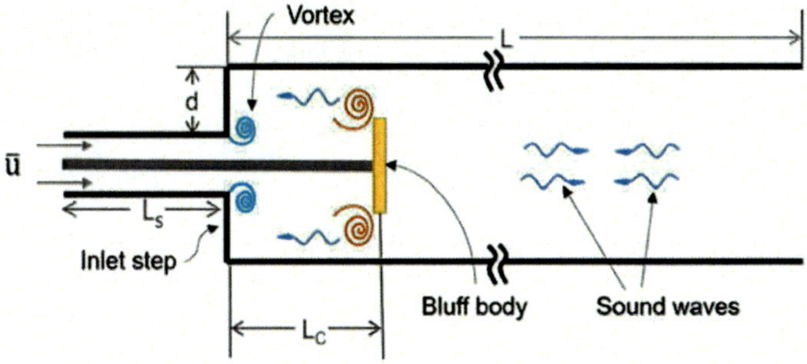

Fig. 9.1 A schematic of the bluff-body stabilized combustor. Here, L is the combustor length, L_c is the location of the bluff body, L_s is the location of the step from the inlet, d is the height of the step (dump plane), and $\overline{u}$ is the mean velocity of the reactant flow entering the combustor. Reproduced with permission from Pawar et al. [18]

Continuity equation: $$\frac{\partial \rho'}{\partial t} + \overline{\rho}\frac{\partial u'}{\partial x} = 0 \tag{9.1}$$

Momentum equation: $$\frac{\partial u'}{\partial t} + \frac{1}{\overline{\rho}}\frac{\partial p'}{\partial x} = 0 \tag{9.2}$$

Energy equation: $$\frac{\partial p'}{\partial t} + \gamma\overline{p}\frac{\partial u'}{\partial x} = (\gamma - 1)\dot{Q}' \tag{9.3}$$

Here, t is the time variable, $\bar{\rho}$ and $\bar{p}$ are the mean density and mean pressure of the flow, γ is the ratio of specific heats, and $\dot{Q}'$ is the heat release rate per unit volume in the system.

Further, we can model the convection velocity of a vortex which is shed from the backward-facing step of the combustor as [13],

$$\frac{dx_j}{dt} = \alpha\overline{u} + u'(x_j, t) \tag{9.4}$$

where α is the fraction of the mean velocity ($\overline{u}$) at which a vortex is convected in the system. The convection of the vortex also depends on the acoustic velocity (u') in the system.

As this shed vortex reaches the bluff-body, we observe an instantaneous heat release rate ($\dot{Q}'$) in the system. We model such impulsive heat release rate generated in the system (shown in Fig. 9.1) due to the breakup of the jth vortex at the bluff-body (L_c) at a time instant t_j as follows [13],

$$\dot{Q}' = \beta\sum_j \Gamma_j\delta(x - L_c)\delta(t - t_j) \tag{9.5}$$

Here, β is a coefficient that relates the vortex impingement to the heat release rate. δ denotes the Dirac-delta function. This means that $\dot{Q}'$ is non-zero only for $x = L_c$ and $t_j = t_c$, and is zero for all other values of x and t.

Since the linearized governing equations mentioned above ((9.1), (9.2), and (9.3)) are partial differential equations, we convert them into ordinary differential equations using the method of Galerkin projection [2, 30]. In this analysis, we need to expand the acoustic variables (p' and u') in terms of basis functions (the natural acoustic modes, M) that satisfy the boundary conditions of the system. Therefore, we can obtain the expression for p' and u' for a combustor geometry shown in Fig. 9.1 as follows,

$$\begin{aligned} p'(x,t) &= \overline{p}\sum_{i=1}^{M}\frac{\dot{\eta}_i(t)}{\omega_i}\cos(k_i x) \\ u'(x,t) &= \frac{\overline{p}}{\overline{\rho}c}\sum_{i=1}^{M}\eta_i(t)\sin(k_i x) \end{aligned} \tag{9.6}$$

where $k_i = \omega_i/c$ is the wavenumber, ω_i is the angular frequency, and c is the speed of sound. $\dot{\eta}$ and η are the time-dependent coefficients in the expansion of p' and u', respectively.

Substituting (9.5) and (9.6) into the linearized energy equation (9.3), we get,

$$\begin{aligned} \frac{\partial}{\partial t}\left(\overline{p}\sum_{i=1}^{M}\frac{\dot{\eta}_i(t)}{\omega_i}\cos(k_i x)\right) + \gamma\overline{p}\frac{\partial}{\partial x}\left(\frac{\overline{p}}{\overline{\rho}c}\sum_{i=1}^{M}\eta_i(t)\sin(k_i x)\right) \\ = (\gamma-1)\beta\sum_j \Gamma_j\delta(x-L_c)\delta(t-t_j) \end{aligned} \tag{9.7}$$

We can simplify the above equation to get,

$$\begin{aligned} \overline{p}\sum_{i=1}^{M}\frac{\ddot{\eta}_i(t)}{\omega_i}\cos(k_i x) + \gamma\frac{\overline{p}^2}{\overline{\rho}c}\sum_{i=1}^{M}\eta_i(t)\cos(k_i x)k_i \\ = (\gamma-1)\beta\sum_j \Gamma_j\delta(x-L_c)\delta(t-t_j) \end{aligned} \tag{9.8}$$

The above equation consists of two independent variables t and x. To eliminate the spatial dependency from this equation, we project the equation along the basis function $\cos(k_n x)$. This is achieved by multiplying both sides of the above equation by $\cos(k_n x)$ and integrating it over x between the limits 0 and L (i.e., taking an inner product with the basis function), we get,

$$\int_0^L \left[\overline{p} \sum_{i=1}^{M} \frac{\ddot{\eta}_i(t)}{\omega_i} \cos(k_i x)\cos(k_n x) + \gamma \frac{\overline{p}^2}{\overline{\rho} c} \sum_{i=1}^{M} \eta_i(t) \cos(k_i x)\cos(k_n x) k_i \right] dx$$

$$= \int_0^L \left((\gamma - 1)\beta \sum_j \Gamma_j \delta(x - L_c)\delta(t - t_j)\cos(k_n x) \right) dx \tag{9.9}$$

Since,

$$\int_0^L \cos\left(\frac{m\pi x}{L}\right)\cos\left(\frac{n\pi x}{L}\right) dx = 0 \quad \text{for} \quad m \neq n$$

$$= L/2 \quad \text{for} \quad m = n$$

and,

$$\int_0^L \cos\left(\frac{m\pi x}{L}\right)\delta(x - L_c) dx = \cos\left(\frac{m\pi L_c}{L}\right)$$

we get,

$$\frac{L}{2}\left(\overline{p}\frac{\ddot{\eta}_n(t)}{\omega_n} + \gamma \frac{\overline{p}^2}{\overline{\rho} c}\eta_n(t) k_n \right) = (\gamma - 1)\cos(k_n L_c) \sum_j \Gamma_j \delta(t - t_j) \tag{9.10}$$

As we know, $k = \omega/c$ and $c^2 = \gamma\overline{p}/\overline{\rho}$, the above equation can be simplified as,

$$\ddot{\eta}_n(t) + \omega_n^2 \eta_n(t) = \frac{2\omega_n(\gamma - 1)}{L\overline{p}} \cos(k_n L_c) \sum_j \Gamma_j \delta(t - t_j) \tag{9.11}$$

Since the thermoacoustic system is a damped oscillator, we add a damping term $\zeta_n = (2n - 1)^2\zeta_1$ in the above equation, where ζ_n is the mode-dependent damping term used to quantify acoustic losses at the boundary [12, 13]. Thus, we get the final second-order differential equation for the acoustic field in a turbulent combustor as,

$$\ddot{\eta}_n(t) + \zeta_n \dot{\eta}_n(t) + \omega_n^2 \eta_n(t) = \frac{2\omega_n(\gamma - 1)}{L\overline{p}} \cos(k_n L_c) \sum_j \Gamma_j \delta(t - t_j) \tag{9.12}$$

Furthermore, the governing equation for the circulation build-up at the backward-facing step of the combustor can be written as follows [13],

$$\dot{\Gamma}_j = \frac{d\Gamma_j}{dt} = \frac{1}{2}u^2_{\text{sep}}(t) \tag{9.13}$$

where $u_{\text{sep}}(t) = \overline{u} + u'(L_s, t)$ is the velocity at the step, $u'(L_s, t)$ is the acoustic velocity at the step, and L_s is the location of the step from the inlet.

The condition for the shedding of a vortex from the step is $\Gamma > \Gamma_{\text{cr}}$, where $\Gamma_{\text{cr}} = \Gamma_0 u_{\text{sep}}$. Here, $\Gamma_0 = d/2St$, and St is a Strouhal number [13]. Thus, when the circulation exceeds this critical value, a vortex is shed, that leads to an immediate fall in the value of circulation to zero. In the subsequent cycle, the circulation grows until it reaches the critical value. Thus, this cycle repeats and, hence, indicates the oscillatory behavior of the circulation [18].

As we mentioned before, the impingement of a fuel-carrying vortex on the bluff-body causes an instantaneous heat release rate in the system. Such an impulse creates an acoustic feedback wave (p'), which travels upstream and perturbs the shear layer at the dump plane; thereby affecting the circulation (Γ) build-up process [23]. This change in the circulation $\Delta\Gamma$ at the dump plane, due to the acoustic pressure fluctuations generated from the burning of a vortex at the bluff-body (L_c), can be modelled as [23],

$$\Delta\Gamma = B p_{L_c} \tag{9.14}$$

where B is the proportionality constant. For matching the experimental results, B is expressed as, $B = B_1 + B_2 S(\overline{u} - B_3)$. Here, S is a step function, B_1 and B_2 are constants indicating the feedback strength, and B_3 is the mean velocity at which the feedback strength changes in the system. Thus, the acoustic feedback is responsible for the change in the time instant at which a vortex is shed in the system. Therefore, we can obtain the modified form of the rate of change of circulation at the step (i.e., (9.13)) as follows [23],

$$\dot{\Gamma}_j = \frac{1}{2}u^2_{\text{sep}}(t) + B \sum_j p'_{L_c}(t_j)\delta(t - t_j - \tau_a) \tag{9.15}$$

where τ_a indicates the acoustic time delay, i.e., time taken by the acoustic wave to reach the dump plane (L_s) from the location of the bluff body (L_c).

Thus, by simultaneously solving (9.4), (9.12), and (9.15), we can study the dynamic properties of the acoustic pressure fluctuations and the circulation build-up in the system. We find that as the mean flow velocity is increased, the system transitions from the state of combustion noise to thermoacoustic instability via intermittency [23]. However, as we discussed in Chap. 8, the state of thermoacoustic instability in turbulent combustors can exhibit two types of limit cycle oscillations: weakly correlated and strongly correlated limit cycle oscillations. In order to capture weakly correlated limit cycle oscillations in the kicked-oscillator model, Pawar et al. [18] added a noise term in (9.14) and (9.15). Such noise terms accounts for the imperfectly closed acoustic boundary condition at the inlet and the underlying turbulent fluctuations that affect the circulation build-up process at the step in practical turbulent combustors. For more details, a reader may refer to [18].

9.2 Intermittency Route to Thermoacoustic Instability in Model

In Fig. 9.2, we show the time series of the acoustic pressure fluctuations obtained from the model during the onset of thermoacoustic instability, as the flow velocity is varied as a control parameter [18]. When the flow velocity is low, we observe the occurrence of low amplitude chaotic fluctuations in the acoustic pressure signal (Fig. 9.2a). The system behavior corresponding to this state is called combustion noise. Once the flow velocity is increased above a certain value, we notice the occurrence of the state of intermittency, where bursts of large amplitude periodic oscillations appear amidst epochs of low amplitude aperiodic oscillations (Fig. 9.2b). A further increase in the flow velocity leads to a corresponding increase in the duration of periodic oscillations observed in the intermittency signal. When the flow velocity is sufficiently high, the system dynamics transitions first to a state of low amplitude thermoacoustic instability (Fig. 9.2c), and then to a state of high amplitude thermoacoustic instability (Fig. 9.2d). Both these states of thermoacoustic instability exhibit limit cycle oscillations. Such a change in the dynamical behavior of the acoustic pressure fluctuations is qualitatively similar to the intermittency route to thermoacoustic instability witnessed in experiments involving a turbulent thermoacoustic system (see Fig. 8.4 in Chap. 8).

The occurrence of the low and high amplitude limit cycle oscillations in the model has different physical reasons. From (9.15), we know that the impingement of a vortex on the bluff-body kicks the acoustic wave due to the instantaneous heat release rate. This acoustic wave then travels upstream and affects the vortex shedding process, which then modifies the subsequent kicking times. Here, kicking time corresponds to the time instance at which a vortex reaches the bluff-body. Thus, kicking times encode physical information about the dynamics of the system [18]. From the temporal variation of kicking times, we can understand the difference between these two types of limit cycle oscillations. If the frequency of a signal is constant in time, the corresponding kicking times would show a single value and, if the frequency of the system exhibits a variation in time, the corresponding kicking times would display a distribution about the period of the dominant frequency of the signal.

In Fig. 9.3, we present the distribution of kicking times (normalized by the period of the corresponding acoustic signal) obtained for the states of low amplitude (Fig. 9.3a) and high amplitude (Fig. 9.3b) limit cycle oscillations. We notice a relatively wide variation of kicking times in the signal for the low amplitude limit cycle oscillations as compared to that of the high amplitude limit cycle oscillations. This, in turn, indicates that the instantaneous frequency of kicking times for low

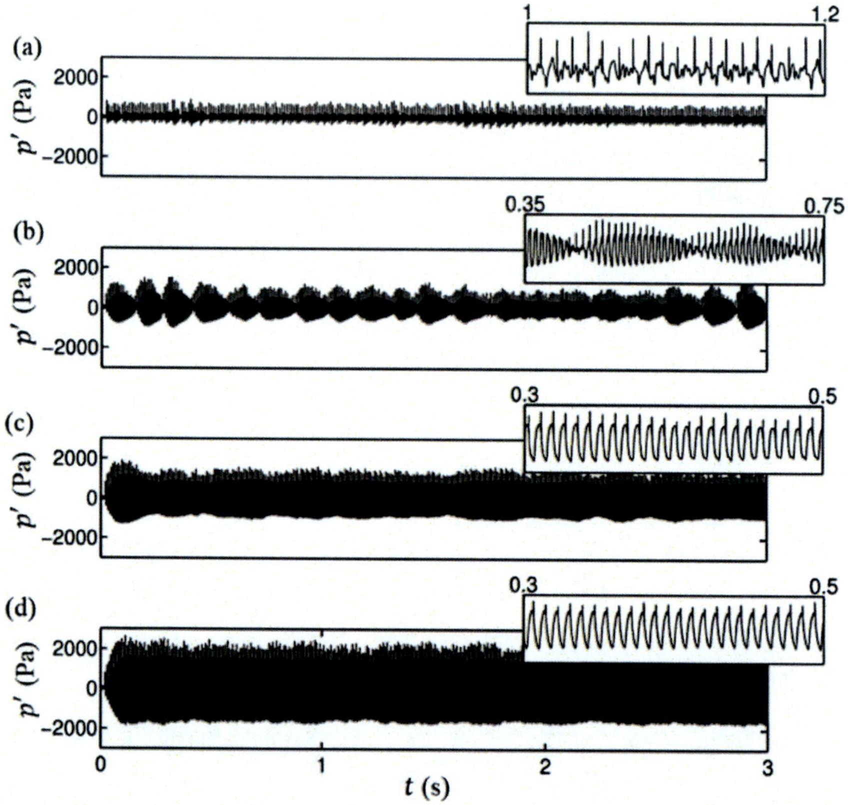

Fig. 9.2 Time series of the acoustic pressure fluctuations (p') for different dynamical states such as (**a**) chaos ($\overline{u} = 9.2$ m/s), (**b**) intermittency ($\overline{u} = 10.2$ m/s), (**c**) low amplitude limit cycle oscillations ($\overline{u} = 10.41$ m/s), and (**d**) high amplitude limit cycle oscillations ($\overline{u} = 10.6$ m/s), during the onset of thermoacoustic instability in the model. Reproduced with permission from Pawar et al. [18]

amplitude limit cycle oscillations shows a small variation about the dominant frequency of the acoustic field, whereas those for the high amplitude limit cycle oscillations exactly match with the acoustic frequency. Thus, feedback between the acoustic fluctuations and the hydrodynamic field is stronger for the later state and it is weaker for the former state of the oscillations.

Next, we study the coupled interaction of the acoustic pressure (p') and the circulation of the step (Γ) for different dynamical states, as illustrated in Fig. 9.2. The oscillation in Γ governs the oscillatory behavior of the heat release rate in the model, as both are directly dependent on each other through a vortex shedding process. Therefore, similar to the experimental investigation of synchronization of coupled acoustic pressure fluctuations and heat release rate fluctuations in a turbulent combustor discussed in Chap. 8, we can also examine synchronization

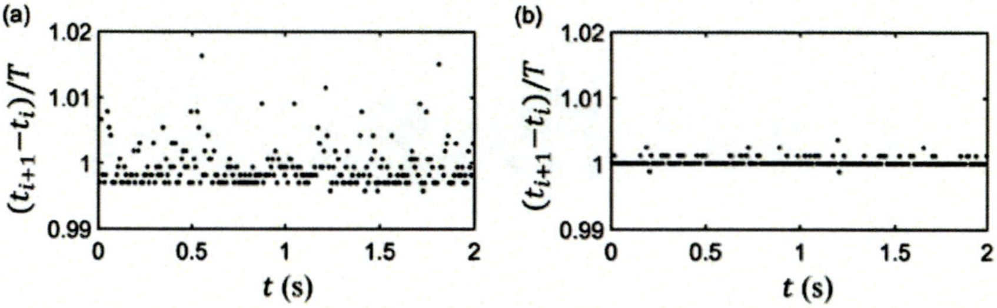

Fig. 9.3 Temporal variation of the difference between kicking times of consecutive oscillation cycles (normalized by the period of the corresponding signal) show (**a**) a wide distribution for the state of low amplitude limit cycle oscillations ($\overline{u}$ = 10.41 m/s) and (**b**) a very small variation for the state of high amplitude limit cycle oscillations ($\overline{u}$ = 10.6 m/s). Reproduced with permission from Pawar et al. [18]

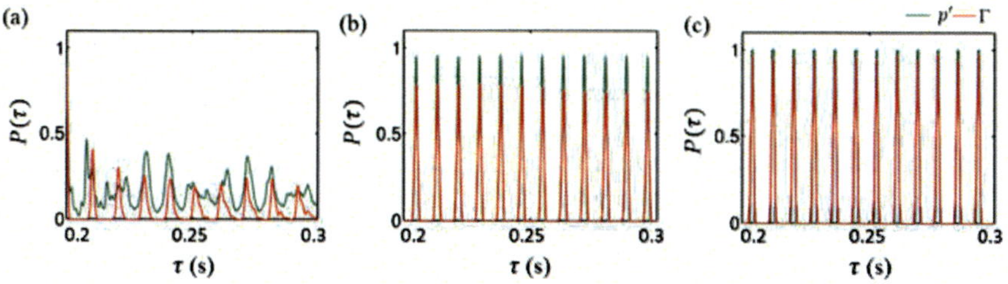

Fig. 9.4 Variation of the probability of recurrence $P(\tau)$ as a function of time delay τ for the acoustic pressure p' and the circulation Γ signals show the property of (**a**) desynchronization during the state of combustion noise ($\overline{u}$ = 9.2 m/s), (**b**) phase synchronization during the state of low amplitude limit cycle oscillations ($\overline{u}$ = 10.41 m/s), and (**c**) generalized synchronization during the state of high amplitude limit cycle oscillations ($\overline{u}$ = 10.6 m/s). Reproduced with permission from Pawar et al. [18]

characteristics of p' and Γ through a statistical measure called the probability of recurrence, $P(\tau)$ (see 8.6).

In Fig. 9.4a–c, we show the variation of the probability of recurrence $P(\tau)$ for different values of time lags (τ) for the signals of p' and Γ observed during the states of combustion noise, low amplitude limit cycle, and high amplitude limit cycle oscillations, respectively. We notice that the plots of $P(\tau)$ for the acoustic pressure and the circulation signals are quite different during the occurrence of combustion noise (Fig. 9.4a). This, in turn, indicates desynchronization of these oscillations in the system [18, 21]. During the occurrence of low amplitude limit cycle oscillations, the $P(\tau)$ plots of p' and Γ signals show the locking of the position of the peaks; however, the heights of these peaks do not show a perfect match (Fig. 9.4b). This indicates the state of phase synchronization (PS) of p' and Γ signals in the model during this state [18, 21]. For the state of high amplitude limit cycle oscillations, we notice the perfect locking of both the positions and the heights of the peaks in $P(\tau)$ plots of p' and Γ signals (Fig. 9.4c), signifying the presence of generalized synchronization (GS) in the system [18, 21].

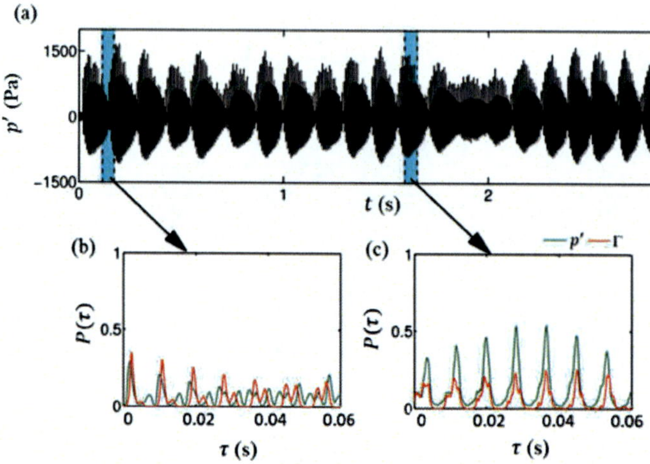

Fig. 9.5 (**a**) Time series of acoustic pressure fluctuations during the state of intermittency ($\overline{u}$ = 10.2 m/s) in the model. (**b, c**) Variation of the probability of recurrence $P(\tau)$ as a function of increasing values of delay time τ, for the acoustic pressure p' and the circulation Γ signals, during epochs of aperiodic fluctuations and periodic fluctuations, respectively. From (**b**) and (**c**), we can confirm the state of intermittent phase synchronization between p' and Γ during the state of intermittency. Reproduced with permission from Pawar et al. [18]

During the occurrence of intermittency, we observe the features of phase synchronization of p' and Γ signals during the bursts of periodic oscillations and their desynchronization during the epochs of aperiodic fluctuations (refer to Fig. 9.5). This is a property of intermittent phase synchronization of p' and Γ oscillations observed in the system [18]. We note that these results shown in Figs. 9.4 and 9.5 are qualitatively similar to the experimental observation of synchronization transition of the acoustic pressure fluctuations and the heat release rate fluctuations observed in a turbulent combustor (Figs. 8.6 and 8.7 in Chap. 8), for the variation of the flow velocity as the control parameter.

To summarize, in this chapter we have discussed a kicked oscillator model for studying the transition to thermoacoustic instability in a turbulent combustor. In this model, the time instant at which a shed vortex from the dump plane impinges on the bluff-body plays an important role in governing the different dynamical states of the combustor. This model is able to qualitatively capture the temporal dynamics of the intermittency route to thermoacoustic instability and also the synchronization characteristics of the acoustic pressure and the heat release rate fluctuations observed in the experiments. In the next chapter, we will introduce the framework of multifractal analysis to characterize the nature of aperiodic fluctuations observed during the occurrence of combustion noise and also to study the phenomenon of the onset of thermoacoustic instability in turbulent combustors.

References

1. Bak, P.: How Nature Works: The Science of Self-organized Criticality. Springer, Berlin (2013)
2. Balasubramanian, K., Sujith, R.I.: Thermoacoustic instability in a Rijke tube: non-normality and nonlinearity. Phys. Fluids **20**(4), 044103 (2008)
3. Chakravarthy, S.R., Shreenivasan, O.J., Boehm, B., Dreizler, A., Janicka, J.: Experimental characterization of onset of acoustic instability in a nonpremixed half-dump combustor. J. Acoust. Soc. Am. **122**(1), 120–127 (2007)
4. Chakravarthy, S.R., Sivakumar, R., Shreenivasan, O.J.: Vortex-acoustic lock-on in bluff-body and backward-facing step combustors. Sadhana **32**(1), 145–154 (2007)
5. Clavin, P., Kim, J.S., Williams, F.A.: Turbulence-induced noise effects on high-frequency combustion instabilities. Combust. Sci. Tech. **96**(1–3), 61–84 (1994)
6. Cohen, J., Anderson, T.: Experimental investigation of near-blowout instabilities in a lean, premixed step combustor. In: 34th Aerospace Sciences Meeting and Exhibit, AIAA 1996-0819 (1996)
7. Dowling, A.P., Mahmoudi, Y.: Combustion noise. Proc. Combust. Inst. **35**(1), 65–100 (2015)
8. George, N.B., Unni, V.R., Raghunathan, M., Sujith, R.I.: Pattern formation during transition from combustion noise to thermoacoustic instability via intermittency. J. Fluid Mech. **849**, 615–644 (2018)
9. Hegde, U.G., Reuter, D., Daniel, B.R., Zinn, B.T.: Flame driving of longitudinal instabilities in dump type ramjet combustors. Combust. Sci. Tech. **55**(4–6), 125–138 (1987)
10. Holland, J.H.: Emergence: From Chaos to Order. OUP Oxford, Oxford (2000)
11. Kendrick, D.W.: An experimental and numerical investigation into reacting vortex structures associated with unstable combustion. Ph.D. Thesis, California Institute of Technology (1995)
12. Matveev, K.I.: Thermoacoustic instabilities in the Rijke tube: experiments and modeling. Ph.D. Thesis, California Institute of Technology (2003)
13. Matveev, K.I., Culick, F.E.C.: A model for combustion instability involving vortex shedding. Combust. Sci. Tech. **175**(6), 1059–1083 (2003)
14. McManus, K.R., Poinsot, T., Candel, S.M.: A review of active control of combustion instabilities. Prog. Energy Combust. Sci. **19**(1), 1–29 (1993)
15. Nair, V., Sujith, R.I.: A reduced-order model for the onset of combustion instability: physical mechanisms for intermittency and precursors. Proc. Combust. Inst. **35**(3), 3193–3200 (2015)
16. Nicoud, F., Wieczorek, K.: About the zero mach number assumption in the calculation of thermoacoustic instabilities. J. Spray Combust. Dyn. **1**(1), 67–111 (2009)
17. Noiray, N., Durox, D., Schuller, T., Candel, S.: A unified framework for nonlinear combustion instability analysis based on the flame describing function. J. Fluid Mech. **615**, 139–167 (2008)
18. Pawar, S.A., Seshadri, A., Unni, V.R., Sujith, R.I.: Thermoacoustic instability as mutual synchronization between the acoustic field of the confinement and turbulent reactive flow. J. Fluid Mech. **827**, 664–693 (2017)
19. Poinsot, T.J., Trouve, A.C., Veynante, D.P., Candel, S.M., Esposito, E.J.: Vortex-driven acoustically coupled combustion instabilities. J. Fluid Mech. **177**, 265–292 (1987)
20. Rogers, D.E., Marble, F.E.: A mechanism for high frequency oscillations in ramjet combustors and afterburners. J. Jet Propuls. **26**(1), 456–462 (1956)
21. Romano, M.C., Thiel, M., Kurths, J., Kiss, I.Z., Hudson, J.: Detection of synchronization for non-phase-coherent and non-stationary data. Europhys. Lett. **71**(3), 466–472 (2005)
22. Schadow, K.C., Gutmark, E., Parr, T.P., Parr, D.M., Wilson, K.J., Crump, J.E.: Large-scale coherent structures as drivers of combustion instability. Combust. Sci. Tech. **64**(4–6), 167–186 (1989)
23. Seshadri, A., Nair, V., Sujith, R.I.: A reduced-order deterministic model describing an intermittency route to combustion instability. Combust. Theor. Model. **20**(3), 441–456 (2016)
24. Seshadri, A., Sujith, R.I.: A bifurcation giving birth to order in an impulsively driven complex system. Chaos **26**(8), 083103 (2016)

25. Smith, D.A., Zukoski, E.E.: Combustion instability sustained by unsteady vortex combustion. In: 21st Joint Propulsion Conference, AIAA 1985-1248 (1985)
26. Sterling, J.D.: Nonlinear analysis and modelling of combustion instabilities in a laboratory combustor. Combust. Sci. Tech. **89**(1–4), 167–179 (1993)
27. Sterling, J.D., Zukoski, E.E.: Nonlinear dynamics of laboratory combustor pressure oscillations. Combust. Sci. Tech. **77**(4–6), 225–238 (1991)
28. Strahle, W.C.: Combustion noise. Prog. Energy Combust. Sci. **4**(3), 157–176 (1978)
29. Yu, H.K., Trouvé, A., Daily, J.W.: Low-frequency pressure oscillations in a model ramjet combustor. J. Fluid Mech. **232**, 47–72 (1991)
30. Zinn, B.T., Lores, M.E.: Application of the Galerkin method in the solution of non-linear axial combustion instability problems in liquid rockets. Combust. Sci. Tech. **4**(1), 269–278 (1971)

Chapter 10
Multifractal Analysis of Turbulent Thermoacoustic Systems

As we discussed in Chap. 7, the dynamics of a turbulent thermoacoustic system during the state of stable operation is complex. Such complexities arise primarily due to the interaction of different subsystems at various temporal and spatial scales in the reaction field of a combustor. In contrast with the traditional viewpoint that a thermoacoustic system is at a fixed point during stable operation, we showed that the dynamics of the acoustic pressure fluctuations during stable operation essentially comprises low amplitude chaotic fluctuations. In the present chapter, we aim to improve our understanding on the dynamical characteristics of irregular fluctuations (or patterns) observed in thermoacoustic systems using the frameworks of fractals and multifractals. After a brief introduction of these methodologies, we will discuss their use to study the temporal dynamics of turbulent thermoacoustic systems during the state of stable operation (combustion noise). Then, we will present a unified characterization of two different dynamical transitions, i.e., thermoacoustic instability and flame blowout, observed in such systems using the framework of multifractality. Finally, we will discuss the use of the multifractal analysis to delineate the change in the behavior of highly wrinkled turbulent flame fronts during the onset of thermoacoustic instability.

10.1 Fractals

Fractals are the geometric representation of natural/man-made objects or time-varying signals that exhibit infinitely many complex patterns or irregularities at various scales. Benoit Mandelbrot [52] was the first to coin the term 'fractal'. According to him, "*A fractal is a shape made of parts similar to the whole in some way*". Since a fractal object owns inherent geometrical irregularities in its shape, overfitting of such an object with a corresponding Euclidean shape does not cover its entire geometric surface. This feature of fractals is nicely put forth by Mandelbrot [52] in his poetic words as,

R. I. Sujith, S. A. Pawar, *Thermoacoustic Instability*, Springer Series in Synergetics,
https://doi.org/10.1007/978-3-030-81135-8_10

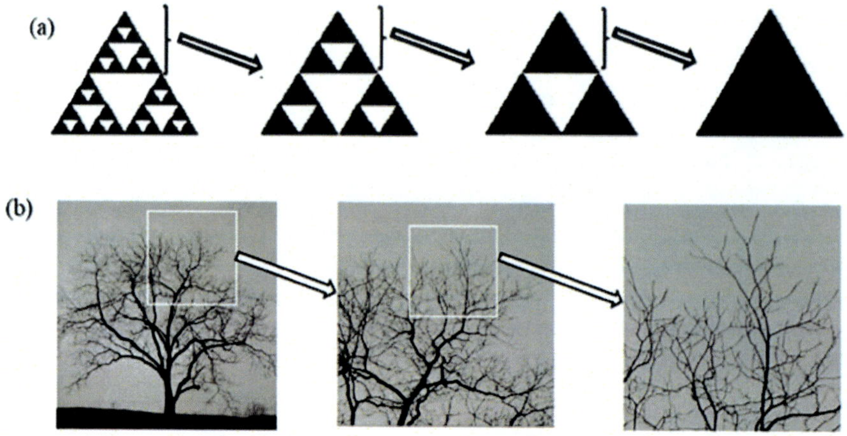

Fig. 10.1 Self-similarity observed in (**a**) a Sierpinski triangle (computer-generated graphic) and (**b**) a tree in winter (natural object). The box in (**b**) highlights the self-similar part of the whole object. These examples reveal that a small part of a fractal object is exactly or nearly similar to the entire object. (**b**) Adapted with permission from Pesin et al. [71]

Clouds are not spheres,
mountains are not cones,
coastlines are not circles,
and bark is not smooth,
nor does lightning travel in a straight line.

We often encounter such objects in our daily lives. Some common examples of fractals include trees, mountains, rivers, coastlines, leaves, ice crystals, seashells, mammalian lungs, turbulence, etc.

All fractals have two key characteristics, namely '*self-similarity*' and '*fractional dimension*' [48], which we elaborate in the following subsections.

10.1.1 Self-Similarity, Scale-Invariance, and Self-Affinity

Self-similarity is a property of an object whose scaled-down parts resemble exactly, or at least to some extent, to the entire object. Thus, self-similarity indicates the recurring behavior of fractals at different scales of observation. The scaling factor of fractals is usually expressed in terms of a percentage of its whole size. For numerical fractals (e.g., Sierpinski triangle in Fig. 10.1a, Mandelbrot set, Koch curve, Cantor set, etc.), the self-similar behavior exists for an infinite number of scales. In contrast, for natural fractals (e.g., trees in Fig. 10.1b, mountains, clouds, lungs, etc.), the self-similar behavior is observed in a certain range of scales with lower and upper cut-offs.

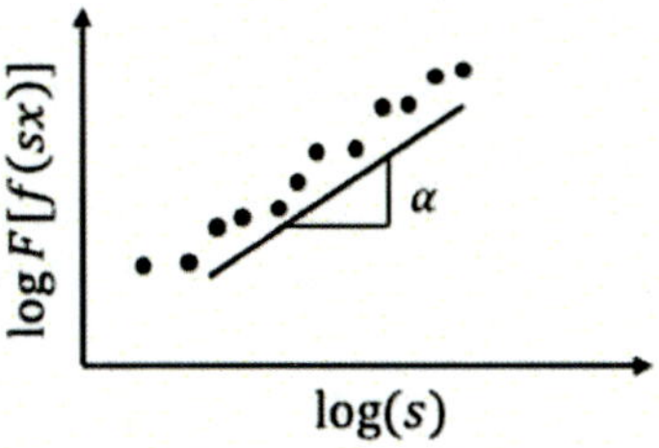

Fig. 10.2 For a fractal signal, the presence of a linear regime in the variation of a measure of fluctuations $F[f(sx)]$ with a scale factor s on the log-log plot indicates its scale invariance property. The fractal dimension corresponds to the slope (α) of this linear regime of the plot

Self-similarity also indicates the exhibition of scale-invariant behavior by fractals. Scale-invariance implies that the characteristics of a fractal remain the same at every scale of observation. To understand the scale invariance property, let us consider a function of variable x as $f(x) = ax^{\alpha}$, where a is a proportionality constant and α is a scaling exponent. If this signal is scaled by a factor s, the scaled version of the signal is obtained as, $f(sx) = a(sx)^{\alpha} = s^{\alpha}(ax^{\alpha}) = s^{\alpha} f(x)$. This indicates that the scaled signal is directly related to the original fractal signal, for all values of the scale factor s. The variation of the properties or a measure of the scaled signal $F[f(sx)]$ with the scale factor s displays a linear relationship on the log-log plot, where the slope of the plot is equal to α (Fig. 10.2). This relationship is also called a '*power-law scaling*'. Note that the power-law scaling (for example, $g(x) = x^{b}$) is different from the exponential scaling (for example, $g(x) = b^{x}$), where the exponent b is a constant in the former case and the exponent x is a variable in the latter case.

Sometimes, we observe that the scaling behavior of fractals change with the direction (for example, x and y coordinates of an object) of magnification. Such fractals with a non-uniform scaling dependence on the direction of the magnification are known as *self-affine fractals*. Such anisotropic scaling is necessary to reveal the self-similar features of most fractals observed in nature [53].

10.1.2 The Fractal Dimension

In simple words, the dimension of an object defines the number of degrees of freedom a point has for its movement on the object or the number of independent coordinates that are needed to indicate the position of the point on the object. However, in mathematics, the dimension has been rigorously defined using a measure theory [20]. Such a description of the dimension is beyond the scope of our discussion in this chapter.

The dimension of Euclidean objects, such as point, line, square, cone, or sphere, is referred to as a topological dimension and its value is always an integer number. For example, a point has a dimension of zero, a line has a dimension one, a plane has a dimension two, a solid has a dimension three, etc. However, the dimension of self-similar fractals is a fractional number and is referred to as the fractal dimension. Many objects in nature possess a non-integer dimension.

Fractal dimensions can be found in different notions, which include the Hausdorff-Besicovitch dimension, box-counting dimension, correlation dimension, information dimension, capacity dimension, generalized dimension, etc. [91]. These dimensions characterize different properties of fractal objects. The mathematical details of these dimensions can be found from [4, 26, 52, 91]. Further, there are various other methods of calculating the fractal dimension of an object. Some of the common methods are box-counting, hand-dividers, prism counting, epsilon-blanket, power spectrum, etc. [91]. We here present the simplest way of calculating the fractal dimension of an object based on the box-counting method [52, 91].

In a box-counting method, we first cover a d-dimensional geometric object (e.g., a curve or a surface) with a set ($s > 0$; $s = s_1, s_2, s_3, \ldots, s_k$) of small equal sized boxes (or d-dimensional hypercubes). Then, we calculate the number of such boxes $N(s)$ of size s that are needed to cover the entire object. Note that by saying a set of boxes cover the object, we mean that every point on the object is present at least in one box. Next, we plot the variation of the number $N(s)$ of boxes of size s that are required to cover the object for the different values of s on a log-log graph. The slope of the linear regime of this plot is obtained as follows [91],

$$D = \lim_{s \to 0} \frac{\log N(s)}{\log(1/s)} \tag{10.1}$$

Equation (10.1) can alternately be written as,

$$N(s) \sim (1/s)^D \quad \text{or} \quad N(s) \sim s^{-D}. \tag{10.2}$$

Here, D is called a box-counting dimension. Thus, there exists a power-law scaling between the number of boxes and the scale factor. This means that, if an object is scaled by a factor s, the corresponding measures (e.g., length, size, volume, mass, etc.) also scale with a power corresponding to the dimension of the object.

For instance, if a line segment of length l is scaled by a factor $s = 1/2$, we get two equal parts of the line (Fig. 10.3a). The dimension of the line is obtained as,

$$D = \frac{\log N(s)}{\log(1/s)} = \frac{\log 2}{\log 2} = 1$$

Similarly, if a square is scaled by a factor of $s = 1/2$, we obtain four small self-similar squares (Fig. 10.3b). The dimension of the square is $D = \log 4/\log 2 = 2$. In this case, the area (a measure of the square) of the newly formed part scales as the power of the dimension. Therefore, the area of the small square is $(1/2)^2$ of the original square. Similarly, we can find the dimension of the cube as 3, and show that the scaling of a cube by half results in the reduction of its volume by $(1/2)^3$ (Fig. 10.3c). This simple calculation further reasserts our previous sentence that the value of the dimension (D) is a positive integer number for Euclidean objects.

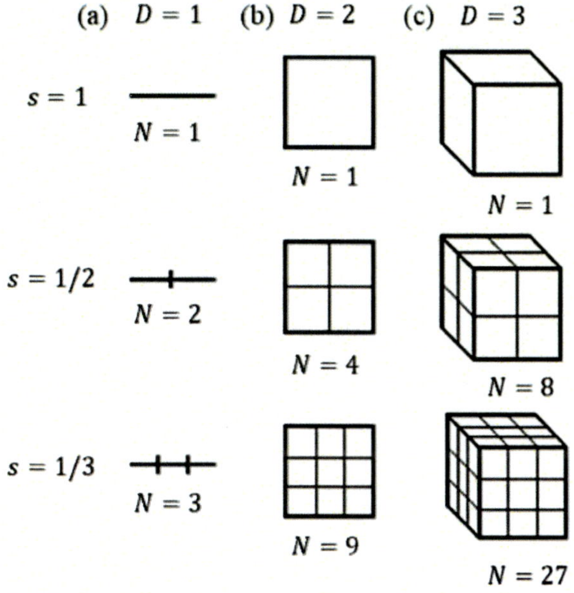

Fig. 10.3 Schematics showing the scaling behavior of different Euclidean geometries. All Euclidean objects have an integer dimension. Adapted from open source https://en.wikipedia.org/wiki/Fractal_dimension

However, this statement is not true for non-Euclidean or fractal objects, which we discuss next from the calculation of the dimension of a famous fractal curve called the Koch snowflake [20, 42].

In order to construct a Koch snowflake curve, we first consider a line segment as indicated by S_0 in Fig. 10.4. In the first iteration, we divide this line into three equal parts ($s = 1/3$) such that we get the length of each new part as $l_1 = (1/3)S_0$. Then, we replace the central $(1/3)$rd division of the line with a two-sided equilateral triangle of size l_1 to get a curve S_1. The length (L) of the new curve is $(4/3)$rd of its original length. In the second iteration, we divide each segment of S_1 by the respective $(1/3)$rd scales such that each new division has a length equal to $l_2 = (1/3)l_1$. Then, we replace the central division of each segment of S_1 with two-sided equilateral triangles of size l_2 so that we get a curve S_2. In this iteration, the length of the curve increases by $(4/3)^2$ of its original length. We then repeat the same procedure for the subsequent n number of iterations to generate the curve of the Koch snowflake (also called the von Koch curve).

In each iteration, every segment of the Koch snowflake curve is scaled down by $(1/3)$rd of its length in the previous iteration and replaced by 4 equal parts of this new length. Thus, every iteration leads to an increase in the length of the curve by $(4/3)$th of its total length in the previous iteration. Therefore, the length of the curve at the nth iteration is equal to $(4/3)^n$ of its original length. If the iteration continues to infinity, the length of this curve also tends to infinity; nevertheless, the size of the curve is still limited to the finite space.

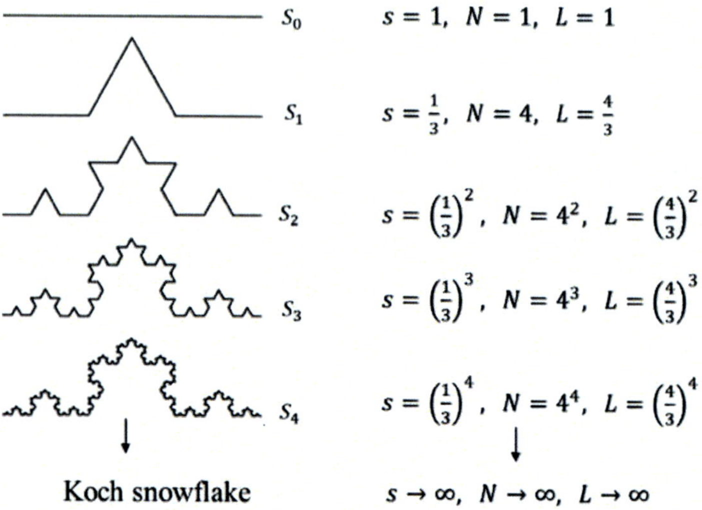

Fig. 10.4 A procedure of constructing a fractal Koch snowflake curve. This curve is self-similar at every 1/3rd portion of its entire scale. Here, $S_0, S_1, \ldots$, etc. indicate the number of iterations, s is the scale factor, N is the number of segments, and L is the length of the curve

From the aforementioned information, we can obtain the dimension of the Koch curve as follows. We notice the generation of four segments of equal length ($N(s) = 4$) in each iteration due to the scaling of the Koch curve by a factor $s = 1/3$. Therefore, the dimension of this curve is calculated as,

$$D = \frac{\log N(s)}{\log(1/s)} = \frac{\log 4}{\log 3} \approx 1.26$$

We note that the dimension of a von Koch curve is a non-integer number.

It is mathematically proved that, for an object to be considered as a fractal, its dimension should strictly exceed the corresponding topological dimension [19, 20, 51]. This also means that a fractal object occupies more space than its topological space. This property of fractals can be understood from different examples shown in Fig. 10.5. We find that the fractal dimension of discrete points objects (e.g., Cantor set in Fig. 10.5a) lies between 0 (topological dimension for a point) and 1 (topological dimension for a line), that of irregular curves (e.g., the von Koch curve in Fig. 10.5b) lies between 1 (topological dimension for a line) and 2 (topological dimension for a plane), and that of broken surfaces (e.g., Menger sponge in Fig. 10.5c) lies between 2 (topological dimension for a plane) and 3 (topological dimension for a cube).

Contrary to the aforementioned exact self-similarity displayed by the synthetic fractals, the fractals observed in nature do not show this behavior of exact self-similarity; however, they exhibit a near statistical resemblance at different scales. One of these natural fractals that have been commonly studied is the length of the

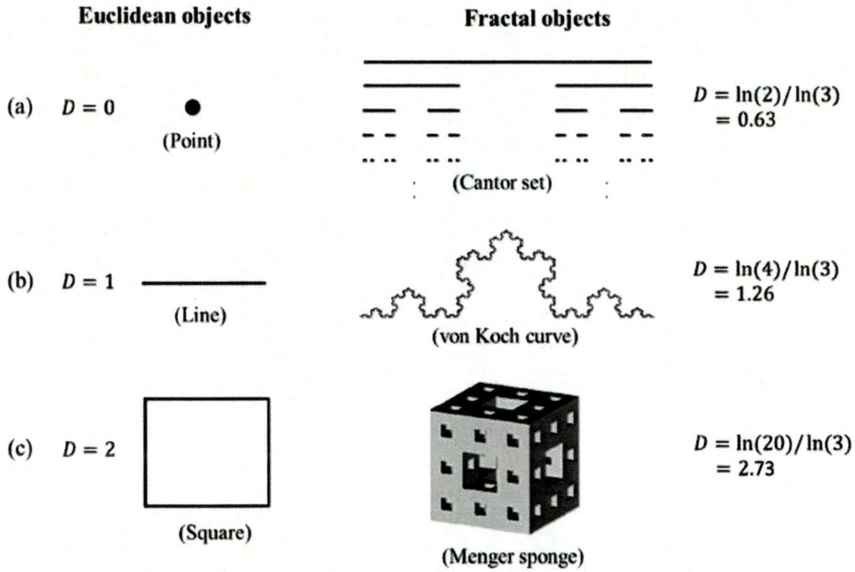

Fig. 10.5 Schematics showing the difference in the dimension D of different Euclidean and non-Euclidean (fractal) objects. The dimension of a fractal object is always greater than the corresponding Euclidean dimension

coastline in the map of Britain [48]. As shown in Fig. 10.6, the coastline of Britain is not a smooth line in its map. As we zoom in on the coastline, we see the existence of more and more wrinkles or finer details along the periphery of the map. Hence, in such a situation the natural question would arise, what is the exact perimeter of Britain? This is because however closer we zoom in, the coastline does not appear to be a straight line. At every magnification of the coastline, the surface roughness exhibits a corresponding increase; hence, the increase in the length of the coastline.

An increase in the magnification of the coastline causes a change in the scale of the measurement. Therefore, the measurement scale of the coastline will change if you measure its length either from a satellite image, from an airplane at different heights relative to the sea level, by car, by cycle, or by a walk down along the shore. In every mode of these measurements, we can expect a change in the length of the coastline, as the number of steps required to measure the length would vary from one mode of transportation to another (Fig. 10.6). Irrespective of these differences in the method of measurement, all these lengths (say, N) exhibit a power-law with the corresponding scale of the measurement (say, r) used in the calculation. The slope of this linear relation is equal to the fractal dimension of the coastline, which is obtained as $D = \log N / \log(1/r)$. The value of D for Britain's coastline is found to be nearly 1.21 [48]. The minimum scale with which the length of the coastline can be measured would be a molecular scale where the distance between the atoms is measured. However, such measurements would then cause the coastline to be of

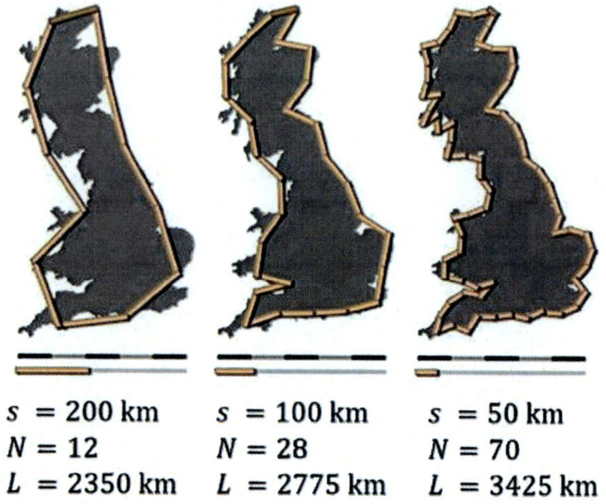

Fig. 10.6 Fractal analysis of coastline of a Britain map, where s indicates the length scale, N indicates the number of segments, and L represents the total length of the coastline measured using a scale s. As the length of the scale becomes smaller and smaller, the total length of the coastline shows a corresponding increase. Adapted from https://en.wikipedia.org/wiki/Fractal_dimension

infinite length, which does not make any sense, as this implies that the finite area of Britain is surrounded by an infinitely long perimeter. Hence, the appropriate choice of a finite scaling range is necessary to obtain physically meaningful results of the fractal dimension.

Similar to the fractal nature of geometrical objects, we encounter several examples of temporal fluctuations in system variables that exhibit fractal characteristics (Fig. 10.7). Such fractal signals exhibit irregular fluctuations resulting from the presence of various ranges of frequencies or amplitudes at different time instances in the signal. The fractal signals are self-similar. Due to the presence of irregular fluctuations in fractal signals, the magnified portion of the signal does not match exactly with the whole signal. In such a situation, fractal signals show statistical self-similarity at different scales of the signal. Fractal signals also display the property of scale-invariance. In order to detect the scale-invariance property of fractal signals, we need to first divide the signal into several non-overlapping time windows. Then, we need to obtain the variation of the root mean square of standard deviation of fluctuations for all such windows with increasing window size on a log-log plot (as elaborated in Sect. 10.1.3). A linear variation of this plot in a particular range of scales confirms the scale-invariance (or the fractal) nature of the signal.

As we discussed before, a fractal dimension measures the self-similarity of an object [52]. It also provides information on how well a fractal geometry occupies the space around it. The larger the fractal dimension, the more the space filled by the fractal object. For example, a series of discrete points along a line has a fractal dimension greater than zero but less than one, as these points occupy more space than a single point but less than the space occupied by a continuous line (Fig. 10.5a). Similarly, a wrinkled curve covers a space larger than a straight line; hence, its dimension is slightly greater than 1 (Fig. 10.8). However, if this curve fills the two-dimensional space, its dimension approaches 2. Consider another example of a two-dimensional image of a tree; if the fractal dimension of the image is 1.1, it indicates

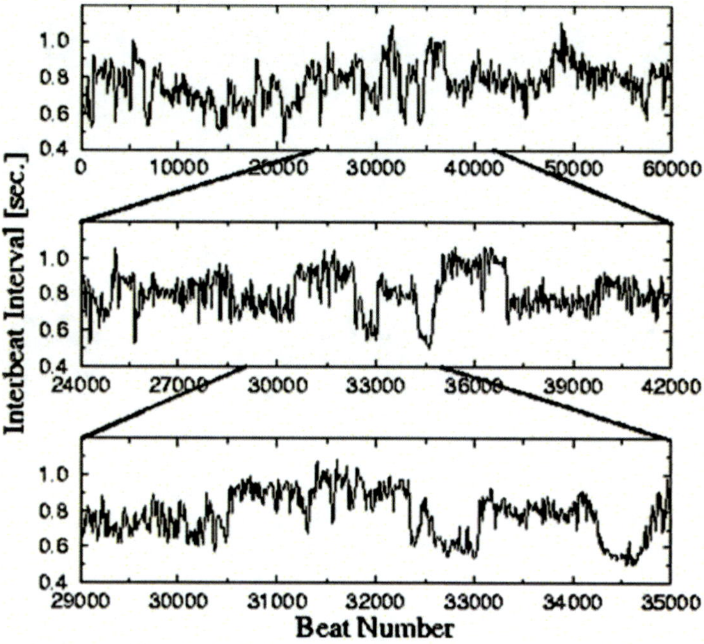

Fig. 10.7 Self-similar property of heartbeat signal, where the magnified portions of the signal show similar trends as that of the original signal. Adapted from https://archive.physionet.org/tutorials/fmnc/node4.html

that the tree sparsely acquires the two-dimensional image (i.e., the image contains a few thin or thick branches of the tree). In contrast, if the fractal dimension of the image is 1.9, it indicates that the tree densely covers the image (i.e. the image contains a large number of both thick and thin branches of the tree).

Next, we discuss the importance of another measure, called the Hurst exponent, to characterize different properties of irregular fluctuations of fractals.

10.1.3 The Hurst Exponent and Properties of Fractal Signals

The Hurst exponent was proposed by a British hydrologist Harold E. Hurst [32] while studying a problem of long term storage capacity of reservoirs on the Nile river. He analyzed the data of fluctuations in the water level of the Nile river and also the influx of water from rainfall taken over nine centuries. Initially, he thought that the water inflow is a random process. However, during his analysis, he found that most such natural phenomena (for example, the river discharge, rainfall, sunspots) follow a behavior of '*biased random walk*' [72], which means the occurrence of

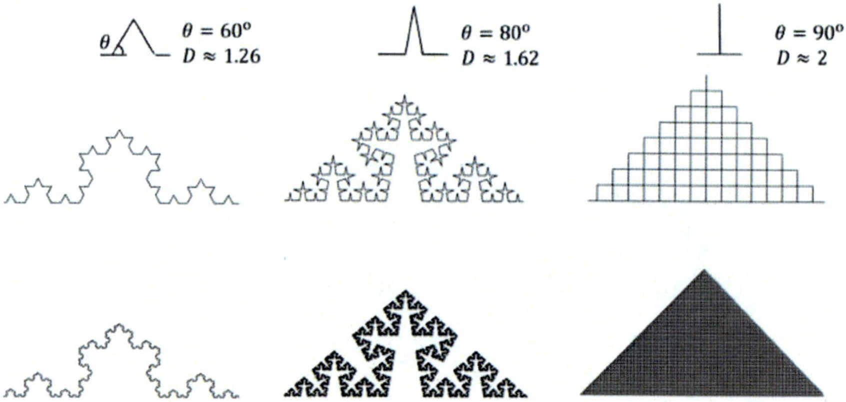

Fig. 10.8 The change in fractal characteristics of a von Koch curve when the base angle of an isosceles triangle (θ) is increased from 60° to 90°. With an increase in the angle, we observe an increase in the fractal dimension of the curve. We notice that the von Koch curve transitions from a smooth fractal having a dimension (D) close to one to a densely filled surface having a dimension of two. The fractal dimension is calculated as $D = \log(2)/\log(2\cos(\theta/2))$. Note that the number iterations used to obtain a von Koch curve are different for the top and bottom rows. Courtesy: Sneha Srikanth (IIT Madras, India)

random events is biased in one direction and the behaviour has a trend with noise. Hurst proposed a measure to quantify the trend and the level of noise present in data. This measure is now popularly known as the Hurst exponent.

Furthermore, Hurst developed a method of rescaled range analysis (R/S analysis) to estimate the Hurst exponent [32, 55]. In the procedure of rescaled analysis of a signal (say, $X_t = X_1, X_2, X_3, \ldots, X_N$), we first divide the signal into N_s non-overlapping segments A_v of size s, where $N_s = N/s$ and $v = 1, 2, 3, \ldots, N_s$ as follows [2],

$$A_1 = \{X(1), X(2), \ldots\ldots\ldots, X(s)\}$$
$$A_2 = \{X(s+1), X(s+2), \ldots\ldots\ldots, X(2s)\}$$
$$\vdots$$
$$A_{N_s} = \{X[(N_s-1)s+1], X[(N_s-1)s+2], \ldots\ldots\ldots, X[N_s s]\}. \quad (10.3)$$

Then, we calculate the mean and the standard deviation of fluctuations in each segment A_v as,

$$\langle X_v \rangle = \frac{1}{s}\sum_{i=1}^{vs} X[(v-1)s+i], \quad (10.4)$$

and

$$S_v = \sqrt{\frac{1}{s}\sum_{i=1}^{vs}\{X[(v-1)s+i] - \langle X_v\rangle\}^2}, \tag{10.5}$$

where $i = 1, 2, \ldots\ldots, N$.

Next, we obtain a cumulative deviate profile for each segment A_v as follows,

$$Y_{vj} = \sum_{i=1}^{j}\{X_{(v-1)s+i} - \langle X_v\rangle\} \tag{10.6}$$

where $j = (v-1)s+1, (v-1)s+2, \ldots.., vs$.

Further, we calculate the range of each segment of the profile Y_{vj} as,

$$R_v(s) = \max(Y_{vj}) - \min(Y_{vj}) \tag{10.7}$$

Finally, we obtain a rescaled range (a dimensionless ratio) by dividing the range for each segment of size s by corresponding standard deviation of fluctuations as follows,

$$F_{RS}(s) = \frac{1}{N_s}\sum_{v=1}^{N_s}\frac{R_v(s)}{S_v(s)} \tag{10.8}$$

For a fractal signal, the log-log plot between $F_{RS}(s)$ and s obeys a power-law such that,

$$F_{RS}(s) \sim s^H \tag{10.9}$$

where the exponent H is called the Hurst exponent.

The Hurst exponent is directly related to the fractal dimension through the following expression [3, 20],

$$D = 2 - H. \tag{10.10}$$

Since the value of H varies between 0 and 1 (i.e., $0 < H < 1$), the value of the fractal dimension D ranges from 2 to 1.

Depending on the value of the Hurst exponent, irregular fluctuations in signals are classified into three groups [34]. For $H = 0.5$, the observations in a signal are independent of each other. Such signals are called uncorrelated or random signals. In contrast, if the value of H deviates from 0.5, each observation in the signal carries a memory of past events. For $H > 0.5$, the signals possess *long-range dependence* and, for $H < 0.5$, the signals possess *short-range dependence*.

Signals that have a value of H between 0 and 0.5 are referred to as *anti-persistent* (anti-correlated) signals. While signals that have a value of H between 0.5 and 1 are referred to as *persistent* (positively correlated) signals. For anti-persistent signals, an increase in value is more likely to be followed by a decrease in value and vice versa. Therefore, anti-persistent signals have a tendency to revert to the mean. Such tendency of mean-reverting increases as H approaches 0. On the other hand, for persistent signals, an increase in value is more likely to be followed by another increase in value and a decrease in value is likely to be followed by another decrease in value. Thus, persistent signals have characteristics of a long memory effect and they display a trend in their temporal variation. The strength of the trend in a signal increases as H approaches 1. Thus, the Hurst exponent provides information about the roughness (fine fluctuations) of the signal. Signals with a higher value of H indicates less roughness (or smoother fluctuations) and clear trends than those observed for signals having lower H (see Fig. 10.9). We can summarise the properties of H as follows,

1. For all signals, $0 < H < 1$
2. For $H = 0.5$, the signal is called random-walk or Brownian motion
3. For $H > 0.5$, the signal possesses persistent (long-term memory, correlated) behavior
4. For $H < 0.5$, the signal possesses anti-persistent (short-term memory, anti-correlated) behavior

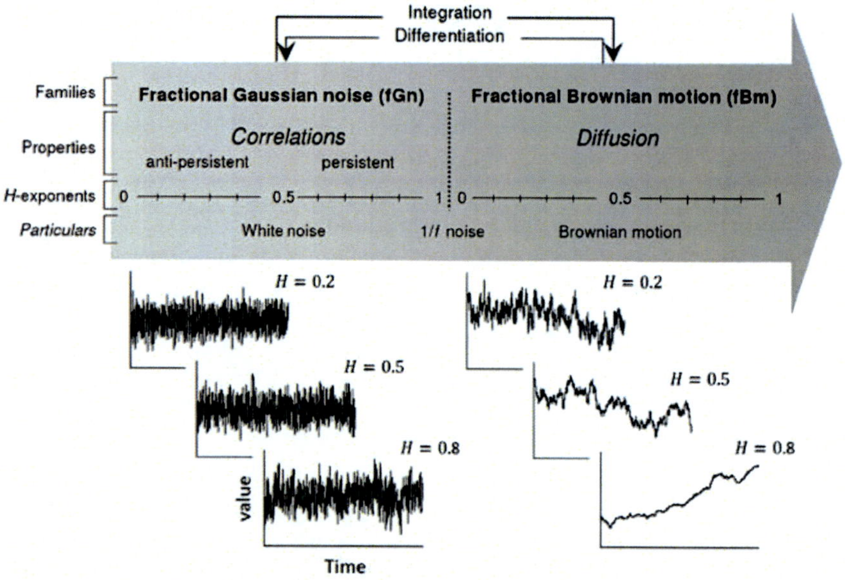

Fig. 10.9 Difference between two families of $1/f$-noise, i.e., fGn and fBm, with respect to different values of the Hurst exponent H. As the value of H increases, we notice that the fluctuations in the fBm decrease and the trend become clearer, which is absent in the fGn. Adapted with permission from Delignieres et al. [15]

We can also understand the difference between persistent and anti-persistent signals using an autocorrelation function, $\rho(k)$. The autocorrelation function calculates the linear statistical dependence between two points of a signal for increasing the spacing (k) between these points, which is obtained as follows [11, 88],

$$\rho(k) = \langle X(t)X(t+k)\rangle \sim C|k|^{-\delta} \tag{10.11}$$

where $\langle\rangle$ denotes average values, C represents a positive constant, and $0 < \delta < 1$.

For persistent signals, the rate of decrease of the autocorrelation function towards zero, for the increasing spacing between the points (k), is very slow (hyperbolic decay), i.e., $\delta \to 1$. Thus, we observe a power-law behavior in the decay of $\rho(k)$, which also indicates the absence of any characteristic scale or the existence of self-similar or fractal behavior of the signal [11]. In contrast, for anti-persistent signals, the autocorrelation function decays at a very fast rate to zero (exponential decay), as the current state in such signals depends only on a few states in its immediate past. This also defines the characteristic scales of the signal. Note that the Hurst exponent explains the nonlinear correlations in the signal, while autocorrelation function highlights linear correlations in the signal.

The power spectral density $S(f)$, obtained as the mean square fluctuations at frequency f, can also be used as a measure to detect fractal characteristics of a signal. For a fractal signal, the variation of $S(f)$ over a certain range of f on the log-log plot shows an inverse power-law scaling [13, 18],

$$S(f) \sim 1/f^{\beta} \tag{10.12}$$

Here, β is a positive integer number and is referred to as a spectral exponent.

The families of irregular fluctuations that exhibit such power-law scaling are called $1/f^{\beta}$ noise [14, 18, 55, 73]. Depending on the value of β, we can further classify $1/f^{\beta}$ noise into two groups: *fractional Gaussian noise* (fGn) or *fractional Brownian motions* (fBm), see Fig. 10.9. Here, fGn and fBm are the generalization of the classical concepts of Gaussian noise and Brownian motion, respectively. fGn is a stationary signal, as the probability distribution and statistical properties (i.e., mean and variance) of such signal are independent of the segment size or its position in the signal. In contrast, fBm is a non-stationary signal, as the variance of such signal is time-dependent and the probability distribution of the large segment is a rescaled version of that of the small segment. Thus, we observe a power-law scaling for the variation of statistical properties over the segment size for fBm [55].

Note that fGn and fBm are interconvertible, and they are characterized by the same Hurst exponent [14, 18], see (10.13). In order to get fBm, we need to cumulatively sum (integrate) the values of fGn. In contrast, in order to obtain fGn, we need to take a difference (differentiate) between the values of fBm.

The characteristics of fBm depend on the value of H, which are also referred to as diffusion properties [15]. Here, diffusion indicates the dispersion of the signal relative to its initial value after a given time interval. For higher values of H, the fBm signal show more diffusive characteristics (Fig. 10.9). Such diffusion property of fBm is not observed for fGn. Therefore, to classify different fGn signals, we can compare their correlation properties, i.e., persistent ($H > 0.5$) and anti-persistent ($H < 0.5$) correlations. From these relations, we also comment that the persistent correlations in fGn lead to strong diffusion in fBm.

The Hurst exponent H is related to the spectral exponent β for fGn and fBm signals as follows [14, 18],

$$H_{\mathrm{fGn}} = \frac{\beta + 1}{2} \quad \text{and} \quad H_{\mathrm{fBm}} = \frac{\beta - 1}{2} \tag{10.13}$$

Depending on the values of H, the different values of β can be obtained for different families of fGn and fBm [6, 7, 17, 18, 80, 82]. As mentioned before, the value of H varies between 0 and 1. Therefore, we have $-1 < \beta < 1$ for fGn and $1 < \beta < 3$ for fBm. A signal corresponding to $\beta = 0$ or $H_{\mathrm{fGn}} = 0.5$ is called Gaussian white noise and that corresponding to $\beta = 2$ or $H_{\mathrm{fBm}} = 0.5$ is called Brownian noise. A signal corresponding to $\beta = 1$ is called pink noise. Thus, $\beta = 1$ is a limiting case that separates two classes of $1/f$ noise.

After the discussion on the characteristics of fractals, we next move our attention to describe the properties of multifractals and the methodologies to study them.

10.2 Multifractals

In general, fractals provide information about the presence of even (homogeneous) distribution of certain measures (or physical properties) on a geometric support. As a result, fractals are indicated by a single non-integer dimension. Here, geometric support could be a line, a plane, or a surface of the sphere. Fractals with a single scaling exponent are called monofractals. However, in some cases, the distribution of measures (or physical properties) is heterogeneous over a geometric support [36]. For example, the spatial distribution of the population in a city fluctuates highly from one part to another part in the space. Another example is the distribution of metals (copper, gold, etc.) on the earth is found in high concentration at a few places, the medium concentration at the majority of the places, and in rare concentrations at remaining places [54].

Such variation in physical properties on different parts of the geometric support is quantified using a normalized measure or probability [36]. In Fig. 10.10, we show an example of a two-dimensional image (geometric support) of a city [82], where streets and parks (unbuilt areas) are indicated by white and buildings (built areas) are indicated by black (Fig. 10.10a). In order to describe the distribution of a measure that quantifies just built or unbuilt areas, we need a single fractal dimension.

On the other hand, in the isometric view of the city, we notice a set of buildings projected outward from the geometric support (Fig. 10.10b). If we consider the height of the building as a measure, then we can quantify the distribution of the heights of buildings in the city. We can characterize the distribution of buildings

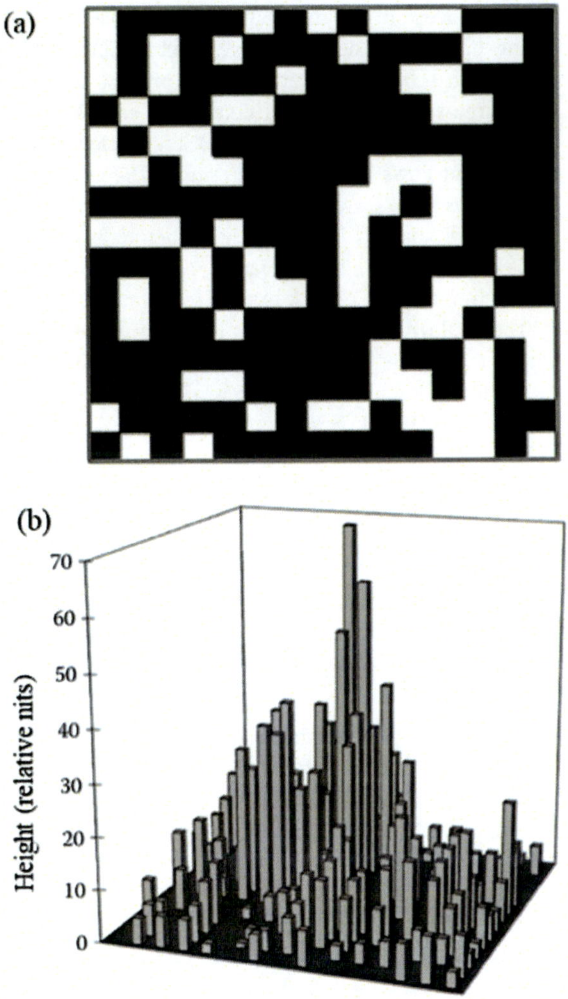

Fig. 10.10 Example of different views, (**a**) top view and (**b**) isometric view, of a city used to illustrate the concept of multifractality. In (**a**), the city is indicated by a set of black and white regions, where black represents buildings and white represents streets and parks. In (**b**), the city is indicated by a set of different height buildings. The distribution of the built (or empty) area in (**a**) can be characterized by a single fractal dimension; hence, it indicates a monofractal. In contrast, as one fractal dimension is associated with the distribution of each building height in (**b**), we need several fractal dimensions to describe the distribution of buildings of different heights; hence, it represents a multifractal. Adapted with permission from Seuront [82]

with similar heights using a fractal dimension. Thus, we need more than one fractal dimension to describe the spread of buildings of different heights in the city. Hence, the distribution of heights of the buildings is multifractal.

In order to quantify such an uneven distribution of measures on the geometric support, we need a framework of multifractality [29]. For multifractal objects or signals, the degree of self-similarity varies with space or in time, respectively. Hence, we need more than one value of fractal dimensions (or scaling exponents) to have a complete description for multifractals. As a result, multifractals are more complex when compared with monofractals, where multifractals can be considered as the interwoven sets of an infinite number of monofractals [93].

The concept of multifractality was originally put forth by Mandelbrot [49, 50] during his studies on fully developed turbulent flows. Such flows are complex and exhibit self-similarity during the process of the transfer of the kinetic energy from large-scale eddies to viscous scale eddies (energy cascade) where energy is dissipated in the form of heat [22, 43, 85]. Richardson [76] described this behavior of turbulent flows as "*Big whorls have little whorls that feed on their velocity, and little whirls have lesser whirls and so on to viscosity*". Recently, the framework of multifractal analysis is applied in numerous problems that include turbulent flows [8, 59], river flows [32, 61], heartbeat patterns [35], internet traffic [56], rainfall [44], ocean circulation [9], stock market [57], brain dynamics [33, 86, 101], etc.

The multifractal properties of the structure are effectively represented using a plot called the singularity spectrum, which we will discuss in detail in the following subsection. The degree of multifractality can be used as a measure of complexity [81, 84]. Here, complexity indicates a notion of "meaningful structural richness" [25]. The broadness of the singularity spectrum has a direct relation with the complexity in the structure; the broader the spectrum, the more complex the structure is. Therefore, the loss of multifractality can be thought of as a loss of complexity in the structure [16].

10.2.1 Methods of Multifractal Analysis

There are many methodologies developed to perform multifractal analysis of temporal and spatial fluctuations. Some of these methods are box-counting [8, 52], rescaled range analysis [5], detrended fluctuation analysis [70], wavelet-based estimation [38], detrended moving average [1, 27], phase space attractor [30], etc. In this section, we will discuss two methods in detail: (1) multifractal detrended fluctuation analysis [39] usually used for analyzing temporal fluctuations (for example, acoustic pressure fluctuations and heat release rate fluctuations [28, 40, 63, 68, 69, 77, 79, 94, 95]) in a thermoacoustic system and (2) box-counting method [8] used for studying both the temporal [23] and the spatial structures (for example, the flame [75]) in a thermoacoustic system.

10.2.1.1 Multifractal Detrended Fluctuation Analysis (MFDFA)

In this section, we systematically present a procedure that is followed to characterize the fractal and multifractal characteristics of temporal fluctuations in a signal using the method of Multifractal Detrended Fluctuation Analysis (MFDFA) [34, 37, 39]. As compared to other methods mentioned before, MFDFA is computationally faster and can be applied to shorter time series.

Let us consider $x(t) = \{x_i\}_{i=1}^{N}$ to be the acoustic pressure signal (or the heat release rate signal) obtained from a thermoacoustic system. In the first step, we subtract the mean $\langle x \rangle$ of the entire signal from its local fluctuations so that fluctuations in the resultant signal are centered at zero. Using this mean-adjusted signal, we next obtain a random-walk profile by the method of the cumulative sum of all values of the signal before point i as follows,

$$Y(i) = \sum_{t=1}^{i} [x(t) - \langle x \rangle], \tag{10.14}$$

where $\langle x \rangle = (1/N) \sum_{t=1}^{N} x(t)$, N is the total number of data points in the signal, and $i = 1, 2, 3, \ldots.., N$.

Next, we divide the random-walk profile $Y(i)$ into non-overlapping segments n_w of equal size w, where $n_w = N/w$. If the number of data points in the profile is not an integer multiple of w, then some data points of the profile will be left over after the last window. In order to account for all such data points, we repeat the same procedure of partitioning the profile starting from the last point towards the first point with the same window size w. This will also lead to the number of segments equal to $n_w = N/w$. Thus, we have in total $2n_w$ windows for the calculation.

After the division of the signal into multiple windows, we perform window-wise detrended fluctuation analysis of the profile $Y(i)$ and calculate the variance for each window v of size w. During detrended fluctuation analysis, we first fit the mth-order polynomial trend, $Y^m(i)$, by the least square method to each window of the profile, $Y(i)$, and then subtract the values of the profile from the polynomial fit. Next, we calculate the variance of this resultant profile obtained from detrended fluctuations as follows,

$$F(w, v) = \frac{1}{w} \sum_{i=1}^{w} \{Y[(v-1)w + i] - Y^m(i)\}^2 \tag{10.15a}$$

for segments $v = 1, 2, 3, \ldots\ldots, n_w$, and

$$F(w, v) = \frac{1}{w} \sum_{i=1}^{w} \{Y[N - (v - n_w)w + i] - Y^m(i)\}^2 \tag{10.15b}$$

for segments $v = n_w + 1, \ldots\ldots\ldots, 2n_w$.

Here, $m = 1, 2, 3, \ldots\ldots$ such that $m = 1$ indicates a linear fit, $m = 2$ indicates a quadratic fit, and so on. The analysis is repeated for different values of m and the best data fit is decided.

Next, from the values of $F(w, v)$, we can calculate a fluctuation function. The fluctuation function for the second-order moment $F_2(w)$ is calculated as the square root of the mean of the detrended variance of all windows of size w,

$$F_2(w) = \left\{ \frac{1}{2n_w} \sum_{v=1}^{2n_w} [F(w, v)] \right\}^{1/2} \tag{10.16}$$

A log-log plot between $F_2(w)$ and the increasing window size, w, helps in characterizing the fractal nature of a signal. For fractal signals, this plot shows a power-law behavior, where the slope of a straight line fitted to the linear regime of the plot is referred to as a scaling exponent or the Hurst exponent (H) such that,

$$F_2(w) \sim w^H \tag{10.17}$$

Thus, the Hurst exponent measures how fast the root mean square of standard deviation of fluctuations in the signal grows with an increase in the segment size. The fluctuation function of second-order moment shown in (10.16) is often used for analyzing monofractal signals, as scaling exponents of such signals remain the same for all moments (Fig. 10.11b). In monofractal signals, we do not observe the presence of extreme (i.e., very small and large magnitude) fluctuations [37].

In the case of multifractal signals, due to the presence of multiple scaling exponents, the detrended fluctuation analysis is performed for different moments (say, q). Thus, the fluctuation function $F_q(w)$ of order q is defined as,

$$F_q(w) = \left\{ \frac{1}{2n_w} \sum_{v=1}^{2n_w} [F(w, v)]^{q/2} \right\}^{1/q} \tag{10.18}$$

where q can take any real value (ranging from negative to positive number) except zero.

For the 0th-order moment (i.e., $q = 0$), the value of $1/q$ diverges to infinity. Therefore, using a logarithmic average approach, we obtain the approximate value of $F_0(w)$ as [34],

$$F_0(w) = \exp \left\{ \frac{1}{4n_w} \sum_{v=1}^{2n_w} \ln[F(w, v)] \right\} \tag{10.19}$$

We repeat the same analysis of the calculation of $F_q(w)$ for different sizes of windows (w). This analysis helps in characterizing the properties of both the fast and the slow scale oscillations present in the signal. Fast oscillations are reflected in the value of the fluctuation function calculated for smaller sized windows, while slow oscillations are reflected in the value of the fluctuation function calculated for larger sized windows.

The value of $F_q(w)$ increases with an increase in the values of w. If there exists a power law behavior between $\log[F_q(w)]$ and $\log(w)$ for different values of qs, we can express the relation between them as,

$$F_q(w) \sim w^{H_q} \tag{10.20}$$

where H_q is referred to as the generalized Hurst exponent [67].

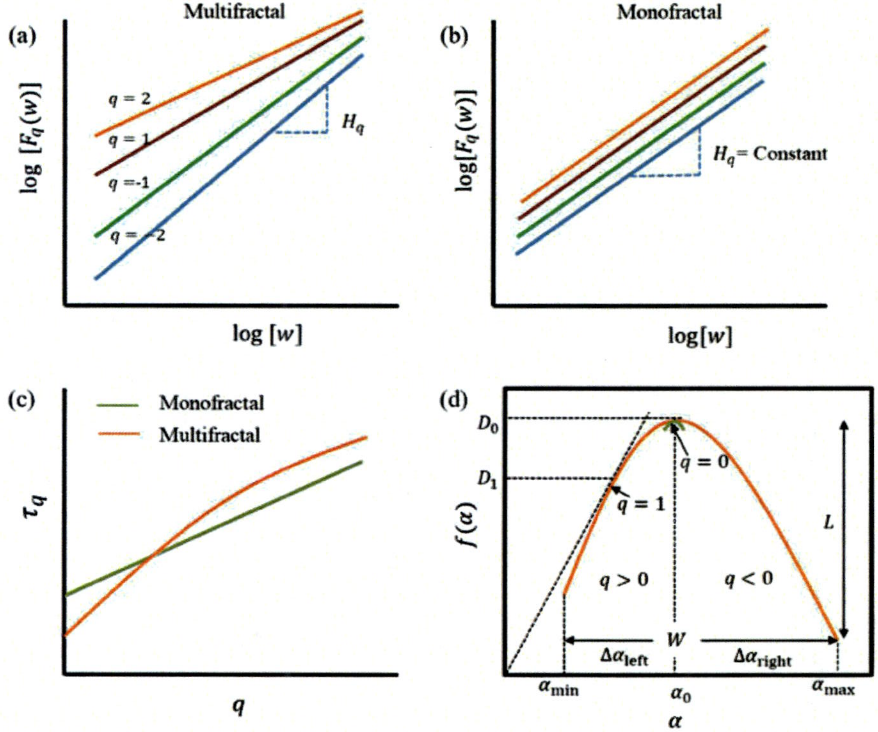

Fig. 10.11 Schematic showing the scale-invariant behavior of (**a**) a multifractal and (**b**) a monofractal signal for different order (q) of moments. (**c**) Variation of the mass exponent (τ_q) for different order (q) of moments, and (**d**) singularity spectrums of a multifractal and a monofractal signal. Various quantifiers of the singularity spectrum are shown in (**d**). Adapted with permission from Ihlen [34]

The value of H_q for $q = 2$ is known as Hurst exponent (H_2). For monofractal signals, the value of H_q is independent of q. This means the scaling exponents of a plot between $\log[F_q(w)]$ and $\log(w)$ are the same for all values of q (Fig. 10.11b). In contrast, for multifractal signals, the value of H_q varies with a change in q. This indicates that the variation of $\log[F_q(w)]$ and $\log(w)$ is non-identical for all values of q (Fig. 10.11a). For positive values of q, H_q describes the scaling behavior of segments with large magnitude fluctuations. In contrast, for negative values of q, H_q indicates the scaling behavior of segments with small magnitude fluctuations.

10.2.1.2 Singularity Spectrum and Generalized Dimension

We can effectively represent the behavior of generalized Hurst exponents observed for different moments of fluctuations using a singularity spectrum (Fig. 10.11d). The singularity spectrum indicates the distribution of sets of Hurst exponents of a signal. The key features of the singularity spectrum further help us in understanding the properties of fractal or multifractal signals. The singularity spectrum usually takes the shape of an inverted parabola. The shape and the position of the spectrum provide information about the range of dimensions required for describing the fractal signal and also the range of scales involved in the signal. We will elaborate on these features of the singularity spectrum in the following discussions.

In order to obtain the singularity spectrum of a signal, we first need to calculate the q-order mass exponent (τ_q), whose value is then used to obtain the singularity exponent α (also called Hölder exponent) and the singularity dimension $f(\alpha)$ through a Legendre transform [20, 39], as given below,

$$\tau_q = qH_q - 1 \tag{10.21}$$

$$\alpha = \partial\tau_q/\partial q \tag{10.22}$$

$$f(\alpha) = q\alpha - \tau_q \tag{10.23}$$

Further, we can obtain the magnitude of a generalized fractal dimension using $\tau(q)$ as follows [31],

$$D_q = \frac{\tau(q)}{q-1}, \quad \text{for} \quad q \neq 1 \tag{10.24}$$

A plot between $f(\alpha)$ and α is called the singularity spectrum. For monofractal signals, as the value of H_q is the same for all values of q, the mass exponent (τ_q) exhibits a linear dependency on q (Fig. 10.11c). Hence, we notice a constant value of singularity exponents (α) for all q. As a consequence, the singularity spectrum compresses to a point for monofractals (Fig. 10.11d). In contrast, for multifractal signals, the variation of τ_q shows a nonlinear dependency on q (Fig. 10.11c). Hence, the value of α is not constant and shows a variation with a change in the values of q. Thus, the singularity spectrum shows a larger sized arc for multifractals (Fig. 10.11d).

We can measure the extent of multifractality present in the signal by calculating the width of the spectrum (W) [47]. It is calculated as the difference between the maximum and the minimum values of α over which the spectrum is spread as follows,

$$W = \Delta\alpha = \alpha_{max} - \alpha_{min} \tag{10.25}$$

Larger the width of the spectrum, the higher the range of scales (i.e., stronger multifractality) present in the signal. For a monofractal signal, the spectrum width is nearly zero. While for a multifractal signal, the spectrum carries a larger width.

Further, the singularity spectrum of multifractal signals is mostly asymmetric about the maximum value of $f(\alpha)$. Hence, the spectrum could be skewed either to the left side (i.e., $\Delta\alpha_{\text{left}} > \Delta\alpha_{\text{right}}$) or the right side (i.e., $\Delta\alpha_{\text{left}} < \Delta\alpha_{\text{right}}$) of the maximum value of $f(\alpha)$. When the spectrum is left-skewed with a long tail on the left side, the variation of H_q with q displays a plateau for negative values of q and nonlinear behavior for positive values of q [34]. This left-skewed behavior of the spectrum, in turn, indicates that the extreme (local large magnitude) fluctuations are dominated in the temporal structures of the multifractal signal [90]. Similarly, for the right-skewed spectrum with a long right tail, the values of H_q are nearly independent of the change in the positive values of q. This right-skewed behavior of the spectrum further points out that the temporal fluctuations in the multifractal signal are dominated by fine (low amplitude) structures.

At the point of the maximum value of $f(\alpha)$, we have $df(\alpha)/d\alpha = 0$. This condition is satisfied only when $q = 0$ (see (10.23)) and it corresponds to $\tau(q) = -1$ (see (10.21)). We denote the value of α corresponding to $f(\alpha)_{\text{max}}$ as α_0, where α_0 indicates the dominant scale present in the multifractal signal. A higher value of α_0 indicates that the signal is less correlated and consists of fine (irregular) structures. In contrast, a lower value of α_0 is observed for a correlated signal having more regular structures.

The height of the spectrum (i.e., $L = f(\alpha)_{\text{max}} - f(\alpha)_{\text{min}}$) represents the range of fractal dimensions required to describe the multifractal signal. The maximum value of $f(\alpha)$ is referred to as a capacity dimension (D_0) and is calculated for $q = 0$. Furthermore, the fractal dimension (D_1) corresponding to $q = 1$ is referred to as an information dimension, and that corresponding to $q = 2$ is called a correlation dimension, D_2 [24].

Let us understand the physical meaning of the aforementioned three dimensions using an example of the distribution of state points of a signal on its attractor in the phase space [64]. In this context, the capacity dimension only measures the number of boxes of a particular size needed to cover all points on the attractor, which is similar to a box-counting method we discussed in Sect. 10.1. This dimension does not account for the number of points present in a box. Thus, the variation of the density of points on the attractor is ignored in the calculation of a capacity dimension. However, we need to consider such density variation of points for strange attractors, where the frequency of re-visits of a trajectory at different regions of the attractor is not uniform and varies significantly from one region to another. Therefore, to resolve this limitation of a capacity dimension, Grassberger [26] and Hentschel and Procaccia [31] introduced two new dimensions, i.e., information dimension and correlation dimension, to account for the distribution of points on the attractor. Information dimension deals with the probability of finding a point in each box, and it is obtained as the ratio between information entropy to the size of a box. Similarly, a correlation dimension takes into account the distribution of points in a box based on the distance between the points in the phase space (as discussed in Chap. 2). If these three dimensions are equal (i.e., $D_0 = D_1 = D_2$), then the corresponding signal is referred to as a monofractal [26]. On the other hand, if these dimensions follow a relation such that $D_0 > D_1 > D_2$, then the corresponding signal is called a multifractal [26].

Note that all these properties of a singularity spectrum discussed so far for a time series are equally applicable to a singularity spectrum obtained from spatial fractal structures, which we discuss next. In this manner, we can detect several features of irregular fluctuations observed in temporal or spatial data through the multifractal analysis.

10.2.1.3 Modified Box-Counting Method

As we discussed previously in Sect. 10.1, the box-counting [78] is one of the most common methods used for the fractal and multifractal analysis of the spatial data (see Fig. 10.12). In this method, we cover a fractal structure using equal-sized boxes. Then, we count the number of boxes, $N(\varepsilon)$, required to cover the fractal structure if the box size (ε) is increased from a small to a large value. We notice that the variation of the number of boxes with the increasing size of the box exhibits a power-law scaling. This means a plot between $\log N(\varepsilon)$ and $\log \varepsilon$ shows a linear variation [58] as given below,

$$N(\varepsilon) \sim \varepsilon^{-D} \tag{10.26}$$

where the scaling exponent D is also called a fractal dimension (or a box-counting dimension). Since the natural structures are multifractal (i.e., they possess several values of D), the box-counting algorithm has been modified further to accommodate it for both fractal and multifractal structures [29].

In the modified box-counting method, we consider a box of size ε to cover the measure (μ) of the fractal structure. Here, the measure (μ) is a normalized quantity that characterizes the property (for example, length, area) of the fractal structure [87]. As we change the box size, the value of the measure also changes. The variation of the measure $\mu(\varepsilon)$ with the box size ε displays a power-law relation [29, 89] as follows,

$$\mu(\varepsilon) \sim \varepsilon^{\alpha} \tag{10.27}$$

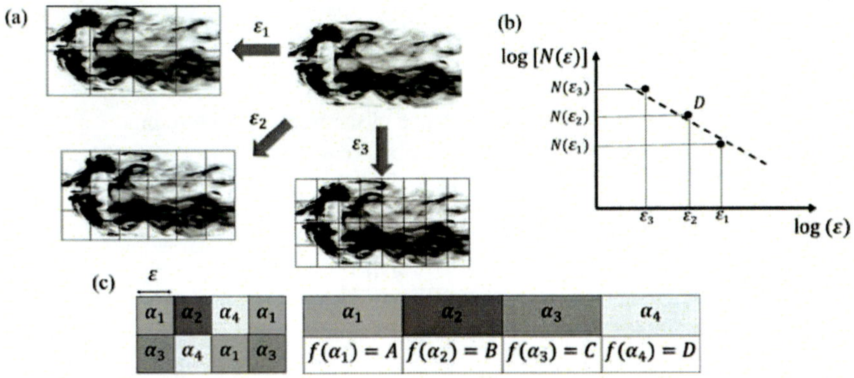

Fig. 10.12 (**a**) Procedure of box-counting method, where an equal-sized (ε) grid is overlapped on the fractal structure (e.g., a turbulent flame) and the number of boxes, $N(\varepsilon)$, that are required to cover the structure are counted. (**b**) Variation of $N(\varepsilon_i)$ with ε_i shows an inverse power-law relation, whose exponent indicates a fractal dimension (D). (**c**) Diagrammatic representation of the distribution of singularity exponent α for a box of size ε over the fractal structure and the classification of sets of α in terms of $f(\alpha)$. Since α_1 appears in a larger number of boxes as compared to other values of α, the fractal dimension corresponding to $f(\alpha_1) = A$ is the highest. In contrast, as α_2 exists only in a single box, the corresponding fractal dimension $f(\alpha_2) = B$ is zero. The concept of singularity spectrum construction in (**c**) is adopted from Murcio et al. [62]

where α is the singularity exponent, also called Lipshitz–Hölder exponent [12].

Note that α is a local property of a fractal structure. Its value remains constant for the monofractal structure and exhibits variation from one location to another for the multifractal structure (Fig. 10.12c). Since many boxes on the multifractal structure may have the same α, its distribution of the structure is considered to be comprised of sets (groups) of α. If $N(\alpha)$ is the number of boxes that have a singularity exponent α, then $f(\alpha)$, an exponent that is loosely related to the fractal dimension or Hausdorff dimension, can be obtained as follows [29],

$$N(\alpha) \sim \varepsilon^{-f(\alpha)} \tag{10.28}$$

In the multifractal analysis of spatial structures of the turbulent flame (as discussed in Sect. 10.3.5), Raghunathan et al. [75] used the box-counting method proposed by Chhabra and Jensen [8]. In this analysis of multifractality, we do not need the Legendre transform to compute the values of the singularity exponent (α) and the fractal dimension, $f(\alpha)$, as required for the MFDFA in the previous section [8, 21, 60, 62, 96, 99, 100]. Instead, we can compute a generalized measure $\mu_i(q, \varepsilon)$ that quantifies the normalized probability of the portion of the fractal structure contained in a box located at i for the qth-order moment. For example, for a binary image of the flame contour (as discussed in Sect. 10.3.5), the measure could be a normalized probability of mass distribution of pixels in the image that contains the flame contour. The value of the generalized measure is obtained as follows [8],

$$\mu_i(q,\varepsilon) = \frac{[P_i(\varepsilon)]^q}{\sum_{i=1}^{N(\varepsilon)}[P_i(\varepsilon)]^q} \tag{10.29}$$

where $P_i(\varepsilon)$ is the probability of occurrence of the fractal structure in a box of size ε. It is calculated as the ratio of number of pixels (M) of the fractal structure in each box to the total number of pixels occupied by the fractal structure. $P_i(\varepsilon)$ is obtained as follows,

$$P_i(\varepsilon) = \frac{M_i(\varepsilon)}{\sum_{i=1}^{N(\varepsilon)} M_i(\varepsilon)} \tag{10.30}$$

Using the expression for the generalized measure $\mu(q)$ (10.29), we can obtain equations for the singularity exponent $\alpha(q)$ and the fractal (Hausdorff) dimension $f(q)$ as follows [8],

$$\alpha(q) = \lim_{\varepsilon\to 0} \frac{\sum_{i=1}^{N(\varepsilon)} \mu_i(q,\varepsilon)\ln[P_i(\varepsilon)]}{\ln\varepsilon} \tag{10.31}$$

$$f(q) = \lim_{\varepsilon\to 0} \frac{\sum_{i=1}^{N(\varepsilon)} \mu_i(q,\varepsilon)\ln[\mu_i(q,\varepsilon)]}{\ln\varepsilon} \tag{10.32}$$

For every value of q, we can obtain the value of α from a linear fit to the plot between $\sum_{i=1}^{N(\varepsilon)} \mu_i(q,\varepsilon)\ln[P_i(\varepsilon)]$ and $\ln\varepsilon$. While we can get the value of f from a linear fit to the plot between $\sum_{i=1}^{N(\varepsilon)} \mu_i(q,\varepsilon)\ln[\mu_i(q,\varepsilon)]$ and $\ln\varepsilon$ [74].

Note that the qth-order statistic helps us in understanding the non-homogeneous distribution of singularity exponents α on the fractal structure [62]. The measure $\mu_i(q)$ amplifies the regions of higher value of singularity exponents for $q > 1$, while it accentuates the regions of lesser value of singularity exponents for $q < 1$. From the plot between f and α, for different values of q, we can construct a singularity spectrum similar to that in Fig. 10.11 and study the multifractal behavior of spatial fractal structures of a system.

Until now, we discussed various properties of fractals and multifractals, and introduced different methodologies required for analyzing such structures. In the next section, we will discuss the application of these fractal/multifractal analyses to reveal hidden complexities underlying the dynamics of turbulent thermoacoustic systems. In Chap. 6, we showed that the dynamical behavior of such systems undergoes a transition from the state of combustion noise to thermoacoustic instability and then to the state of flame blowout, as we decrease the equivalence ratio to a fuel lean condition. In this transition, a state of intermittency is observed prior to and post the state of thermoacoustic instability. Using the methods of fractal/multifractal analysis, we will describe the features of aperiodic fluctuations observed during the state of combustion noise (stable operation), study the transition between different dynamical states, and characterize the structure of the turbulent flame.

10.3 Multifractal Analysis of Thermoacoustic Systems

10.3.1 Loss of Multifractality Prior to Thermoacoustic Instability

As we discussed in previous chapters, the dynamical behavior of a thermoacoustic system is usually characterized by analyzing its independent variables, which include the acoustic pressure, the acoustic velocity, and the heat release rate fluctuations in the flame. A coupled interaction between these variables under certain operating conditions leads to the onset of thermoacoustic instability. Therefore, the dynamical changes in the system can be tracked by analyzing the temporal properties of these variables as the system approaches thermoacoustic instability.

In Fig. 10.13a, b, we show the time series of acoustic pressure fluctuations obtained during the states of combustion noise and thermoacoustic instability from a turbulent bluff-body stabilized dump combustor, respectively [63]. We notice that the dynamics of the acoustic pressure fluctuations during the state of combustion noise exhibits low amplitude aperiodic fluctuations (Fig. 10.13a). The power spectrum of these fluctuations shows a broadband of frequencies with shallow peaks around the acoustic modes of the combustor (Fig. 10.13c). Traditionally, the dynamics of such acoustic pressure fluctuations were considered as stochastic and modeled using ad-hoc terms of additive or multiplicative noises in acoustic equations [10, 45, 65]. In contrast, in Chap. 6, we discussed that the acoustic pressure fluctuations observed during the state of combustion noise comprise high-dimensional chaotic oscillations.

On the other hand, the dynamics of the combustor during thermoacoustic instability shows self-sustained, high amplitude periodic oscillations (Fig. 10.13b). The power spectrum of these fluctuations shows sharp peaks at the frequencies corresponding to integer multiples of the fundamental mode of the combustor (Fig. 10.13d). Thus, we notice the emergence of ordered oscillations from chaos during the occurrence of thermoacoustic instability. Therefore, in order to accurately predict the onset of thermoacoustic instability, we need to have a clear understanding of the dynamics of the state of combustion noise. In this section, we discuss the fractal and multifractal properties of different dynamical states observed in a turbulent thermoacoustic system.

For fractal analysis of thermoacoustic oscillations, we need to choose a scale ratio in a specific range of timescales (or segment lengths) of the signal. As a rule of thumb, it is often selected between two-four cycles of the dominant frequency of thermoacoustic instability [63]. If the scaling window size is less than one acoustic cycle, we cannot detect the periodicity in the signal correctly. In contrast, if the scaling window size is greater than the four acoustic cycles, we end up averaging out the fluctuations. This, in turn, results in the inappropriate calculation of the fractal behavior of the signal. Furthermore, if there are two dominant modes (say, f_1 and f_2 such that $f_2 > f_1$) present in the signal, we need to consider a scale range from

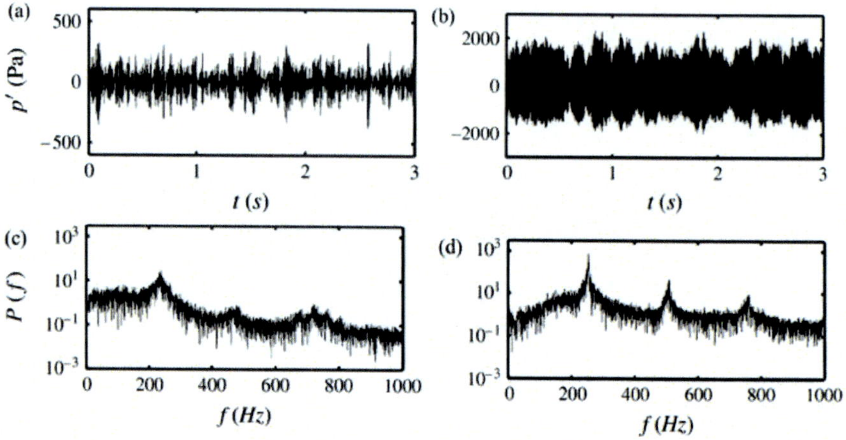

Fig. 10.13 (**a, b**) The time series of the acoustic pressure fluctuations and (**c, d**) the corresponding power spectrum obtained for the states of (**a**) combustion noise and (**c**) thermoacoustic instability, respectively, in a turbulent combustor. Adapted with permission from Nair and Sujith [63]

Fig. 10.14 Variation of the fluctuation function of order two $F_2(w)$ with the change in the sample size w for signals of Gaussian white noise, periodic oscillations, and acoustic pressure fluctuations obtained experimentally during the state of combustion noise. We notice different values of the Hurst exponent (H) for these signals. Adapted with permission from Nair and Sujith [63]

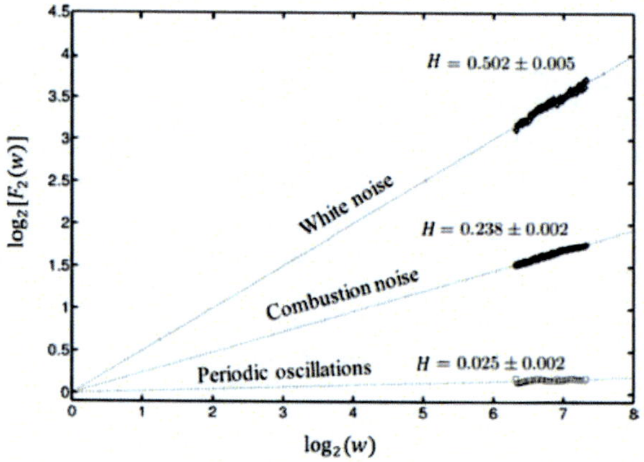

two cycles of the shortest timescale ($2/f_2$) to the four cycles of the longest timescale ($4/f_1$), to account for the effects of both the frequencies [41].

In Fig. 10.14, we show the comparison of the fractal properties of the experimentally obtained acoustic pressure fluctuations using MFDFA algorithm during the state of combustion noise with that of a numerically generated Gaussian white noise signal and a periodic signal (i.e., thermoacoustic instability). The fractal behavior of these signals is characterized through the variation of the second-order fluctuation function, $F_2(w)$, with the increasing window size, w, of the signal on a log-log plot (see (10.16)). This plot shows a power-law scaling for all the signals. However, we observe that the slope of the plot (i.e., the Hurst exponent H) is different for each signals.

As expected, we notice the value of the Hurst exponent close to 0.5 for the Gaussian white noise signal. For the periodic signal, its value is observed to be close to zero, due to the presence of a single timescale throughout the signal. In contrast, the Hurst exponent shows a value near 0.24 for the acoustic pressure fluctuations observed during the state of combustion noise. This value of the Hurst exponent lies in between the two extreme values of the Hurst exponents observed for the periodic and the random oscillations. Thus, we confirm that the dynamics of the acoustic pressure during the state of combustion noise is fractal. As the value of H lies between 0 and 0.5 (i.e., $0 < H < 0.5$), the acoustic pressure oscillations observed during this state show the characteristics of the anti-persistent behavior, which indicates that the signal does not possess any trend and has a tendency to revert to the mean.

As seen in Fig. 10.13a, the acoustic pressure signal observed during the state of combustion noise consists of fluctuations of various scales at different time intervals. In order to study such signals, we need to analyze their multifractal properties for a range of orders (q) of fluctuations. Next, we discuss the multifractal properties of the aforementioned signals by calculating variation of the qth-order fluctuation function $F_q(w)$ with increasing window size w (10.18), see Fig. 10.15I. We also plot the singularity spectrum (a plot between $f(\alpha)$ and α) corresponding to these signals (see Fig. 10.15II). As we mentioned previously, the singularity spectrum helps us in differentiating a monofractal signal from a multifractal signal. For monofractal signals, the value of the Hurst exponent remains invariant for the variation of q, whereas for multifractal signals, its value changes with a change in q. Hence, the singularity spectrum of a monofractal signal shows a point, while that of a multifractal signal displays a spread over a wider range of α.

For the state of combustion noise (Fig. 10.15b-I), we notice a change in the slope of the plot between $F_q(w)$ and w for different values of q. Furthermore, we observe that the singularity spectrum of this state is broad, exhibiting a spread over a large range of α (Fig. 10.15b-II). This behavior of the spectrum further reasserts that the dynamics of acoustic pressure fluctuations during the state of combustion noise is multifractal, having fluctuations at both large and small amplitudes in the signal. We also notice that the spectrum is right-skewed, i.e., the spectrum is asymmetric on both the halves and has a long tail on the right side. This right-skewed property of the spectrum indicates that the multifractal characteristics of the signal are strongly dependent on the local small amplitude fluctuations than the local large amplitude fluctuations.

On the other hand, we notice that the slope of the plot between $F_q(w)$ and w for different values of q remains constant for both the Gaussian white noise (Fig. 10.15a-I) and the periodic signal (Fig. 10.15c-I). The singularity spectrums for these signals display a small arc centered around $\alpha \approx 0.5$ and 0, respectively (Fig. 10.15a-II and c-II). As the spectrum width of these signals is very small, we can conclude that these signals consist of nearly uniform amplitude fluctuations throughout the signal. From close inspection, we notice that the singularity spectrum of the acoustic pressure fluctuations during the state of thermoacoustic instability is not a point but has a small width that is spread over a small range of α. This

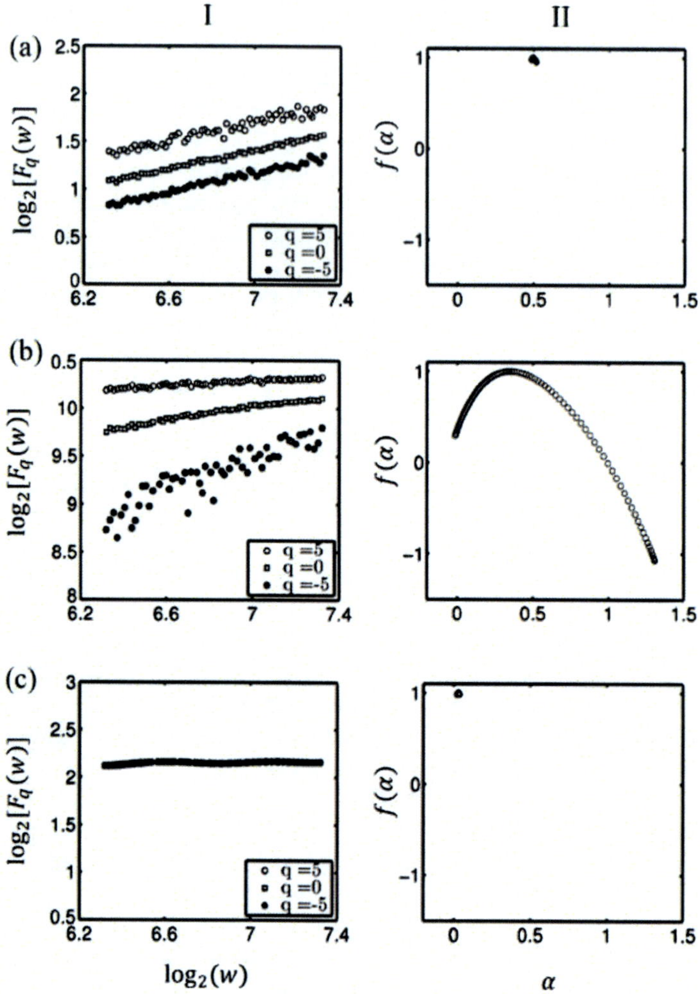

Fig. 10.15 (I) Variation of the qth-order fluctuation function $F_q(w)$ with an increase in the sample size w and (II) the corresponding singularity spectrums for the signals of (**a**) Gaussian white noise, (**b**) aperiodic acoustic pressure fluctuations observed during the state of combustion noise, and (**c**) periodic oscillations observed during thermoacoustic instability. The singularity spectrums of signals of white noise and periodic oscillations indicate a very small arc located around $\alpha = 0.5$ and 0, respectively. In contrast, the singularity spectrum of acoustic pressure fluctuations during the state of combustion noise is broad and exhibits features of multifractality. Adapted with permission from Nair and Sujith [63]

small spread in the spectrum could be a result of small modulations observed in cycle-to-cycle amplitudes of the signal during this state (see Fig. 10.13b). These amplitude modulations occur primarily due to the presence of underlying turbulent perturbations in the flow [46]. Thus, the dynamics of the Gaussian white noise and the periodic thermoacoustic oscillations are not multifractal.

The comparison of the singularity spectrums of acoustic pressure fluctuations for the states of combustion noise and thermoacoustic instability, indicates the collapse of the spectrum from a broad distribution to nearly a point during the onset of thermoacoustic instability. This transition of the combustor dynamics from the state of combustion noise to thermoacoustic instability can be alternately viewed as a *loss of multifractality* in the system. Such a loss of multifractality is associated with the diminishment of the multiple spectral scales observed during the state of combustion noise to a single scale (or at most a few discrete scales) during the state of thermoacoustic instability. This narrowing behavior of the broadband frequencies to a single frequency in the spectrum can also be referred to as a '*loss of spectral reverse*' [68, 97].

Note that, in case we ignore the multifractal features of acoustic fluctuations observed during the state of combustion noise, we indirectly remove the contribution of several scales involved in the built-up of the thermoacoustic phenomenon [63]. As a consequence of the removal of such complexities existing in the combustion dynamics, we further limit our capability to accurately predict the onset of thermoacoustic instability in a system.

10.3.2 Detection of Correlations Through Surrogate Analysis and Singularity Spectrum

As we mentioned before, the acoustic pressure signal exhibits irregular (aperiodic) fluctuations during the state of combustion noise. Traditionally, such irregular fluctuations have been confirmed to be stochastic through the observation of their unimodal (or Gaussian) probability distribution [46, 65]. However, according to Kantelhardt et al. [39], a multifractal signal can also exhibit a regular distribution with finite moments (e.g., Gaussian distribution). Since multifractal signals occur due to two processes: (1) broad probability distribution of the data points (for example, Levy distribution) and (2) long-range correlations in both high and low amplitude fluctuations, the confirmation of the signal properties only through their probability distribution could be misleading.

One simple way to detect the presence of correlations in a signal is to use a surrogate test of random shuffling [98]. The surrogate analysis is a statistical test often used to detect nonlinear determinism (correlations) in a signal [92]. Here, a surrogate indicates analogous data that is generated from the original measured signal using linear models. In the method of the random shuffling surrogate, the null hypothesis is formulated such that the data is generated from a random process (Gaussian white noise). During random shuffling, the data points in a signal are

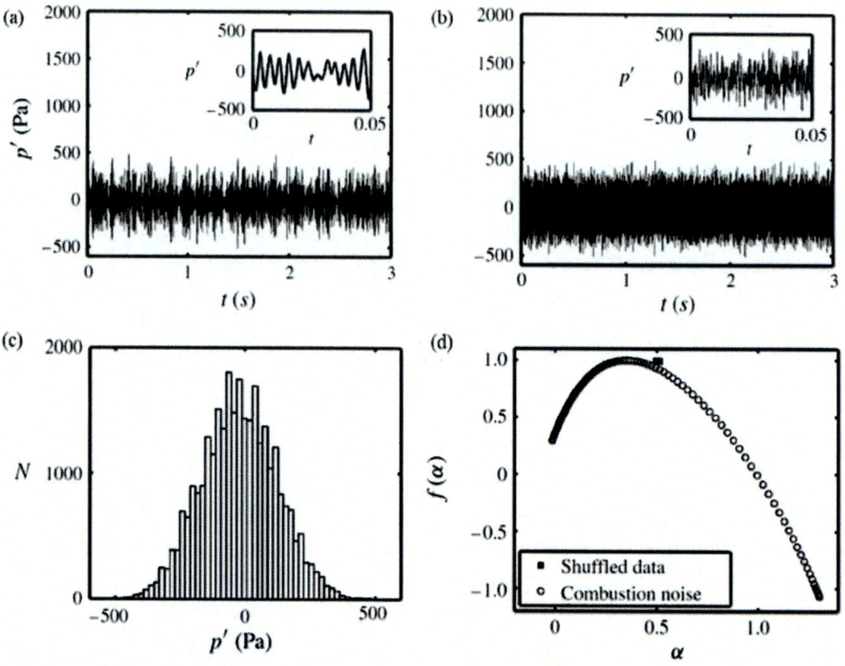

Fig. 10.16 The time series of (**a**) acoustic pressure fluctuations observed during the state of combustion noise and (**b**) a corresponding randomly shuffled data. (**c**) Histogram of these two signals, where N is the number of data points. (**d**) The singularity spectrums of the experimental data and the shuffled data. We observe that although the histogram of these signal appears to have a Gaussian distribution, their singularity spectrums show a significant difference. A wider spectrum observed for the experimental data indicates the presence of low level of correlations in the signal. Such correlations are completely absent in the shuffled data. Reproduced with permission from Nair and Sujith [63]

randomly swapped so that the linear correlations between the points are destroyed while preserving the amplitude distribution of the original signal. Therefore, the loss of correlation causes the newly regenerated surrogate signals to appear memoryless and display the features of a random process. Then, a measure that computes linear correlations in the signal is calculated for both the original signal and for all surrogates. If the computed measure is significantly different for both the original signal and surrogates, then the proposed null hypothesis of the absence of linear temporal correlations in a given signal is rejected.

In practical combustion systems, fluctuations occurring due to different sources such as measurement noise, experimental noise, underlying flow fluctuations, or due to interaction of different processes in the combustor are inevitable. In such situations, surrogate analysis provides a powerful tool to determine the existence of determinism in apparently stochastic data.

In Fig. 10.16a, b, we show the time series of acoustic pressure fluctuations obtained from experiments for the state of combustion noise and the corresponding surrogate signal obtained from the application of the random shuffling test, respectively [63]. We notice a loss of weak correlations present in the original experimental signal (inset of Fig. 10.16a) when the data points are randomly shuffled (inset of Fig. 10.16b), although the distribution of both the signals conforms to the Gaussian distribution (Fig. 10.16c). Further, we notice that the singularity spectrum of the shuffled data is collapsed to a point around $\alpha = 0.5$ (Fig. 10.16d). In contrast, the singularity spectrum of the experimental data is observed to be broad and shows a peak away from $\alpha = 0.5$ (Fig. 10.16d). These behaviors of the singularity spectrums further indicate the presence of noise-like (random) fluctuations in the shuffled data and the existence of multifractality in the experimental data. Thus, we can confirm that the dynamics of the acoustic pressure fluctuations is not stochastic and consists of a low level of correlations between data points in the signal. From this analysis, we understood that although the probability distribution of the acoustic pressure fluctuations during the state of combustion noise being Gaussian or its power spectrum being broadband may indicate a noise-like behavior in the signal, the underlying dynamics of this state could be complex, deterministic, and multifractal.

10.3.3 Fractional-Order Van der Pol Model to Study Multifractality

As we see in Fig. 10.15b-II, the experimental data of acoustic pressure fluctuations for the state of combustion noise in a turbulent thermoacoustic system displays a wide right-skewed multifractal spectrum. In order to capture these features of the multifractal spectrum, Varghese et al. [95] developed a fractional-order model. This model was obtained by considering a minimal fractional generalization of the classical Van der Pol oscillator forced with additive white noise,

$$\frac{d^{\mu}\eta}{dt^{\mu}} + (\beta - \kappa\eta^2)\frac{d\eta}{dt} + \omega^{\mu}\eta = \xi \quad \text{for} \quad 1 < \mu \leq 2 \tag{10.33}$$

where μ is the order of differentiation, η is the time-dependent coefficient of the acoustic pressure, β and κ are the linear damping and nonlinear driving of the system, respectively, ω is the angular frequency of the oscillator, and ξ is the white Gaussian noise. For a detailed description of the model, the reader may refer to Varghese et al. [95].

In Fig. 10.17, we show the time series of the acoustic pressure fluctuations during the states of combustion noise and thermoacoustic instability, and the corresponding singularity spectrum obtained using the method of MFDFA from the fractional Van der Pol model [95]. The time series from this model during the occurrence of combustion noise (Fig. 10.17b) shows aperiodic fluctuations and has features similar to that shown in Fig. 10.13a for the experimental data. The multifractal spectrum

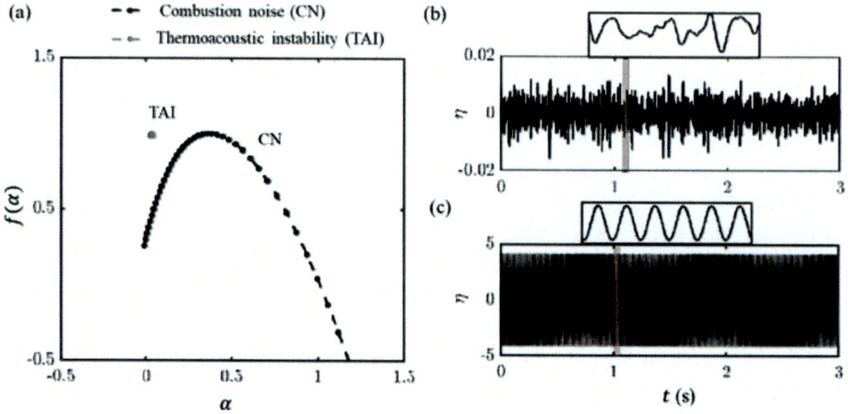

Fig. 10.17 (**a**) The singularity spectrum and (**b, c**) the time series of acoustic pressure fluctuations for the states of combustion noise and thermoacoustic instability, respectively, from the fractional Van der Pol model. The spectrum is observed to be broad for the state of combustion noise and a point for the state of thermoacoustic instability. Adapted with permission from Varghese et al. [95]

of the pressure fluctuations corresponding to this state is broad (Fig. 10.17a) and closely matches with that observed for the experimental data of the thermoacoustic system (Fig. 10.15b). Furthermore, the Hurst exponent of the pressure fluctuations during combustion noise from the fractional Van der Pol model was observed to be 0.254, which is close to that of the experimental data ($H = 0.238$) (Fig. 10.14). We also notice the loss of multifractality during the occurrence of thermoacoustic instability in the model (Fig. 10.17a), similar to that observed in the experimental data (refer to singularity spectrum in Fig. 10.15b-II, c-II).

According to Nair and Sujith [63], the multifractal nature of the experimental pressure fluctuations is due to the effect of memory rather than the distribution of the data (as discussed in Sect. 10.3.2). Fractional differential operators introduce long range-correlations in the pressure fluctuations and consequently result in a wider multifractal spectrum. However, more work needs to be done to find the physical basis for the proposed fractional Van der Pol model (10.33). The fractional Van der Pol equation could possibly be obtained from considering the fractional form of the Helmholtz equation and will be a topic of future studies. Moreover, a system identification method to estimate the various adjustable parameters used in the fractional-order model needs to be developed.

10.3.4 Unified Multifractal Analysis During the Transition to Thermoacoustic Instability and Flame Blowout

As we discussed in Sect. 6.5 of Chap. 6, similar to thermoacoustic instability, lean flame blowout is also an important dynamic phenomenon exhibited by turbulent thermoacoustic systems. The dynamics of the acoustic pressure prior to flame blowout shows low amplitude aperiodic oscillations. Gotoda et al. [23] used the framework of multifractal analysis to detect the properties of such acoustic pressure oscillations prior to flame blowout in a turbulent thermoacoustic system. Subsequently, Unni et al. [94] provided a unified framework of multifractality to characterize the properties of acoustic pressure fluctuations during the state of combustion noise and that observed prior to flame blowout.

In Fig. 10.18, we show the singularity spectrums of acoustic pressure fluctuations, obtained using the method of MFDFA, for the states of combustion noise, thermoacoustic instability, and low amplitude aperiodic fluctuations observed prior to flame blowout. As we discussed earlier, a broad singularity spectrum is observed for the acoustic pressure fluctuations during the state of combustion noise, while the spectrum for the state of thermoacoustic instability (periodic oscillations) shows a small cluster of points near $\alpha = 0$. The singularity spectrum of the acoustic pressure fluctuations observed just before flame blowout displays a broad distribution. This behavior of the spectrum further indicates that the aperiodic oscillations observed prior to flame blowout are multifractal. We also notice a shift in the position of the spectrum during the transition from the state of combustion noise to thermoacoustic instability and then to flame blowout due to a change in the underlying fractal characteristics of the pressure signal during these states.

So far in this chapter, we presented the multifractal analysis of temporal oscillations of the acoustic pressure for different states of the combustor dynamics, such as combustion noise, thermoacoustic instability, and a state prior to lean flame blowout. We showed that the dynamics of aperiodic oscillations observed in thermoacoustic systems are indeed complex and display a multifractal behavior. The loss of multifractality is observed at the onset of regular (periodic) oscillations in the system. Next, we will discuss the application of the multifractal analysis to study the characteristics of spatial structures of the flame (reacting flow field) during different dynamical states in a turbulent thermoacoustic system.

10.3.5 Multifractal Behavior of Flame Surfaces

In Chap. 7, we discussed that the temporal dynamics of a thermoacoustic system is the result of spatiotemporal interactions happening over a range of scales between different subsystems such as the flow, the flame, and the acoustics of a turbulent combustor. Furthermore, as we mentioned before, the temporal dynamics of the acoustic pressure during the state of combustion noise is multifractal and a loss of

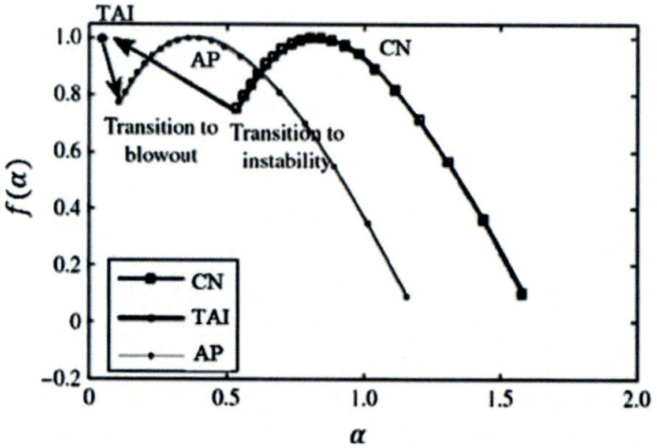

Fig. 10.18 The singularity spectrum of the acoustic pressure fluctuations obtained for the states of combustion noise (CN), thermoacoustic instability (TAI), and aperiodic oscillations observed prior to flame blowout (AP) in a turbulent bluff-body stabilized dump combustor. The structure and the position of the singularity spectrum vary during the transition from combustion noise to thermoacoustic instability and then to flame blowout (as indicated by arrows). Adapted with permission from Unni and Sujith [94]

multifractality happens during the occurrence of thermoacoustic instability. Then the natural questions that arise are what is the source of such multifractality in the temporal dynamics of the system? Is the flame multifractal? How do we reconcile the loss of multifractality in the temporal dynamics of acoustic pressure to a highly turbulent flame? Do we see the loss of multifractality in the spatial structures of the flame during the onset of thermoacoustic instability, even though the flame is highly turbulent? We will try to address these questions in this section. We will discuss the application of the multifractal analysis to study the spatiotemporal interactions of the flow and the flame, and also to analyze the flame topology, for different dynamical states observed during the onset of thermoacoustic instability [75].

In Figs. 10.19, 10.20, 10.21, and 10.22, we show the experimental data of the simultaneously acquired acoustic pressure fluctuations, global heat release rate fluctuations, and 2-D planer Mie scattering images of the reacting flow from a turbulent bluff-body-stabilized dump combustor. As discussed in Chap. 7, in a dump (backward-facing step) combustor (Fig. 7.3), the formation of a shear layer takes place at the step. This, in turn, helps in the stabilization of the flame in the system. The flame surface is influenced by vortical structures formed along the shear layers at the step. These vortical structures wrinkle (roll-up) the flame topology, where the size of a wrinkle depends on the size of the vortex. Furthermore, such vortical structures enhance the mixing of hot products with cold reactants; thus maintain continuous combustion in the system.

From the Mie scattering images of the reaction field, we can extract the flame contours (structures) using a method of edge detection from an image processing technique [66, 83]. In this method, the Mie scattering image is converted into a binary image using a suitable choice of the threshold. These binary flame images

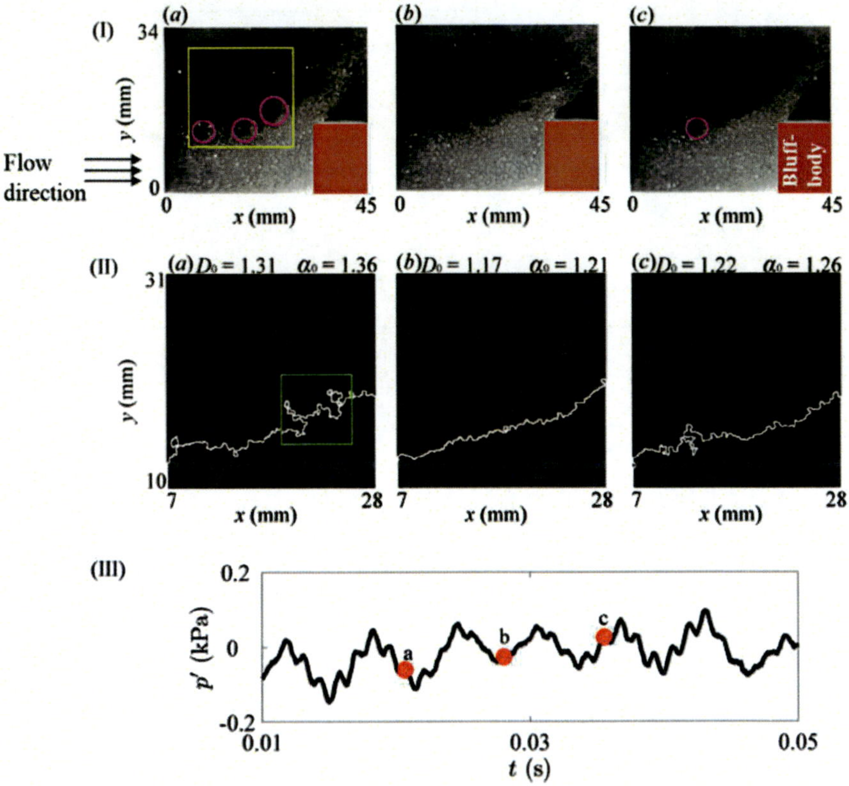

Fig. 10.19 (I) Mie scattering images of the reacting flow field and (II) the flame contour images obtained from the edge detection method for a specific region indicated by the yellow box in (Ia). These images correspond to time instances a, b, and c of the acoustic pressure signal (III) obtained during the state of combustion noise. The red box in (I) indicates the location of the bluff-body in the combustor. The presence of small-scale vortical structures is marked by purple circles in (I). The flow direction is from the left to right. During this state, we observe that the flame topology fills the smaller space and wrinkles are heterogeneously distributed (see green box in IIa). Reproduced with permission from Raghunathan et al. [75]

exhibit the flame contours that have several wrinkles of different sizes, distributed heterogeneously in the reaction zone of the combustor. By using the box-counting method (as detailed in Sect. 10.2.1.3), we can characterize the multifractal properties of such wrinkles on the flame topology. Further, we can compute several measures of multifractality, such as the capacity dimension (D_0), the dominant singularity exponent (α_0), and the singularity spectrum, to quantify features of the turbulent flame surfaces during different dynamical states of the combustor.

In Fig. 10.19I, we show the Mie scattering images of the flame obtained at different time instances of the chaotic oscillations in the acoustic pressure signal (Fig. 10.19III) during the state of combustion noise. We notice the presence of small-

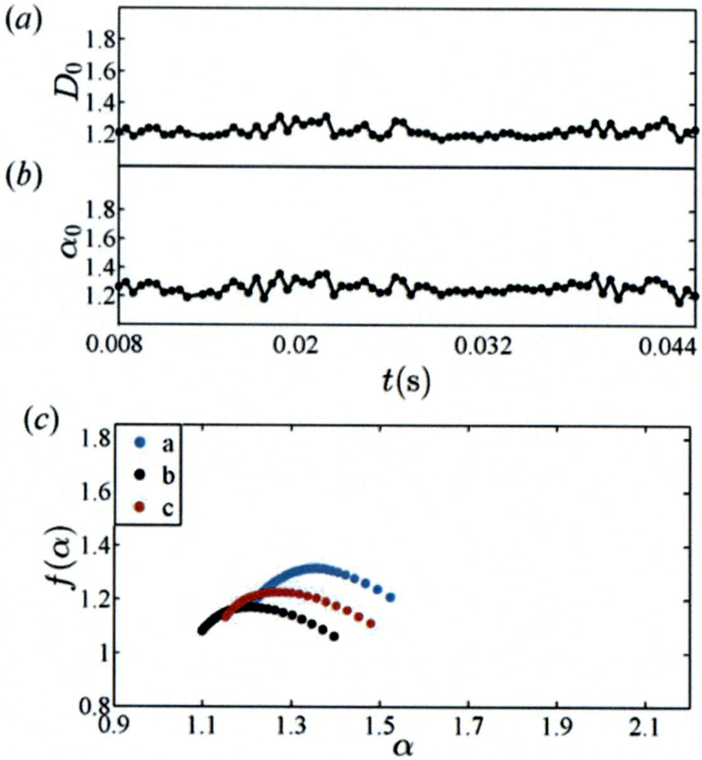

Fig. 10.20 The temporal variation of (**a**) the capacity dimension (D_0) and (**b**) the dominant singularity exponent (α_0) observed during the state of combustion noise. (**c**) The singularity spectrums corresponding to flame contours shown in Fig. 10.19IIa–c. The singularity spectrum is wider, right-skewed, and exhibits small aperiodic oscillations. Adapted with permission from Raghunathan et al. [75]

scale vortical structures (marked by in purple circles) along the shear layer, which causes a small-scale roll-up of the flame surface. The flame contour obtained after image processing shows a heterogeneous distribution of wrinkles along the flame (see Fig. 10.19II). These wrinkles are the results of the underlying perturbations induced by the flow.

Figures 10.20 show the multifractal analysis of the flame contours (shown in the yellow square of Fig. 10.19Ia) observed during the state of combustion noise. The multifractal analysis is performed using a generalized box-counting method discussed in Sect. 10.2.1.3 [75]. The temporal variation of D_0, α_0, and the singularity spectrum of flame contours in Fig. 10.20a–c, respectively. The values of D_0 and α_0 are indicated on the top of each image in Fig. 10.19II. We observe that the value of D_0 fluctuates around a mean of 1.22 (Fig. 10.20a). As the value of D_0 lies near 1, we observe that the flame surface does not fill the entire surface of interest shown by a yellow square in Fig. 10.19Ia. Furthermore, a higher value of D_0 in Fig. 10.19IIa as compared to that observed in Fig. 10.19IIb, c suggests the presence of increased irregularities/wrinkling on the flame contour (indicated by a green box in Fig. 10.19IIa.

Further, we notice that the value of α_0 fluctuates around a mean of 1.26, having a maximum value of 1.36 and a minimum value of 1.16 (Fig. 10.20b). A higher value of α_0 indicates the existence of higher flame wrinkles in specific locations of the flame contour, as compared to that observed for lower values of α_0. This behavior of the flame structure is reflected in Fig. 10.19IIa where we observe that the wrinkles are not homogeneously distributed on the flame topology as compared to the near homogeneous distribution of wrinkles observed on the flame topology in Fig. 10.19IIb, c.

The singularity spectrum of all flame contours lies at different locations of α_0 and exhibits a right-skewed structure (Fig. 10.20c). Here, right-skewed signifies the dominance of small-scale wrinkles on the flame topology. To summarize, the flame topology observed during the occurrence of combustion noise is less space-filling and consists of heterogeneous distribution of small-scale wrinkles along the flame surface.

On the other hand, during the occurrence of thermoacoustic instability (Fig. 10.21III), we observe the periodic emergence of large-scale structures in the reaction field of the combustor, as seen from the Mie scattering images in Fig. 10.21I. The flame contours obtained from the corresponding processed Mie scattering images are shown in Fig. 10.21II. We notice that the large-scale vortical structures induce a large roll-up on the flame contour in one acoustic cycle. We also note the existence of small-scale wrinkles within the large-scale structures of the flow and flame topologies. These small-scale vortices cause increased irregularities in the flame topology. Thus, the flame structure consists of a wider range of spatial scales during the state of thermoacoustic instability as compared to that observed during the occurrence of combustion noise (Fig. 10.19).

In Fig. 10.22a, we notice the oscillatory behavior in the temporal variation of D_0 and α_0. Such oscillations are the result of the periodic shedding of large-scale structures in the system. We observe that the value of D_0 varies between 1.81 and 1.28 with a mean value of 1.59. Higher values of D_0 indicate the space-filling nature of the flame contour, i.e., the flame occupies a larger area (see Fig. 10.21IIb–d) as compared to the less space-filling flame contour observed at lower values of D_0 (see Fig. 10.21IIa, e, f). The space-filling is a result of the existence of both the large-scale and small-scale structures in the flame.

We observe a similar variation in α_0 as that observed for D_0. The value of α_0 fluctuates between 1.87 and 1.31 with a mean value of 1.64. A higher value of α_0 indicates that the flame contour is highly wrinkled and it fills the maximum portion of the region of interest in the combustor. Moreover, the maximum values of both D_0 and α_0 are observed when the large-scale structures reach their maximum size, before impinging on the bluff-body and break down into small-scale structures.

Figure 10.22c shows the singularity spectrums of all flame contours presented in Fig. 10.21II. We observe periodic oscillations in the movement of the spectrum in an oscillation cycle. The spectrum moves towards the upper right corner during

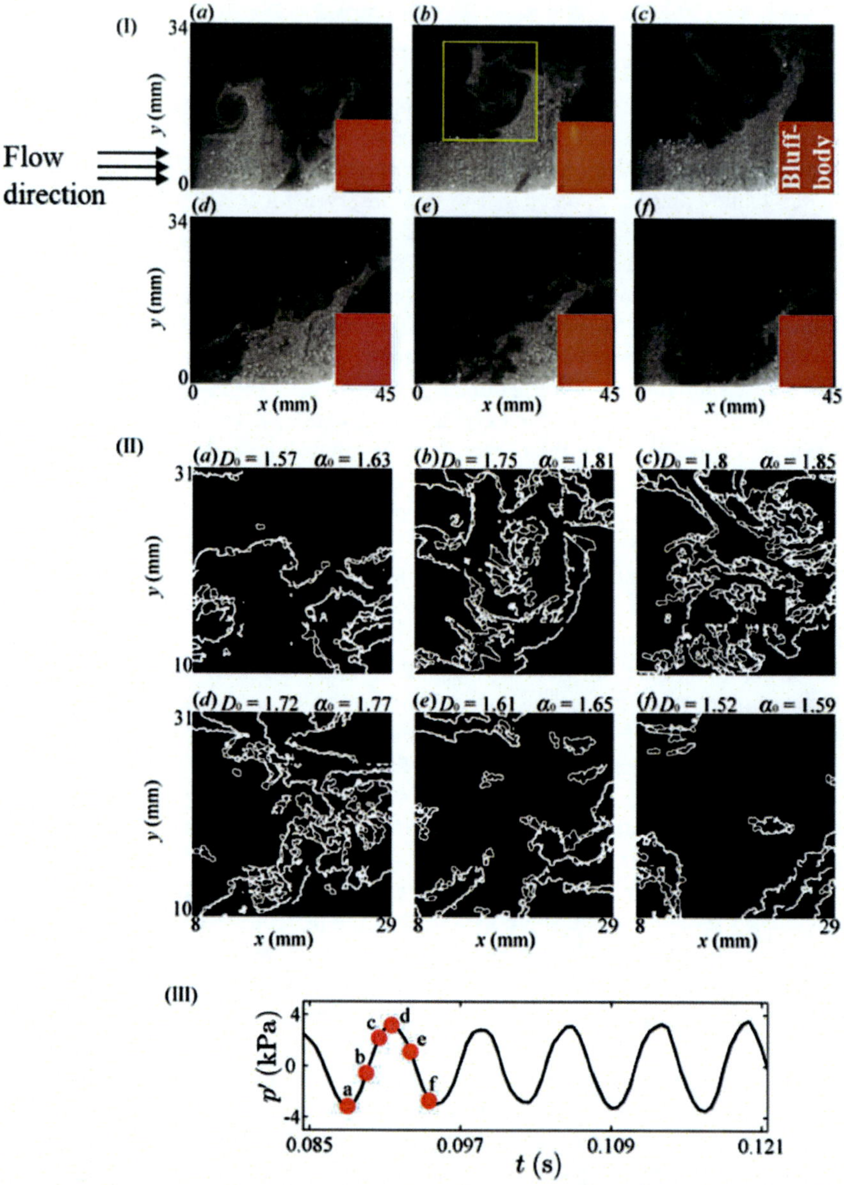

Fig. 10.21 (I) Mie scattering images of the reacting flow field and (II) the flame contour images obtained from the edge detection method for a specific region indicated by the yellow box in (Ib). These images correspond to time instances a to f of the acoustic pressure signal (III) obtained for the state of thermoacoustic instability. We observe a wide range of small and large-scale wrinkles in the flame topology. Reproduced with permission from Raghunathan et al. [75]

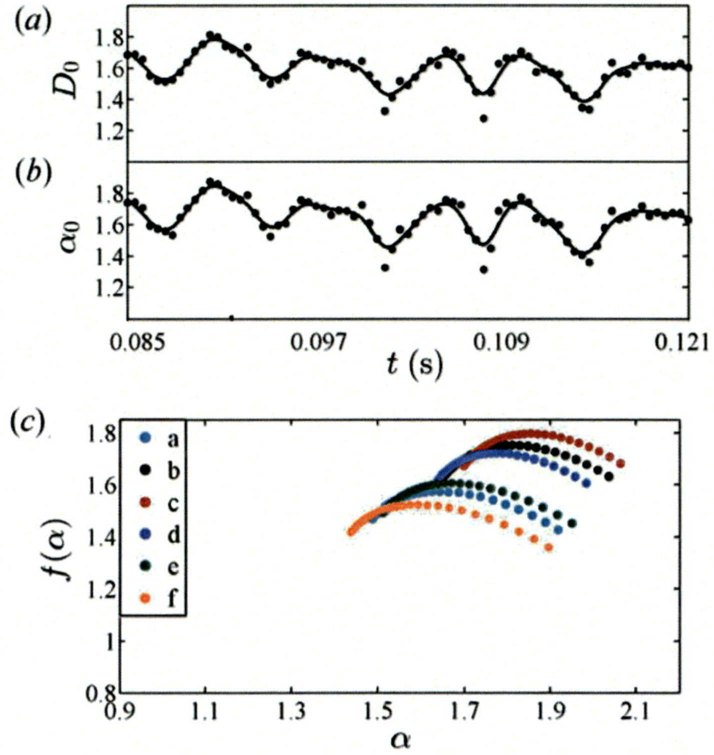

Fig. 10.22 The temporal variation of (**a**) the capacity dimension (D_0) and (**b**) the dominant singularity exponent (α_0) observed during the state of thermoacoustic instability. (**c**) The singularity spectrums corresponding to flame contours shown in Fig. 10.21IIa–f. During this state, we observe that the instantaneous flame contours of a turbulent flame are multifractal and their singularity spectrum exhibits periodic oscillations. Adapted with permission from Raghunathan et al. [75]

the emergence of the large-scale structures in the flow field (or during the growth of the amplitude in the acoustic pressure signal) and returns in the same path after the destruction of the large-scale structures when impinged on the bluff-body (i.e., during the decay of the amplitude in the acoustic pressure signal). The singularity spectrum is broad and has a width of 0.35. This width is larger than the spectrum width of 0.28 observed during the state of combustion noise. Furthermore, the spectrum is right-skewed, indicating the dominance of small-scale structures in the flame topology as compared to the large-scale structures.

From this analysis, we infer that although the temporal dynamics of the acoustic pressure fluctuations shows a loss of multifractality during the onset of thermoacoustic instability (refer to Fig. 10.15), the spatial structures of the reaction field in the combustor are still complex and preserves the multifractal characteristics during all the dynamical states.

To summarize, in this chapter, we discussed the application of the frameworks of fractals and multifractals to characterize the aperiodic pressure fluctuations observed during the state of combustion noise, and also to study the onset of thermoacoustic instability in turbulent thermoacoustic systems. We showed that such aperiodic oscillations observed during the occurrence of combustion noise are complex and multifractal due to the presence of a range of scaling exponents. We also observed the loss of multifractality during the onset of thermoacoustic instability. Further, we introduced a measure of fractality of a signal called the Hurst exponent. This measure can be used for detecting the dynamical transitions of the system variables from one state to another, as the control parameter is changed. We can use this measure as a precursor for the occurrence of the undesired states of thermoacoustic instability and flame blowout in a turbulent thermoacoustic system, which we will elaborate in Chap. 12. From the multifractal analysis of spatial structures of the reaction field during different states of flow-flame interaction, we found that the turbulent flame is multifractal and it retains its multifractal features even during the state of thermoacoustic instability. The singularity spectrum corresponding to the state of thermoacoustic instability displays periodic oscillations.

In the next chapter, we will study the temporal and spatiotemporal dynamics of a turbulent combustor during the onset of thermoacoustic instability using a complex network approach. We will learn the construction of complex networks from the temporal and spatiotemporal data of turbulent thermoacoustic systems using different methodologies, and the interpretation of results obtained through different measures of the complex network analysis.

References

1. Alessio, E., Carbone, A., Castelli, G., Frappietro, V.: Second-order moving average and scaling of stochastic time series. Eur. Phys. J. B **27**(2), 197–200 (2002)
2. Banerjee, S., Hassan, M.K., Mukherjee, S., Gowrisankar, A.: Fractal Patterns in Nonlinear Dynamics and Applications: Patterns in Nonlinear Dynamics and Applications. CRC Press, Boca Raton (2020)
3. Bassingthwaighte, J.B., Liebovitch, L.S., West, B.J.: Fractal Physiology. Springer, Berlin (2013)
4. Bolotin, Y., Tur, A., Yanovsky, V.: Chaos: Concepts, Control and Constructive Use. Springer, Berlin (2009)
5. Bunde, A., Havlin, S.: Fractals in Science. Springer, Berlin (2013)
6. Caccia, D.C., Percival, D., Cannon, M.J., Raymond, G., Bassingthwaighte, J.B.: Analyzing exact fractal time series: evaluating dispersional analysis and rescaled range methods. Physica A **246**(3–4), 609–632 (1997)
7. Cannon, M.J., Percival, D.B., Caccia, D.C., Raymond, G.M., Bassingthwaighte, J.B.: Evaluating scaled windowed variance methods for estimating the Hurst coefficient of time series. Physica A **241**(3–4), 606–626 (1997)
8. Chhabra, A., Jensen, R.V.: Direct determination of the $f(\alpha)$ singularity spectrum. Phys. Rev. Lett. **62**(12), 1327–1330 (1989)
9. Chu, P.C.: Multi-fractal thermal characteristics of the southwestern GIN sea upper layer. Chaos Solitons Fractals **19**(2), 275–284 (2004)

10. Clavin, P., Kim, J.S., Williams, F.A.: Turbulence-induced noise effects on high-frequency combustion instabilities. Combust. Sci. Tech. **96**(1–3), 61–84 (1994)
11. Coronado, A.V., Carpena, P.: Size effects on correlation measures. J. Biol. Phys. **31**(1), 121–133 (2005)
12. De Bartolo, S.G., Gabriele, S., Gaudio, R.: Multifractal behaviour of river networks. Hydrol. Earth Syst. Sci. Discuss. **4**(1), 105–112 (2000)
13. Delignieres, D., Ramdani, S., Lemoine, L., Torre, K., Fortes, M., Ninot, G.: Fractal analyses for 'short' time series: a re-assessment of classical methods. J. Math. Psychol. **50**(6), 525–544 (2006)
14. Delignieres, D., Torre, K.: Fractal dynamics of human gait: a reassessment of the 1996 data of Hausdorff et al. J. Appl. Physiol. **106**(4), 1272–1279 (2009)
15. Delignières, D., Torre, K., Bernard, P.L.: Transition from persistent to anti-persistent correlations in postural sway indicates velocity-based control. PLoS Comput. Biol. **7**(2), e1001089 (2011)
16. Diosdado, A.M.: Analysis of the relation between complexity and multifractality in cardiac interbeat intervals time series. In: World Congress on Medical Physics and Biomedical Engineering, 7–12 Sept 2009, Munich, pp. 1506–1509. Springer, Berlin (2009)
17. Eke, A., Herman, P., Bassingthwaighte, J., Raymond, G., Percival, D., Cannon, M., Balla, I., Ikrényi, C.: Physiological time series: distinguishing fractal noises from motions. Eur. J. Physiol. **439**(4), 403–415 (2000)
18. Eke, A., Herman, P., Kocsis, L., Kozak, L.R.: Fractal characterization of complexity in temporal physiological signals. Physiol. Meas. **23**(1), R1 (2002)
19. Falconer, K.: Fractal Geometry: Mathematical Foundations and Applications. Wiley, New York (1990)
20. Feder, J.: Fractals. Springer Science & Business Media, Berlin (1988)
21. França, L.G.S., Montoya, P., Miranda, J.G.V.: On multifractals: a non-linear study of actigraphy data. Physica A **514**, 612–619 (2019)
22. Frisch, U., Kolmogorov, A.: Turbulence: the legacy of A N Kolmogorov. Cambridge University Press, Cambridge (1995)
23. Gotoda, H., Amano, M., Miyano, T., Ikawa, T., Maki, K., Tachibana, S.: Characterization of complexities in combustion instability in a lean premixed gas-turbine model combustor. Chaos **22**(4), 043128 (2012)
24. Grassberger, P.: Generalizations of the Hausdorff dimension of fractal measures. Phys. Lett. A **107**(3), 101–105 (1985)
25. Grassberger, P.: Information and complexity measures in dynamical systems. In: Information Dynamics, pp. 15–33. Springer, Berlin (1991)
26. Grassberger, P., Procaccia, I.: Characterization of strange attractors. Phys. Rev. Lett. **50**(5), 346–349 (1983)
27. Gu, G.F., Zhou, W.X.: Detrending moving average algorithm for multifractals. Phys. Rev. E **82**(1), 011136 (2010)
28. Guan, Y., Li, L.K.B., Ahn, B., Kim, K.T.: Chaos, synchronization, and desynchronization in a liquid-fueled diffusion-flame combustor with an intrinsic hydrodynamic mode. Chaos **29**(5), 053124 (2019)
29. Halsey, T.C., Jensen, M.H., Kadanoff, L.P., Procaccia, I., Shraiman, B.I.: Fractal measures and their singularities: the characterization of strange sets. Phys. Rev. A **33**(2), 1141–1151 (1986)
30. Harikrishnan, K.P., Misra, R., Ambika, G., Amritkar, R.E.: Computing the multifractal spectrum from time series: an algorithmic approach. Chaos **19**(4), 043129 (2009)
31. Hentschel, H.G.E., Procaccia, I.: The infinite number of generalized dimensions of fractals and strange attractors. Physica D **8**(3), 435–444 (1983)
32. Hurst, H.E.: Long-term storage capacity of reservoirs. Trans. Am. Soc. Civil Eng. **116**, 770–799 (1951)
33. Ihlen, E.A.F., Vereijken, B.: Interaction-dominant dynamics in human cognition: beyond $1/^{\alpha}$ fluctuation. J. Exp. Psychol. Gen. **139**(3), 436 (2010)

34. Ihlen, E.A.F.E.: Introduction to multifractal detrended fluctuation analysis in Matlab. Front. Physiol. **3**, 141 (2012)
35. Ivanov, P.C., Amaral, L.A.N., Goldberger, A.L., Havlin, S., Rosenblum, M.G., Struzik, Z.R., Stanley, H.E.: Multifractality in human heartbeat dynamics. Nature **399**(6735), 461–465 (1999)
36. Jestczemski, F., Sernetz, M.: Multifractal approach to inhomogeneous fractals. Physica A **223**(3-4), 275–282 (1996)
37. Kantelhardt, J.W.: Fractal and multifractal time series. In: Encyclopedia of Complexity and Systems Science, pp. 3754–3778. Springer, Berlin (2009)
38. Kantelhardt, J.W., Roman, H.E., Greiner, M.: Discrete wavelet approach to multifractality. Physica A **220**(3-4), 219–238 (1995)
39. Kantelhardt, J.W., Zschiegner, S.A., Koscielny-Bunde, E., Havlin, S., Bunde, A., Stanley, H.E.: Multifractal detrended fluctuation analysis of nonstationary time series. Physica A **316**(1-4), 87–114 (2002)
40. Kasthuri, P., Pavithran, I., Pawar, S.A., Sujith, R.I., Gejji, R., Anderson, W.: Dynamical systems approach to study thermoacoustic transitions in a liquid rocket combustor. Chaos **29**(10), 103115 (2019)
41. Kerres, B., Nair, V., Cronhjort, A., Mihaescu, M.: Analysis of the turbocharger compressor surge margin using a Hurst-exponent-based criterion. SAE Int. J. Engines **9**(3), 1795–1806 (2016)
42. Koch, H.: Sur une courbe continue sans tangente, obtenue par une construction géométrique élémentaire. Ark. Mat. Astr. Fys. **1**, 681–704 (1904)
43. Kolmogorov, A.N.: A refinement of previous hypotheses concerning the local structure of turbulence in a viscous incompressible fluid at high Reynolds number. J. Fluid Mech. **13**(1), 82–85 (1962)
44. Labat, D., Mangin, A., Ababou, R.: Rainfall–runoff relations for karstic springs: multifractal analyses. J. Hydrol. **256**(3–4), 176–195 (2002)
45. Lieuwen, T., Banaszuk, A.: Background noise effects on combustor stability. J. Propuls. Power **21**(1), 25–31 (2005)
46. Lieuwen, T.C.: Experimental investigation of limit-cycle oscillations in an unstable gas turbine combustor. J. Propuls. Power **18**(1), 61–67 (2002)
47. Mali, P.: Multifractal characterization of global temperature anomalies. Theor. Appl. Climatol. **121**(3-4), 641–648 (2015)
48. Mandelbrot, B.: How long is the coast of Britain? Statistical self-similarity and fractional dimension. Science **156**(3775), 636–638 (1967)
49. Mandelbrot, B.B.: Possible refinement of the lognormal hypothesis concerning the distribution of energy dissipation in intermittent turbulence. In: Statistical Models and Turbulence, pp. 333–351. Springer, Berlin (1972)
50. Mandelbrot, B.B.: Intermittent turbulence in self-similar cascades: divergence of high moments and dimension of the carrier. J. Fluid Mech. **62**(2), 331–358 (1974)
51. Mandelbrot, B.B.: Fractals - Form, Chance and Dimension. Freeman, San Francisco (1977)
52. Mandelbrot, B.B.: The Fractal Geometry of Nature. W H Freeman, New York (1982)
53. Mandelbrot, B.B.: Self-affine fractals and fractal dimension. Phys. Scr. **32**(4), 257 (1985)
54. Mandelbrot, B.B.: Multifractal measures, especially for the geophysicist. In: Fractals in Geophysics, pp. 5–42. Springer, Berlin (1989)
55. Mandelbrot, B.B., Van Ness, J.W.: Fractional Brownian motions, fractional noises and applications. SIAM Rev. **10**(4), 422–437 (1968)
56. Masugi, M., Takuma, T.: Multi-fractal analysis of IP-network traffic for assessing time variations in scaling properties. Physica D **225**(2), 119–126 (2007)
57. Matia, K., Ashkenazy, Y., Stanley, H.E.: Multifractal properties of price fluctuations of stocks and commodities. Europhys. Lett. **61**(3), 422 (2003)
58. Meakin, P.: Fractals, Scaling and Growth Far from Equilibrium, vol. 5. Cambridge University Press, Cambridge (1998)
59. Meneveau, C., Sreenivasan, K.R.: Simple multifractal cascade model for fully developed turbulence. Phys. Rev. Lett. **59**(13), 1424–1427 (1987)

60. Miranda, J.G.V., Montero, E., Alves, M.C., González, A.P., Vázquez, E.V.: Multifractal characterization of saprolite particle-size distributions after topsoil removal. Geoderma **134**(3–4), 373–385 (2006)
61. Movahed, M.S., Hermanis, E.: Fractal analysis of river flow fluctuations. Physica A **387**(4), 915–932 (2008)
62. Murcio, R., Masucci, A.P., Arcaute, E., Batty, M.: Multifractal to monofractal evolution of the London street network. Phys. Rev. E **92**(6), 062130 (2015)
63. Nair, V., Sujith, R.I.: Multifractality in combustion noise: predicting an impending combustion instability. J. Fluid Mech. **747**, 635–655 (2014)
64. Nayfeh, A.H., Balachandran, B.: Applied Nonlinear Dynamics: Analytical, Computational, and Experimental Methods. Wiley, New York (2008)
65. Noiray, N., Schuermans, B.: Deterministic quantities characterizing noise driven Hopf bifurcations in gas turbine combustors. Int. J. Nonlin. Mech. **50**, 152–163 (2013)
66. Otsu, N.: A threshold selection method from gray-level histograms. IEEE Trans. Syst. Man Cybern. **9**(1), 62–66 (1979)
67. Parisi, G., Frisch, U.: A multifractal model of intermittency. In: Turbulence and Predictability in Geophysical Fluid Dynamics and Climate Dynamics, pp. 84–88. North-Holland, Amsterdam (1985)
68. Pavithran, I., Unni, V.R., Varghese, A.J., Sujith, R.I., Saha, A., Marwan, N., Kurths, J.: Universality in the emergence of oscillatory instabilities in turbulent flows. Europhys. Lett. **129**(2), 24004 (2020)
69. Pawar, S.A., Raghunathan, M., Reeja, K.V., Midhun, P.R., Sujith, R.I.: Effect of preheating of the reactants on the transition to thermoacoustic instability in a bluff-body stabilized dump combustor. Proc. Combust. Inst. **38**(4), 6193–6201 (2021)
70. Peng, C.K., Buldyrev, S.V., Havlin, S., Simons, M., Stanley, H.E., Goldberger, A.L.: Mosaic organization of DNA nucleotides. Phys. Rev. E **49**(2), 1685–1689 (1994)
71. Pesin, Y.B., Pesin, Y.B., Climenhaga, V.: Lectures on Fractal Geometry and Dynamical Systems, vol. 52. American Mathematical Society, Providence (2009)
72. Peters, E.E.: Chaos and Order in the Capital Markets: A New View of Cycles, Prices, and Market Volatility. Wiley, New York (1996)
73. Pilgram, B., Kaplan, D.T.: A comparison of estimators for $1/f$ noise. Physica D **114**(1–2), 108–122 (1998)
74. Puthenveettil, B.A., Arakeri, J.H.: Plume structure in high-Rayleigh-number convection. J. Fluid Mech. **542**, 217–249 (2005)
75. Raghunathan, M., George, N.B., Unni, V.R., Midhun, P.R., Reeja, K.V., Sujith, R.I.: Multifractal analysis of flame dynamics during transition to thermoacoustic instability in a turbulent combustor. J. Fluid Mech. **888**, A14 (2020)
76. Richardson, L.F.: Weather Prediction by Numerical Process. Cambridge University Press, Cambridge (1922)
77. Roy, A., Premchand, C.P., Raghunathan, M., Krishnan, A., Nair, V., Sujith, R.I.: Critical region in the spatiotemporal dynamics of a turbulent thermoacoustic system and smart passive control. Combust. Flame **226**, 274–284 (2021)
78. Russell, D.A., Hanson, J.D., Ott, E.: Dimension of strange attractors. Phys. Rev. Lett. **45**(14), 1175–1178 (1980)
79. Saurabh, A., Imran, H., Nawroth, H., Paschereit, C.O., Kabiraj, L.: Fractal characteristics of combustion noise. J. Eng. Gas Turbine Power **140**(12) (2018)
80. Schaefer, A., Brach, J.S., Perera, S., Sejdić, E.: A comparative analysis of spectral exponent estimation techniques for $1/f^{\beta}$ processes with applications to the analysis of stride interval time series. J. Neurosci. Methods **222**, 118–130 (2014)
81. Sen, A.K.: Multifractality as a measure of complexity in solar flare activity. Sol. Phys. **241**(1), 67–76 (2007)
82. Seuront, L.: Fractals and Multifractals in Ecology and Aquatic Science. CRC Press, Boca Raton (2009)

83. Shepherd, I.G., Cheng, R.K., Talbot, L.: Experimental criteria for the determination of fractal parameters of premixed turbulent flames. Exp. Fluids **13**(6), 386–392 (1992)
84. Shimizu, Y.U., Thurner, S., Ehrenberger, K.: Multifractal spectra as a measure of complexity in human posture. Fractals **10**(01), 103–116 (2002)
85. Sreenivasan, K.R., Antonia, R.A.: The phenomenology of small-scale turbulence. Annu. Rev. Fluid Mech. **29**(1), 435–472 (1997)
86. Suckling, J., Wink, A.M., Bernard, F.A., Barnes, A., Bullmore, E.: Endogenous multifractal brain dynamics are modulated by age, cholinergic blockade and cognitive performance. J. Neurosci. Methods **174**(2), 292–300 (2008)
87. Tao, T.: An introduction to measure theory. In: Graduate Studies in Mathematics, vol. 126. American Mathematical Society, Providence (2011)
88. Tarnopolski, M.: On the relationship between the Hurst exponent, the ratio of the mean square successive difference to the variance, and the number of turning points. Physica A **461**, 662–673 (2016)
89. Tél, T., Fülöp, Á., Vicsek, T.: Determination of fractal dimensions for geometrical multifractals. Physica A **159**(2), 155–166 (1989)
90. Telesca, L., Lovallo, M.: Analysis of the time dynamics in wind records by means of multifractal detrended fluctuation analysis and the Fisher–Shannon information plane. J. Stat. Mech: Theory Exp. **2011**(07), P07001 (2011)
91. Theiler, J.: Estimating fractal dimension. J. Opt. Soc. Am. A **7**(6), 1055–1073 (1990)
92. Theiler, J., Eubank, S., Longtin, A., Galdrikian, B., Farmer, J.D.: Testing for nonlinearity in time series: the method of surrogate data. Physica D **58**(1–4), 77–94 (1992)
93. Thompson, J.R., Wilson, J.R.: Multifractal detrended fluctuation analysis: Practical applications to financial time series. Math. Comput. Simul. **126**, 63–88 (2016)
94. Unni, V.R., Sujith, R.I.: Multifractal characteristics of combustor dynamics close to lean blowout. J. Fluid Mech. **784**, 30–50 (2015)
95. Varghese, A.J., Chechkin, A., Metzler, R., Sujith, R.I.: Capturing multifractality of pressure fluctuations in thermoacoustic systems using fractional-order derivatives. Chaos **31**(3), 033108 (2021)
96. Vázquez, E.V., Ferreiro, J.P., Miranda, J.G.V., González, A.P.: Multifractal analysis of pore size distributions as affected by simulated rainfall. Vadose Zone J. **7**(2), 500–511 (2008)
97. West, B.J., Goldberger, A.L.: Physiology in fractal dimensions. Am. Sci. **75**(4), 354–365 (1987)
98. West, B.J., Latka, M., Glaubic-Latka, M., Latka, D.: Multifractality of cerebral blood flow. Physica A **318**(3–4), 453–460 (2003)
99. Xu, H.C., Gu, G.F., Zhou, W.X.: Direct determination approach for the multifractal detrending moving average analysis. Phys. Rev. E **96**(5), 052201 (2017)
100. Zeleke, T.B., Si, B.C.: Characterizing scale-dependent spatial relationships between soil properties using multifractal techniques. Geoderma **134**(3–4), 440–452 (2006)
101. Zorick, T., Mandelkern, M.A.: Multifractal detrended fluctuation analysis of human EEG: preliminary investigation and comparison with the wavelet transform modulus maxima technique. PLoS One **8**(7), e68360 (2013)

Chapter 11
Complex Network Approach to Thermoacoustic Systems

In the previous chapters, we discussed the phenomenon of the onset of thermoacoustic instability in turbulent combustors. In such systems, the transition of the state of combustion noise to thermoacoustic instability happens via intermittency. The state of combustion noise consists of low amplitude chaotic fluctuations and shows features of multifractality. A collective interaction between small-scale structures leads to the formation of large-scale coherent structures in the combustor. Furthermore, the coupling between the acoustic pressure and the heat release rate fluctuations exhibits different forms of synchronization. The existence of all these rich dynamical behaviors during the onset of thermoacoustic instability indicates the complex nature of turbulent thermoacoustic systems. Therefore, we can consider such systems as 'complex systems'. In this chapter, we will discuss the application of complex network theory to study the onset of thermoacoustic instability in turbulent combustion systems. We will cover the application of different methods of complex network analysis, such as visibility algorithm, cycle networks, and recurrence networks, to study temporal properties of the states of combustion noise, intermittency, and thermoacoustic instability. We will also discuss the application of correlation networks, acoustic power production and vorticity networks, and turbulence networks to identify critical regions existing in the combustor flow field.

As we discussed in Chap. 5, the global dynamical behavior of a complex system cannot be inferred from the individual behavior of its interacting subsystems. Therefore, in the case of thermoacoustic systems, the traditional reductionist approach of studying the individual components is no longer adequate to describe the emergent behavior of the system during the occurrence of thermoacoustic oscillations. Complex systems approach is a novel approach that has recently emerged in science to study how the interaction of constituents of a system leads to a collective global dynamical behavior [4]. This approach has widely been employed in several fields such as physics, biology, ecology, social science, and engineering [2, 5, 11, 12, 18, 21, 23, 31, 34, 36, 84, 94, 108]. An efficient way to implement this approach to real-world data or simulation data is to use tools based

R. I. Sujith, S. A. Pawar, *Thermoacoustic Instability*, Springer Series in Synergetics,
https://doi.org/10.1007/978-3-030-81135-8_11

on complex network theory [8]. The application of complex networks has found its importance in thermoacoustic literature as well. We come across many studies [16, 49–51, 53, 58–61, 77, 79, 80, 93, 113, 118] that have used complex network approach to investigate temporal, spatiotemporal, and coupled dynamics of different dynamic variables of a thermoacoustic system. In the following section, we will introduce the basics of complex networks and subsequently, we will discuss the application of complex networks to thermoacoustic systems.

11.1 An Introduction to Complex Networks

Nature has an abundance of networks. Examples of such networks include the neuron network in brain [46], vascular network [112], protein-protein interaction network [101, 105], internet topology network [24, 43], transportation (airline and road) networks [15], social (Facebook, Instagram, and Researchgate) networks [103, 124], World Wide Web (WWW) [3, 55], etc. Every network is made by the interaction amongst its components. The interaction of a large number of components in a network sometimes results in an emergence of collective behavior [122]. An example of collective behavior in nature is the emergence of synchronized motion in a flock of birds or a school of fish [107]. A system consisting of such emergent behavior arising due to the interaction of the parts is usually referred to as a complex system [63, 89, 91, 121].

A complex system possess the following key properties [63],

- Such a system consists of many interacting subparts.
- Each part consists of own internal structure and has a specific behavior.
- The interaction between subparts is nonlinear.

Therefore, the global property of a complex system is more than simply a sum of the properties of the individual parts. The complex network approach has been found to be an efficient way to study such complex systems. Albert Barabási, a pioneer in the field of the complex network, has mentioned in one of his articles, "*Therefore, if we are ever to have a theory of complexity, it will sit on the shoulders of network theory*" [8].

The concept of a complex network is primarily based on graph theory and statistical physics. A network consists of two main components, which are the nodes (or vertices) and the links (or edges). In a network, the node indicates the individual element of the network and the link represents an interaction between two nodes. For example, in a social network, nodes are people and links between them could be based on different factors such as friendship, family relationship, or scientific collaboration.

Networks can be classified as (i) directed and undirected, and (ii) weighted and unweighted [9, 42, 73]. In an undirected network (Fig. 11.1a, c), the nodes are bidirectionally connected. In directed networks (Fig. 11.1b, d), we observe a directionality in the connection of nodes and it is indicated by arrows. In the case of

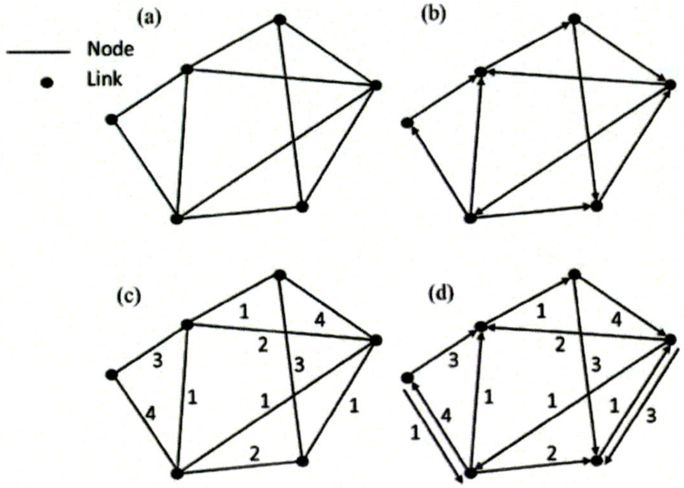

Fig. 11.1 Schematics of **(a)** the unweighted, undirected network, **(b)** the unweighted, directed network, **(c)** the weighted, undirected network, and **(d)** the weighted, directed network. Arrows in the links indicate the directionality of the connectivity between two nodes and numbers indicate the weights for the links

unweighted networks, all pairs of nodes in the network are connected with the same strength or weights (Fig. 11.1a, b). However, for weighted networks (Fig. 11.1c, d), the strength of the interaction between the nodes is also taken into account.

Typically, in the network analysis, multiple connections between a pair of nodes and connections of a node onto itself (self-loops) are not considered. A graph with such connections is called multigraph [124] and such studies are not a focus of our discussion in this chapter.

Mathematically, the interaction (link) between the nodes in a network is represented in the form of an adjacency matrix [84]. For unweighted and undirected networks, the adjacency matrix consists of 0s and 1s (i.e., a binary matrix). Such a matrix simply indicates whether a connection between a pair of nodes exists or not. Thus, we can obtain the adjacency matrix of unweighted network as follows,

$$A_{ij} = \begin{cases} 1 & \text{if } i \neq j \text{ or if } i \text{ and } j \text{ are connected} \\ 0 & \text{if } i = j \text{ or if } i \text{ and } j \text{ are not connected} \end{cases} \tag{11.1}$$

Here, $i, j = 1, 2, \ldots.., N$ are the nodes of a network. The components of A_{ij} indicate the connection from a node j to node i. When two nodes are connected in a network, we mark the element corresponding to this pair in A_{ij} as 1 and, if the nodes are not connected in a network, we mark the element corresponding to this pair in A_{ij} as 0 (Fig. 11.2). In order to eliminate self-connections in the network, we need to assign a value of 0 to the diagonal elements of A_{ij}.

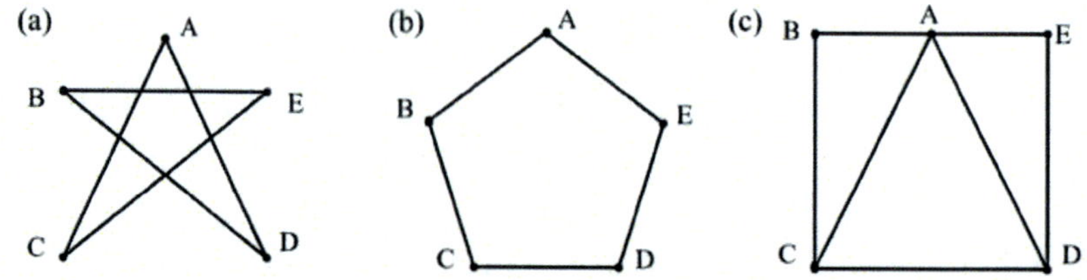

$$
\begin{array}{c} \\ A \\ B \\ C \\ D \\ E \end{array}
\begin{array}{c} \begin{array}{ccccc} A & B & C & D & E \end{array} \\ \begin{pmatrix} 0 & 0 & 1 & 1 & 0 \\ 0 & 0 & 0 & 1 & 1 \\ 1 & 0 & 0 & 0 & 1 \\ 1 & 1 & 0 & 0 & 0 \\ 0 & 1 & 1 & 0 & 0 \end{pmatrix} \end{array}
\quad
\begin{array}{c} \begin{array}{ccccc} A & B & C & D & E \end{array} \\ \begin{pmatrix} 0 & 1 & 0 & 0 & 1 \\ 1 & 0 & 1 & 0 & 0 \\ 0 & 1 & 0 & 1 & 0 \\ 0 & 0 & 1 & 0 & 1 \\ 1 & 0 & 0 & 1 & 0 \end{pmatrix} \end{array}
\quad
\begin{array}{c} \begin{array}{ccccc} A & B & C & D & E \end{array} \\ \begin{pmatrix} 0 & 1 & 1 & 1 & 1 \\ 1 & 0 & 1 & 0 & 0 \\ 1 & 1 & 0 & 1 & 0 \\ 1 & 0 & 1 & 0 & 1 \\ 1 & 0 & 0 & 1 & 0 \end{pmatrix} \end{array}
$$

Fig. 11.2 **(a)–(c)** Examples of the adjacency matrices $A_{i,j}$ constructed from different network topologies of nodes A, B, C, D, and E

For weighted networks, we need to assign the connection between different pair of nodes with given weights (strengths), where weights could be any real number. The adjacency matrix of the undirected networks is symmetric, while that of the directed network need not be symmetric.

11.2 Measures of Complex Networks

In the complex network analysis, the networks are constructed using different methods, while the topologies of these networks are quantified using various statistical measures [9, 42, 73]. These measures are directly or indirectly derived from the adjacency matrix of the network. In this section, we summarize a few measures that are commonly used in the network analysis.

Since nodes form a main component in the network, we can quantify the importance of a node in a network using centrality measures. In a network, the central nodes have the highest importance and play a significant role in the spread of information. Intuitively, the center could have several meanings. For example, a center of a star network can either be considered as the node with the largest degree (as explained next) or the node through which most of the shortest paths pass or the node that is closest to other nodes. Therefore, depending on the importance of a node in the network, we can develop different centrality measures. In the following subsections, we will discuss some of the key centrality measures for an unweighted undirected network, such as degree, betweenness centrality, and closeness centrality [9, 42, 73]. These measures are commonly used in the analysis of thermoacoustic oscillations.

(a) **Degree**

The number of connections a node (say, i) has to any given node in the network is called a degree of that node. We can calculate the degree as,

$$k_i = \sum_{j=1}^{N} A_{ij} \tag{11.2}$$

where N is the total number of nodes in a network. For an isolated node, k_i takes a value of zero. The maximum possible degree any node could have is $N-1$. A network with degree $N-1$ for all nodes is called a complete graph (or a clique), as all nodes are connected to each other. A node with the highest degree exerts a stronger influence on the functioning of the network, as this node interacts with a large number of nodes in a network.

We can detect a scaling behavior (or power-law) of a network from its degree distribution. Degree distribution $P(k)$ of a network is calculated as,

$$P(k) = n(k)/N \tag{11.3}$$

where $P(k)$ is the probability that a node has k number of connections and $n(k)$ is the number of nodes having degree k.

The degree distribution, i.e., a plot showing the variation of $P(k)$ with k, is different for different types of networks. For example, regular networks show delta distribution, random networks show Gaussian distribution, small world networks show skewed distribution, and scale-free networks show power-law distribution, as shown in Fig. 11.3a–d, respectively. We will discuss the properties of these networks in the next section.

In Fig. 11.4, we present an example of an unweighted undirected network and calculation of degree of its nodes. In this network, the total number of nodes is 5. As node A is connected to nodes B and C, its degree is 2. Similarly, we can evaluate the degree of nodes B, C, D, and E in the network as 4, 2, 1, and 1, respectively. Further, in this network, we observe that two nodes have a degree

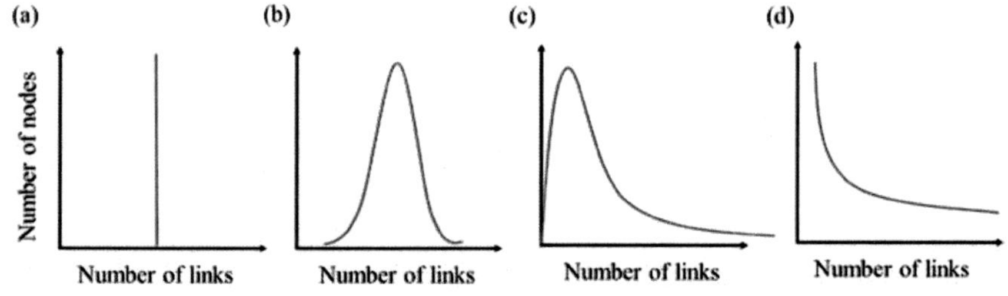

Fig. 11.3 Schematics showing the distribution of links between a pair of nodes (i.e., the degree distribution) of different networks on a linear scale. (**a**) Regular network. (**b**) Random network. (**c**) Small-world network. (**d**) Scale-free network

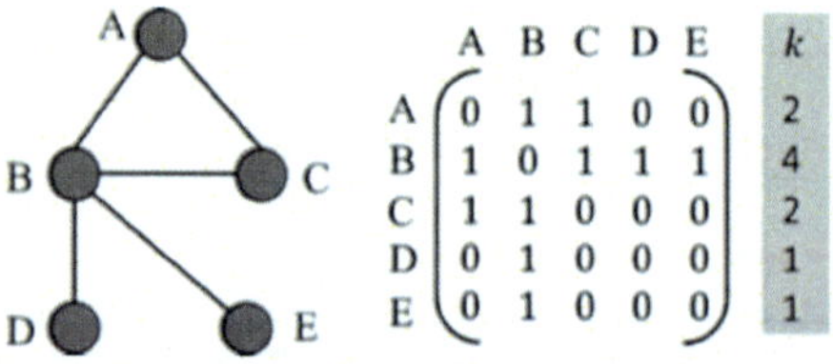

Fig. 11.4 Schematic showing the network topology, the corresponding adjacency matrix, and the calculation of degree (k) of each node in the network

of 1, two nodes have a degree of 2, and one node has a degree of 4. Therefore, the probability of occurrence of degree 1 in the network is, $P(1) = n(1)/N = 2/5 = 0.4$. Similarly, we can obtain the probability of the occurrence of degree 2 and 4 in the network as 0.4 and 0.2, respectively.

(b) **Betweenness centrality**

Betweenness centrality quantifies how many of the shortest paths between any two nodes in the network contain a particular node as compared to the total number of such paths. We can calculate the betweenness centrality of a node as,

$$b_i = \sum_{j,k \epsilon n, j \neq k} \frac{d_{(j,k)}(i)}{d_{(j,k)}} \tag{11.4}$$

where $d_{(j,k)}$ indicates the number of the shortest paths between nodes j and k, $d_{(j,k)}(i)$ indicates how many such paths pass through node i. The shortest path represents the minimum number of links that are needed to travel from a node j to node k. Higher the value of b_i, the more the importance the ith node has. This measure helps to identify the nodes that have the largest influence on the transfer of information through a network.

The calculation of betweenness centrality of a node is illustrated in Fig. 11.5. In order to measure the betweenness centrality of a node B in the network, we first calculate the total number of shortest paths between a pair of nodes in the network such that the node B does not appear at the end of the path. These paths are 6, namely AC, AD, AE, CD, CE, and DE. From these paths, the paths that pass through node B are 5, namely AD, AE, CD, CE, and DE. Therefore, the betweenness centrality of node B is obtained as $5/6 = 0.833$. For rest of the nodes in the network (i.e., A, C, D, and E), the betweenness centrality is zero, as none of the shortest paths between a pair of nodes in the network pass through these nodes.

(c) **Closeness centrality**

Closeness centrality quantifies the average shortest distance of a node with all other nodes in a network. We can express closeness centrality as the normalized inverse of the sum of the shortest paths between a node and the rest of the nodes,

$$c_i = \frac{N-1}{\sum_j d_{(i,j)}} \tag{11.5}$$

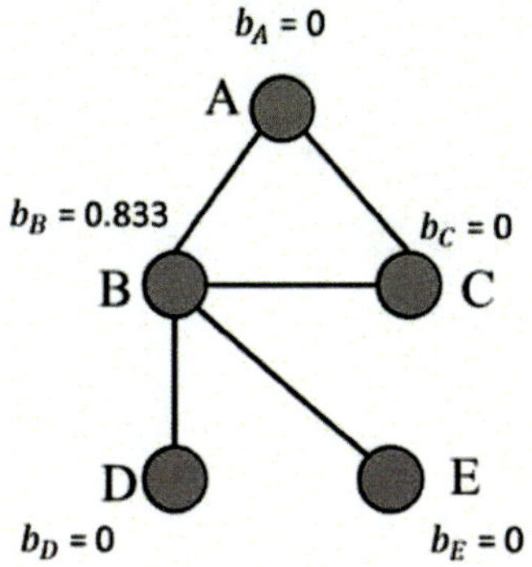

Betweeness centrality of node B

(j, k)	$d_{j,k}$	$d_{j,k}(B)$
(A, C)	1	0
(A, D)	1	1
(A, E)	1	1
(C, D)	1	1
(C, E)	1	1
(D, E)	1	1

$$b_B = \sum \frac{d_{j,k}(B)}{d_{j,k}} = \frac{5}{6} = 0.833$$

Fig. 11.5 Calculation of betweenness centrality of each node in the network. The example for calculation of betweenness centrality of node B is shown in the table

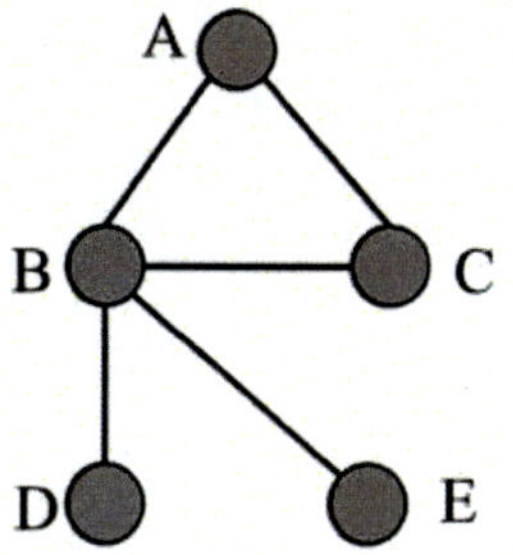

	A	B	C	D	E	$\sum d_i$	c_i
A	0	1	1	2	2	6	0.67
B	1	0	1	1	1	4	1
C	1	1	0	2	2	6	0.67
D	2	1	2	0	2	7	0.57
E	2	1	2	2	0	7	0.57

$$c_A = \frac{5 - 1}{6} \approx 0.67$$

Fig. 11.6 Calculation of closeness centrality of each node in a network with $N = 5$. The matrix illustrates the number of shortest paths between two nodes

where $d_{(i,j)}$ is the length of the shortest path of node i with node j.

We can refer to the inverse of closeness as farness in a network [14]. Nodes with high closeness centrality are the nodes that are in overall proximity to all the nodes. This measure also helps us understand how fast the information can pass through a given node to all other nodes. Transfer of information is efficient when passed through nodes with high closeness centrality.

The example for the measurement of closeness centrality of a node is shown in Fig. 11.6. From node A, the shortest paths to other nodes in the network are AB, AC, AD, and AE. In these paths, we can reach from node A to nodes B and C in one step, but to reach nodes D and E, we need at least two steps. Hence, closeness centrality of node A is $(5-1)/6 \approx 0.67$. We can compute the closeness centrality for other nodes in a similar manner.

(d) **Average path length or characteristic path length**

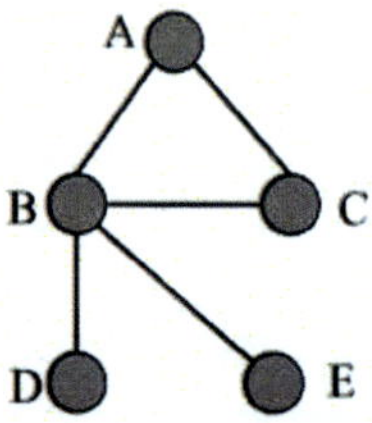

$d_{A,B} = 1 \quad d_{B,D} = 1$

$d_{A,C} = 1 \quad d_{B,E} = 1$

$d_{A,D} = 2 \quad d_{C,D} = 2$

$d_{A,E} = 2 \quad d_{C,E} = 2$

$d_{B,C} = 1 \quad d_{D,E} = 2$

$$L = \frac{d_{A,B} + d_{A,C} + d_{A,D} + d_{A,E} + d_{B,C} + d_{B,D} + d_{B,E} + d_{C,D} + d_{C,E} + d_{D,E}}{N(N-1)}$$

$$= \frac{1+1+2+2+1+1+1+2+2+2}{5(5-1)} = 15/20 = 0.75$$

Fig. 11.7 Calculation of the average path length of the network

This measure gives the average of the shortest path between every pair of nodes in a network. A network with a smaller average shortest path length is considered to be more efficient for information transfer from one node to another. We can calculate the value of the average path length of any network as,

$$L = \frac{1}{N(N-1)} \sum_{i,j=1,i\neq j}^{N} d_{(i,j)} \tag{11.6}$$

We can refer to the maximum value of the shortest path $d_{(i,j)}$ between node i and node j as the diameter of a network. The diameter indicates the shortest distance between the two most distant nodes in a network. In many real-world networks, it is preferred to have a short average path length for the quick transfer of information from one place to another. In Fig. 11.7, we present an example for the calculation of the average path length of a network.

(e) **Local clustering coefficient**

Unlike the centrality measures discussed above, the local clustering coefficient quantifies the probability that the neighbors of a particular node are connected and form a clique (a complete graph) in a network. In other words, the local clustering coefficient measures a ratio of the number of connected pairs of neighbors of a node i (i.e., neighbors of a node i that form triangles) to the total number of connections possible between neighbors of the node i (i.e., total number of triplets at node i), see Fig. 11.8. Here, a triplet represents a group of three nodes where the neighbors of a centered node (say, i) are unconnected (i.e., these three nodes form an open triangle). In contrast, a triangle indicates a group of three nodes that are connected to each other (i.e., a closed triplet), see Fig. 11.8.

The local clustering coefficient is calculated as follows,

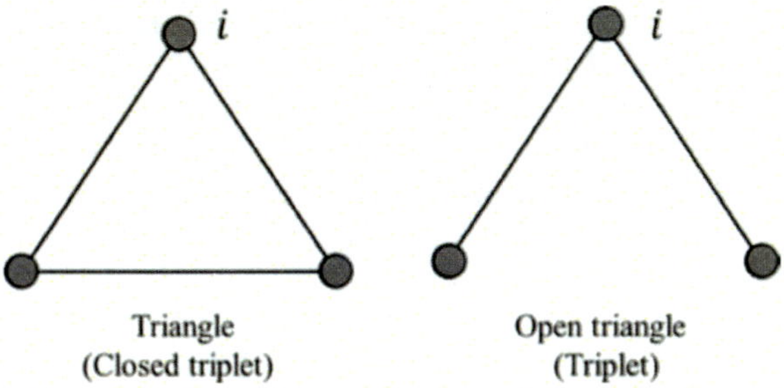

Fig. 11.8 Schematics illustrating a triangle (close triplet) and an open triangle (triplet) in a network

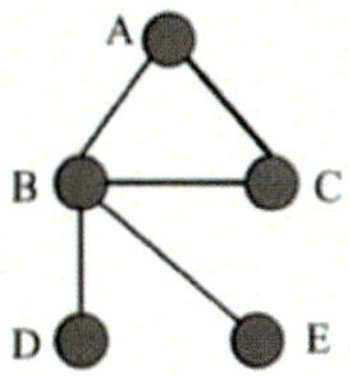

	k_i	e_i	C_i
A	2	1	1
B	4	1	0.167
C	2	1	1
D	1	0	0
E	1	0	0

Fig. 11.9 Calculation of the clustering coefficient C_i of each node in the network

$$C_i = \frac{e_i}{k_i(k_i - 1)/2} \tag{11.7}$$

where e_i is the number of links between the neighbors of a node i, k_i is the total number of possible neighbors of a node i, and $k_i(k_i - 1)/2$ is the maximum possible links amongst the neighbors.

The value of C_i lies between 0 and 1. For $C_i = 0$, all neighbors of a node are disconnected, which is seen in the case of a tree network or a square lattice. On the other hand, $C_i = 1$ suggests that all neighbors of a node are interconnected, which happens for the complete graph.

For example, in the network shown in Fig. 11.9, B and C are the neighbors of a node A and they are connected. Thus, the set of nodes A, B, and C form a triangle in the network. Thus, the clustering coefficient of node A is 1. The node D has only one neighbor (i.e., B); hence, it does not have a possibility to form a triangle. Therefore, the clustering coefficient of node D is zero. Node B has four neighbors (such as A, C, D, and E). Among these nodes, only nodes A and C are connected, while others nodes D and E are not connected with each other and also with the nodes A and C. Therefore, we observe only one triangle and five triplets for node B. As a result, the clustering coefficient of node B is very low (Fig. 11.9).

A high value of clustering coefficient indicates a locally dense network. Since the clustering coefficient is based on the connections of the neighbors of a node and not just the node, the concept of clustering coefficient is introduced to identify and classify small world networks and scale free networks.

(f) **Transitivity**

Transitivity is a global measure of the local clustering coefficient. Transitivity of a network gives the information of the likelihood of modularity or the presence of community structure in a network. Transitivity can be quantified as the ratio between the number of closed triplets and the total number of possible triplets (i.e., a sum of closed triplets and open triplets) in a network. Therefore, we can calculate the network transitivity as follows,

$$\mathcal{T} = \frac{\text{Total number of closed triplets}}{\text{Total number of triplets in a network}} \tag{11.8}$$

In an undirected network with three nodes (i, j and k), if the nodes j and i are connected and the nodes j and k are connected, then the possibility of the nodes i and k being connected is called transitivity closure (or partial transitivity) (Fig. 11.10a). In contrast, if all three nodes are connected to each other, then the network demonstrates a perfect transitivity (or a clique).

Transitivity can be understood from a simple example of a friend circle, where a friend of my friend need not be my friend. However, there is a high chance that my friend's friend is likely to be my friend rather than being a completely unknown (or random) person.

In Fig. 11.10b, we show the example of the calculation of the transitivity of a network. In this network, the number of closed triplets are three, each one is centred at node A, B, and C (i.e., BAC, CBA, ACB). The number of possible triplets in the network are five (i.e., three closed triplets: BAC, CBA, ACB, and two open triplets: DBC and DBA). Hence, the transitivity of the network is $\mathcal{T} = 3/5 = 0.6$.

11.3 Types of Complex Networks

Complex networks are primarily classified into different topologies [9, 64]. The two extremes of such topologies are: (1) a *random network*, where the connection between nodes are made in a random fashion, and (2) a *regular network*, where all the nodes have the same interaction properties. However, most real-world networks do not fall into these two categories. Two other topologies of networks have been discovered recently, which are: (3) *scale-free network* and (4) *small-world network*. The classification of these networks is based on the connectivity of nodes and the degree distribution of connections in the network.

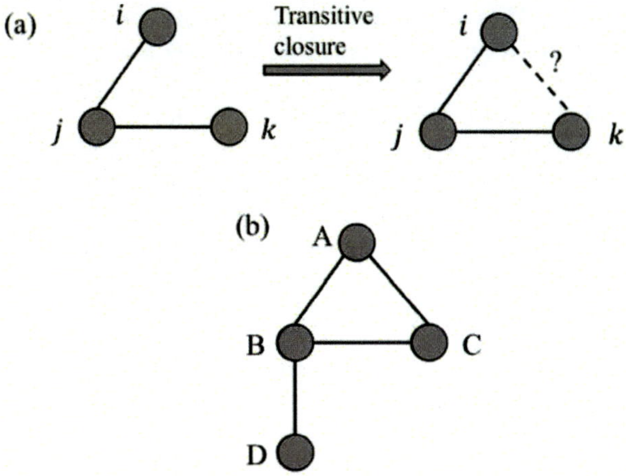

$$\text{Transitivity} = \frac{\text{Total number of closed triplets}}{\text{Total number of triplets in a network}} = \frac{3}{5} = 0.6$$

Fig. 11.10 (**a**) An example for understanding the transitive closure of a network. In a network with three nodes (i, j and k), transitivity closure indicates that if a node j is connected to its neighboring nodes i and k, then there is a high chance for nodes i and k to be connected. (**b**) An example demonstrating the calculation of transitivity of a network containing 4 nodes

11.3.1 Regular Networks

In regular networks, every node in the network has an equal degree, i.e., the same number of links [26, 52]. Hence, the degree distribution of regular networks shows point. These networks are highly ordered. Examples of regular networks include a ring network (Fig. 11.11a), a square lattice (Fig. 11.11b), and a globally connected network (Fig. 11.11c). In a ring network, every node in the network is connected to a few of its neighbors with an equal number of links (Fig. 11.11a). In a square lattice, each node is connected to all of its nearest neighbors; hence, every node in such a network has a degree of four (Fig. 11.11b). In a globally connected network (also called a complete graph), every node is connected to all nodes in the network (Fig. 11.11c). Therefore, for a globally connected network, the value of the shortest path length is one and the clustering coefficient is highest, i.e., unity [25]. Since most real-world networks appear to be sparse (i.e., not fully connected); they usually do not possess the property of a regular network.

11.3.2 Random Network

Random network is the opposite of the regular network, where the nodes in a network are randomly connected. This type of network was initially proposed by

Erdos and Rényi [40, 41] and was the major part of the network analysis until the recent discovery of small-world networks and scale-free networks. The connection between a pair of nodes is established through a link selected from the uniform distribution, while avoiding multiple and self-connections. On an average, every node in the network carries the same degree, whose value is $k = p(N-1) \approx pN$, where p is the probability of the selection of a node and N is the total number of nodes in the network ($N >> k$). The degree distribution of such networks follows a Poison's distribution. The presence of random connections leads to a low average path length and a nearly zero clustering coefficient of the network.

11.3.3 Small-World Networks

Regular and random networks do not capture most features of real-world networks. Most real-world networks possess the property of a high clustering coefficient and a low shortest path length. These properties imply that most nodes in a network need not be neighbors of each another (i.e., the existence of triangles); however, the neighbors of a node in the network have the possibility of being neighbors of each other (a property of high local clustering coefficient). Further, a connection between most nodes can be achieved through a few steps, mostly through hubs (a property of low shortest path). A network with these properties is referred to as a small-world network and the cause of their formulation is referred to as the '*small-world effect*'.

A first well-known small-world experiment on the social network was conducted by Stanley Milgram [74, 115]. In the experiment, he asked several people to post letters to some random persons they did not know, and whose name and details were given to them but not their address. The idea was that a person A, who needs to send the letter to B, will send it actually to someone he knows (say, C), and C will send it further to a known person D and so on until the letter reaches B. Through this experiment he found that any two randomly chosen people are connected through

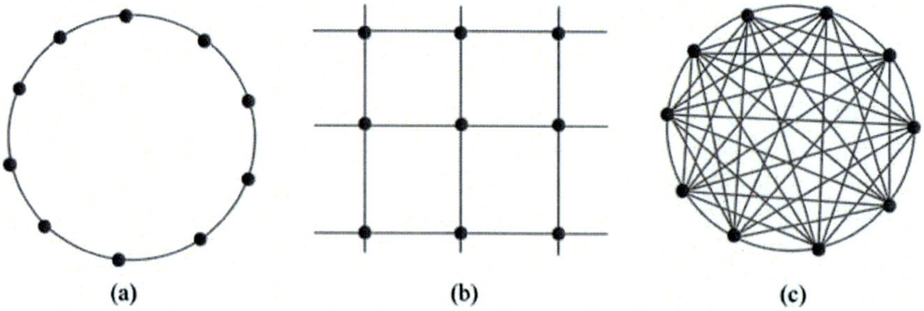

Fig. 11.11 Different topologies of regular networks. (**a**) Ring network has local connectivity to nearest neighbors, (**b**) square lattice has local connectivity with four nearest neighbors, and (**c**) the complete graph has global connectivity between all nodes

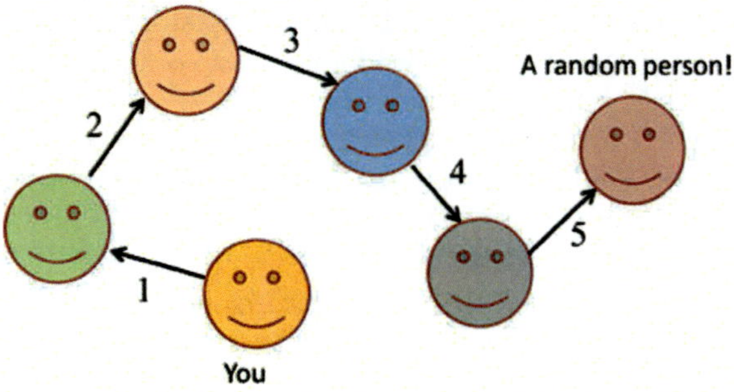

Fig. 11.12 Pictorial representation of six-degrees of separation observed in small-world networks

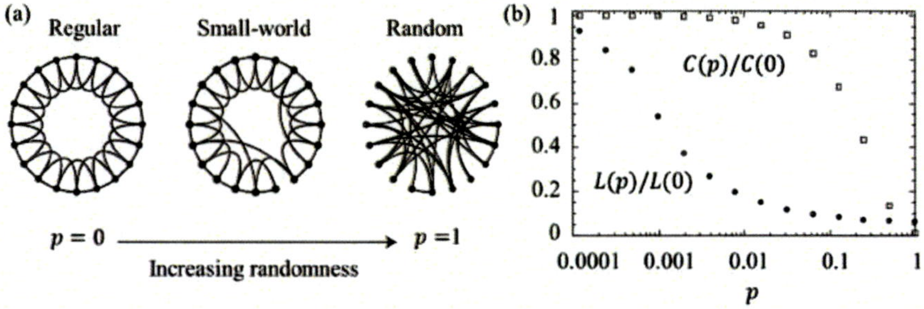

Fig. 11.13 (**a**) Conversion of the topology of a locally coupled regular network to a random network through a small-world network by changing the probability (p) of the rewiring of links in the network. (**b**) The variation of the clustering coefficient (C) and the short path length (L) with the probability (p) of rewiring of links in the network. Here, a filled circle indicates L and hollow square represents C. Reproduced with permission from Watts and Strogatz [126]

six other people on an average (Fig. 11.12). This phenomenon is often referred to as the '*six degrees of separation*' between people. The network constructed based on this idea has nodes as people and links are established if two people know each other. In a world with a population of nearly 8 billion, why would two people know each other with a just short chain of acquaintances?

To understand this small-world phenomenon, Watts and Strogatz [126] developed a simple model based on two small-world properties namely, high clustering and low short path length (Fig. 11.13). Consider a regular network of N number of nodes each with a degree k. Therefore, the number of links in the network are Nk. In order to induce randomness in the network, disconnect a number of links in the regular network with a probability p and rewire the network to new distant nodes such that these new pairs of nodes cannot have more than one connection and also the self-connections. Here, $p = 0$ for a regular network and $p = 1$ for a random network. With an increase in p, we observe that the value of the clustering coefficient of the network remains high for a large range of p and falls rapidly as $p \to 1$. On the other hand, the average path length of the network falls rapidly at lower values of p and stays near zero for higher values of p. A small-world network falls between random and regular networks. It consists of a large clustering coefficient (a property of regular network) and short average path length (a property of random network). This model is commonly referred to as the Watts–Strogatz (WS) model.

Subsequently, Newman and Watts [87] proposed a new model (NW small-world model) based on the modification in the WS model. In their model, instead of removing the links between nodes and rewiring them to another node as done in the WS model, new links are added with probability p in the network such that no two pairs of nodes in a network are connected more than once. Thus, for $p = 0$, the network is regular. In contrast, for $p = 1$, it shows the properties of a globally connected network.

A network with small-world property is highly efficient in information transfer, signal processing, and synchronization as compared to regular networks. Examples of such small-world networks include social connections, brain networks, power grids, metabolic reactions, protein-protein interactions, research collaboration, etc [125].

11.3.4 Scale-Free Networks

As we have seen previously, the degree distribution of nodes for the random (Fig. 11.3b) and small-world (Fig. 11.3c) networks are nearly homogeneous, showing a bell-shaped distribution with a peak near the average value and an exponential decay near tails. Therefore, these networks are referred to as *exponential networks*. Further, these networks assume a fixed number of nodes, whose connectivity can only be rearranged or newly established. According to Barabási and Albert [6], many practical networks do not possess a uniform degree for all the nodes, but have a large variation in their degree such that a few nodes have a high degree and the remaining nodes have a low degree. The networks with these properties are *scale-free* and display a *power-law* behavior of their degree distribution (Fig. 11.3d).

To understand this power-law nature of the networks, Barabási and Albert [6] proposed a simple model based on two main ingredients: *growth* and *preferential attachment*. Here, a growth indicates a continuous addition of nodes in the network and a preferential attachment signifies that the added node has a tendency to get connected to an existing node with a higher degree. Thus, the degree of the nodes with a higher degree progressively increases, while the degree of the nodes with a lower degree remains nearly the same all the time. This process in which higher degree nodes get more new connections is called '*rich gets richer*' process. Eventually, the network evolves into a scale-invariant state and the addition of new nodes does not alter the shape of the degree distribution. A log-log plot of the degree distribution for a scale-free network shows a power-law behavior, such that $P(k) = k^{-\gamma}$. Here, γ is the slope of the linear variation in the plot between the probability, $P(k)$, of a node having a degree k as a function of the degree of a node (k). For scale-free networks, the value of γ lies between 2 and 3 [9].

In scale-free networks, the most vital parts are the '*hubs*', also called '*Achilles' heel*'. Hub is a node in the network that has the highest degree (i.e., a large number

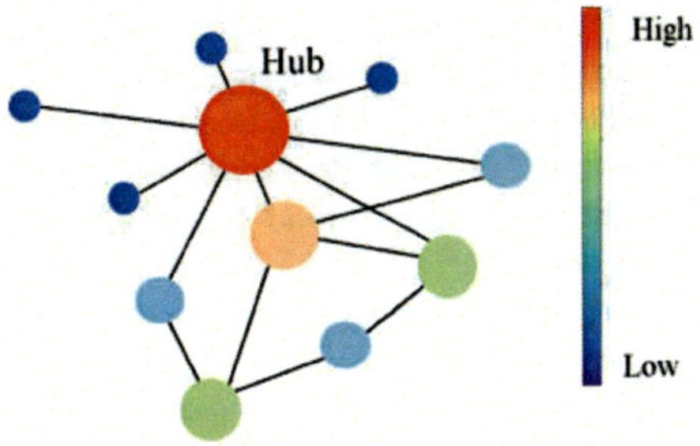

Fig. 11.14 A network topology showing the presence of nodes (filled circles) and links (black lines). Size of the nodes indicates the number of connections associated it. A node with highest connections (shown in red) is called a hub

of links pass through this node) and it plays a major role in the functioning or robustness of the network (Fig. 11.14). The removal of such hubs in the network may most likely disintegrate the network into disconnected parts, whereas the removal of nodes with a low degree may not affect the functioning of the network. For example, in a network of U.S. airlines [7], the nodes are airports and links indicate the presence of a direct flight between two airports. As the airports at Chicago, New York, Denver, and Atlanta have direct flights to nearly all U.S. airports, these networks can be referred to as hubs. Therefore, a shutdown of these airports due to some reasons will adversely affect the global airline connectivity as compared to the shutdown of airports that have less connectivity. Identifying hubs in a scale-free network is thus very critical. Examples of scale-free networks in the real-world are internet, citation network, a network of Hollywood actors, research collaborations, etc. [5].

So far, we presented the basics of complex networks, their types, and measures derived from the connectivity of nodes in a network. In the next section, we study the application of the acquired knowledge of the network theory to analyze the data obtained from experiments and numerical simulations of thermoacoustic systems. We classify this analysis into two parts: (i) temporal networks and (ii) spatiotemporal networks. In temporal networks, the time series of one or more dynamical variables of the system is used for the construction of a network. On the other hand, in spatiotemporal networks, network construction is based on the inter-connections between spatially distributed points in the combustor flow field.

11.4 Complex Network Approach to Study Temporal Dynamics of Thermoacoustic Systems

As we discussed in Chaps. 6–8, the temporal dynamics of a thermoacoustic system with turbulent flow field is complex. The dynamics of such systems changes from the state of combustion noise to thermoacoustic instability via intermittency. To understand the underlying features of temporal fluctuations in one or more dynamic variables of a system, we need to first map these fluctuations onto a complex network and then access the hidden information using different measures from network theory [48, 109, 132].

There are several ways to construct a network from a single time series data. These methods include phase space coarse-graining method [123], cycle network [130], correlation method [128], visibility graph [62], horizontal visibility [69], partitioning the phase space [88], neighborhoods in phase space [47, 93, 127], and recurrence network [35, 56, 71]. During the network construction, we need to first obtain an adjacency matrix by applying one of the aforementioned network construction algorithms on the temporal data of a system variable and then use any network visualization software (for example, Gephi software, https://gephi.org/ and [13, 27]) to picturize the connectivity of nodes in a network.

The application of complex network approach to thermoacoustic systems is very recent. Murugesan and Sujith [79] were the first to study the onset of thermoacoustic instability in the framework of complex network theory. Subsequently, several researchers have used this approach to map a time series of the acoustic pressure fluctuations into complex network through different methods of network construction. Using the complex systems approach, we can study different features of the temporal dynamics of the states of combustion noise [79] and thermoacoustic instability [93], and also the properties of the route to thermoacoustic instability [50, 79, 80, 113]. We can also perform the multivariate analysis of both the acoustic pressure and the heat release rate fluctuations during the transition from combustion noise to thermoacoustic instability via intermittency [49]. Furthermore, we can develop numerous early warning measures (precursors) for the onset of thermoacoustic instability, using different statistical properties of complex networks that are constructed from the time traces of acoustic pressure fluctuations [49–51, 77, 78, 80, 113]. In the following subsections, we will elaborate on the characterization of different features of temporal dynamics of thermoacoustic systems using the complex network approach.

11.4.1 Combustion Noise Is Scale-Free

In this section, we characterize the complexities underlying the dynamics of the acoustic pressure fluctuations during the state of combustion noise in a turbulent thermoacoustic system. In Chaps. 6 and 10, we discussed in detail a few novel properties of this state using different techniques from nonlinear time series analysis and multifractal analysis of the acoustic pressure signal. These properties are different from the traditional steady state (or stochastic) description of this state. The acoustic pressure fluctuations observed during the state of combustion noise consist of low amplitude chaotic fluctuations [83, 114]. Further, these chaotic oscillations are multifractal, having a range of scaling exponents associated with different orders of fluctuation function of the pressure signal [82]. Similarly, we can reveal such a scale-invariance property of acoustic pressure fluctuations during the state of combustion noise, using the approach of complex networks.

In the case of turbulent flames in an open environment, the spectrum of acoustic pressure fluctuations observed during the state of combustion noise exhibits a power-law scaling. Such a power-law behavior is an indication of the scale-invariance property of the physical system [32, 65, 68]. Abugov and Obrezkov [1] reported that open turbulent flames display power-law scaling with an exponent of $-5/2$ in the spectrum of acoustic pressure fluctuations for a frequency range of 2–10 kHz. In contrast, Rajaram and Lieuwen [98] reported that such power spectrum exhibits a scaling of f^{β} for a low frequency range and a scaling of $f^{-\alpha}$ for the intermediate and high-frequency range, where α and β are positive numbers and f is the frequency. Furthermore, Rajaram [97] identified the relation between the power-law scaling of the acoustic pressure (p') and the heat release rate ($\dot{q}'$) fluctuations of an open turbulent flame, in the high frequency (ω) range, as $p'(\omega) \sim \langle|\dot{q}'(\omega)|^2\rangle$. Clavin and Siggia [30] and Clavin [29] related the power-law scaling of acoustic pressure fluctuations of such open flames to the scaling in the Kolmogorov spectrum of turbulence.

On the other hand, the power spectrum of the acoustic pressure fluctuations during the occurrence of stable operation in confined turbulent thermoacoustic systems is broadband and, consists of shallow peaks around the unstable acoustic mode of the combustor [28, 54, 106]. These shallow peaks in the spectrum are a result of the preferential amplification of the acoustic waves generated by turbulent flames near the natural modes of the combustion chamber. The presence of such multiple peaks in the spectrum makes it difficult to discern the scale-invariance property of this state, just from the analysis of the power spectrum. Therefore, Murugesan and Sujith [79] proposed a complex network approach, based on the visibility algorithm, to study the scale invariance characteristics of acoustic pressure fluctuations observed during the state of combustion noise in confined turbulent flame combustors.

While constructing a complex network using the visibility algorithm, we first consider every local positive peak in the time series of acoustic pressure fluctuations (say, P) as a node in the network [92]. Then, to construct a visibility plot, we represent these data points of the signal as vertical bars (Fig. 11.15). We can express the condition of visibility (connectivity) for any two nodes (say, i and j in a vector P) in the network through the following equation [62],

$$A_{ij} = \begin{cases} 1 & \text{if } P_k < P_i + (P_j - P_i)\frac{t_k - t_i}{t_j - t_i} \\ 0 & \text{otherwise} \end{cases} \tag{11.9}$$

where $A_{i,j}$ is the adjacency matrix. The visibility condition indicates that the nodes i and j can be connected by a straight line (or "visibility line") given that the connection is not interrupted by an intermediate node k, where $i < k < j$. If nodes are connected, their connection is marked as 1 in the adjacency matrix, and if nodes are disconnected, their connection is represented as 0 in the adjacency matrix. This method of network construction is efficient and fast, as compared to the methods that

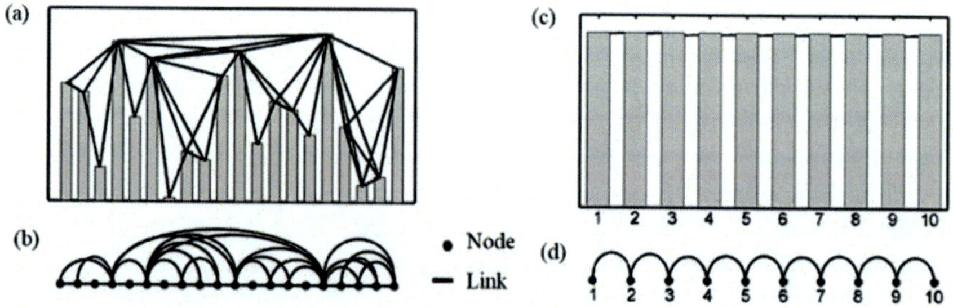

Fig. 11.15 Schematic showing the construction of a complex network using a visibility graph from the time series data of (**a**) irregular and (**c**) regular fluctuations. In (**a**) and (**c**), vertical bars indicate the local amplitudes of the signal and solid lines signify their connection (links). In (**b**) and (**d**), a dot indicates a node corresponding to each amplitude and a solid line represents the connectivity (links) between two nodes. For irregular fluctuations, each node can have more than two connections, while for regular fluctuations, each node is connected to the same number of nearest neighbors. Adapted with permission from Murugesan and Sujith [79]

are based on the construction of the phase space which require the identification of embedding parameters [48].

In Fig. 11.16a, b, we show the time series and the corresponding power spectrum of the acoustic pressure fluctuations for the state of combustion noise [79]. The dynamics of the acoustic pressure shows low amplitude aperiodic fluctuations and its power spectrum shows a low amplitude broadband peak near the fundamental mode of the combustor. Furthermore, in Fig. 11.16c, we present a complex network constructed by identifying the nodes and the connectivity between a pair of nodes using the visibility algorithm . The complex network is visualized using a Force Atlas layout in a Gephi software [13, 27]. In this network, the nodes are indicated as the dots, nodes of different degrees are represented by different colors, and the size of each dot indicates the degree of a particular node. The nodes with a large degree are called hubs of the network. In Fig. 11.16c, we notice that the color distribution across the nodes of the network of combustion noise is heterogeneous (various colors and sizes of nodes), as there is no single characteristic degree in the network for all the nodes.

The degree distribution of the complex network is shown in Fig. 11.16d. We notice a power-law (scale invariance) behavior for the degree distribution such that,

$$P(k) \sim k^{-\gamma},$$

where the exponent of this power-law is $\gamma = 2.7$ [79]. From the presence of such a power-law variation, we can confirm the existence of a scale-free network [7] in the dynamics of acoustic pressure fluctuations during the state of combustion noise.

Murugesan and Sujith [79] derived insights about the dynamics of the underlying turbulent reacting flow from the heterogeneity of nodes observed in the complex network (Fig. 11.16c). In such flows, the length scales range from a largest-scale governed by the characteristic dimension of the combustor to the smallest-scale

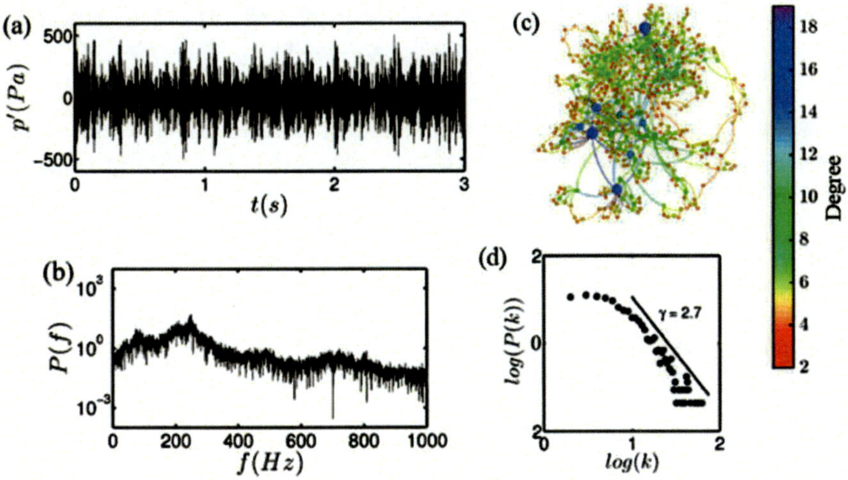

Fig. 11.16 During the state of combustion noise in a turbulent combustor, (**a**) the acoustic pressure signal (p') shows low amplitude chaos and (**b**) the corresponding power spectrum shows a broadband of frequencies. (**c**) The complex network constructed using the visibility algorithm exhibits a wide distribution of degree of nodes and (**d**) the corresponding degree distribution on a log-log plot shows a power-law behavior. The color bar indicates the variation of the degree of a node. Adapted with permission from Sujith and Unni [109]

governed by the Kolmogorov scales. The large-scale vortices produce the large magnitude fluctuations in the flow, which reflect in the form of the large amplitude fluctuations in the acoustic pressure signal. On the other hand, the intermediate and the small-scale vortices produce the small fluctuations in the flow field and, hence, govern the small amplitude oscillations in the acoustic pressure signal. In the network, we observe that the nodes corresponding to the large amplitude fluctuations have visibility to nodes corresponding to small amplitude fluctuations. Therefore, such nodes have a larger number of connections and, hence, we can refer to them as hubs. From Fig. 11.16c, we notice that the number of such hubs in the network is less, as we observe only a few large sized nodes colored by blue and pink in the network. In contrast, the number of nodes with a smaller degree is large as these nodes possess very few connections, mostly with their neighbors. From this information of the degree distribution of nodes, we can infer that the presence of combustion noise is the result of the cascade of turbulent scales and, hence, the dynamics of this state possess the property of scale invariance.

11.4.2 The Onset of Thermoacoustic Instability as a Transition from Scale-Free to Regular Networks

As we know, the dynamical transition of turbulent thermoacoustic systems from the state of combustion noise (stable operation) to thermoacoustic instability (unstable operation) happens via intermittency. Here, we present the change in characteristics of the complex networks for these states of the combustor operation. Figure 11.17-I illustrates the time series of the acoustic pressure fluctuations corresponding to the states of combustion noise (Fig. 11.17a), intermittency (Fig. 11.17d), and thermoacoustic instability (Fig. 11.17g). The corresponding degree distributions and complex networks are shown in columns II and III of Fig. 11.17, respectively.

The acoustic pressure fluctuations observed during the state of combustion noise show the features of low amplitude chaotic fluctuations (Fig. 11.17a). During the state of intermittency, the acoustic pressure fluctuations show the intermittent occurrence of bursts of large amplitude periodic oscillations amidst epochs of low amplitude chaotic oscillations (Fig. 11.17d). The state of thermoacoustic instability shows the presence of large, nearly constant amplitude, periodic oscillations in the system (Fig. 11.17g). Thus, the occurrence of thermoacoustic instability from the state of combustion noise is reflected as the transition of the system dynamics that is dominated by multiple time scales to that dominated by a single (or a few discrete) time scale.

Further, we observe a power-law behavior in the degree distribution for the states of combustion noise (Fig. 11.17b) and intermittency (Fig. 11.17e). Hence, the corresponding complex networks exhibit a property of scale-free networks [79]. In contrast, the degree distribution of the state of thermoacoustic instability (Fig. 11.17h) shows a single dot, i.e., the absence of power-law scaling. Thus, we observe the loss of the scale-free characteristics of the complex network at the onset of thermoacoustic instability.

We observe a wide range of degrees, as indicated by different colors, in the networks constructed for the states of combustion noise (Fig. 11.17c) and intermittency (Fig. 11.17f). However, we notice a single color in the complex network constructed for the state of thermoacoustic instability (Fig. 11.17i). A network with the same degree for all nodes is called a regular network [96]. In this network, as each node is connected to two of its nearest neighbors, the degree of each node in the network is two. Thus, we can view the transition of the system dynamics from the state of combustion noise to thermoacoustic instability as a loss of scale-free behavior of the complex network, leading to an emergence of a regular network [79]. This emergence of regular networks is similar to the emergence of order observed, due to the formation of large-scale coherent structures, during the onset of thermoacoustic instability in a turbulent combustor, as we discussed in Chap. 7.

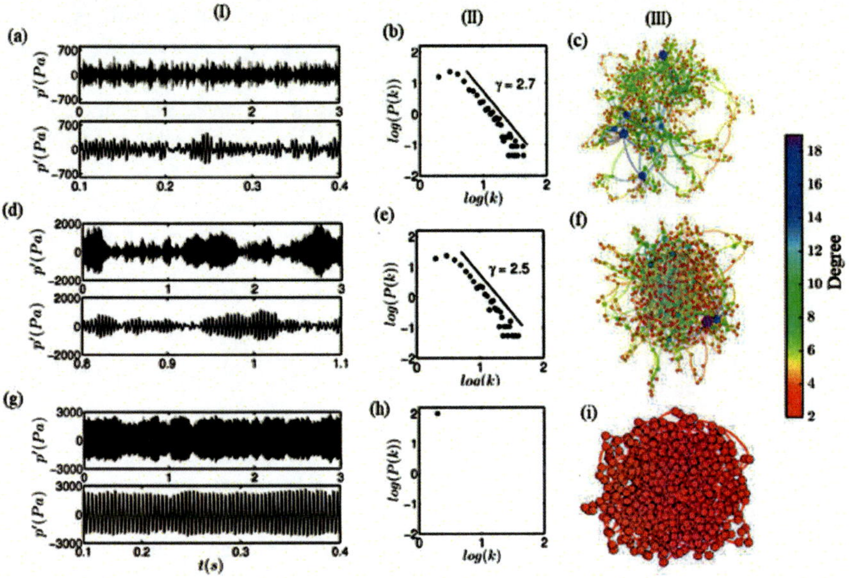

Fig. 11.17 (I) The time series of the acoustic pressure fluctuations along with their zoomed portions for the states of (**a**) combustion noise, (**d**) intermittency, and (**g**) thermoacoustic instability. (II) The degree distribution and (III) complex networks obtained using the visibility algorithm corresponding to these states of the combustor dynamics. We observe the emergence of regular network from the scale-free network during the transition from combustion noise to thermoacoustic instability. Plots in (I) are reproduced with permission from Murugesan and Sujith [79] and plots in (II) and (III) are reproduced with permission from Sujith and Unni [109]

11.4.3 Small-World-Like Behavior of Thermoacoustic Instability Using Cycle Network

In practical turbulent combustors, sometimes the limit cycle oscillations observed during the state of thermoacoustic instability are not perfectly periodic, while they exhibit a wide variation in the cycle-to-cycle amplitude of oscillations in the signal [66, 67, 90, 95]. This observation is similar to that we discussed for the state of phase synchronization in Chap. 8. Lieuwen [67] surmised that such fluctuations in the limit cycle amplitude arise due to the background random perturbations present in the turbulent system. In contrast, Pawar et al. [95] showed that the variation in amplitudes of the limit cycle oscillations is the result of a relatively wide distribution in the instantaneous frequency of the vortex shedding with respect to the frequency of the unstable acoustic mode of the combustor. Such signals with varying amplitude of periodic oscillations are sometimes referred to as "pseudoperiodic signals" [104]. Here, pseudoperiodic indicates the oscillatory behavior of a signal that has components of both strong periodicity and some level of irregular or chaotic oscillations. Due to the oscillatory nature, these signals contain repeating patterns in every consecutive cycle of the oscillation. Zhang and Small [130] proposed a method of cycle networks to analyze the properties of pseudoperiodic signals. Okuno et al. [93] used this network approach to construct a weighted complex network and reveal the hidden properties of the acoustic pressure fluctuations observed during the state of thermoacoustic instability in a swirl-stabilized turbulent combustor.

In order to construct a network from a pseudoperiodic time series (x) containing N observations $\{x_i\}_1^N$, we first need to divide the signal into m disjoint cycles. These cycles are obtained by identifying the events of local maximum (crest) or local minimum (trough) in the signal. Then, a portion of the signal between two consecutive events is considered as a cycle (Fig. 11.18a, b). Thus, we obtain a vector of cycles from such partitioning of the signal as C_i ($i = 1, 2, \ldots\ldots, m$), where m represents the number of cycles in the signal.

In the network construction, each cycle corresponds to a node of the network. The connection between the nodes is obtained from the strength of temporal correlation (similarity index) observed between the cycles. In this way, we can encode the temporal dynamics into a network topology. The correlation between a pair of cycles (say, C_i and C_j, where $i, j = 1, 2, 3, \ldots, m,\ i \neq j$) can be obtained as follows [129],

$$\rho_{ij} = \max_{l=0,1,\ldots,l_j-l_i} \frac{\mathrm{Cov}[C_i(1:l_i), C_j(1+l:l_i+l)]}{\sqrt{\mathrm{Var}[C_i(1:l_i)]}\sqrt{\mathrm{Var}[C_j(1+l:l_i+l)]}} \tag{11.10}$$

where l_i and l_j indicate the length of cycles C_i and C_j, respectively. $C_i(a : b)$ represents the portion of the cycle between ath and bth elements, Cov denotes the covariance, and Var indicates the variance. If two cycles are not of the same length (i.e., $l_i \neq l_j$), we need to overlap the shorter cycle onto a longer cycle and obtain the correlation for $(l_j - l_i)$ steps. Then, the maximum correlation (ρ_{ij}) between cycles is obtained by varying this overlap. The larger the value of ρ_{ij}, the higher the level of similarity between cycles.

Furthermore, the correlation coefficient has a direct relation with the closeness of cycles in the phase space. We can obtain such closeness from the distance between the phase space points of cycles as follows [129],

$$D_{ij} = \min_{l=0,1,\ldots,l_j-l_i} \frac{1}{l} \sum_{k=1}^{l_i} \|X_k - Y_{k+1}\| \tag{11.11}$$

where X_k and Y_k are the kth state points of cycles C_i and C_j in the reconstructed phase space, respectively, and $l_i < l_j$. Higher correlation between cycles indicates the presence of a lower phase space distance between them. Thus, the distance between the cycles is inversely proportional to their correlation (Fig. 11.18c,d). As practical signals are usually influenced by the background fluctuations, their phase space trajectories possess some randomness. Therefore, we can use the correlation coefficient for the network construction of signals obtained from experiments and

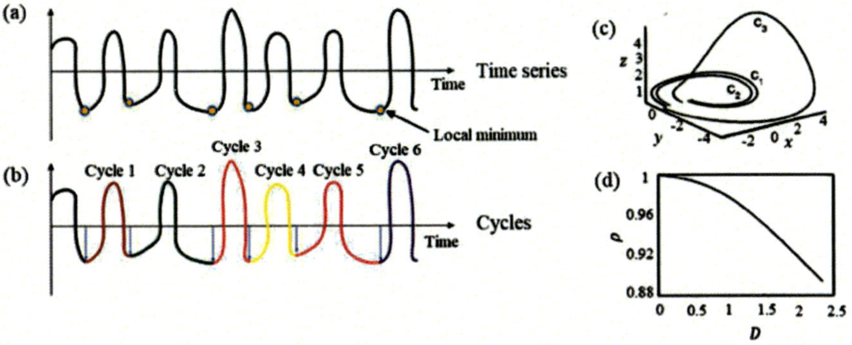

Fig. 11.18 Schematic showing (**a**) a time series and (**b**) a partition of the consecutive cycles based on the detection of local minimum in the signal. (**c**) The reconstructed phase space of the Rössler oscillator showing the evolution of three cycles C_1, C_2, and C_3. The value of the correlation coefficient between cycles is $\rho_{12} = 0.9961$ and $\rho_{13} = 0.9192$. Since $\rho_{12} > \rho_{13}$, cycle C_2 is closer to C_1 as compared to C_3. (**d**) Variation of the correlation (ρ) with the increasing distance (D) between the points in a pair of cycles. (**a, b**) Adapted with permission from Emmert-Streib [39] and (**c, d**) adapted with permission from Zhang et al. [129]

the phase space distance for the network construction of signals obtained from models [130].

From the information of correlation or phase space distance between the cycles, we can construct undirected and unweighted/weighted networks. These networks are always undirected, as the values of ρ or D do not have any information about a directionality. To construct a weighted network, we can apply a global threshold for the values of ρ or D, and obtain a resultant adjacency matrix containing binary elements. In the case of the weighted network, the value of the weight is equal to the actual value of ρ or D observed in the analysis. Once the adjacency matrix is constructed from the signal, we can obtain several network properties that discussed in Sec. 11.2.

In Fig. 11.19a, b, we show the variation of two basic properties of a cycle network, such as the average path length (L) and the average clustering coefficient (C_a), respectively, as a function of the threshold of the correlation coefficient (ρ), for the acoustic pressure fluctuations observed during the state of thermoacoustic instability in a swirl-stabilized turbulent combustor [93]. The network construction is based on the correlation coefficient and a correlation threshold of 0.9943 is used for obtaining the adjacency matrix. For $\rho = 0.9943$, we observe the values of L and C_a as 1.43 and 0.71, respectively. The simultaneous occurrence of a sufficiently high value of C_a and a low value of L is a property of a small-world network [126]. This behavior suggests that, the dynamics of the acoustic pressure fluctuations during the state of thermoacoustic instability, consisting of wide cycle-to-cycle variation in the amplitude (or pseudoperiodicity), may have a small-world property [93].

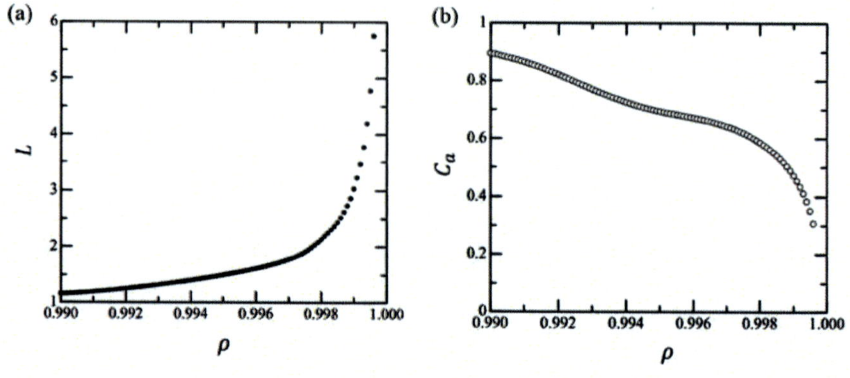

Fig. 11.19 Variation of the network properties (**a**) the characteristic path length L and (**b**) the average clustering coefficient C_a as a function of the correlation threshold ρ for the state of thermoacoustic instability observed in a swirl-stabilized turbulent combustor. Reproduced with permission from Okuno et al. [93]

11.4.4 Thermoacoustic Instability as a Condensation Transition in the Phase Space

The various dynamical states observed en route to thermoacoustic instability can be analyzed from the trajectories of these states in the phase space. The skeletal structure comprising the phase space topology is a set of periodic orbits that attract and trap the trajectory for certain epochs. For distinct realizations of the system during a particular dynamical state, the trajectory is attracted by several periodic orbits in the phase space and the order in which it visits these orbits is not unique. However, the set of periodic orbits that govern such an evolution of the trajectory in the phase space remain unaltered, and are thus said to be dynamically invariant. The trajectory may approach an orbit along the stable manifold and get ejected along the unstable manifold as shown in Fig. 11.20. These phase space orbits may have distinct stability properties and are also known as "*unstable periodic orbits*" (UPOs). The amount of time a trajectory spends around an orbit characterizes the stability of that orbit. For example, UPO-2 in Fig. 11.20 is comparatively more stable than UPO-1 as evident by the greater number of cycles of the trajectory around UPO-2 as compared to UPO-1.

By constructing and analyzing complex networks from phase space cycles, as discussed in the previous section, one can infer the topology of the phase space during various dynamical states. Using unweighted cycle networks, Tandon and Sujith [113] discussed the transformations in the phase space topology and the corresponding cycle networks during the transition from combustion noise to thermoacoustic instability via the route of intermittency in two systems, namely bluff-body and swirl stabilized turbulent combustors.

In Fig. 11.21(I–III), we show the acoustic pressure time series, the corresponding phase space diagram, and the cycle networks during various dynamical states observed in a bluff-body stabilized turbulent combustor. The phase space is reconstructed from the acoustic pressure time series using Takens' delay embedding method [111].

During the state of combustion noise (Fig. 11.21a), low-amplitude chaotic fluctuations in the acoustic pressure time series translate to a chaotic arrangement of the cycles of the trajectory in the phase space that are weakly correlated. As a result, the corresponding cycle network topology comprises mostly very low-degree nodes ($\approx$ degree 10) along with a few central nodes having degree $\approx$ 100.

During the state of intermittency, alternate epochs of periodicity and aperiodicity are observed in the signal (Fig. 11.21b). As a result, we observe both chaotic motion of the trajectory in the central region of the phase space, and several concentric rings of the trajectory characterizing its periodic motion. Furthermore, the corresponding network consists of both high-degree nodes ($\approx$ degree 250) and low-degree nodes ($\approx$ degree 50). The fraction of low degree nodes in the cycle network reduces as the system dynamics approaches the state of thermoacoustic instability in the combustor.

As we discussed in Chap. 8, during the state of weakly correlated limit cycle oscillations, which correspond to the state of phase synchronization of acoustic pressure and heat release rate fluctuations, the cycle-to-cycle amplitude of pressure fluctuations varies in time, as shown in Fig. 11.21c. During this state, the phase space trajectory is oriented in the form of a thick band of concentric rings while no chaotic motion of the trajectory is observed. The cycle network obtained from such dynamics comprises several high degree nodes ($\approx$ degree 300) and very few low degree nodes ($\approx$ degree 100) owing to the amplitude modulation of acoustic pressure fluctuations. Finally, during the state of strongly correlated limit cycle oscillations, which corresponds to generalized synchronization of pressure and heat release rate fluctuations (see Chap. 8), the phase space trajectory seems to condense into a thin ring-like structure characterizing the presence of a stable periodic orbit.

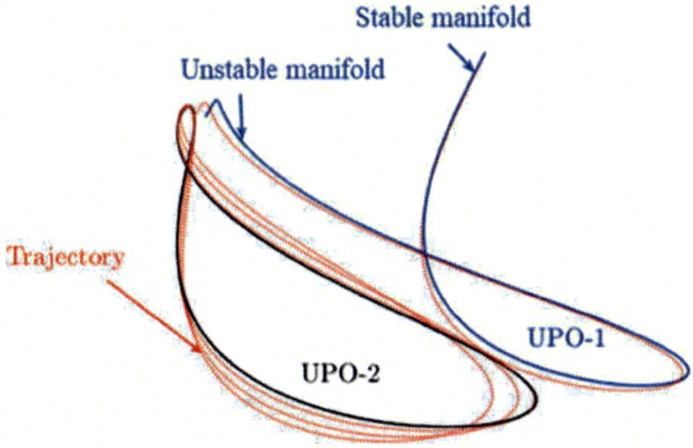

Fig. 11.20 Schematic diagram showing trajectory around two unstable periodic orbits in the phase space. Reproduced with permission from Tandon and Sujith [113]

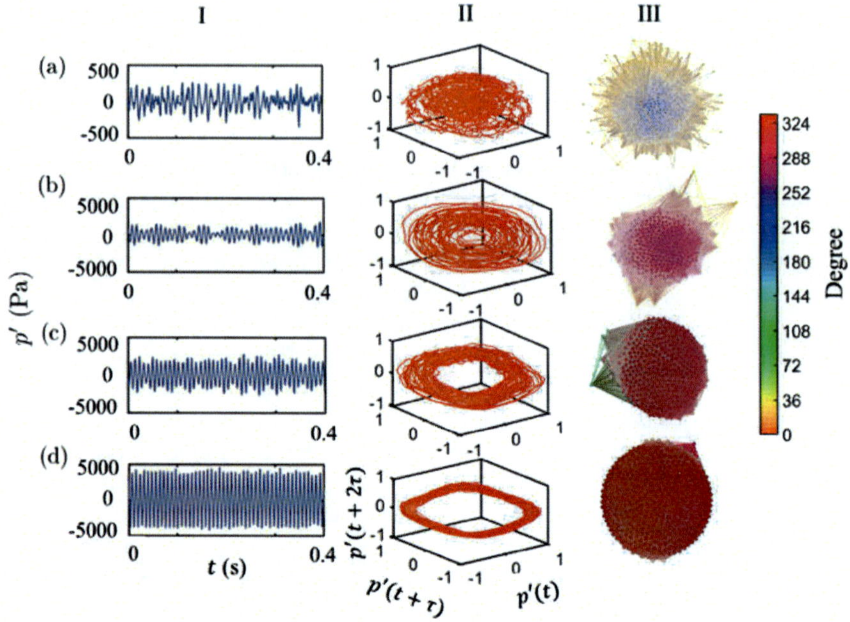

Fig. 11.21 (I) Time series of acoustic pressure fluctuations, (II) corresponding normalized phase space, and (III) cycle network during various dynamical states observed in a bluff-body stabilized turbulent combustor when $\rho = 0.92$. The dynamical states shown are (**a**) combustion noise, (**b**) intermittency, (**c**) weakly correlated and (**d**) strongly-correlated limit cycle oscillations observed during thermoacoustic instability. Reproduced with permission from Tandon and Sujith [113]

The corresponding cycle network forms a globular structure which is almost fully connected and almost all the nodes have a very high degree ($\approx$ degree 350).

Next, we discuss the use of network centrality measures derived from cycle networks to identify the occurrence of various dynamical states in the combustor and the corresponding phase space topology. If the phase space topology comprises several UPOs, the trajectory jumps often between these UPOs. The average phase space distance between these cycles is large, while the average correlation is small. When the cycles of the trajectory are weakly correlated in phase space, the corresponding nodes in the cycle network are sparsely connected. As a result, the average closeness centrality $\langle C_{\text{close}} \rangle$ tends to be low in value. Further, since the trajectory jumps often between two different UPOs, the trajectory will consist of certain 'transition cycles' which connect the path of the trajectory between these two orbits. These transition cycles can be associated with the cycles around both the orbits and the corresponding nodes will have a high betweenness centrality. Therefore, average betweenness centrality $\langle C_{\text{BC}} \rangle$ captures the tendency of the transition between different UPOs and thus helps quantify the stability of periodic orbits in the phase space. Finally, since the various cycles of the trajectory form clusters around several UPOs, we expect the average clustering coefficient $\langle CC \rangle$ to

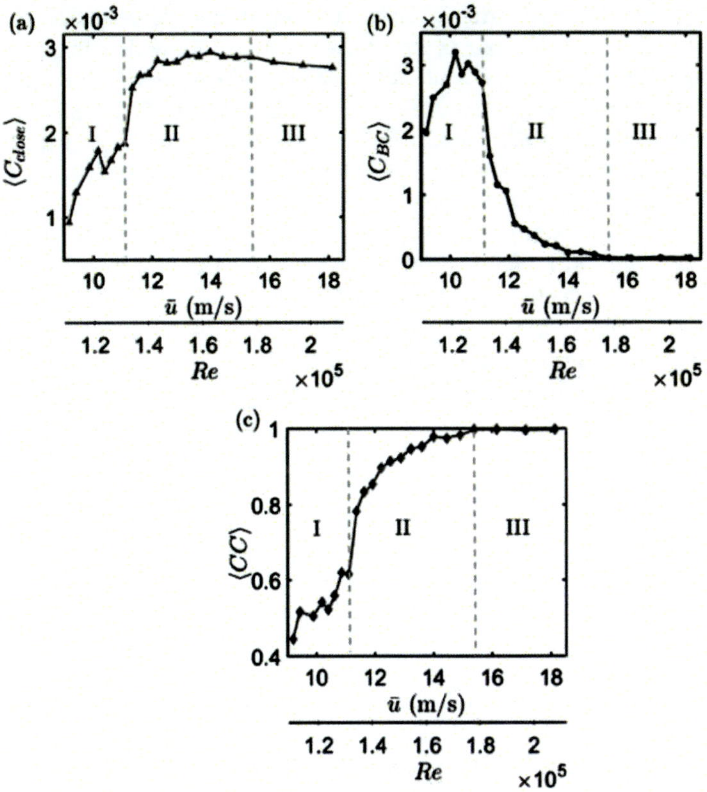

Fig. 11.22 Variation of the average (**a**) closeness centrality $\langle C_{\text{close}} \rangle$, (**b**) betweenness centrality $\langle C_{\text{BC}} \rangle$, and (**c**) clustering coefficient $\langle CC \rangle$ as a function of the control parameter (mean flow velocity $\overline{u}$ or the corresponding Reynolds number Re) when the cycle networks are derived with $\rho = 0.92$. The regions (I), (II), and (III) demarcate the occurrence of combustion noise, intermittency, and thermoacoustic instability, respectively, in a bluff-body stabilized turbulent combustor. Adapted with permission from Tandon and Sujith [113]

be small during chaotic dynamics. Thus, $\langle CC \rangle$ characterizes the number of periodic orbits in the phase space.

On the other hand, when the phase space consists of a highly stable periodic orbit, all cycles of the trajectory are located close to each other on that orbit and the corresponding nodes in the cycle network have high closeness centrality. Moreover, the trajectory clusters around a dominant and stable periodic orbit, and the number of transition cycles in the phase space is negligible. As a result, the average betweenness centrality is expected to be very small, while the average clustering coefficient is expected to be very high.

In region I of Fig. 11.22, we observe that $\langle C_{\text{close}} \rangle$ and $\langle CC \rangle$ are very low and $\langle C_{\text{BC}} \rangle$ is very high and all these three measures increase during the state of combustion noise. We, therefore, infer that the phase space comprises several highly unstable periodic orbits during the state of combustion noise, leading to chaotic dynamics. Then, with increase in the control parameter ($\overline{u}$), we find that $\langle C_{\text{close}} \rangle$ and $\langle CC \rangle$ increases significantly, while $\langle C_{\text{BC}} \rangle$ begins to decrease, which is identified as the onset of the state of intermittency (region II in Fig. 11.22). We infer that the number of periodic orbits decreases while their stability increases with $\overline{u}$ during intermittency. As a result, the trajectory is trapped for longer epochs around the more stable UPOs, leading to epochs of periodic oscillations amidst aperiodic fluctuations in the acoustic pressure signal. Finally, in region III of Fig. 11.22, we find that $\langle C_{\text{close}} \rangle$ saturates to a very high value, $\langle CC \rangle \rightarrow 1$ and $\langle C_{\text{BC}} \rangle \rightarrow 0$, indicating the presence of a dominant and highly stable periodic orbit in the phase space. We identify the point at which all the three centrality measures saturate as the onset point of thermoacoustic instability in the combustor.

We noticed that as order (periodicity) emerges in the combustor dynamics, the cycles in the phase space tend to concentrate around fewer and more stable periodic orbits. At the onset of thermoacoustic instability, all phase space cycles "*condense*" around a single periodic orbit which is the stable limit cycle oscillator. Tandon and Sujith [113] drew a parallel between the well-known Bose–Einstein condensation [20, 38] and the transformation of phase space topology as order emerges from chaos in the dynamics of the combustor (Fig. 11.21), where the motivation was to cast the transition from chaotic to periodic dynamics in a combustor as a "*phase transition*" [99].

Bose–Einstein condensation is a well-known phenomenon in quantum physics, where typical quantum particles called '***Bosons***' that are initially distributed across several high-energy levels condense to the lowest and most stable energy state with the decrease in temperature of the system [20, 38, 99]. In statistical physics, a system is said to show a condensation transition when a large fraction of particles of the system occupies a single energy level referred to as the 'ground state'.

In order to draw an analogy with Bose–Einstein condensation, we can consider phase space cycles to be analogous to Boson particles and periodic orbits are equivalent to energy levels. But how do we define the energy of periodic orbits in the phase space? We observed earlier that, higher the stability of a periodic orbit, higher the number of phase space cycles associated with it. All these cycles will be strongly correlated and the corresponding nodes in the cycle network will be connected. Therefore, higher the stability of an orbit, higher the degree of nodes corresponding to the cycles associated with that orbit. We can define energy (E) of a node such that higher the stability, lower the energy of the periodic orbit with which that node is associated. Therefore, we can define E as,

$$E = 1 - k/k_{\text{max}} \tag{11.12}$$

where k is the degree of the node and k_{max} is the maximum possible degree of the cycle network. Note that $k_{\text{max}} = N - 1$, where N is the total number of nodes in the cycle network.

In Fig. 11.23, we show how the particles are distributed in different energy levels for various dynamical states observed in a turbulent combustor when $\rho = 0.92$.

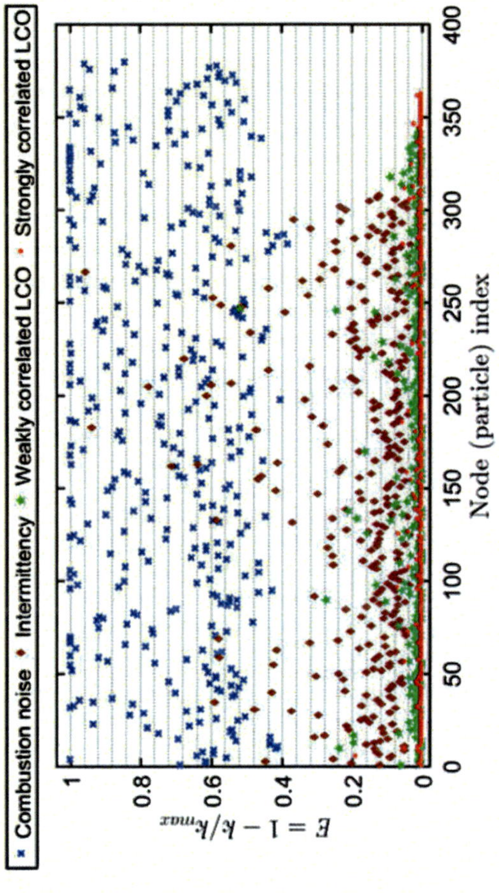

Fig. 11.23 Distribution of particles (nodes) in energy levels (E) during various dynamical states observed in a bluff-body stabilized turbulent combustor for $\rho = 0.92$. Reproduced with permission from Tandon and Sujith [113]

During the state of combustion noise, we find that the particles are distributed across several high energy levels during the state of combustion noise. We infer that the phase space cycles are associated with several highly unstable periodic orbits. During the state of intermittency, the particles are spread across a smaller number of energy levels, indicating that the number of such unstable periodic orbits in the phase space decreases during this state. Further, we also observe that the particles are associated with lower energy levels during intermittency as compared to combustion noise, which implies that on average, the stability of periodic orbits is comparatively higher during the occurrence of intermittency. During the state of weakly correlated limit cycle oscillations, we observe that a significant fraction of particles occupy a very few and very low energy levels. Whereas, during the state of strongly correlated limit cycle oscillations almost all the particles condense to the lowest energy level, i.e., the ground state. Such condensation implies that the phase space topology comprises a highly stable periodic orbit during the occurrence of thermoacoustic instability.

Note that such an analogy to Bose–Einstein condensation provides a novel insight and simultaneously portrays how the number and the stability of periodic orbits in the phase space morphs when order emerges in the dynamics of a turbulent combustor. We, therefore, propose that the onset of thermoacoustic instability can be viewed as the onset of a condensation transition. This new perspective can potentially allow cross-fertilization of concepts (or borrowing tools) from the study of phase transitions in statistical physics to predict the onset of oscillatory instabilities. Further, a spatio-temporal analysis along these lines can possibly help infer the flow physics of the combustor.

11.4.5 Recurrence Network Topologies of Different Dynamical States

In this section, we discuss another popular method of mapping a time series into a complex network, called ϵ-recurrence network [35, 71]. This method of network construction is based on the analysis of recurrence properties of state points in the reconstructed phase space (Fig. 11.24). Every state point in the phase space is considered as a node. Based on the recurrence behavior (or the mutual closeness) of these state points, we can establish links between the nodes. A recurrence matrix provides the information for the recurrence behavior of phase space trajectory, which is obtained as follows, [72],

$$R_{i,j}(\epsilon) = \Theta(\epsilon - \|x_i - x_j\|) \tag{11.13}$$

where Θ is a Heaviside function, x_i and x_j are the state space points, ϵ is the recurrence threshold, and $\|.\|$ is the Euclidean norm. $i, j = 1, 2, ..., N - (d - 1)\tau$, where N is the total number of points in the signal, τ is the optimum time delay,

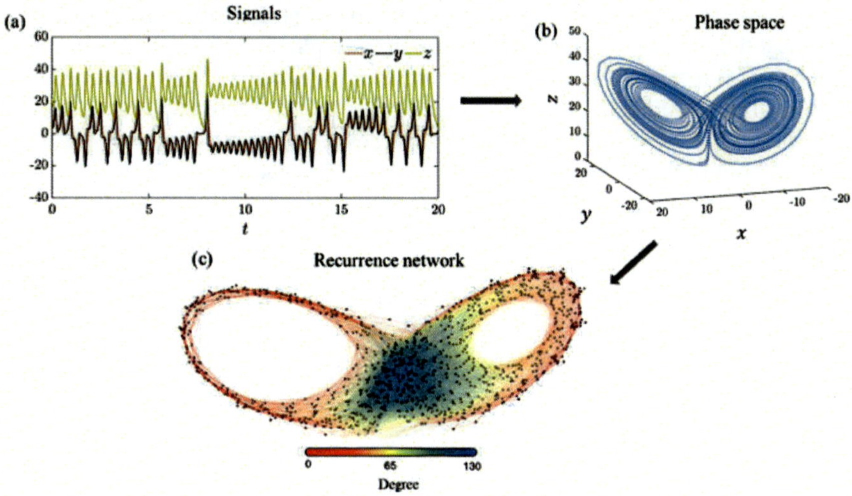

Fig. 11.24 (**a**) Time series of three variables of a Lorenz model and a corresponding (**b**) phase space and (**c**) recurrence network. The recurrence network preserves the topology of a three-dimensional phase space of a Lorenz attractor in a two-dimensional space. Courtesy: Vedasri Godavarthi (University of California, USA)

and d is the embedding dimension. When the state points are within ϵ distance from each other (i.e., $\|x_i - x_j\| < \epsilon$), we consider them as recurred. The recurring state points are indicated by 1 and non-recurring states points are represented by 0 in the recurrence matrix.

We can derive the adjacency matrix from the recurrence matrix. To remove self-connections of a node, we obtain an adjacency matrix by subtracting a Kronecker delta (identity) matrix ($\delta_{i,j}$) from a recurrence matrix as follows,

$$A_{i,j}(\epsilon) = R_{i,j}(\epsilon) - \delta_{i,j} \tag{11.14}$$

where $\delta_{i,j} = 1$, if $i = j$ and, $\delta_{i,j} = 0$, if $i \neq j$.

The choice of threshold ϵ plays a crucial role in the construction of a network. A larger threshold would lead to a very dense network containing too many links, whereas a smaller size threshold would result in a disconnected network with a few links [56]. There are multiple ways of choosing a threshold for the construction of an ϵ-recurrence network. Jacob et al. [56] developed a unified framework to choose an optimum threshold for the construction of a recurrence network. Other methods include a fixed recurrence rate or a fraction of the diameter of the attractor, etc. We can derive all complex network measures, as discussed in Sec. 11.2, directly from the adjacency matrix of the ϵ—recurrence network and analyze different properties of a signal [71].

Since the adjacency matrix consists of only 0s and 1s (see (11.14)) and no information for the direction of the connection between the nodes, we can refer to the corresponding recurrence network as an unweighted, undirected network. We note that the measures derived from the recurrence plot of a signal help us in characterizing its temporal properties (i.e., regular or irregular behaviors). On the other hand, as the construction of recurrence networks is solely based on the closeness of state points in the phase space and does not account for time information, measures derived from recurrence networks aid in describing the geometrical properties and not the dynamic properties of the underlying attractor [35].

The advantage of the recurrence network analysis is that, in order to get useful information of underlying features of an attractor, it does not require high temporal sampling of the data; we can construct a network from a lesser number of data points as compared to that required in other methods of the network construction [131]. Furthermore, we can use this approach for noisy and non-uniformly sampled signals as well [35].

In Figs. 11.25a–c, we show topologies of recurrence networks, obtained from the time series of the acoustic pressure fluctuations, corresponding to the states of combustion noise, intermittency, and thermoacoustic instability, respectively [50]. These networks show the distribution of the degree of nodes. The adjacency matrix is obtained from the recurrence matrix and the network topology is visualized using a Gephi software.

As the dynamics of the state of combustion noise is aperiodic, its network topology shows a cluttered behavior (Fig. 11.25a). Furthermore, since the state of thermoacoustic instability consists of periodic oscillations, the corresponding network topology shows a ring-like structure (Fig. 11.25c), similar to that of a limit cycle attractor. During the state of intermittency (Fig. 11.25b), as the dynamics alternately switches between periodic and aperiodic oscillations, the corresponding network topology displays a combination of a ring-like structure and a cluttered structure. Furthermore, as the number of recurrences of the phase space trajectories is lesser for the state of combustion noise than that for intermittency, the degree of nodes in the network of combustion noise is very low as compared to that of intermittency. As every state point in the phase space possesses nearly equal recurrences during the state of thermoacoustic instability, the network topology of this state shows very high value for a single degree. All these observations also confirm that the network topology preserves the geometric properties of an attractor of a particular dynamical state.

11.4.6 Directional Dependence Between the Coupled Acoustic Pressure and Heat Release Rate Fluctuations Using Recurrence Networks

In the previous section, we discussed the construction of ϵ–recurrence networks using a single time series data, to analyze network topologies of different dynamical states of a thermoacoustic system. Next, we extend this analysis on recurrence

Fig. 11.25 Different topologies of ϵ-recurrence network obtained from acoustic pressure signals corresponding to the states of **(a)** combustion noise, **(b)** intermittency, and **(c)** thermoacoustic instability observed in a bluff-body stabilized turbulent thermoacoustic system. These networks show the distribution of degree of the nodes, as indicated in the color bar. A recurrence threshold used for the network construction is based on the fraction of the attractor size. Adapted with permission from Godavarthi et al. [50]

networks to study the coupled interaction between two oscillators, i.e. the acoustic pressure and the heat release rate, of a turbulent thermoacoustic system, similar to that presented in Chap. 8. Similar to the method of construction of joint and cross recurrence plots (as discussed in Chap. 8), we can also use this multivariate recurrence analysis to construct joint and cross recurrence networks. Measures derived from these networks, such as the joint transitivity and the cross transitivity [45], can then be used to study the coupled behavior and causality of two variables of a system. To obtain the joint and the cross transitivities of a network, we first need to compute the adjacency matrices from joint and cross recurrence matrices, respectively.

As we discussed in Chap. 2, the joint recurrence matrix measures the occurrence of simultaneous recurrence of two signals (say, x and y) in their individual phase spaces. The joint recurrence matrix is calculated through the product of the recurrence matrices of the individual signals, as given below [45, 100],

$$JR_{ij}(\epsilon_x, \epsilon_y) = \Theta(\epsilon_x - \|x_i - x_j\|)\Theta(\epsilon_y - \|y_i - y_j\|) \tag{11.15}$$

where x_i and y_i are state vectors of two signals, $i, j = 1, 2, ..., N_1$, and N_1 is the number of states points in the phase space.

The cross recurrence matrix measures the recurrence of trajectories of two normalized signals in the same neighborhood, i.e., it measures the similarity of phase spaces as follows,

$$CR_{ij}(\epsilon) = \Theta(\epsilon - \|x_i - y_j\|) \tag{11.16}$$

where $i = 1, 2, ..., N_x$, $j = 1, 2, ..., N_y$ and, N_x and N_y are the number of states points in the phase space for x and y signals, respectively.

Furthermore, we can obtain the adjacency matrix for joint recurrence networks as,

$$A_{ij}(\epsilon_x, \epsilon_y) = JR_{ij}(\epsilon_x, \epsilon_y) - \delta_{ij} \tag{11.17}$$

and the adjacency matrix for cross recurrence networks as,

$$A_{ij}(\epsilon) = CR_{ij}(\epsilon) - \delta_{ij} \tag{11.18}$$

Unlike the recurrence matrix and the joint recurrence matrix, the cross recurrence matrix is not symmetric. Using this asymmetric property of CR, we can quantify the directional dependence in the coupling of two systems.

As we discussed in Sect. 11.2, the network transitivity computes the fraction of triplets that form triangles in a network. Here, a triplet comprises a center node that is connected to two other nodes which are not connected to each other, and a triangle consists of three nodes that are connected to each other [85]. The transitivity of two networks is calculated as [45],

$$\mathcal{T} = \frac{\sum_{i,j,k=1}^{N} A_{ij} A_{jk} A_{ki}}{\sum_{i,j,k=1}^{N} A_{ij} A_{ki}} \tag{11.19}$$

where i, j and k are three nodes, A is the adjacency matrix [45]. If nodes j and k are connected to node i, i.e., $A_{ij} = 1$, $A_{ik} = 1$, then transitivity indicates nodes j and k are also connected, i.e., $A_{jk} = 1$.

The value of joint transitivity ($\mathcal{T}_J$) quantifies the simultaneous existence of closed triangles in the networks of two systems. In contrast, cross transitivity ($\mathcal{T}_{XY}$) quantifies the probability of existence of two connected nodes in a network of one system that form a cross-triangle with a node in the network of another system. Thus, cross transitivity measures the inter-system connectivity of two networks.

When two systems are uncoupled, we have $\mathcal{T}_x$, $\mathcal{T}_y > \mathcal{T}_J$. Here, $\mathcal{T}_x$ and $\mathcal{T}_y$ are transitivity of networks of individual systems X and Y, respectively. During the state of generalized synchronization, $\mathcal{T}_J \to \mathcal{T}_x, \mathcal{T}_y$ and the value of $\mathcal{T}_J$ becomes maximum due to an increase in the simultaneous connectivity in the network [44]. Feldhoff et al. [45] proposed a normalized measure, called the transitivity ratio, to detect the onset of generalized synchronization in the system of coupled oscillators. It is a ratio of the joint transitivity ($\mathcal{T}_J$) and the arithmetic mean of transitivity of individual oscillators (i.e. $\mathcal{T}_x$, and $\mathcal{T}_y$) as given below,

$$Q_{\mathcal{T}} = \frac{\mathcal{T}_J}{(\mathcal{T}_x + \mathcal{T}_y)/2} \tag{11.20}$$

At the onset of generalized synchronization, individual transitivity of both the systems becomes the same as their joint transitivity. Hence, $Q_{\mathcal{T}}$ attains a value equal to 1.

Furthermore, if a cross recurrence matrix of two systems is not symmetric (i.e., CR_{ij} is not the same as CR_{ji}), the corresponding cross transitivity $\mathcal{T}_{xy}$ for both the systems also shows asymmetry, i.e., $\mathcal{T}_{xy} \neq \mathcal{T}_{yx}$. Therefore, we can use this asymmetry property of the cross transitivity ($\mathcal{T}_{xy}$) to detect the directionality in the coupling of two systems. For unidirectionally coupled systems, i.e., if X is a driver and Y is a driven system, $\mathcal{T}_{xy} < \mathcal{T}_{yx}$. This is because, a driver system X attracts the trajectory of a driven system Y; hence, triplets in X are more likely to have connections with a node in Y and form closed cross-triangles, than vice versa. In contrast, for mutually coupled X and Y systems, as there is an equal probability for triplets in X to form cross-triangles with nodes in Y and vice versa, implying $\mathcal{T}_{xy} \approx \mathcal{T}_{yx}$. Thus, based on the difference between the cross transitivities, we can estimate the directional dependence between the coupled systems.

In Fig. 11.26, we show recurrence networks constructed from joint (Fig. 11.26a–d) and cross (Fig. 11.26e–h) recurrence matrices of the acoustic pressure (p') and heat release rate ($\dot{q}'$) fluctuations, for different dynamical states observed during the onset of thermoacoustic instability in a turbulent thermoacoustic system [49]. In a network, nodes are colored based on their degree. During the state of combustion noise (Fig. 11.26a), as both p' and $\dot{q}'$ signals are desynchronized (as we discussed

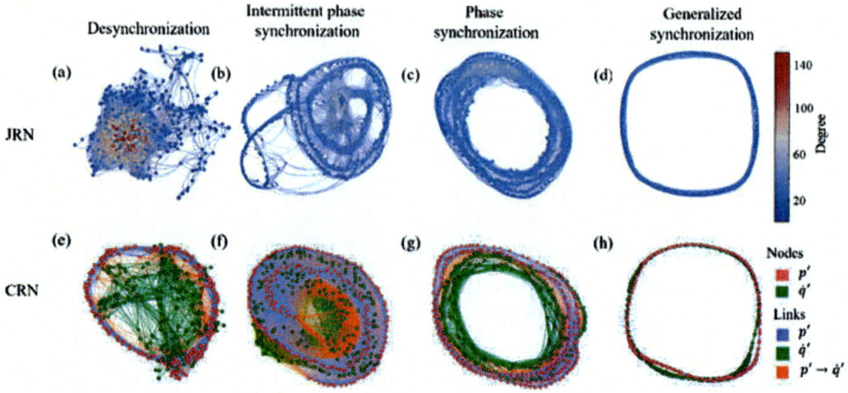

Fig. 11.26 Different topologies of the multivariate ϵ-recurrence networks, constructed using (**a**–**d**) joint recurrence networks (JRN) and (**e**–**h**) cross recurrence networks (CRN), from the signals of acoustic pressure (p') and heat release rate ($\dot{q}'$) fluctuations. These networks are obtained for the states of (**a, e**) desynchronization, (**b, f**) intermittent phase synchronization, (**c, g**) phase synchronization, and (**d, h**) generalized synchronization observed in a bluff-body stabilized turbulent thermoacoustic system. The color bar indicates the variation of the degree of the nodes in the network. The networks are constructed by fixing the recurrence rate as a threshold. Reproduced with permission from Godavarthi et al. [49]

in Chap. 8), their joint network topology shows cluttered behavior, consisting nodes with a wide range of degrees (from low to high).

In contrast, for the states of thermoacoustic instability, i.e., weakly correlated (Fig. 11.26c) and strongly correlated (Fig. 11.26d) limit cycles, as both p' and $\dot{q}'$ signals are periodic and synchronized, their joint network topologies exhibit a similar topology as that of the limit cycle attractor. Due to the simultaneous recurrence of phase space trajectories during these states, the degree of each node in the network attains nearly the same value. As we discussed in Chap. 8, since p' and $\dot{q}'$ signals are phase synchronized and not amplitude synchronized during the occurrence of weakly correlated thermoacoustic instability, their joint network shows a thick ring topology (Fig. 11.26c). In contrast, as p' and $\dot{q}'$ signals show generalized synchronization (i.e., both phase and amplitude synchronization) during the occurrence of strongly correlated thermoacoustic instability, their joint network topology shows a thin ring (Fig. 11.26d). During the occurrence of intermittency, the joint network topology consists of a combination of topologies which are observed for states of weakly correlated thermoacoustic instability and combustion noise (Fig. 11.26b).

In Figs. 11.26e–h, we show network topologies obtained from the cross recurrence matrix and also from the individual recurrence matrices of the normalized acoustic pressure (p') and heat release rate ($\dot{q}'$) signals, for different dynamical states of a turbulent thermoacoustic system [49]. The cross recurrence of states in the phase space of p' to the states in the phase space of $\dot{q}'$ are interconnected through orange colored links in the network topology.

During the state of combustion noise (Fig. 11.26e), we observe random connections in the cross-recurrence network of p' and $\dot{q}'$ signals. Furthermore, we notice that individual networks of these signals do not overlap with each other, indicating the desynchrony between them. For the state of intermittency (Fig. 11.26f), a portion of nodes (indicated in red and green) appears closer and other nodes appear farther apart in the network. The presence of synchronization between p' and $\dot{q}'$ signals leads to the occurrence of similar states in their cross recurrence plot. This, in turn, brings the nodes in a network closer and also causes the overlap of individual networks of these signals on each other. Thus, we notice the property of intermittent synchronization of p' and $\dot{q}'$ signals during the occurrence of intermittency. For the state of weakly correlated thermoacoustic instability (Fig. 11.26g), we observe thick ring-type network topologies for both p' and $\dot{q}'$ signals and their interconnections. In contrast, the network topologies observed for p' and $\dot{q}'$ signals during the state of strongly correlated thermoacoustic instability (Fig. 11.26h) exhibit thin rings, and individual networks appear very close to each other due to the presence of similar states. This behavior further confirms the existence of phase synchronization in the former state (Fig. 11.26g) and generalized synchronization in the later state (Fig. 11.26h), which we will confirm next through quantitative measures.

In Fig. 11.27, we show the variation of different network measures of transitivity, derived from both the joint and the cross recurrence matrices, as a function of the flow velocity ($\bar{u}$) to quantify the synchronization transition in the system. Figure 11.27a shows the variation of joint transitivity (τ_J) and transitivity ratio (Q_τ) with $\bar{u}$ during the transition from combustion noise to thermoacoustic instability via intermittency. The values of these measures first show an increase and then saturation at the onset of thermoacoustic instability. During the state of generalized synchronization of thermoacoustic instability, the presence of similar states in the recurrence plots causes the individual transitivity of p' and $\dot{q}'$ signals to become equal to their joint transitivity. As Q_τ is a normalized quantity, its value reaches 1 during the occurrence of thermoacoustic instability. Therefore, we can use Q_τ to detect the onset of the generalized synchronization between the acoustic field and the heat release rate fluctuations in the system. A gradual increase of both τ_J and Q_τ indicate the increase in the coupling between p' and $\dot{q}'$ oscillations during the onset of thermoacoustic instability.

In Fig. 11.27b, we show the variation of cross transitivity $\tau_{p'\dot{q}'}$ and $\tau_{\dot{q}'p'}$ with the flow velocity ($\overline{u}$). Here, $\tau_{p'\dot{q}'}$ indicates the probability that two nodes in the recurrence network of $\dot{q}'$ are connected, given that they are neighbors to a node in the recurrence network of p'. The interconnection between the acoustic pressure and the heat release rate signals is obtained from the cross recurrence plot.

As we discussed before, the cross recurrence matrix need not be symmetric and interconnections between the nodes in the network depend on whether the distance matrix is calculated from $p' \rightarrow \dot{q}'$ or from $\dot{q}' \rightarrow p'$. The difference in interconnections between p' and $\dot{q}'$ makes the cross transitivity asymmetric. Thus, we can use this measure as an indicator of the directionality of the coupling between the systems [44]. The values of $\tau_{p'\dot{q}'}$ and $\tau_{\dot{q}'p'}$ show nearly the same values for the state of combustion noise and intermittency as a result of weak

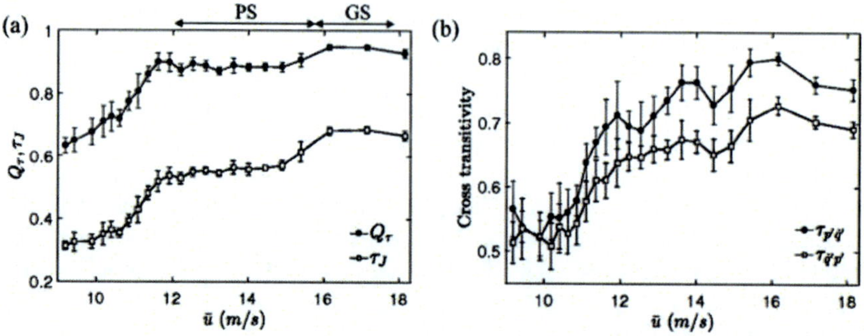

Fig. 11.27 The variation of different measures of the multivariate ϵ-recurrence networks, (**a**) joint transitivity (τ_J) and the transitivity ratio (Q_τ), and (**b**) cross transitivity ($\tau_{p'\dot{q}'}$ and $\tau_{\dot{q}'p'}$), as a function of the flow velocity, $\overline{u}$ (control parameter). A plot in (**a**) helps in detecting the onset of phase synchronization (PS) and generalized synchronization (GS) states and a plot in (**b**) helps in detecting the possible directionality of coupling between the pressure and heat release rate fluctuations in a turbulent thermoacoustic system. Reproduced with permission from Godavarthi et al. [49]

coupling between the signals during these states. As the system dynamics approach the onset of thermoacoustic instability, we notice a divergence in the variation of these plots, where the variation of $\tau_{p'\dot{q}'}$ is higher than that observed for $\tau_{\dot{q}'p'}$. This difference could be a result of the stronger influence of $\dot{q}'$ on p' than vice versa. This difference in the cross transitivity of p' and $\dot{q}'$ signals further indicates the presence of asymmetric bidirectional coupling between them.

Until now, we discussed the methodologies of complex network construction from the temporal data of a thermoacoustic system. We also shed light on analyzing the temporal dynamics of coupled acoustic pressure and heat release rate fluctuations during the onset of thermoacoustic instability. Next, we will discuss the spatiotemporal analysis of a turbulent thermoacoustic system using the complex network approach.

11.5 Complex Network Approach to Study Spatial Dynamics of Thermoacoustic Systems

As we discussed in the introduction of this chapter, a turbulent thermoacoustic system is a complex system. In this system, the local interaction between elements of the reaction field, under the influence of global acoustic field, leads to the periodic occurrence of self-organized large-scale emergent structures during the state of thermoacoustic instability (also refer to Chap. 7).

A few studies have exploited the approach of complex networks to study spatial interaction of different parts in the combustor [53, 59–61, 76, 77, 118]. Several measures have been developed to construct spatial networks and interpret the results of spatiotemporal dynamics of such systems during the onset of thermoacoustic instability. The approach of complex networks is also used to identify critical regions existing in the reaction field of the combustor [59, 61, 76, 119]. These critical regions play a crucial role in controlling the dynamics of the system during the state of thermoacoustic instability. Hence, we need to destroy these regions, through some means of external perturbations, to quench such instabilities in the system. We will be summarizing the control strategies developed using the complex network approach in Chap. 12. In the following subsections, we will discuss in detail the application of complex network theory to study the spatiotemporal properties of thermoacoustic systems.

There are different measures, such as Pearson correlation, mutual information, and event synchronization [34], that are used to construct spatial networks in different areas of climate [19, 70, 116, 117], ocean [75], and fluid mechanics [81, 102, 110]. In a turbulent thermoacoustic system, we can construct networks from spatiotemporal measurements of the reaction field (heat release rate fluctuations) and the hydrodynamic field (velocity fluctuations) in the combustor. Using different network measures, we can disclose the hidden information of the combustion dynamics. We can construct the acoustic power networks from information of both the local heat release rate fluctuations (obtained from high-speed chemiluminescence imaging of the flame) and the acoustic pressure fluctuations (obtained from pressure transducers) in the system. Furthermore, we can obtain the velocity networks from information of the local velocity vectors (obtained from the measurements of particle image velocimetry, PIV) of the hydrodynamic field in the combustor.

During the construction of spatial networks, every image of the spatial field of a combustor is divided into a rectangle grid, where each grid point is considered as the node of the network. The connection between two nodes is established through some specific criteria. In the following subsection, we present a few methods of constructing networks from the spatiotemporal data of the turbulent reactive field in a combustor.

11.5.1 Unweighted Spatial Correlation Networks of the Time-Averaged Flow Field

We can construct unweighted, undirected correlation networks using the information of the velocity field, obtained from PIV measurements in a turbulent combustor [118]. There are two different ways to obtain these networks: (1) the Pearson correlation coefficient and (2) the dot product .

11.5.1.1 Construction of Spatial Networks Using the Pearson Correlation

The Pearson correlation coefficient between two variables (say, x and y) is obtained as follows [118],

$$R_p = \frac{\sum_{i=1}^{N}(x_i - \bar{x})(y_i - \bar{y})}{\sqrt{\sum_{i=1}^{N}(x_i - \bar{x})^2}\sqrt{\sum_{i=1}^{N}(y_i - \bar{y})^2}} \tag{11.21}$$

Here, x_i and y_i are the velocity information for the ith time instant at two different locations of the grid. $\bar{x}$ and $\bar{y}$ indicate the arithmetic mean of x and y time series, respectively. N corresponds to the total number of data points in the signal.

The value of R_p lies between -1 and 1. The correlation between two signals attains a maximum value when these signals vary in a similar fashion, despite them having different amplitudes. For an unweighted network, we can establish the connection between the nodes when the value of the Pearson correlation coefficient is above a chosen threshold (R_{pt}). We can select the value of R_{pt} from the analysis of the link density (ρ) of the network. The link density measures the fraction of number of links present in a network. It is obtained as the ratio of the actual number of links (say, E) to the maximum possible number of links (say, $N(N-1)/2$) in a network [12],

$$\rho = \frac{2E}{N(N-1)} \tag{11.22}$$

We need to choose the value of R_{pt} such that the variation in the link density is maximum for the states of combustion noise, intermittency, and thermoacoustic instability [118]. A maximum variation in the link density accounts for the maximum variability of the network topology across different dynamical states of the combustor operation.

11.5.1.2 Networks Based on Dot Products

The spatial network can be constructed using the dot product of two signals as follows [118],

$$R_s = \sum_{i=1}^{n} x_i y_i \tag{11.23}$$

The value of R_s depends on the magnitude of the signals. It takes a maximum value when two signals follow each other in a similar fashion and possess similar amplitudes. As a result, the value of R_s is not bounded and carries any value depending on the amplitude of the signals. For example, to construct a network based on this measure, we can fix the link density of the network at 50% [118]. It also means that the network corresponding to every dynamical state contains 50% of the total possible links.

The adjacency matrix (A_{ij}) of the spatial network is obtained as follows,

$$A_{ij} = \begin{cases} 1 & \text{if } i \neq j \text{ or if } i \text{ and } j \text{ are connected} \\ 0 & \text{if } i = j \text{ or if } i \text{ and } j \text{ are not connected} \end{cases} \tag{11.24}$$

where, $i, j = 1, 2, \ldots.., N$ are the nodes of a network.

The adjacency matrix takes a value of unity when there exists a connection between the nodes i and j, and a value of zero in the absence of any link between these nodes. The adjacency matrix of correlation network is always symmetric due to the absence of any directional dependence between the nodes (i.e., if node i is connected to node j, it implies that node j is also connected to node i).

Depending on the aforementioned methodologies employed to construct a network, the meaning associated with the properties of the corresponding network also varies. For example, in a turbulent reactive flow, a correlation between two points decays as the distance between these points increases and it dies down to zero once the distance is increase above a critical value. Nevertheless, as the system dynamics reach the state of thermoacoustic instability, the correlation is established between different regions of the combustor due to the emergence of the large-scale coherent structures in the flow field. In the case where the correlation between two points is measured using a Pearson correlation coefficient, irrespective of the distance between the points, if the velocity fluctuations between them vary similarly, the Pearson correlation coefficient exhibits very high value. Thus, we can detect the correlation between farther points, which is only possible in the presence of large-scale coherent structures in the system. Therefore, the construction of the network through Pearson correlation coefficient helps in detecting both the existence of large-scale coherent structures in the flow field and the local correlation between spatial locations of the system. On the other hand, if the network is constructed using a dot product, velocities of comparable amplitude at two locations in the reaction field exhibit high correlation only for neighboring points alone. Thus, we can detect the local correlation between spatial points using the network obtained from the method of dot product.

In Fig. 11.28, we show the spatial networks constructed from the velocity field of a turbulent bluff-body stabilized dump combustor using the Pearson correlation coefficient [118]. The spatial distribution of several network properties (as discussed in Sec. 11.2) of such networks is shown for the states of combustion noise, intermittency, and thermoacoustic instability.

Figure 11.28a–c presents the spatial networks formed using the degree of a node in the network for the aforementioned states of the combustor dynamics, respectively. The degree provides information about the number of neighbors of a given node in the network. A very high value of degree indicates that the correlation between the velocity fluctuations observed between a pair of points (nodes) is greater than the correlation threshold (R_{pt}). For the state of combustion

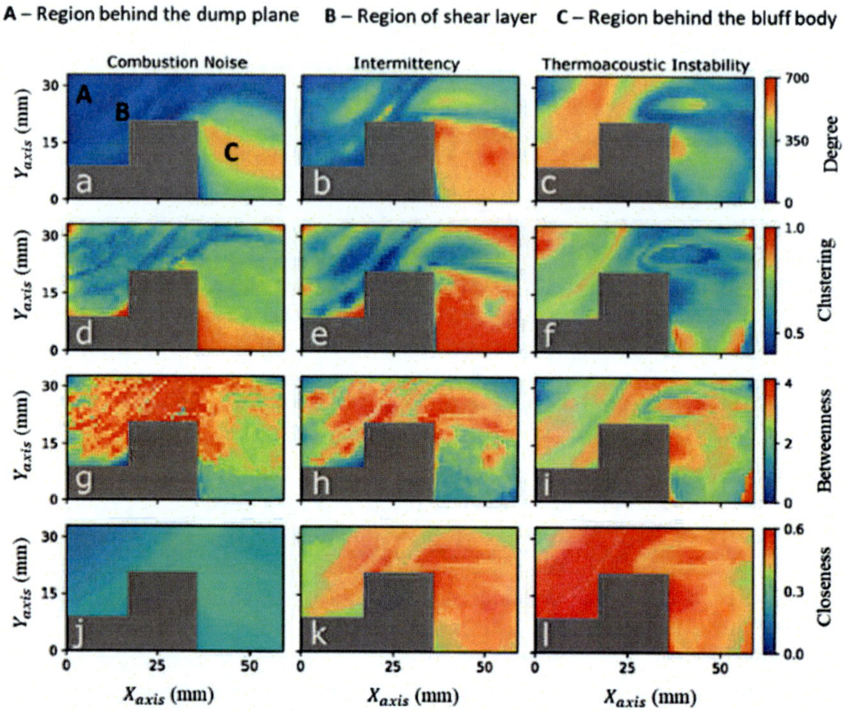

Fig. 11.28 Spatial distribution of different network properties, such as **(a–c)** degree, **(d–f)** clustering coefficient, **(g–i)** betweenness centrality, and **(j–l)** closeness centrality, for the states of combustion noise, intermittency, and thermoacoustic instability observed in a turbulent bluff-body stabilized dump combustor. These averaged spatial networks are constructed using a Pearson correlation coefficient for a correlation threshold of 0.25. In **(a)**, A, B, and C indicate the regions downstream of the dump plane, the shear layer, and the wake of the bluff-body, respectively. Higher values of these measures indicate the presence of critical regions in the reaction flow field of a turbulent combustor. The masked region in gray in all plots indicate the location of the bluff-body in the combustor. Adapted with permission from Unni et al. [118]

noise (Fig. 11.28a), we observe that the degree is highest in the wake region of the bluff-body (indicated as region C in Fig. 11.28a) as compared to other regions in the combustor flow field. Thus, the region of high degree coincides with the recirculation zone of the bluff-body. During the state of intermittency (Fig. 11.28b), a change in the degree distribution is observed in the shear region (indicated as region B in Fig. 11.28a) existing between the dump plane and the tip of the bluff body. As the system behavior transitions to the state of thermoacoustic instability (Fig. 11.28c), the degree distribution becomes maximum in the shear region and shows a corresponding decrease in its value in the wake of the bluff-body. Thus, we observe a shift in the region of the high degree distribution in the spatial network during the onset of thermoacoustic instability.

Figure 11.28d–f show the distribution of the local clustering coefficient in the spatial networks of the velocity field during the occurrence of combustion noise, intermittency, and thermoacoustic instability, respectively. The local clustering coefficient (C_i) provides information about the inter-connectivity of a node in the network. Higher the value of C_i, larger the connectivity amongst the nodes in a network. Thus, if a region in the network has higher values of degree and C_i, the network forms a cluster of connected nodes. In other words, clusters indicate the presence of a spatially coherent region of the velocity fluctuations in the system [118].

For the state of combustion noise (Fig. 11.28d), we observe that the wake of the bluff-body has higher values of C_i as compared to that seen downstream of the dump plane. During the state of intermittency (Fig. 11.28e), the region of the interconnected nodes is observed to increase in the wake of the bluff-body as compared to that noticed for the state of combustion noise. In the case of thermoacoustic instability (Fig. 11.28c), the wake of the bluff-body shows lower values of C_i, whereas the region downstream of the dump plane exhibits higher values of C_i. Thus, we notice a shift in the region of clustering of the nodes in the network towards the upstream flow field in the combustor, as the system dynamics transitions to the state of thermoacoustic instability.

Further, Fig. 11.28g–i demonstrate the distribution of the betweenness centrality in the spatial networks constructed from the velocity field for different states of the combustor dynamics. Betweenness centrality (b_i) measures the fraction of shortest paths between a pair of nodes that pass through a given node in the network. The region of high b_i indicates the locations that connect different parts of the flow field, which are otherwise uncorrelated to each other. Thus, it provides information of main pathways that correspond to fastest information transfer in the network.

In the case of combustion noise (Fig. 11.28g), we observe a larger region of the reaction field, i.e., both the shear zone and the top of the bluff-body, possesses high values of b_i. In contrast, we notice a reduction in the values of b_i in these regions during the state of intermittency (Fig. 11.28h). During the occurrence of thermoacoustic instability (Fig. 11.28i), nearly the entire reaction field shows moderately high value of b_i due to increased spatial correlation in the system. During this state, the presence of high values of degree and b_i along with minimal values of C_i in the shear layer imply that the nodes in this region have larger number of links, which are spread across the entire flow field, but possess a low value of interconnections. Thus, the shear zone possibly acts as a bridge that transports the flow fluctuations between the dump plane and the bluff-body wake during the state of thermoacoustic instability.

In Figs. 11.28j–l, we present the distribution of the closeness centrality in the spatial networks of the velocity field for the states of combustion noise, intermittency, and thermoacoustic instability, respectively. The closeness centrality (c_i) measures the closeness of a node to all other nodes in the network. It is proportional to the reciprocal of the sum of shortest path lengths between a node and all other nodes in the network. Therefore, any perturbation provided to a node with the highest c_i requires the least time to travel to all other nodes in the network. During the occurrence of combustion noise (Fig. 11.28j), the value of c_i is medium in the regions on the top and the wake of the bluff-body. For the state of intermittency (Fig. 11.28k), we observe an increase in the value of c_i in these regions. In contrast, in the case of thermoacoustic instability (Fig. 11.28l), we notice a very high value of c_i in the shear zone as compared to that observed in the bluff-body wake.

All these analyses suggest that regions corresponding to the higher values of different network measures in the spatial network indicate the existence of critical regions in the system. We observe the region of the bluff-body wake as a critical region during the state of combustion noise, while the shear zone appears as a critical region for the state of thermoacoustic instability. Therefore, the destruction of such critical regions through some external means (e.g. micro-jets, flow restrictors, vortex generators) would provide an effective way to control the occurrence of thermoacoustic instability in the system [59]. These external perturbations spread locally in the flow field and affect the global dynamics of the system.

Similar to the aforementioned spatial network analysis based on the Pearson correlation coefficient, we can also identify local correlations between the nodes using networks constructed through dot products (as shown in (11.23)). Using this method, Unni et al. [118] found the maximum value of betweenness centrality in the region on top of the bluff-body during the state of thermoacoustic instability. In this region, a higher heat release rate is observed due to the impingement of vortices on the bluff-body. The maximum heat release rate results in the maximum value of acoustic power production. Therefore, we can also consider this region on the top of the bluff-body as a critical region.

Having studied the process of identification of critical regions in the reaction field using the unweighted, undirected averaged spatial networks corresponding to different states of the combustor dynamics, we next discuss the use of time-varying spatial networks to study the occurrence of thermoacoustic instability in a combustor. Using time-varying network analysis, we can obtain the instantaneous information of the dynamical changes happening in the spatial network of the system. Such an approach of network analysis is extensively used in brain research to understand the dynamical functional connectivity of brain networks [22, 33, 37, 120].

11.5.2 Weighted Time-Varying Spatial Networks Obtained Though Acoustic Power and Vorticity Fields

Using the approach of time-varying networks, we can examine the spatiotemporal dynamics of a turbulent combustor during the onset of thermoacoustic instability [60]. Such a network can be constructed using the information from the production rate of acoustic power sources or the vorticity field in the combustor. We can obtain the networks for the acoustic power sources ($p'\dot{q}'$) by multiplying the local heat release rate fluctuations $\dot{q}'(x, y, t)$, obtained from OH* chemiluminescence images of the flame, with the acoustic pressure fluctuations $p'(t)$, measured at a point in the combustor. Since the length of the reaction field is much smaller than the wavelength of acoustic oscillations developed during the state of thermoacoustic instability (see Fig. 7.3 in Chap. 7), we can assume the acoustic pressure fluctuations measured at a point to remain the same in the reaction zone. In a network, every pixel of the flame image is considered as a node. The nodes corresponding to positive values of $p'\dot{q}'$ are called *acoustic power sources* and the nodes with negative values of $p'\dot{q}'$ are called *acoustic power sinks*. Further, we can assume that the connections between two nodes are local and happen due to the underlying convecting fluid flow. Hence, every node in the reaction field links to only its nearest neighbors. Since the flame image forms a square lattice structure, every node can have a maximum of eight local links in the network. We can assign the average value of $p'\dot{q}'$ between a pair of connected nodes as the weight to the link.

On the other hand, we can construct vorticity (ω) networks from the information of the local velocity field in the combustor. This information can be obtained from the PIV measurements of the flow field. In this network, a node corresponds to the interrogation window chosen for analyzing the vector field of the flow direction. The connection between nodes is established when the vorticities at these nodes are both positive ($\omega_i, \omega_j > 0$, anti-clockwise vorticity) and both negative ($\omega_i, \omega_j < 0$, clockwise vorticity). We can assign the value of average vorticity, $(\omega_i + \omega_j)/2$, as the weight to a link between the connected nodes.

In Figs. 11.29 and 11.30, we show the time series of the acoustic pressure fluctuations (p') and the instantaneous spatial distribution of node strength (s_i) obtained from the weighted networks during the states of combustion noise and thermoacoustic instability, respectively. These spatial networks are constructed from the information of the local acoustic power sources ($p'\dot{q}'$) and the vorticity field (ω) of the combustor. We can define the node strength of a particular node i as the sum of the weights of its all links as follows [10, 18],

$$s_i = \sum_{j \in n_i} w_{ij} \tag{11.25}$$

where i and j indicate nodes in a network, w is the weight of the link connecting nodes i and j, and n_i is the neighborhood of the node i. Higher the value of the node strength, the higher the weights carried by that node.

Figure 11.29b shows the spatial networks constructed by connecting the nodes with positive $p'\dot{q}'$ alone, i.e., tracking the spatiotemporal evolution of the acoustic power sources [60]. In these networks, the acoustic power production is observed to happen in small fragmented clusters distributed in the large region of the acoustic power sink (shown by blue color). These small clusters convect downstream in time. Figure 11.29c shows the plot of velocity vectors obtained from PIV, overlaid with the vorticity network. In these networks, the higher value of positive node strength is indicated in red, representing the existence of a sequence of small size vortices shed from the dump plane of the combustor. On the other hand, the presence of the

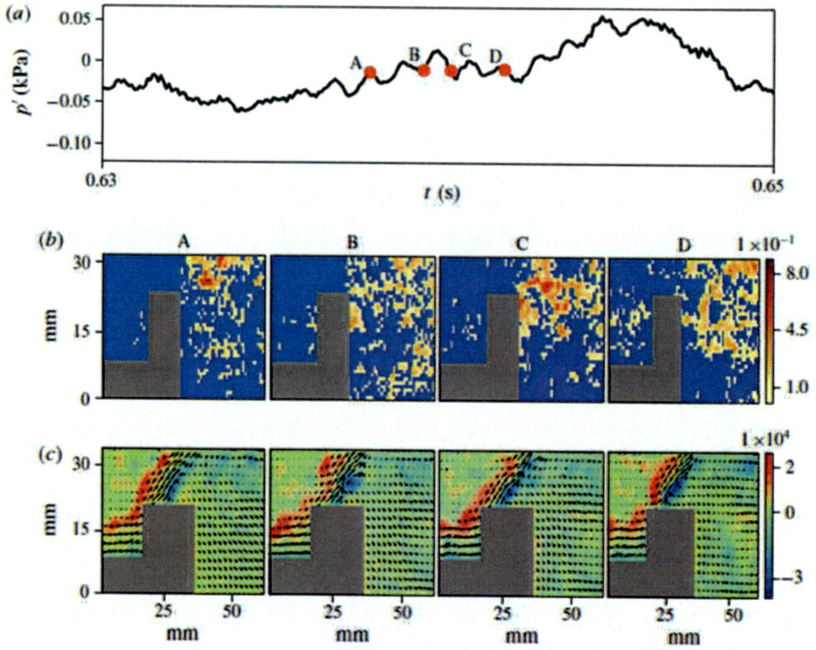

Fig. 11.29 (**a**) Time series of the acoustic pressure fluctuations during the state of combustion noise shows low amplitude aperiodic fluctuations. (**b, c**) The time-varying weighted spatial networks of acoustic power production ($p'\dot{q}'$) and the vorticity (ω) fields, respectively, constructed using node strength (s) for time instances A, B, C, and D shown in (**a**). During this state, we observe the distribution of small fragmented clusters of acoustic power production over a large region of the reaction field. The masked region in gray in all plots indicates the location of the bluff-body in the combustor. Reproduced with permission from Krishnan et al. [60]

negative node strengths (indicated in blue) denotes the presence of small vortices generated from the tip of the bluff body.

In Fig. 11.30, we show the time-varying networks of the acoustic power production and the vorticity, over an oscillation cycle of the acoustic pressure fluctuations, for the state of thermoacoustic instability. During this state, the acoustic power production happens in large clusters (Fig. 11.30b) and the magnitude of the node strength is almost three orders higher than that observed for the state of combustion noise. Further, these clusters occupy most regions of the acoustic power sink in the network. When the acoustic pressure is at a lower value, we notice the formation of a large-scale vortex in the vorticity network (Fig. 11.30c). As the acoustic pressure amplitude increases, the size of the vortex grows and its impingement on the bluff-body leads to the maximum acoustic power production in the system.

We can perform a similar analysis of network construction for the state of intermittency [60]. The dynamics of the network observed during epochs of aperiodicity

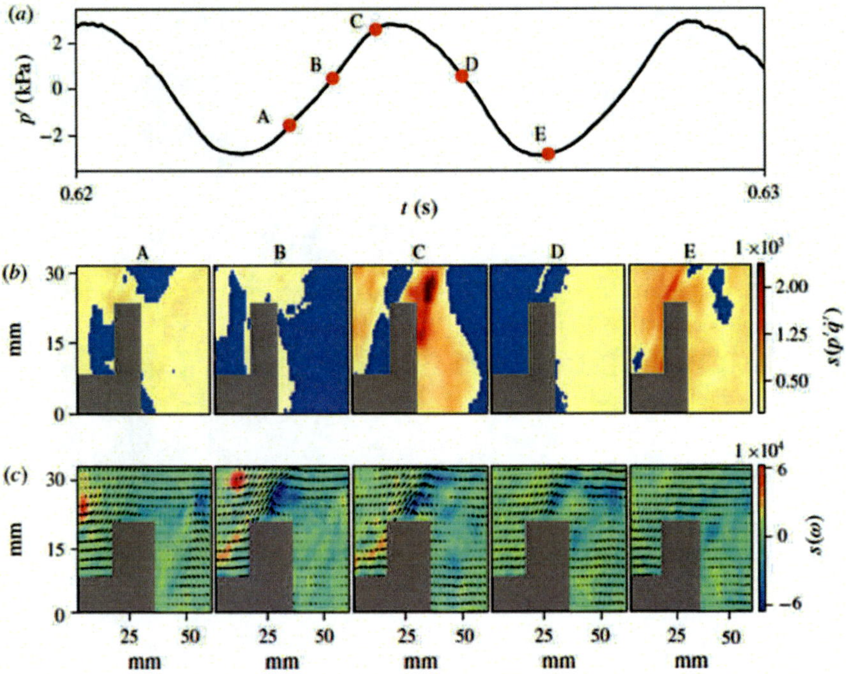

Fig. 11.30 (**a**) Time series of the acoustic pressure fluctuations during the state of thermoacoustic instability shows large amplitude periodic oscillations. (**b, c**) The time-varying weighted spatial networks of acoustic power production ($p'\dot{q}'$) and vorticity (ω) fields, respectively, constructed using node strength (s) for points A, B, C, D, and E shown in (**a**). During this state, we observe the formation of large size clusters of the acoustic power production from small fragmented clusters due to nucleation, coalescence, and growth processes in the reaction field. Reproduced with permission from Krishnan et al. [60]

of the intermittency signal is similar to that seen for the state of combustion noise (in Fig. 11.29). In contrast, during the epochs of periodic oscillation of intermittency, the size of clusters of the acoustic power production and also the size of the large-scale vortices are observed to be smaller than that for the state of thermoacoustic instability (Fig. 11.30). For epochs of the growth of acoustic pressure oscillations, observed during the transition from aperiodic to periodic oscillations at the onset of a burst, Krishnan et al. [60] witnessed the coalescence of small fragmented clusters of the acoustic power production leading to the formation of large-size clusters via nucleation, coalescence, and growth processes. They showed that these large size clusters occur due to the merging of small-scale vortices present along the shear layer.

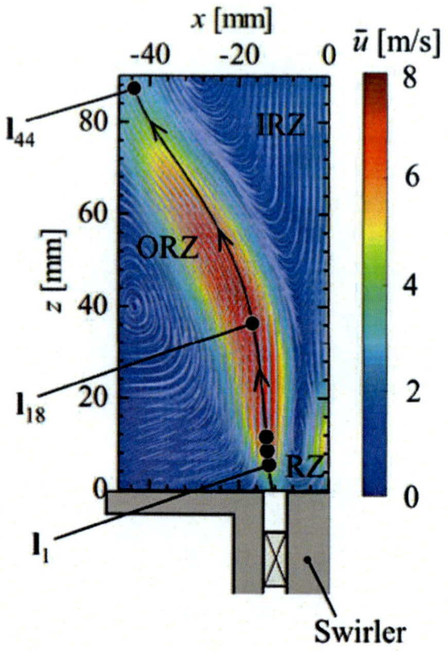

Fig. 11.31 The mean velocity field of left half of a swirl-stabilized turbulent combustor during the state of thermoacoustic instability. I_1, I_{18}, and I_{44} indicate the spatial locations along the shear layer. The vortex formation occurs along the shear layer existing between inner recirculation zone (IRZ) and outer recirculation zone (ORZ). Reproduced with permission from Murayama et al. [77]

11.5.3 Weighted Time-Varying Turbulence Networks Obtained Though Vorticity Fields

In this section, we discuss the approach of time-varying complex networks to study vorticity interactions in the reaction field of a turbulent combustor during the state of thermoacoustic instability [61, 77]. In Fig. 11.31, we show the averaged velocity field on the left-half of the swirl-stabilized combustion chamber. In this system, the flame is stabilized due to three recirculation zones named, inner recirculation zone (IRZ), recirculation zone (RZ), and outer recirculation zone (ORZ). The vortical structures are formed along the shear layer that exists between IRZ and ORZ (as denoted by a black line). We can map this vorticity information of the reaction field in the combustor into complex networks using the approach of turbulence networks.

Turbulence networks were first proposed by Taira et al. [110] to investigate the interaction between vortices in turbulent flows. In this network analysis, they considered a spatial domain of two-dimensional isotropic turbulence with periodic boundary conditions (Fig. 11.32). In a network, every point vortex is viewed as a node and the interaction between vortices is governed by specific rules [81]. Every vortex in the two-dimensional flow field advances due to the velocity induced by other vortices and not by its own velocity (i.e., a potential vortex cannot induce a velocity on itself). Therefore, the network does not have any self-connections. Furthermore, every vortex is influenced by all the vortices present in the flow field; hence, a connection between the nodes is established if the velocity of a vortex is

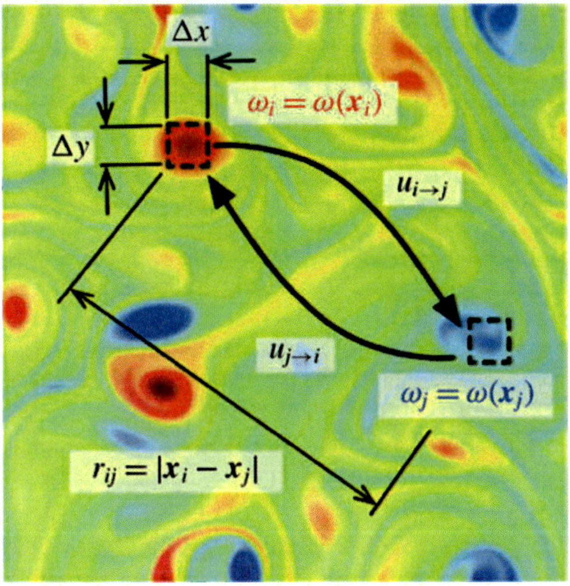

Fig. 11.32 Pictorial representation of the interaction of fluid elements (point vortices) in a two-dimensional turbulence network. This network indicates that the strength of the interaction between vortices at r_i and r_j having vorticities ω_i and ω_j depends on the respective induced velocities $u_{i\to j}$ and $u_{j\to i}$. Colors indicate the contour of the vorticity field. Reproduced with permission from Taira et al. [110]

influenced by another vortex. The velocity of the vortex is directly proportional to the strengths (circulations) of individual vortices and inversely proportional to the separation distance between them. Thus, closer vortices exhibit higher interaction between each other and stronger vortices have influence over a larger range in the flow field.

In Fig. 11.32, we show the interaction between vortices observed in a discrete cartesian coordinate system of line vortices. We can obtain the magnitude of the velocity induced in the fluid element located at r_j due to the fluid element located at r_i from the Biot–Savart law as follows [110],

$$u_{i\to j} = \frac{|\Omega_i|}{2\pi |r_i - r_j|} \tag{11.26}$$

where $\Omega_i = \omega(r_i)\Delta x \Delta y$ is the circulation of the fluid element of size $\Delta x \Delta y$, $r = (x, y)$ is the position vector, and ω is the vorticity. This velocity is induced perpendicular to the line of action. We can obtain the resultant velocity of the fluid element at r_i by the superposition of induced velocities from all fluid elements present in the spatial domain. We can represent such vortex interactions (or connectivity in the network) in the form of the adjacency matrix, whose elements indicate the induced velocities in the system. Thus, the adjacency matrix of a turbulence network is obtained as,

$$A_{ij} = \begin{cases} W_{ij} & \text{if } i \neq j \\ 0 & \text{if } i = j \end{cases} \tag{11.27}$$

where W_{ij} is the weight of the link between the vortices. Note that, since $u_{i\to j}$ is not the same as $u_{j\to i}$, we can construct directed networks using turbulence network approach. We can assign a weight for the connection as the average of the induced velocities at these vortices [110],

$$W_{ij} = \frac{1}{2}(u_{i\to j} + u_{j\to i}) \tag{11.28}$$

The distribution of the vortex strength can be obtained as $s_i = \sum_j A_{ij}$.

Murayama et al. [77] investigated the probability density function (PDF) of the vortex strength $p(s)$ in the instantaneous vorticity network of a swirl-stabilized combustor during the state of thermoacoustic instability (Fig. 11.33a). The PDF shows a power-law behavior with a scaling exponent of -3.4. This suggests the possibility of the presence of a scale-free network in the vorticity field of the system. Furthermore, they observed that the lifetime of such scale-free structures is very short (around 21% of the total length of the signal). This means scale-free structures intermittently appear and disappear in the vorticity field of the combustor. A power-law with an exponent of -2.1 is observed in the PDF of the lifetime of the scale-free structures (Fig. 11.33b). Such a characterization of the vorticity field into a scale-free network further indicates the presence of hubs in the network. Similar observations have been made by Krishnan et al. [61] for the states of combustion noise, intermittency and thermoacoustic instability observed in a turbulent bluff-body stabilized dump combustor.

Figures 11.34a, b show the time-varying spatial distribution of the vortex strength and community structure of the vorticity field over an acoustic pressure cycle during the state of thermoacoustic instability [77]. In each plot of Fig. 11.34b, open circles indicate the locations of communities and the thickness of black lines indicates the degree of interaction between communities. Here, a community indicates a subset of nodes in a network that are densely connected to each other, but less densely connected with other nodes in the rest of the network [86]. We can detect such communities in a network using the widely used Louvain method based on modularity optimization [17, 57].

When the acoustic pressure is at its minimum value ($t = 0.501$ s), the shedding of a ring of vortices in noticed from the injector rim. The distribution of vortex strength is observed to be the highest at the locations of these ring vortices (Fig. 11.34a). Thus, ring vortices act as hubs in the network. The community structure of the vorticity field suggests the presence of a strong link between communities formed at the locations of primary hubs (Fig. 11.34b). As time passes, these communities convect downstream and the link between them becomes weak. At the instance of acoustic pressure maximum (i.e., $t = 0.505$ s in Fig. 11.34b), the vortex strength is very low, the formation of hubs in the network is suppressed, and the links between communities appear to be weak and randomly distributed. Thus, construction of turbulence networks helps us in understanding the vortex interactions and also locating critical regions (i.e., hubs) in the spatial field of the combustor.

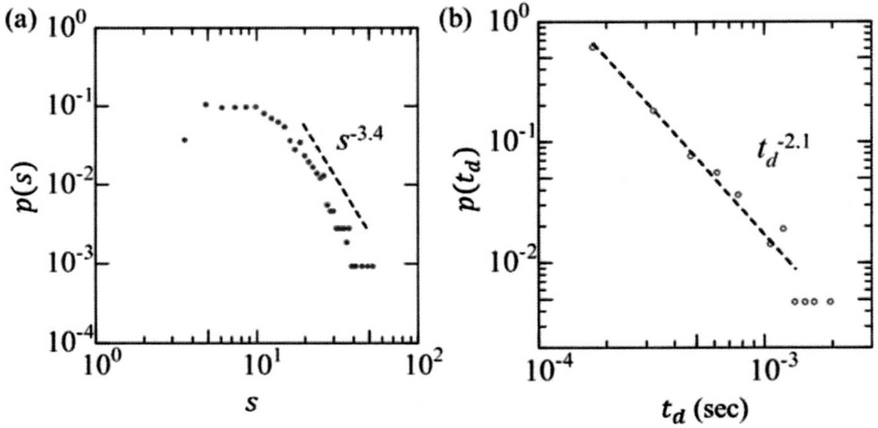

Fig. 11.33 The probability density functions of (**a**) the vortex strength s and (**b**) lifetime t_d show a power-law behavior during the state of thermoacoustic instability. These power-law behaviors further indicate the presence of scale-free networks in the system. Adapted with permission from Murayama et al. [77]

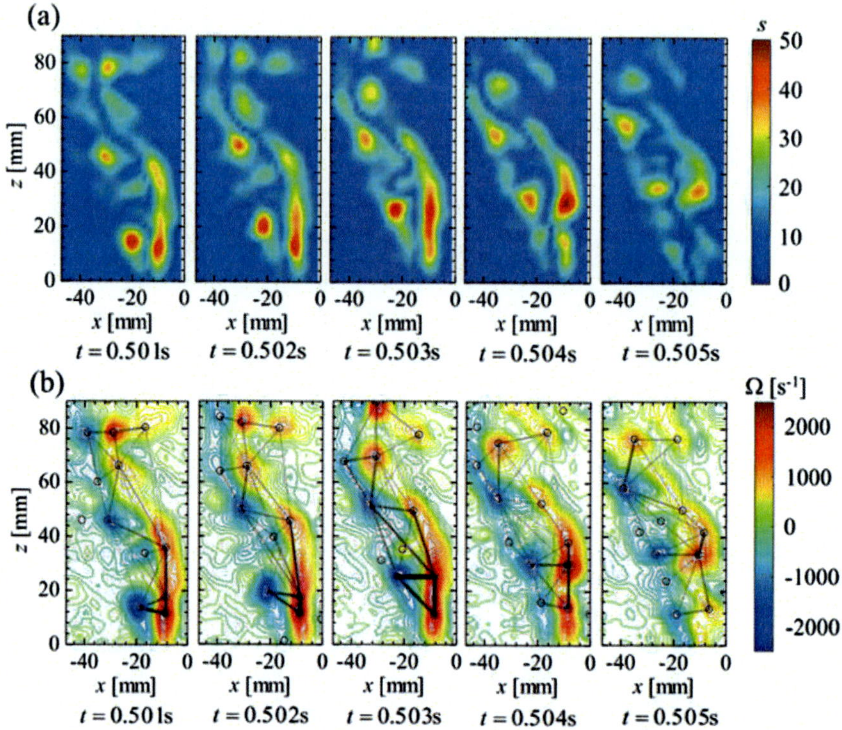

Fig. 11.34 Time-varying spatial distribution of (**a**) the vortex strength (s) in turbulence networks and (**b**) community structure of the vorticity field for different instances of the acoustic pressure cycle during the state of thermoacoustic instability, where $t = 0.501$ s corresponds to $\theta = 270°$ (pressure minima) and $t = 0.505$ s corresponds to $\theta = 90°$ (pressure maxima). We observe that the vortex interactions happen through the formation of communities in turbulence networks. Reproduced with permission from Murayama et al. [77]

In summary, in this chapter, we demonstrated that complex network analysis provides an emerging platform to characterize both the temporal and spatiotemporal dynamics of thermoacoustic systems. Using different methods of network construction, we revealed several hidden properties, such as scale-free, small-world, pseudoperiodicity, synchronization, etc., of complex thermoacoustic oscillations observed in turbulent combustors. Further, using different measures of network topology, we studied spatiotemporal interaction between different regions of the combustor flow field. We found that complex network analysis can help in identifying critical regions in the reaction field, studying interactions of vortices, and improving our understanding of the onset of thermoacoustic instability.

In the next chapter, we will discuss various novel precursor technologies and mitigation strategies that are developed for thermoacoustic instability using different tools from dynamical systems theory and complex systems theory.

References

1. Abugov, D.I., Obrezkov, O.I.: Acoustic noise in turbulent flames. Combust. Explos. Shock Waves **14**(5), 606–612 (1978)
2. Albert, R., Barabási, A.L.: Statistical mechanics of complex networks. Rev. Mod. Phys. **74**(1), 47 (2002)
3. Albert, R., Jeong, H., Barabási, A.L.: Diameter of the world-wide web. Nature **401**, 130–131 (1999)
4. Bar-Yam, Y.: Dynamics of Complex Systems. CRC Press, Boca Raton (1997)
5. Barabási, A.L.: Linked: The New Science of Networks. Perseus, Cambridge (2002)
6. Barabási, A.L., Albert, R.: Emergence of scaling in random networks. Science **286**(5439), 509–512 (1999)
7. Barabási, A.L., Bonabeau, E.: Scale-free networks. Sci. Am. **288**(5), 60–69 (2003)
8. Barabási, A.L., Gulbahce, N., Loscalzo, J.: Network medicine: a network-based approach to human disease. Nat. Rev. Genet. **12**(1), 56–68 (2011)
9. Barabási, A.L., et al.: Network Science. Cambridge University Press, Cambridge (2016)
10. Barrat, A., Barthelemy, M., Pastor-Satorras, R., Vespignani, A.: The architecture of complex weighted networks. Proc. Natl. Acad. Sci. **101**(11), 3747–3752 (2004)
11. Barrat, A., Barthelemy, M., Vespignani, A.: Dynamical Processes on Complex Networks. Cambridge University Press, Cambridge (2008)
12. Barthélemy, M.: Spatial networks. Phys. Rep. **499**(1–3), 1–101 (2011)
13. Bastian, M., Heymann, S., Jacomy, M.: Gephi: an open source software for exploring and manipulating networks. In: Proceedings of the International AAAI Conference on Web and Social Media, vol. 3 (2009)
14. Bavelas, A.: Communication patterns in task-oriented groups. J. Acoust. Soc. Am. **22**(6), 725–730 (1950)
15. Bell, M.G., Iida, Y.: Transportation Network Analysis. Wiley, Chichester (1997)
16. Bhattacharya, A., De, S., Mondal, S., Mukhopadhyay, A., Sen, S.: Recurrence network analysis exploring the routes to thermoacoustic instability in a Rijke tube with inverse diffusion flame. Chaos **31**(3), 033117 (2021)
17. Blondel, V.D., Guillaume, J.L., Lambiotte, R., Lefebvre, E.: Fast unfolding of communities in large networks. J. Stat. Mech: Theory Exp. **2008**, P10008 (2008)
18. Boccaletti, S., Latora, V., Moreno, Y., Chavez, M., Hwang, D.U.: Complex networks: structure and dynamics. Phys. Rep. **424**(4–5), 175–308 (2006)
19. Boers, N., Bookhagen, B., Barbosa, H.M.J., Marwan, N., Kurths, J., Marengo, J.A.: Prediction of extreme floods in the eastern central andes based on a complex networks approach. Nat. Commun. **5**(1), 1–7 (2014)
20. Bose, S.N.: Plancks Gesetz und Lichtquantenhypothese. Z. Phys. **26**, 178–181 (1924)
21. Bullmore, E., Sporns, O.: Complex brain networks: graph theoretical analysis of structural and functional systems. Nat. Rev. Neurosci. **10**(3), 186–198 (2009)
22. Cabral, J., Kringelbach, M.L., Deco, G.: Functional connectivity dynamically evolves on multiple time-scales over a static structural connectome: models and mechanisms. NeuroImage **160**, 84–96 (2017)
23. Caldarelli, G.: Scale-free Networks: Complex Webs in Nature and Technology. Oxford University Press, Oxford (2007)
24. Calvert, K.L., Doar, M.B., Zegura, E.W.: Modeling internet topology. IEEE Commun. Mag. **35**(6), 160–163 (1997)
25. Chen, G., Wang, X., Li, X.: Fundamentals of Complex Networks: Models, Structures and Dynamics. Wiley, Hoboken (2014)
26. Chen, W.K.: Graph Theory and its Engineering Applications, vol. 5. World Scientific, Singapore (1997)
27. Cherven, K.: Mastering Gephi Network Visualization. Packt, Birmingham (2015)

28. Chiu, H.H., Summerfield, M.: Theory of combustion noise. Acta Astronaut. **1**(7–8), 967–984 (1974)
29. Clavin, P.: Dynamics of combustion fronts in premixed gases: from flames to detonations. Proc. Combust. Inst. **28**(1), 569–585 (2000)
30. Clavin, P., Siggia, E.D.: Turbulent premixed flames and sound generation. Combust. Sci. Tech. **78**(1–3), 147–155 (1991)
31. Cohen, R., Havlin, S.: Complex Networks: Structure, Robustness and Function. Cambridge University Press, Cambridge (2010)
32. Davis, A., Marshak, A., Wiscombe, W., Cahalan, R.: Scale invariance of liquid water distributions in marine stratocumulus. Part I: spectral properties and stationarity issues. J. Atmos. Sci. **53**(11), 1538–1558 (1996)
33. Dimitriadis, S.I., Laskaris, N.A., Tsirka, V., Vourkas, M., Micheloyannis, S., Fotopoulos, S.: Tracking brain dynamics via time-dependent network analysis. J. Neurosci. Methods **193**(1), 145–155 (2010)
34. Donges, J.F., Zou, Y., Marwan, N., Kurths, J.: Complex networks in climate dynamics. Eur. Phys. J. Spec. Top. **174**(1), 157–179 (2009)
35. Donner, R.V., Zou, Y., Donges, J.F., Marwan, N., Kurths, J.: Recurrence networks—a novel paradigm for nonlinear time series analysis. New J. Phys. **12**(3), 033025 (2010)
36. Dorogovtsev, S.N., Mendes, J.F.: Evolution of Networks: From Biological Nets to the Internet and WWW. OUP Oxford, Oxford (2013)
37. Doron, K.W., Bassett, D.S., Gazzaniga, M.S.: Dynamic network structure of interhemispheric coordination. Proc. Natl. Acad. Sci. **109**(46), 18661–18668 (2012)
38. Einstein, A.: Quantum theory of the monatomic ideal gas. Sitzungsber. Preuss. Akad. Wiss. Phys. Math. Klasse 261–267 (1924)
39. Emmert-Streib, F.: Parametric construction of episode networks from pseudoperiodic time series based on mutual information. PLoS One **6**(12), e27733 (2011)
40. Erdős, P., Rényi, A.: On random graphs I. Publ. Math. Debrecen **6**(18), 290–297 (1959)
41. Erdős, P., Rényi, A.: On the evolution of random graphs. Publ. Math. Inst. Hung. Acad. Sci. **5**(1), 17–60 (1960)
42. Estrada, E., Knight, P.A.: A first Course in Network Theory. Oxford University Press, Oxford (2015)
43. Faloutsos, M., Faloutsos, P., Faloutsos, C.: On power-law relationships of the internet topology. ACM SIGCOMM Comput. Commun. Rev. **29**(4), 251–262 (1999)
44. Feldhoff, J.H., Donner, R.V., Donges, J.F., Marwan, N., Kurths, J.: Geometric detection of coupling directions by means of inter-system recurrence networks. Phys. Lett. A **376**(46), 3504–3513 (2012)
45. Feldhoff, J.H., Donner, R.V., Donges, J.F., Marwan, N., Kurths, J.: Geometric signature of complex synchronisation scenarios. Europhys. Lett. **102**(3), 30007 (2013)
46. Fornito, A., Zalesky, A., Bullmore, E.: Fundamentals of Brain Network Analysis. Academic Press, Cambridge (2016)
47. Gao, Z., Jin, N.: Complex network from time series based on phase space reconstruction. Chaos **19**(3), 033137 (2009)
48. Gao, Z.K., Small, M., Kurths, J.: Complex network analysis of time series. Europhys. Lett. **116**(5), 50001 (2017)
49. Godavarthi, V., Pawar, S.A., Unni, V.R., Sujith, R.I., Marwan, N., Kurths, J.: Coupled interaction between unsteady flame dynamics and acoustic field in a turbulent combustor. Chaos **28**(11), 113111 (2018)
50. Godavarthi, V., Unni, V.R., Gopalakrishnan, E.A., Sujith, R.I.: Recurrence networks to study dynamical transitions in a turbulent combustor. Chaos **27**(6), 063113 (2017)
51. Gotoda, H., Kinugawa, H., Tsujimoto, R., Domen, S., Okuno, Y.: Characterization of combustion dynamics, detection, and prevention of an unstable combustion state based on a complex-network theory. Phys. Rev. Appl. **7**(4), 044027 (2017)
52. Harary, F.: Graph Theory. Addison-Wesley, Reading (1969)

53. Hashimoto, T., Shibuya, H., Gotoda, H., Ohmichi, Y., Matsuyama, S.: Spatiotemporal dynamics and early detection of thermoacoustic combustion instability in a model rocket combustor. Phys. Rev. E **99**(3), 032208 (2019)
54. Hegde, U.G., Reuter, D., Zinn, B.T.: Sound generation by ducted flames. AIAA J. **26**(5), 532–537 (1988)
55. Huberman, B.A., Adamic, L.A.: Growth dynamics of the world-wide web. Nature **401**(6749), 131–131 (1999)
56. Jacob, R., Harikrishnan, K.P., Misra, R., Ambika, G.: Uniform framework for the recurrence-network analysis of chaotic time series. Phys. Rev. E **93**(1), 012202 (2016)
57. Jutla, I.S., Jeub, L.G.S., Mucha, P.J.: A generalized louvain method for community detection implemented in MATLAB (2011). http://netwiki.amath.unc.edu/GenLouvain
58. Kobayashi, T., Murayama, S., Hachijo, T., Gotoda, H.: Early detection of thermoacoustic combustion instability using a methodology combining complex networks and machine learning. Phys. Rev. Appl. **11**(6), 064034 (2019)
59. Krishnan, A., Manikandan, R., Midhun, P.R., Reeja, K.V., Unni, V.R., Sujith, R.I., Marwan, N., Kurths, J.: Mitigation of oscillatory instability in turbulent reactive flows: a novel approach using complex networks. Europhys. Lett. **128**(1), 14003 (2019)
60. Krishnan, A., Sujith, R.I., Marwan, N., Kurths, J.: On the emergence of large clusters of acoustic power sources at the onset of thermoacoustic instability in a turbulent combustor. J. Fluid Mech. **874**, 455–482 (2019)
61. Krishnan, A., Sujith, R.I., Marwan, N., Kurths, J.: Suppression of thermoacoustic instability by targeting the hubs of the turbulent networks in a bluff body stabilized combustor. J. Fluid Mech. **916**, A20 (2021)
62. Lacasa, L., Luque, B., Ballesteros, F., Luque, J., Nuno, J.C.: From time series to complex networks: the visibility graph. Proc. Natl. Acad. Sci. **105**(13), 4972–4975 (2008)
63. Ladyman, J., Lambert, J., Wiesner, K.: What is a complex system? Eur. J. Philos. Sci. **3**(1), 33–67 (2013)
64. Latora, V., Nicosia, V., Russo, G.: Complex Networks: Principles, Methods and Applications. Cambridge University Press, Cambridge (2017)
65. Lesne, A., Laguës, M.: Scale Invariance: From phase Transitions to Turbulence. Springer, Berlin (2011)
66. Lieuwen, T., Banaszuk, A.: Background noise effects on combustor stability. J. Propuls. Power **21**(1), 25–31 (2005)
67. Lieuwen, T.C.: Experimental investigation of limit-cycle oscillations in an unstable gas turbine combustor. J. Propuls. Power **18**(1), 61–67 (2002)
68. Lovejoy, S., Schertzer, D.: Scale invariance, symmetries, fractals, and stochastic simulations of atmospheric phenomena. Bull. Am. Meteorol. Soc. **67**(1), 21–32 (1986)
69. Luque, B., Lacasa, L., Ballesteros, F., Luque, J.: Horizontal visibility graphs: exact results for random time series. Phys. Rev. E **80**(4), 046103 (2009)
70. Malik, N., Bookhagen, B., Marwan, N., Kurths, J.: Analysis of spatial and temporal extreme monsoonal rainfall over South Asia using complex networks. Clim. Dyn. **39**(3–4), 971–987 (2012)
71. Marwan, N., Donges, J.F., Zou, Y., Donner, R.V., Kurths, J.: Complex network approach for recurrence analysis of time series. Phys. Lett. A **373**(46), 4246–4254 (2009)
72. Marwan, N., Romano, M.C., Thiel, M., Kurths, J.: Recurrence plots for the analysis of complex systems. Phys. Rep. **438**(5), 237–329 (2007)
73. Menczer, F., Fortunato, S., Davis, C.A.: A First Course in Network Science. Cambridge University Press, Cambridge (2020)
74. Milgram, S.: The small world problem. Psychology Today **2**(1), 60–67 (1967)
75. Molkenthin, N., Rehfeld, K., Marwan, N., Kurths, J.: Networks from flows-from dynamics to topology. Sci. Rep. **4**, 4119 (2014)
76. Murayama, S., Gotoda, H.: Attenuation behavior of thermoacoustic combustion instability analyzed by a complex-network-and synchronization-based approach. Phys. Rev. E **99**(5), 052222 (2019)

77. Murayama, S., Kinugawa, H., Tokuda, I.T., Gotoda, H.: Characterization and detection of thermoacoustic combustion oscillations based on statistical complexity and complex-network theory. Phys. Rev. E **97**(2), 022223 (2018)
78. Murugesan, M., Nair, V., Sujith, R.I.: System and method for early detection of onset of instabilities using complex networks. US Patent US10337414B2, Priority date: April 29 (2014)
79. Murugesan, M., Sujith, R.I.: Combustion noise is scale-free: transition from scale-free to order at the onset of thermoacoustic instability. J. Fluid Mech. **772**, 225–245 (2015)
80. Murugesan, M., Sujith, R.I.: Detecting the onset of an impending thermoacoustic instability using complex networks. J. Propuls. Power **32**(3), 707–712 (2016)
81. Nair, A.G., Taira, K.: Network-theoretic approach to sparsified discrete vortex dynamics. J. Fluid Mech. **768**, 549–571 (2015)
82. Nair, V., Sujith, R.I.: Multifractality in combustion noise: predicting an impending combustion instability. J. Fluid Mech. **747**, 635–655 (2014)
83. Nair, V., Thampi, G., Karuppusamy, S., Gopalan, S., Sujith, R.I.: Loss of chaos in combustion noise as a precursor of impending combustion instability. J. Spray Combust. Dyn. **5**(4), 273–290 (2013)
84. Newman, M.: Networks. Oxford University Press, Oxford (2010)
85. Newman, M.E.J.: The structure and function of complex networks. SIAM Rev. **45**(2), 167–256 (2003)
86. Newman, M.E.J.: Analysis of weighted networks. Phys. Rev. E **70**(5), 056131 (2004)
87. Newman, M.E.J., Watts, D.J.: Renormalization group analysis of the small-world network model. Phys. Lett. A **263**(4–6), 341–346 (1999)
88. Nicolis, G., Cantu, A.G., Nicolis, C.: Dynamical aspects of interaction networks. Int. J. Bifurc. Chaos **15**(11), 3467–3480 (2005)
89. Nicolis, G., Nicolis, C.: Foundations of complex systems. Eur. Rev. **17**(2), 237–248 (2009)
90. Noiray, N., Durox, D., Schuller, T., Candel, S.: A unified framework for nonlinear combustion instability analysis based on the flame describing function. J. Fluid Mech. **615**, 139–167 (2008)
91. Northrop, R.B.: Introduction to Complexity and Complex Systems. CRC Press, Boca Raton (2014)
92. Nuñez, A.M., Lacasa, L., Gomez, J.P., Luque, B.: Visibility algorithms: a short review. In: New Frontiers in Graph Theory, pp. 119–152. InTech, Rijeka (2012)
93. Okuno, Y., Small, M., Gotoda, H.: Dynamics of self-excited thermoacoustic instability in a combustion system: pseudo-periodic and high-dimensional nature. Chaos **25**(4), 043107 (2015)
94. Pastor-Satorras, R., Vespignani, A.: Evolution and Structure of the Internet: A Statistical Physics Approach. Cambridge University Press, Cambridge (2007)
95. Pawar, S.A., Seshadri, A., Unni, V.R., Sujith, R.I.: Thermoacoustic instability as mutual synchronization between the acoustic field of the confinement and turbulent reactive flow. J. Fluid Mech. **827**, 664–693 (2017)
96. Quenell, G.: Spectral diameter estimates for *k*-regular graphs. Adv. Math. **106**(1), 122–148 (1994)
97. Rajaram, R.: Characteristics of sound radiation from turbulent premixed flames. Ph.D. Thesis, Georgia Institute of Technology (2007)
98. Rajaram, R., Lieuwen, T.: Acoustic radiation from turbulent premixed flames. J. Fluid Mech. **637**, 357–385 (2009)
99. Reif, F.: Fundamentals of Statistical and Thermal Physics. Waveland Press, Long Grove (2009)
100. Romano, M.C., Thiel, M., Kurths, J., Kiss, I.Z., Hudson, J.: Detection of synchronization for non-phase-coherent and non-stationary data. Europhys. Lett. **71**(3), 466–472 (2005)
101. Rual, J.F., Venkatesan, K., Hao, T., Hirozane-Kishikawa, T., Dricot, A., Li, N., Berriz, G.F., Gibbons, F.D., Dreze, M., Ayivi-Guedehoussou, N., et al.: Towards a proteome-scale map of the human protein–protein interaction network. Nature **437**(7062), 1173–1178 (2005)

102. Scarsoglio, S., Iacobello, G., Ridolfi, L.: Complex networks unveiling spatial patterns in turbulence. Int. J. Bifurc. Chaos **26**(13), 1650223 (2016)
103. Scott, J.: Social network analysis. Sociology **22**(1), 109–127 (1988)
104. Small, M., Yu, D., Harrison, R.G.: Surrogate test for pseudoperiodic time series data. Phys. Rev. Lett. **87**(18), 188101 (2001)
105. Stelzl, U., Worm, U., Lalowski, M., Haenig, C., Brembeck, F.H., Goehler, H., Stroedicke, M., Zenkner, M., Schoenherr, A., Koeppen, S., et al.: A human protein-protein interaction network: a resource for annotating the proteome. Cell **122**(6), 957–968 (2005)
106. Strahle, W.C.: Combustion noise. Prog. Energy Combust. Sci. **4**(3), 157–176 (1978)
107. Strogatz, S.: Sync: The Emerging Science of Spontaneous Order. Penguin, London (2004)
108. Strogatz, S.H.: Exploring complex networks. Nature **410**(6825), 268–276 (2001)
109. Sujith, R.I., Unni, V.R.: Complex system approach to investigate and mitigate thermoacoustic instability in turbulent combustors. Phys. Fluids **32**(6), 061401 (2020)
110. Taira, K., Nair, A.G., Brunton, S.L.: Network structure of two-dimensional decaying isotropic turbulence. J. Fluid Mech. **795**, R2 (2016)
111. Takens, F.: Detecting strange attractors in turbulence. Lect. Notes Math. **898**(1), 366–381 (1981)
112. Tammela, T., Zarkada, G., Wallgard, E., Murtomäki, A., Suchting, S., Wirzenius, M., Waltari, M., Hellström, M., Schomber, T., Peltonen, R., et al.: Blocking VEGFR-3 suppresses angiogenic sprouting and vascular network formation. Nature **454**(7204), 656–660 (2008)
113. Tandon, S., Sujith, R.I.: Condensation in the phase space and network topology during transition from chaos to order in turbulent thermoacoustic systems. Chaos **31**(4), 043126 (2021)
114. Tony, J., Gopalakrishnan, E.A., Sreelekha, E., Sujith, R.I.: Detecting deterministic nature of pressure measurements from a turbulent combustor. Phys. Rev. E **92**(6), 062902 (2015)
115. Travers, J., Milgram, S.: An experimental study of the small world problem. In: Social Networks, pp. 179–197. Elsevier, Amsterdam (1977)
116. Tsonis, A.A., Roebber, P.J.: The architecture of the climate network. Phys. A **333**, 497–504 (2004)
117. Tupikina, L., Rehfeld, K., Molkenthin, N., Stolbova, V., Marwan, N., Kurths, J.: Characterizing the evolution of climate networks. Nonlinear Processes Geophys. **21**, 705–711 (2014)
118. Unni, V.R., Krishnan, A., Manikandan, R., George, N.B., Sujith, R.I., Marwan, N., Kurths, J.: On the emergence of critical regions at the onset of thermoacoustic instability in a turbulent combustor. Chaos **28**(6), 063125 (2018)
119. Unni, V.R., Sujith, R.I., Krishnan, A., Marwan, N., Kurths, J.: System and method for optimizing passive control of oscillatory instabilities in turbulent flows. US Patent US10895382B2, Priority date: February 27 (2018)
120. Valencia, M., Martinerie, J., Dupont, S., Chavez, M.: Dynamic small-world behavior in functional brain networks unveiled by an event-related networks approach. Phys. Rev. E **77**(5), 050905 (2008)
121. Vicsek, T.: Complexity: the bigger picture. Nature **418**(6894), 131–131 (2002)
122. Vicsek, T., Zafeiris, A.: Collective motion. Phys. Rep. **517**(3–4), 71–140 (2012)
123. Wang, M., Tian, L.: From time series to complex networks: the phase space coarse graining. Phys. A **461**, 456–468 (2016)
124. Wasserman, S., Faust, K., et al.: Social Network Analysis: Methods and Applications, vol. 8. Cambridge University Press, Cambridge (1994)
125. Watts, D.J.: Small Worlds. Princeton University Press, Princeton (2018)
126. Watts, D.J., Strogatz, S.H.: Collective dynamics of 'small-world' networks. Nature **393**(6684), 440–442 (1998)
127. Xu, X., Zhang, J., Small, M.: Superfamily phenomena and motifs of networks induced from time series. Proc. Natl. Acad. Sci. **105**(50), 19601–19605 (2008)
128. Yang, Y., Yang, H.: Complex network-based time series analysis. Phys. A **387**(5–6), 1381–1386 (2008)

129. Zhang, J., Luo, X., Small, M.: Detecting chaos in pseudoperiodic time series without embedding. Phys. Rev. E **73**(1), 016216 (2006)
130. Zhang, J., Small, M.: Complex network from pseudoperiodic time series: topology versus dynamics. Phys. Rev. Lett. **96**(23), 238701 (2006)
131. Zou, Y., Donner, R.V., Donges, J.F., Marwan, N., Kurths, J.: Identifying complex periodic windows in continuous-time dynamical systems using recurrence-based methods. Chaos **20**(4), 043130 (2010)
132. Zou, Y., Donner, R.V., Marwan, N., Donges, J.F., Kurths, J.: Complex network approaches to nonlinear time series analysis. Phys. Rep. **787**, 1–97 (2019)

Chapter 12
Early Warning and Mitigation Strategies for Thermoacoustic Instability

The presence of thermoacoustic instability has been a great concern in several continuous combustion systems used in propulsion (e.g., aeroengines, rocket engines) and power generation (e.g., land-based gas turbine engines) applications. The large amplitude self-sustained oscillations generated due to thermoacoustic instability can reduce the lifetime of structural components of an engine. It can increase the heat transfer from combustor surfaces and overwhelm thermal protection systems. In gas turbine engines, the presence of thermoacoustic instability leads to immediate shut down of systems, causing an interruption in power generation. Further, in rocket engines, the destruction of system components due to thermoacoustic instability can result in mission failure. Hence, the prediction and control of thermoacoustic instabilities in practical combustion systems are important. In this chapter, we provide an overview of prediction and control strategies that have been traditionally used and recently developed using dynamical systems and complex systems approaches. We first present the application of precursor technology to provide early warning signals for the occurrence of thermoacoustic instability. Subsequently, we discuss both active and passive control strategies developed for mitigation of thermoacoustic instabilities.

12.1 Precursors for the Onset of Impending Thermoacoustic Instability

In this section, we will discuss recent developments in the precursor technology related to the online monitoring of the combustor operation. As discussed before, the presence of thermoacoustic instability is hazardous to the structural integrity and the performance of combustion systems. Such instabilities, therefore, need to be avoided or suppressed before they emerge in the system. One way to evade the onset of thermoacoustic instabilities is to develop reliable and robust precursors that

R. I. Sujith, S. A. Pawar, *Thermoacoustic Instability*, Springer Series in Synergetics,
https://doi.org/10.1007/978-3-030-81135-8_12

can provide early warning signals to the onset of thermoacoustic instability. In such a case, appropriate modifications of the operational parameters (for example, fuel and air flow rates, secondary injections, fuel staging, etc.), based on the information from the combustion chamber, can help in avoiding thermoacoustic instability.

Traditional methodologies [49, 60] detect an incipient thermoacoustic instability (i.e., after its onset) and not when it is impending (i.e., prior to its onset). Hence, controls based on the detection of incipient thermoacoustic instability lack effectiveness, as the controls activate only after the instability is established in the system. In this case, the system may be exposed to the undesired large amplitude oscillations observed during thermoacoustic instability before the application of the control, incurring some damage to system components. In order to avoid such problems, it is desirable to have early warning measures for impending instabilities, where based on the "live" information from the combustor dynamics, appropriate health monitoring of the system is performed much prior to the occurrence of thermoacoustic instability [83]. In this manner, we can avoid the problems arising from the presence of these instabilities and, thereby, extend the lifetime of the combustor.

We will now highlight the difference between measures that detect an 'incipient instability' and an 'impending instability'. In Fig. 12.1, we show the variation of measures used for the detection of an incipient instability (Fig. 12.1a) and an impending instability (Fig. 12.1b) [80]. Figure 12.1a demonstrates the change in the amplitude levels of the acoustic pressure fluctuations, measured in terms of root mean square (rms) value (p_{rms}) and the magnitude of the dominant frequency in the fast Fourier transform (p_{FFT}) of the signal, with the increasing values of Reynolds number of the flow (Re). These measures register a significant increase in their values only when thermoacoustic instability has occurred in the combustor, i.e., an incipient instability. They do not provide information about the proximity of the system to the onset of thermoacoustic instability.

In contrast, the variation of the Hurst exponent (H), a measure of fractal characteristics of a signal, with the Reynolds number shows a smooth decrease towards zero, when the system behavior starts changing from chaotic to periodic dynamics via intermittency (Fig. 12.1b). As we have seen in Chap. 10, H takes a value between 0 and 0.5 for the anti-persistent fractal signals and 0 for the periodic signals. A close inspection of Figs. 12.1(a,b) suggests that the decrease in the value of H starts much earlier than the value of Re at which the pressure amplitudes show a visible change. Thus, by providing a suitable threshold for H (for example, $H = 0.1$ marked by a dotted line Fig. 12.1b) that is far from zero, we can effectively detect the proximity for the onset of thermoacoustic instability [83]. Further, we can take specific control actions before the system actually transitions to such instability. Therefore, we can use a feature of the loss of fractal nature of the acoustic pressure signal as an early warning of thermoacoustic instability in turbulent combustors.

There were a few attempts in the past to develop precursors for thermoacoustic instability based on tools from linear theory. Hobson et al. [49] measured vibrations of the combustor casing and oscillations in acoustic pressure, and analysed the change in bandwidth of the dominant frequency of these fluctuations. They showed

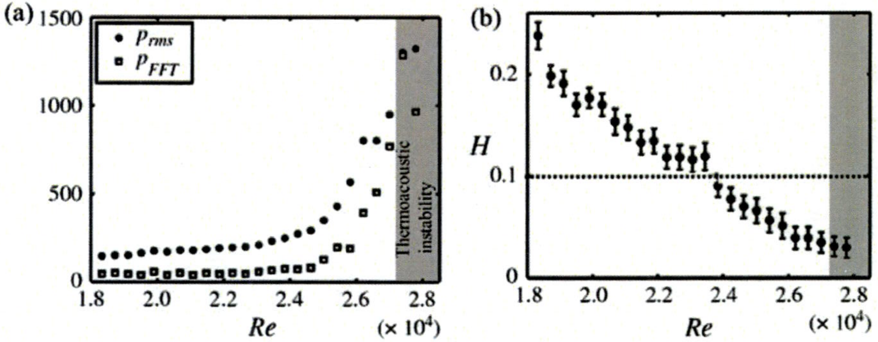

Fig. 12.1 Variation of measures **(a)** the amplitudes and **(b)** the Hurst exponent (H) of the acoustic pressure signal as a function of the Reynolds number of the flow (Re). In **(a)**, p_{rms} is the root mean square of the pressure signal and p_{FFT} is the amplitude of the dominant frequency in the spectrum of the pressure signal. The value of H shows a decreasing trend much prior to the value of Re at which the pressure amplitudes exhibit an increase. Thus, the pressure amplitudes measure incipient thermoacoustic instability and the Hurst exponent measures impending thermoacoustic instability. The horizontal dotted line in **(b)** indicates the threshold value of H that can be used as an indicator of the onset of thermoacoustic instability. Adapted with permission from Nair and Sujith [80]

that the bandwidth is indicative of the damping in the system and its value decreases to zero as the system dynamics approaches unstable operation (thermoacoustic instability). Lieuwen [60] measured damping of the system using a linear measure called autocorrelation, and monitored the variation of acoustic pressure fluctuations with the change in operating parameters as the damping approaches zero. Since these measures do not exhibit a significant change much prior to the rise in the pressure amplitude, they work only as a marker for incipient thermoacoustic instability.

Recently, we have seen an increased interest in developing precursors (early warning measures) based on different tools from dynamical systems theory and complex systems approach [89]. These precursors successfully help in forewarning the onset of thermoacoustic instability in turbulent combustors where the state of intermittency is observed prior to the onset of thermoacoustic instability. In the following discussion, we summarize various studies dedicated to develop different precursors for the occurrence of thermoacoustic instability. For technical details of these measures, the readers may refer to the corresponding research papers.

Nair et al. [81] showed that the state of combustion noise consists of the features of high dimensional chaos. The presence of chaos (low determinism) during this state, in turn, provides sufficient information for predicting the occurrence of thermoacoustic instability in a combustor. They applied the 0–1 test for chaos on the dynamics of the acoustic pressure fluctuations and devised a measure that can provide early warning for the proximity of thermoacoustic instability (refer to Fig. 6.2 in Chap. 6). During the state of combustion noise, this measure shows a value of unity. As the system dynamics approaches thermoacoustic instability, this measure shows a decrease to zero. Thus, the loss of chaos in a system can be used as an indicator for the occurrence of thermoacoustic instability.

During the transition from combustion noise to thermoacoustic instability, the system behavior changes from the state of chaotic to periodic oscillations. This dynamic change, in turn, causes the change in phase space characteristics of a pressure signal. Therefore, we can develop different precursors that will characterize such a dynamic change in the phase space trajectories. By analyzing the recurrence property of the phase space trajectory, Nair et al. [82] proposed various measures such as recurrence rate, entropy, and trapping time as precursors for thermoacoustic instability. Unni et al. [138, 140] and Ramanan et al. [102] used a measure based on the symbolic time series approach, called anomaly detection [103], as a precursor for the occurrence of thermoacoustic instability. Pavithran et al. [90, 92] developed a methodology to predict the amplitude of thermoacoustic instability at the onset using the inverse power law scaling between the spectral amplitude of the dominant mode and the Hurst exponent [91].

Murugesan and Sujith [77] employed a novel approach of complex network analysis to predict the occurrence of thermoacoustic instability in turbulent thermoacoustic systems. They suggested measures that characterize the network topology, such as clustering coefficient, characteristic path length, network diameter, and global efficiency, as good predictors for the onset of thermoacoustic instability [78]. On the other hand, Godavarthi et al. [36] used measures based on the analysis of the topology of the recurrence network, i.e., characteristic path length and betweenness centrality, to detect the transition to thermoacoustic instability. Chiocchini et al. [20] used a chaotic synchronization measure, called the interdependence index, as an online precursor for an impending thermoacoustic instability. Rouwenhorst et al. [107] developed a low-order model for the online detection of azimuthal instability in the annular combustor, where the decay rates and the mode shapes are instantaneously monitored.

Gotoda and co-workers [39, 75] introduced approaches from complex network (such as visibility graph and motifs) for online detection of thermoacoustic instability in turbulent combustors. Furthermore, they incorporated machine learning (standard support vector machine) into ordinal partition transition network [56] and permutation based complexity analysis [48] to develop precursors for thermoacoustic instability. Ray and co-workers [13, 70, 113–115, 144] have developed different data-driven approaches to predict the occurrence of thermoacoustic instability from experimental signals of the system variable.

Gopalakrishnan et al. [37] applied a novel approach of critical slowing down [122, 135] to provide early warning measures for the onset of thermoacoustic instability in a Rijke tube system. Here, critical slowing down indicates a slow recovery of the dynamical behaviour of a system from any perturbations provided in the vicinity of the bifurcation point (also called tipping point). The signature of critical slowing down of the system behavior can be observed from the increase in the value of measures such as variance, lag-1 autocorrelation, and conditional heteroskedasticity before the occurrence of critical transition in the system [120, 121]. For studying the behavior of critical slowing down in the system, we need to acquire data of the acoustic pressure fluctuations for the continuous variation of the control parameter (e.g., heater power in the Rijke tube) during the transition from a stable steady state to stable limit cycle oscillations [37]. Then, we need to calculate the variation of the aforementioned measures of critical slowing down for a sliding window along the pressure signal. Gopalakrishnan et al. [37] observed a continuous increase in the values of variance and conditional heteroskedasticity much before the onset of thermoacoustic oscillations in the system. These measures showed robustness in the presence of noisy fluctuations in the underlying flow field as well. Thus, we can use variance and conditional heteroskedasticity as precursors for thermoacoustic instability in the system where the control parameter is continuously changed at a specific rate. Pavithran et al. [88] found lag-1 autocorrelation and Hurst exponent to be good measures to predict the transition well before the occurrence of thermoacoustic instability in rate-dependent studies. They also observed that the warning time obtained from these measures to predict the onset of thermoacoustic instability decreases with the rate of change of control parameter, following an inverse power law relation. Hence, for very fast rates, the warning time may be too short to perform any control action. Although such studies of critical slowing down are conducted in a simple thermoacoustic system of Rijke tube, careful examination of the working of these measures still needs to be investigated in turbulent combustors.

Similar to the prediction of the onset of thermoacoustic instability, many studies have also focused on forewarning the dynamic event of lean flame blowout in turbulent combustors. Nair and Lieuwen [79] developed several tools from time series analysis of the acoustic pressure data such as the spectral, statistical, wavelet-based, and threshold-based approaches to detect blowout event in the combustor. Muruganandam et al. [76] analyzed the data of OH* chemiluminescence from the flame and used threshold-based tools to forewarn the blowout event. Yi and Gutmark [145] devised indicators such as a normalized root mean square of chemiluminescence in the flame and a normalized cumulative duration of local extinctions of the flame as real-time predictors for lean flame blowout. Unni and Sujith [141] used the Hurst exponent of the acoustic pressure signal to detect flame blowout [139]. A group from Jadavpur University have used different data-driven approaches from nonlinear time series analysis [53] and symbolic time series analysis [25] to characterize the dynamic behavior of the flame and established different precursors for lean blowout [19, 26, 27, 29, 72, 116, 117, 123, 124]. On the other hand, Gotoda and co-workers developed several online detectors for flame blowout based on dynamical systems theory [30, 41] and complex network theory [40, 74]. Premraj et al. [102] discovered the presence of Dragon-king extreme events just prior to flame blowout and proposed the identification of such events as a precursor for the catastrophic transition such as flame blowout.

12.2 Traditional Approaches for Passive and Active Controls of Thermoacoustic Instability

The attenuation methodologies of thermoacoustic instability are primarily divided into two types, which include active and passive controls [23, 32, 62, 68, 94]. As thermoacoustic instability is a closed-loop interaction between the acoustic pressure fluctuations and the heat release rate fluctuations, the main aim of these control strategies is to break the coupling between these fluctuations and, thus, curb thermoacoustic instability when it has occurred in the system.

Passive control methods focus on modifications in the hardware of the combustor, where these changes made in the design are immovable and they do not possess any dynamic component. These modifications may change the flow dynamics and also the acoustic standing-wave patterns in the combustor. Some examples of passive controls that have been adopted by engine manufacturing companies include, fixing of baffles on the injector plate of the rocket engine, changes in fuel manifolds, installation of acoustic liners, Helmholtz resonators (acoustic damper), quarter-wave and half-wave tubes in gas turbine combustors, modification of the combustor length, etc. [62, 106, 146].

Baffles are solid components located in the combustor (especially, on the injector head of a rocket engine) to disrupt the acoustic standing wave patterns that emerge during thermoacoustic instability [23]. The Helmholtz resonator is a large cavity with a small neck usually located on the combustor surface, where it absorbs acoustic energy at the undesired frequencies that are within the range of the resonant frequency of the Helmholtz resonator [12, 47]. Acoustic liners are the perforated casings located along the combustor length, where secondary dilution air jets enter through small holes in the combustor [33]. The vortex shedding that occurs along the rim of these perforations dampens the acoustic oscillations developed in the system [50]. The main purpose of dampers or liners is to weaken the acoustic field by increasing the damping or dissipation of acoustic energy in the combustor.

Another method is to change the length of the combustor that modifies the natural frequency of the acoustic field or to change the size of flame holding devices that alters the vortex shedding frequency in the system. The modification of the frequency, in turn, helps in avoiding the excitation of unstable acoustic modes or modifying the frequency-locking characteristics of the acoustic pressure and the heat release rate fluctuations in the combustor. Fuel-staging is another way of mitigation of thermoacoustic instability, where the total fuel consumed in the combustor is injected at different rates in different parts of the combustor [16, 17, 112]. Such fuel distribution changes the heat release rate field in the combustors and, hence, reduces the sound pressure level observed during the state of thermoacoustic instability.

There are a few disadvantages of using passive controls in practical combustion systems [23, 50]. All passive control devices are effective only in a specific range of frequencies, i.e., they have a limited range of operation. In order to mitigate instabilities over a wider range, we need to have several dampers installed on the combustor. This, in turn, increases the overall weight of the combustor. Further, the accommodation of these large size devices in the limited space of the combustor is unfeasible. The response of these devices in controlling low-frequency thermoacoustic instability is very poor. The modification of these parts is time-consuming and also costly. The presence of these devices also results in the loss of total pressure in the system which, in turn, affects the system performance. Despite these disadvantages, passive controls are considered to be the most viable option of damping thermoacoustic instability by engine manufacturers over the use of active controls [62].

Active control of thermoacoustic instability relies on the instantaneous monitoring of the combustor operation [32, 68]. In this control strategy, the acoustic pressure field or the fuel injection rates in the combustor are altered through actuators. The acoustic pressure field in the combustor is modified using acoustic drivers (e.g., loudspeakers, sirens, electropneumatic actuator), while fuel distribution in the combustor is altered through servo (e.g., solenoid) valves. These modifications in operating conditions, in turn, change the heat release rate fluctuations in the combustor. This further modifies the coupling of the heat release rate field with the acoustic field, leading to the mitigation of thermoacoustic instability.

Active controls are often classified into two types: open-loop and closed-loop [32, 68]. In open-loop controls, we do not use feedback information from the combustor dynamics for controlling thermoacoustic instability (Fig. 12.2a). The input to the combustor is predefined and fixed in time. We need to supply an

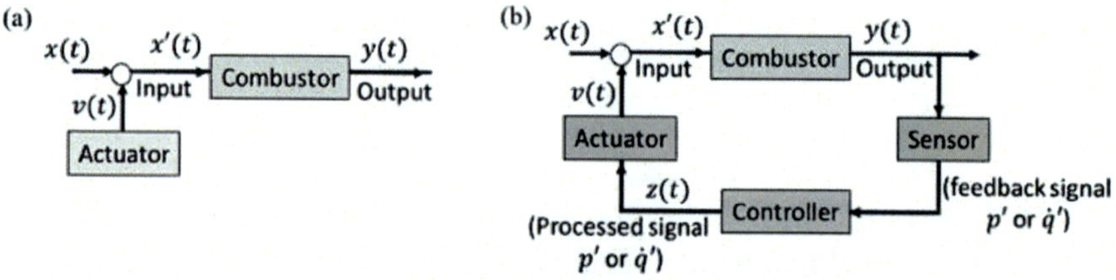

Fig. 12.2 Schematics of (**a**) open-loop control and (**b**) closed-loop control strategies of thermoacoustic instability. In open-loop control, the input of a combustor is independent of its output. In contrast, in closed loop control, the input to a combustor is feedback coupled with its output

oscillatory signal of constant amplitude and frequency through the actuator to affect the acoustic pressure field or the fuel supply rate in the combustor. The effectiveness of this control strategy in the mitigation of thermoacoustic instability depends on the choice of these forcing parameters, which we need to obtain from several trials or through models [11, 42, 46, 64, 69, 71, 108].

In contrast, in closed-loop controls, the control action required for the mitigation of thermoacoustic instability is completely dependent on the input (feedback) information acquired from the combustor [32]. In Fig. 12.2b, we show a typical closed-loop control arrangement used in combustion systems. In this method, using appropriate sensors, we feed the output information of the combustor dynamics to the controller. Often, we measure the system response in terms of the acoustic pressure fluctuations (using a pressure transducer) or the heat release rate fluctuations (using a photodiode tube). We then supply this acquired information from the sensor to the controller, where the signal is compared with a pre-set or user-defined operating condition, and accordingly the actuator unit is driven. The actuator then supplies a time-varying input to the combustor and makes changes in the system operating conditions. This change, in turn, breaks the coupling between the heat release rate and the acoustic field developed during the state of thermoacoustic instability and stabilizes the system to stable operation.

Closed-loop controllers are further categorized as fixed-parameter control and adaptive control [32]. A fixed-parameter control is based on the rule of single-input single-output, where the response of the transfer function in the controller does not change in time. In contrast, for adaptive control, the output of the controller is automatically modified using an internal algorithm based on the information of both instantaneous and time-averaged dynamics of the combustor. In the controller, the input signal is usually phase-shifted and amplified as per the requirement before supplying it to the actuator. The details of traditional active closed-loop control strategies developed over the years by various researchers can be found from several review papers [3, 18, 32, 50, 68].

Compared to passive control methodologies, active control methodologies are advantageous in the sense that they are simpler in design, have fast response, wide operation range, easy to replace, cost-effective, and eliminate design changes in the combustor [8, 23, 62]. Despite these advantages, active control strategies have several disadvantages as well [98]. As practical combustors work in a harsh environment of high operating temperature, pressure, and turbulent flow field, the reliability of sensors has always been a problem. As acoustic drivers have a very good high-frequency response, their use in controlling thermoacoustic instabilities occurring at low frequency is usually ineffective. The installation of such actuators on large-scale gas turbine combustors is a big concern due to the limited space, the requirement of cooling, and the need for a huge power supply to drive these units. Fuel (solenoid) valves work well only in the low-frequency range; they are costlier and have a short lifetime. Furthermore, the control of thermoacoustic instabilities that occur due to the interaction of multiple modes has yet to be fully explored, with only a few studies having been done so far [43, 54, 118].

Having summarized the traditional methods used for controlling thermoacoustic instability, we next discuss the recent developments in mitigating such instabilities using approaches from synchronization theory.

12.3 Control of Thermoacoustic Instability Using Methodologies from Synchronization Theory

As we discussed in Chap. 8, synchronization is the phenomenon of coupled interaction of two or more oscillators, where the coupling of oscillators in a particular range of parameters leads to the locking of their instantaneous phases and frequencies to a common value [93]. Synchronization of oscillators can be achieved through the mechanism of mutual coupling (Fig. 12.3a) or through the application of external forcing (Fig. 12.3b). The presence of coupling modifies the frequencies of these oscillators to a common value, while the application of forcing locks the natural frequency of the oscillator with the forcing frequency [5].

In addition to synchronization, the coupling of oscillators sometimes leads to an important phenomenon known as "oscillation quenching" [57, 119, 147] where the oscillators attain a steady state. The phenomenon of oscillation quenching is further classified into two types: *amplitude death* and *oscillation death*. During the occurrence of amplitude death, all coupled oscillators achieve the same steady state

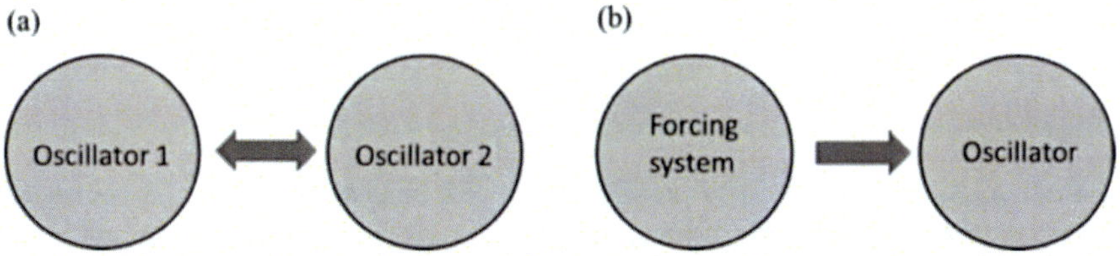

Fig. 12.3 Schematic of different mechanisms of synchronization of oscillators, where arrows indicate the direction of coupling between the oscillators. (**a**) Mutual synchronization. (**b**) Forced synchronization

(homogeneous steady state). While during the occurrence of oscillation death, all coupled oscillators attain different steady states (inhomogeneous steady states). In some situations, the coexistence of an oscillatory state and a steady state is also observed in a system of coupled oscillators and this state of coupled dynamics is referred to as *partial amplitude death* [4, 63]. In short, the coupling of oscillators can affect the phase, frequency, and amplitude of their signals.

Similarly, forcing an oscillator using an external source can also change the phase, frequency, and amplitude characteristics of the forced system (refer to Sect. 8.3 in Chap. 8). The state of locking of instantaneous phases of the forced oscillator with the forcing signal is called forced synchronization [5]. The process of forced synchronization of an oscillator can occur through the states of phase-drifting, intermittent phase-locking, phase trapping, and phase locking. The forcing also affects the amplitude of originally self-excited oscillations. The forcing of such self-excited oscillations at a frequency far from the natural frequency can completely suppress these oscillations in the system. This phenomenon of suppression of self-excited oscillations due to non-resonant forcing is called *asynchronous quenching* [15, 28, 55, 86, 126]. During asynchronous quenching, the amplitude of the forced system decreases to a minimum value corresponding to the amplitude of the forcing applied to the system. When the forcing is applied close to the natural frequency of the oscillator, forcing leads to the amplification of forced oscillations to a very high value, referred to as *resonant amplification* or *synchronance* [71].

In the following sections, we discuss the methodologies of quenching of thermoacoustic instability through the phenomena of amplitude death and asynchronous quenching. The occurrence of these phenomena has been investigated in laboratory-scale laminar thermoacoustic systems. However, their systematic study for the mitigation of thermoacoustic oscillations in practical turbulent combustors is yet to be executed.

12.3.1 Mitigation of Thermoacoustic Instability Using Amplitude Death Phenomenon

To study the occurrence of amplitude death phenomenon, oscillators exhibiting self-sustained (periodic or chaotic) oscillations are coupled through appropriate coupling mechanisms. The types of coupling schemes that are commonly used in achieving amplitude death are time-delay, dissipative, conjugate, environmental, etc. [119, 147]

The phenomenon of amplitude death was discovered by Rayleigh [104] in a system of coupled organ pipes. He observed that when two organ pipes of the same pitch are kept close to each other, the interaction of these pipes leads to quenching of oscillations in both these pipes. Abel et al. [1, 2] performed experimental investigation and mathematical modeling of organ pipes and showed that the decrease in the amplitude of sound waves in the organ pipes is due to a complex interaction of aerodynamic jets emitting from the mouth of pipes, leading to anti-phase radiation of sound waves.

Amplitude death has been observed in several systems, which include electronic circuits [105], chemical oscillators [9, 22], lasers [97, 142], candle-flame oscillators [65–67], etc. The existence of amplitude death has also been investigated both numerically and experimentally in a system of two coupled thermoacoustic oscillators, exhibiting limit cycle oscillations during the state of thermoacoustic instability [14, 24, 51, 52, 99, 111, 125, 132, 133].

In the following discussion, we will first present a theoretical study of achieving amplitude death in a model of two coupled thermoacoustic Rijke tube systems. Subsequently, we will present the experimental studies pertaining to amplitude death in physical systems.

12.3.1.1 Model for Studying Amplitude Death in Coupled Thermoacoustic Systems

Balasubramanian and Sujith [6] developed a mathematical model to study the bifurcation characteristics of a single Rijke tube system comprised of an electrically heated wire mesh as a compact heat source, which is given as follows,

$$\frac{d\eta_j}{dt} = \dot{\eta}_j$$

$$\frac{d\dot{\eta}_j}{dt} + 2\zeta_j\omega_j\dot{\eta}_j + \omega_j^2\eta_j = -j\pi K\left[\sqrt{\left|\frac{1}{3} + u'_f(t-\tau_1)\right|} - \sqrt{\frac{1}{3}}\right]\sin(j\pi x_f) \tag{12.1}$$

Here, $\dot{\eta}$ indicates the time-varying component of the jth mode of the acoustic pressure in the duct, t is the time, ζ is the damping coefficient, ω is the natural frequency of the oscillator, K is the non-dimensional heater power, u'_f is the acoustic velocity at the heater location (x_f), and τ_1 is the delay of thermal inertia.

Thomas et al. [133] modified equation (12.1) to investigate the effect of time-delay and dissipative coupling mechanisms on the occurrence of amplitude death in two coupled Rijke tube systems. We denote these systems as a and b. The governing equations for the system a in the coupled state is given below [125, 133],

$$\frac{d\eta_j^a}{dt} = \dot{\eta}_j^a$$

$$\begin{aligned}\frac{d\dot{\eta}_j^a}{dt} + 2\zeta_j\omega_j^a\dot{\eta}_j^a + (\omega_j^a)^2\eta_j^a = -j\pi K^a\left[\sqrt{\left|\frac{1}{3} + u'^a_f(t-\tau_1^a)\right|} - \sqrt{\frac{1}{3}}\right]\sin(j\pi x_f^a) \\ + \underbrace{\mathcal{K}_d(\dot{\eta}_j^b - \dot{\eta}_j^a)}_{\text{Dissipative coupling}} + \underbrace{\mathcal{K}_\tau\left[\dot{\eta}_j^b(t-\tau) - \dot{\eta}_j^a(t)\right]}_{\text{Time-delay coupling}}\end{aligned} \tag{12.2}$$

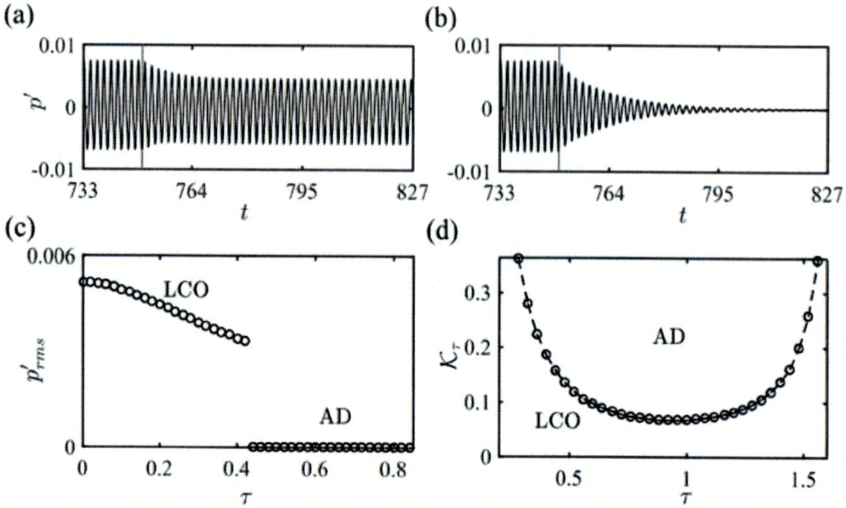

Fig. 12.4 The time series of the acoustic pressure signal when the coupling strength is (**a**) low and (**b**) sufficiently high to cause amplitude death in the model of time-delay coupled limit cycle oscillations. (**c**) A one-parameter bifurcation diagram between the rms (root mean square value) of the acoustic pressure signal (p'_{rms}) with the coupling delay (τ). (**d**) A two-parameter bifurcation diagram between the coupling strength ($\mathcal{K}_\tau$) and the delay time (τ). LCO and AD indicate the regions of limit cycle oscillations and amplitude death, respectively. We need to have finite values of $\mathcal{K}_\tau$ and τ to achieve AD in the system. Reproduced with permission from Thomas et al. [133]

where τ is the coupling delay, $\mathcal{K}_d$ and $\mathcal{K}_\tau$ are the coupling strengths associated with the terms of dissipative and time-delay coupling, respectively. We can obtain a similar equation for the system b by swapping a and b in (12.2).

To study the interaction of coupled thermoacoustic systems, we first need to establish both the systems in a state of limit cycle oscillations, by varying a bifurcation parameter (i.e., the heater power or mass flow rate [38]). In this state of limit cycle oscillations, we can refer to thermoacoustic systems as thermoacoustic oscillators. Once both the oscillators exhibit limit cycle oscillations, the coupling between them is switched on.

In (12.2), $\mathcal{K}_d = 0$ indicates the presence of only time-delay coupling, and $\mathcal{K}_\tau = 0$ represents the presence of only dissipative coupling in the system. In Figs. 12.4 and 12.5, we show the effect of the individual application of these coupling schemes on the coupled behavior of thermoacoustic oscillators. The coupled response of the system can be shown in terms of bifurcation diagrams. A one-parameter bifurcation diagram indicates the transition of the system behavior from the state of limit cycle oscillations to amplitude death. On the other hand, a two-parameter bifurcation diagram distinguishes the regions of coupled dynamics (i.e., limit cycle and amplitude death) observed in the parameter space of coupling constants.

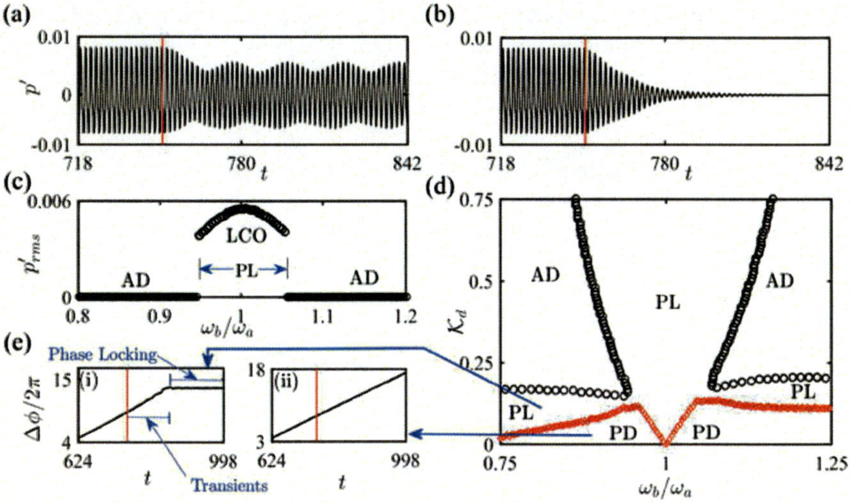

Fig. 12.5 The time series of the acoustic pressure signal when dissipative coupling strength is (**a**) low and (**b**) sufficiently high to cause amplitude death (AD) in the model of dissipatively coupled limit cycle oscillations. (**c**) A one-parameter bifurcation diagram between the rms of the acoustic pressure signal (p'_{rms}) with the frequency ratio between the oscillators (ω_b/ω_a). (**d**) A two-parameter bifurcation diagram between the coupling strength ($\mathcal{K}_d$) and the frequency ratio (ω_b/ω_a). (**e**) The temporal variation of the relative phase between the oscillators for regions of phase-locking (PL) and phase-drifting (PD). We need to have finite values of frequency detuning (ω_b/ω_a) and $\mathcal{K}_d$ to achieve AD in the system. Reproduced with permission from Thomas et al. [133]

Figure 12.4 shows the effect of time-delay coupling ($\mathcal{K}_d = 0$) on the coupled behavior of limit cycle oscillations observed in identical thermoacoustic oscillators (i.e., $\omega_a = \omega_b$). Here, the term 'identical' signifies that limit cycle oscillations in both the oscillators have nearly the same amplitude and frequency in the uncoupled state, and the term 'non-identical' indicates a mismatch in these parameters of oscillators in the uncoupled state. Further, delay coupling arises due to the finite time required for the acoustic wave to transfer from one oscillator to another once they are coupled, and dissipative coupling accounts for the interaction that arises due to mass transfer between two systems through the coupling tube [14] or the acoustic losses.

In Fig. 12.4a, b, we observe that, depending upon the value of the coupling parameters, the introduction of coupling between the oscillators can either reduce the amplitude of limit cycle oscillations or lead to complete quenching of these oscillations to a steady state, respectively (a vertical line in Fig. 12.4a, b indicates the time instant at which the coupling is switched on in the system). Figure 12.4c shows the one-parameter bifurcation diagram exhibiting the variation of the root mean square of the acoustic pressure fluctuations (p'_{rms}) with delay time (τ) for a fixed value of the coupling strength. We notice that the increase in delay monotonically reduces the amplitude of the limit cycle oscillations. As the delay is increased above a critical value, the transition from limit cycle oscillations to a steady state is observed to be explosive (i.e., abrupt) [125]. We refer to this steady state as amplitude death of coupled thermoacoustic oscillators.

Furthermore, in Fig. 12.4d, we present a two-parameter bifurcation diagram of coupled thermoacoustic oscillators, showing the variation of the coupling strength and delay time. We notice a U-shaped region of amplitude death (AD) in the plot. Outside the AD region, oscillators exhibit limit cycle oscillations at the reduced amplitude than their amplitude value observed in the uncoupled state. Furthermore, we also notice that, to achieve AD in coupled oscillators, we need finite (non-zero) values of both the coupling strength and the coupling delay in the system.

In Fig. 12.5, we demonstrate the coupled behavior of limit cycle oscillations ($\mathcal{K}_\tau = 0$) in two dissipatively coupled non-identical thermoacoustic oscillators ($\omega_a \neq \omega_b$). For a low value of the coupling strength, we notice a small decrease in the amplitude of limit cycle oscillations (Fig. 12.5a). In this state, we observe amplitude modulation in the signals of coupled oscillators, where the frequency of these modulations is equal to the difference in natural frequencies of these oscillators. In contrast, when the value of the coupling strength is sufficiently high, we notice a complete quenching of limit cycle oscillations in the system (Fig. 12.5b).

The one-parameter bifurcation diagram in Fig. 12.5c indicates that, we need to have a small frequency mismatch ($\omega_b/\omega_a \neq 1$) between the dissipatively coupled oscillators to achieve amplitude death. For a low value of the frequency mismatch, we observe a small decrease in the amplitude of limit cycle oscillations, where these oscillations are synchronized (phase-locked) in time (indicated as PL in the Fig. 12.5c). From the two-parameter bifurcation diagram (Fig. 12.5d) between the coupling strength ($\mathcal{K}_d$) and the ratio of dominant frequencies (ω_b/ω_a) of the oscillators, we notice that the values of $\mathcal{K}_d$ and $\Delta\omega$ ($= |\omega_b - \omega_a|$) need to be sufficiently large to achieve amplitude death in the system (boundaries are indicated by black circles in Fig. 12.5d). Furthermore, the oscillators are desynchronized (or show phase-drifting behavior, PD) for low values of $\mathcal{K}_d$ and are synchronized when $\mathcal{K}_d$ is sufficiently high.

Thomas et al. [133] also showed that when both the time-delay and dissipative couplings are applied simultaneously, the region of amplitude death increases in the parameter space. This also means that in such a case, it is easy to attain amplitude death of oscillators. Furthermore, they observed that the increase in the amplitude of the uncoupled limit cycle oscillations reduces the region of amplitude death in the parameter space of coupled oscillators. In other words, for a given set of coupling parameters, although the quenching of large amplitude limit cycle oscillations might not be possible, it may be easily achieved when the amplitude of limit cycle oscillations is low [99].

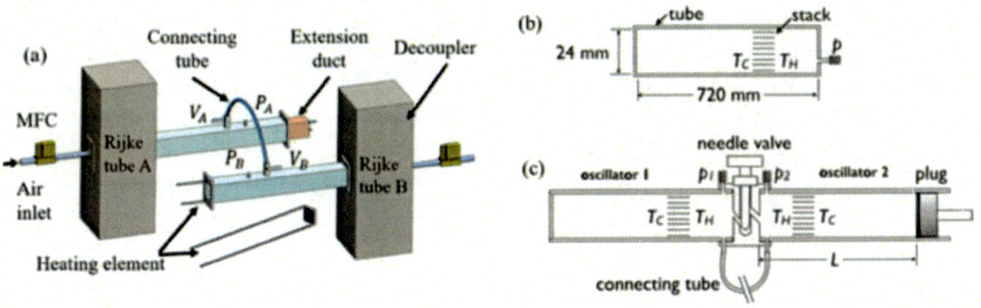

Fig. 12.6 Schematics of (**a**) a system of time-delay coupled two horizontal Rijke tube oscillators, (**b**) a single thermoacoustic engine, and (**c**) a system of time-delay and dissipatively coupled two thermoacoustic engines. In (**a**) and (**c**), the time-delay coupling between the oscillators is achieved through a connecting tube. In (**c**), the dissipative coupling between the engines is obtained through a needle valve. A frequency detuning between the oscillators is achieved by changing the length of one of the oscillators. Towards this purpose, an extension duct is used in (**a**) and a plunger arrangement is provided in (**c**). (**a**) Reproduced with permission from Dange et al. [24] and (**b, c**) reproduced with permission from Biwa et al. [14]

12.3.1.2 Experiments for Studying Amplitude Death in Coupled Thermoacoustic Systems

Apart from the aforementioned theoretical study, a few experimental studies have also been conducted to illustrate the occurrence of amplitude death in lab-scale thermoacoustic systems [24, 51] and thermoacoustic engines [14]. A typical thermoacoustic engine is different from a thermoacoustic system that is discussed in the present book. A thermoacoustic engine consists of a long cylindrical tube closed at both ends, wherein the working fluid (air) is filled at ambient conditions [129]. A porous stack is sandwiched between the hot and cold regions (Fig. 12.6b). When the temperature gradient across the stack is greater than a critical value, self-sustained large amplitude acoustic oscillations are developed in the system. On the other hand, a typical thermoacoustic system is an open system, where reactants continuously enter the system and hot gases leave the system [23, 62]. The interaction of flow fluctuations with the heat source (a flame or a compact electric heater) in the system leads to local fluctuations in the heat release rate. When these heat release rate fluctuations happen to be in-phase with the acoustic fluctuations, self-sustained large amplitude thermoacoustic instability is established in the system.

In Figs. 12.7, 12.8, and 12.9, we show the experimental characterization of amplitude death phenomenon when two horizontal Rijke tubes (Fig. 12.6a), i.e., prototypical thermoacoustic systems, are connected [24]. To study the occurrence of amplitude death phenomenon, we couple the two Rijke tubes, both exhibiting self-sustained limit cycle oscillations, using a flexible tube of length (L') and diameter (d). The change in the length of the coupling tube changes the delay time in the coupling of oscillators. We can express this delay time as $\phi = 2\pi L'/\lambda$, where λ is the wavelength of acoustic oscillations in the system [24]. Further, a change in the diameter of the coupling tube varies its admittance characteristics ($Y = S/\rho c$,

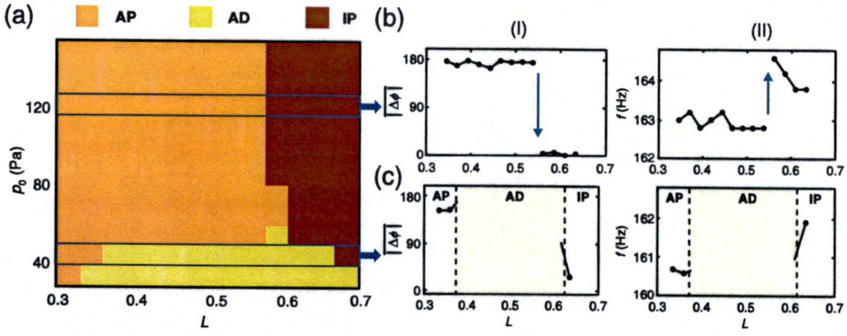

Fig. 12.7 (**a**) The two-parameter bifurcation diagram between the amplitude of limit cycle oscillations (LCO) in the uncoupled state (p_0) and the normalized length of the coupling tube (L), obtained from experiments involving two coupled Rijke tubes. Variation of (**b**) the mean phase difference ($\Delta\phi$) between LCOs and (**c**) the dominant frequency (f) of LCO with the change in L. For lower values of p_0, we notice the occurrence of a region of amplitude death (AD) between anti-phase synchronization (AP) and in-phase synchronization (IP). For higher value of p_0, we notice a direct transition from AP to IP, corresponding to phase-flip bifurcation. Reproduced with permission from Dange et al. [24]

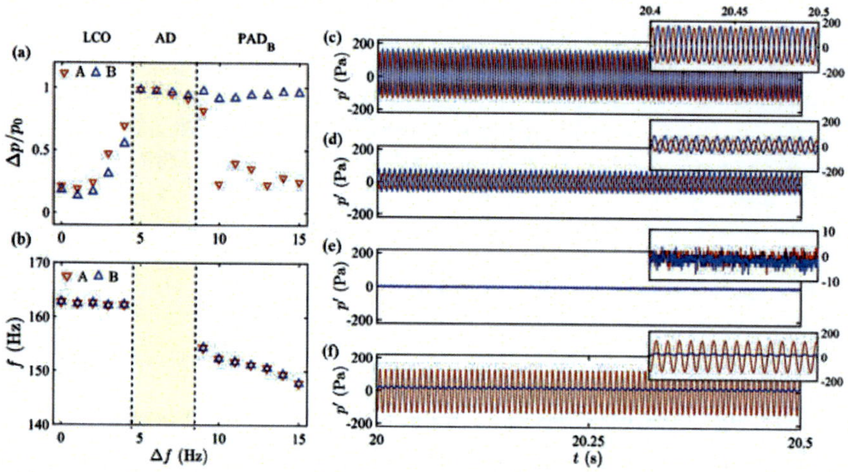

Fig. 12.8 Variation of the normalized change in the amplitude of acoustic pressure fluctuations ($\Delta p/p_0$) and the dominant frequency (f) of limit cycle oscillations (LCO) with the change in frequency difference between them (Δf), obtained from experiments in the system of coupled Rijke tube systems. The time series of acoustic pressure fluctuations in both the systems during the states of (**c, d**) limit cycle oscillations (for $\Delta f = 0$ and 5 Hz), (**e**) amplitude death AD (for $\Delta f = 7$ Hz), and (**f**) partial amplitude death PAD (for $\Delta f = 15$ Hz). A and B correspond to notations of two Rijke tube systems. We need a finite frequency detuning to achieve AD and even higher values of Δf to obtain PAD of large amplitude LCOs in time-delay coupled systems. Reproduced with permission from Dange et al. [24]

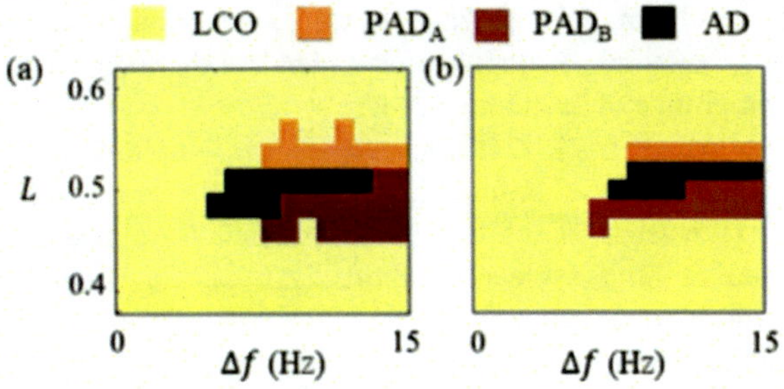

Fig. 12.9 **(a, b)** Two parameter bifurcation diagrams between the length of the coupling tube (L) and the frequency difference between the oscillators (Δf) for low (1 cm) and high (1.2 cm) values of coupling tube diameters, respectively. We need to have a finite frequency detuning and a specific range of L to obtain AD and PAD of large amplitude limit cycle oscillations in the experiments on time-delay coupled Rijke tube systems. Adapted with permission from Dange et al. [24]

where S is the cross-sectional area of the coupling tube, c is the speed of sound, and ρ is the density of air) . This change in the diameter, in turn, changes the amount of acoustic energy transmitted or reflected in the tube; thus determining the strength of the coupling between the oscillators [24].

Furthermore, we present the effect of the change in system parameters (such as the amplitude and the natural frequency of limit cycle oscillations in the uncoupled state) on the coupled behavior of Rijke tube oscillators [24]. Frequency detuning can be achieved in the system by varying the length of one of the oscillators, which in turn changes its natural frequency ($f_n = c/2L'$). Moreover, the amplitude of limit cycle oscillations can be varied by changing the electric power supplied to the heater in the system.

In Fig. 12.7, we show the coupled behavior of identical thermoacoustic systems obtained from experiments. Figure 12.7a shows a two-parameter bifurcation diagram exhibiting the variation of the amplitude of limit cycle oscillations in the uncoupled state (p_0) as a function of the normalized length of the connecting tube ($L = L'/\lambda$). We observe that when the value of p_0 is low, the coupled behaviour of oscillators changes from the state of anti-phase synchronization (AP) to in-phase synchronization (IP) via amplitude death (AD) (Figs. 12.7a, c-I). In contrast, when the value of p_0 is large, we notice a direct transition from the state of anti-phase synchronization to in-phase synchronization (Figs. 12.7a and 12.7b-I). We refer to this sudden change in the nature of synchronization of the oscillators due to a change in the coupling parameter as *phase-flip bifurcation* [66, 95, 96]. We observe that the critical length of the coupling tube corresponding to this bifurcation point nearly equals half the wavelength of the acoustic field (i.e., $L'_{cr} \sim \lambda/2$ or $L \sim 0.5$) in the system. For this value of the tube length, the phase angle between the oscillators changes as $\phi = 2\pi L'/\lambda = 2\pi(\lambda/2)/\lambda = \pi$. Therefore, we observe a change in the phase angle between the oscillators by π when the length of the coupling tube crosses L_{cr}. Furthermore, this transition from the states of anti-phase to in-phase synchronization is associated with a sudden jump in the dominant frequency

(f) of coupled oscillators (Fig. 12.7b-II). From these results, we understand that the occurrence of amplitude death in coupled identical thermoacoustic oscillators is difficult if the amplitude of oscillators is large.

According to Dange et al. [24], we can achieve amplitude death in coupled oscillators exhibiting large amplitude limit cycle oscillations, if a finite frequency detuning (Δf) is present in the system. Figure 12.8a depicts the suppression of the amplitude of limit cycle oscillations ($\Delta p/p_0$) in coupled non-identical thermoacoustic oscillators (represented as A and B) for different values of frequency detuning, when the length and the diameter of the coupling tube are fixed. Here, Δp is the change in the root mean square of the acoustic pressure fluctuations before and after the coupling is introduced. Therefore, $\Delta p/p_0 = 0$ indicates no suppression and $\Delta p/p_0 = 1$ signifies a complete quenching (amplitude death) of limit cycle oscillations in the system.

We observe that the increase in the frequency detuning shows a continuous decrease in the amplitude of limit cycle oscillations (Fig. 12.8a, c, and d). Once the frequency detuning is increased above a critical value, we notice the occurrence of amplitude death in the system. During the state of amplitude death, limit cycle oscillations are completely suppressed, and the acoustic pressure signals observed in the form of very low amplitude noisy fluctuations in both the systems (refer to AD region in Fig. 12.8a, e). A further increase in the frequency detuning engenders an emergence of periodic oscillations in one of the oscillators (oscillator A), while other oscillator (oscillator B) exhibits very small amplitude periodic oscillations (see PAD_B region of Fig. 12.8a, f). We refer to this state of coexistence of oscillatory and nearly quenched pressure dynamics in the system as partial amplitude death, PAD [4, 63, 134, 143]. Thus, we can infer that the introduction of frequency detuning can lead to quenching of large amplitude thermoacoustic oscillations in either one or both the time-delay coupled oscillators. Furthermore, in Fig. 12.8b, we notice that despite the presence of frequency detuning in the system, both coupled oscillators are frequency-locked in the regions of LCO.

In Fig. 12.9a, b, we show two-parameter bifurcation diagrams between the length of the coupling tube (L) and the frequency detuning (Δf) between the oscillators, for two different diameters (such as 1 and 1.2 cm, respectively) of the coupling tube. We notice that the increase in the diameter of the coupling tube affects the occurrence of AD and PAD states in the system. We observe that the regions of oscillation quenching (i.e. both AD and PAD) are concentrated over a finite range of the length of the coupling tube. Furthermore, these regions of quenching are observed only when a finite frequency detuning is present between the coupled oscillators. We observe the presence of finite amplitude limit cycle oscillations outside these quenching regions in the system.

Similarly, the occurrence of the amplitude death phenomenon is also studied in coupled thermoacoustic engines (Fig. 12.6c). In Fig. 12.10, we show the coupled behavior of limit cycle oscillations in two thermoacoustic engines under the influence of dissipative and time-delay coupling schemes [14]. Here, we can achieve dissipative coupling by connecting two systems directly through a needle value. The strength of the dissipative coupling can be varied by changing the opening angle of

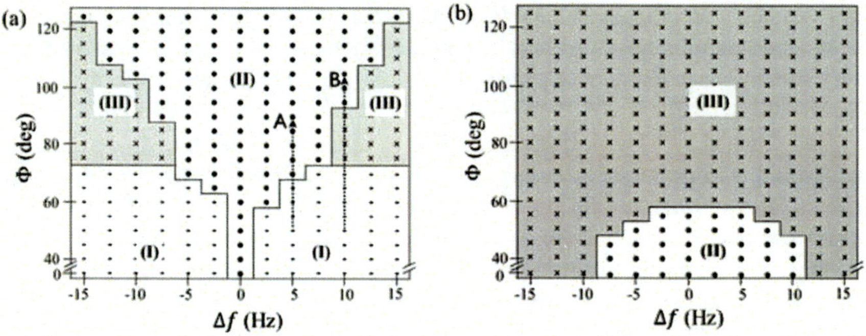

Fig. 12.10 Two-parameter bifurcation diagrams between the opening angle of the value (Φ) and the frequency difference (Δf) when two thermoacoustic engines are (**a**) only dissipatively coupled and (**b**) both dissipative and time-delay coupled. Regions (I), (II), and (III) indicate the states of desynchronization, synchronization, and amplitude death, respectively. We observe that the presence of both dissipative and time-delay coupling enhances the occurrence of AD region in the system. Adapted with permission from Biwa et al. [14]

the valve (Φ). As explained before, time-delay coupling is achieved by connecting two oscillators through a flexible tube, whose length is used to change the delay time (τ) in the coupling of oscillators. The coupled response of both the systems can be measured in terms of acoustic pressure fluctuations for different values of coupling parameters.

Figure 12.10 shows two-parameter bifurcation diagrams, presenting the variation of the valve opening angle (Φ) as a function of the frequency detuning (Δf) between the dissipatively coupled thermoacoustic engines. In these plots, regions I to III indicate desynchronized quasiperiodic oscillations, synchronized limit cycle oscillations, and amplitude death, respectively. In the absence of time-delay coupling (Fig. 12.10a), we observe that the dissipative coupling between the oscillators is insufficient in achieving amplitude death for a small value of Δf (i.e., nearly identical oscillators). In the range of small frequency detuning (path A), the coupled behavior of oscillators transitions from the state of desynchronization to synchronization with an increase in the valve angle, Φ. In contrast, for a larger value of frequency detuning (path B), their coupled behaviour exhibits a transition from the states of desynchronization to synchronization via amplitude death.

In the presence of time-delay coupling, the dissipatively coupled thermoacoustic engines show a significant change in their coupled behavior (Fig. 12.10b). With these coupling arrangements, the amplitude death region is observed over a larger range of Φ albeit the frequency detuning in the system is zero. Furthermore, we observe that the region of phase-locking is concentrated around a zero frequency detuning at lower values of Φ in the system. Thus, the simultaneous presence of both time-delay and dissipative couplings is useful in enhancing the occurrence of amplitude death in coupled thermoacoustic engines [14].

Until now, we discussed the coupled interaction of limit cycle oscillations observed in laminar thermoacoustic systems. However, in practical thermoacoustic systems, the underlying flow field is highly turbulent. The phenomena of amplitude death and synchronization of limit cycle oscillations observed for laminar systems may provide a stepping stone on understanding the coupled behavior of turbulent thermoacoustic systems.

Thomas et al. [132] investigated the effect of Gaussian white noise on the coupled behavior of limit cycle oscillations in the model of horizontal Rijke tube shown in (12.2). They showed that the presence of noise in the system significantly affects the coupled behavior of thermoacoustic systems as compared to that observed in noise-free systems. In such systems, complete quenching of thermoacoustic oscillations (i.e., the transition from limit cycle oscillations to steady state) may not be possible due to the presence of background fluctuations. In the presence of noise, an abrupt transition of limit cycle oscillations to the steady state (as seen in the noise-free condition in Fig. 12.4c) becomes a continuous transition. This gradual transition is a result of the phenomenon called *pre-bifurcation noise amplification* happening in noisy systems near the bifurcation point [128]. During the transition from limit cycle to noisy steady state, we observe a change in the amplitude distribution of the acoustic pressure fluctuations from bimodal to unimodal [132]. We can use this change in the nature of the distribution of pressure oscillations as an indicator to detect the occurrence of amplitude death in coupled thermoacoustic systems in the presence of noise.

Although we have achieved some level of understanding on the occurrence of amplitude death in laminar thermoacoustic systems, much work needs to be done to characterize this experimentally and theoretically in turbulent thermoacoustic systems. Hence, a systematic investigation of the amplitude death phenomenon in lab-scale turbulent combustors [52] is necessary before we can implement this approach in practical combustors.

12.3.2 Open-Loop Control of Thermoacoustic Instability Through Asynchronous Quenching

In the last section, we discussed that complete quenching (AD) or partial quenching (PAD) of limit cycle oscillations in coupled thermoacoustic systems can be achieved by the process of mutual interaction. Similarly, we can quench thermoacoustic instabilities developed in a single system [42, 71] or coupled thermoacoustic systems [111] through the applications of external forcing. In this control methodology, a unidirectional coupling is developed between the oscillator and the forcing signal (as shown in Fig. 12.3b), where the forcing affects the limit cycle oscillations, and not vice versa. This method of controlling thermoacoustic instability has been traditionally referred to as open-loop control [68].

As mentioned in Sec. 12.1, there are different ways to externally perturb thermoacoustic instabilities in a combustor, which include the acoustic perturbations generated using loudspeakers [11, 46] or siren [7], pulsations induced in the inlet air-fuel flow through actuators or valves [64], secondary injection of low-frequency modulated high-momentum air jet [137], etc. Through these methodologies, thermoacoustic oscillations are periodically forced at different amplitudes and frequencies. When the forcing frequency is sufficiently away from the natural frequency, i.e., non-resonant forcing [21], forcing can suppress the amplitude of limit cycle oscillations through a phenomenon called asynchronous quenching [28, 42, 55, 71, 108, 111].

Many studies in the past have employed the methodology of active open-loop control to mitigate thermoacoustic instability in laboratory combustion systems [11, 21, 46, 61, 64, 87, 137]. These studies have shown that external perturbations break the positive coupling developed between the acoustic field and the heat release rate fluctuations during the state of thermoacoustic instability. As a result, the overall acoustic power in the system approaches a value near zero. In the absence of any driving from the flame to the acoustic field, thermoacoustic instabilities are quenched in the system. Recent studies [7, 42–45, 54, 71, 108, 111] have shown the usefulness of the framework of forced synchronization [5] to systematically investigate the forced response of thermoacoustic instabilities (as discussed in Chap. 8) and also to explain the basis for the mitigation of these oscillations. In the rest of this section, we will discuss studies pertaining to the application of this framework for quenching of thermoacoustic instability in laminar thermoacoustic systems [42, 71]. Similar approaches can be used for the mitigation of thermoacoustic instability through open-loop control in turbulent thermoacoustic systems [7].

In Fig. 12.11, we show the comparison of the amplitude response of limit cycle oscillations of a thermoacoustic system for different frequencies of the external harmonic forcing. The amplitude response is obtained in terms of the variation in the root mean square (rms) of the acoustic pressure signal (P_{rms}) as a function of the increasing forcing amplitude (A_f), keeping the forcing frequency (f_f) constant. The frequency of unforced limit cycle oscillations is referred to as the natural frequency (f_{n0}).

When the forcing frequency is close to the natural frequency (i.e., $f_f = 155$ and 175 Hz, where f_{n0} = 168.8 Hz), we observe that the increase in forcing amplitude initially decreases the amplitude of limit cycle oscillations to a minimum value and then exhibits almost a steep linear increase (Fig. 12.11b,c). The minimum amplitude achieved through forcing is more for 155 Hz as compared to 175 Hz. On the other hand, when the forcing frequency is far from the natural frequency ($f_f = 100$ and 200 Hz), we notice a significant difference in the forced response of limit cycle oscillations in the system (see Fig. 12.11a, d). For $f_f = 100$ Hz, the amplitude of limit cycle oscillations displays a significant decrease at lower values of the forcing amplitude and a minimal increase at higher values of the forcing amplitude (Fig. 12.11a). In contrast, for $f_f = 200$ Hz, we observe a continuous increase in the amplitude of limit cycle oscillations for the entire range of the forcing amplitude (Fig. 12.11d). This behavior further suggests that the response of limit cycle oscillations to forcing is asymmetric in the horizontal Rijke tube system, where the application of forcing at a value higher than the natural frequency leads

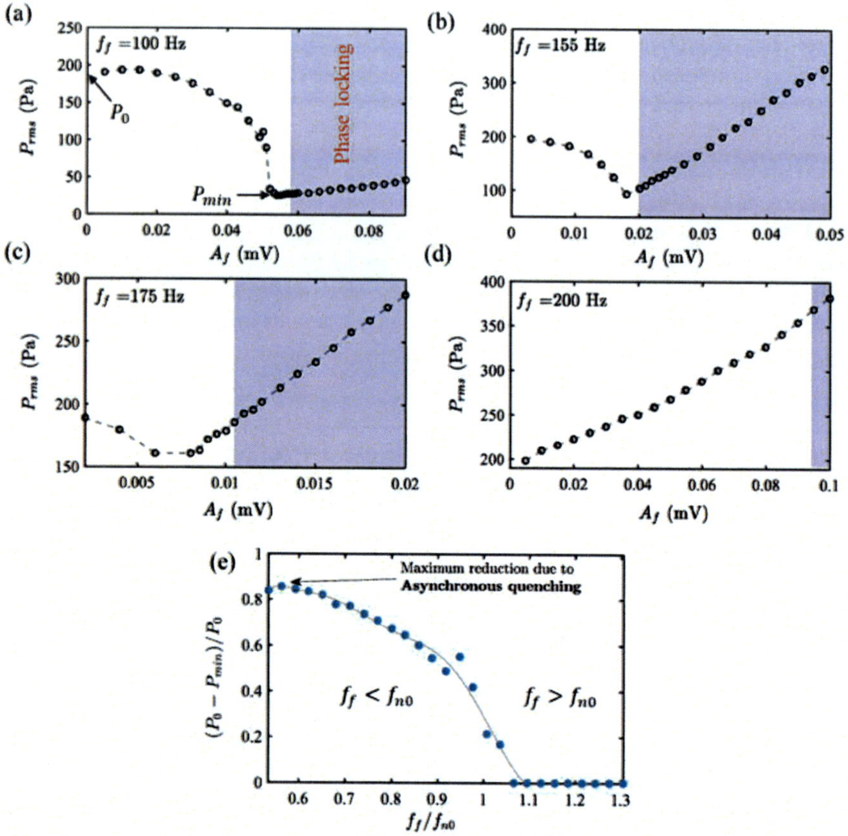

Fig. 12.11 (**a–d**) The forced response of limit cycle oscillations (LCO) showing the variation of the root mean square of the acoustic pressure fluctuations (P_{rms}) with the amplitude of forcing (A_f) for different values of the forcing frequency (f_f), from experiments in a horizontal Rijke tube. The highlighted region indicates forced synchronization of LCO. (**e**) Variation of the maximum decrease in the amplitude of LCO (i.e., $P_0 - P_{min}$), normalized by the unforced amplitude (P_0) of LCO, with change in the normalized forcing frequency, f_f/f_{n0}. We notice the synchronous amplification of LCO in the region of forced synchronization (highlighted region in **a–d**) and asynchronous quenching of LCO only at frequencies much lower than its natural frequency (see **e**). Adapted with permission from Mondal et al. [71]

to the amplification of limit cycle oscillations above their unforced amplitude. Such amplification of limit cycle oscillations is undesirable in a practical scenario.

Figure 12.11e represents the overall suppression in the amplitude of limit cycle oscillations achieved through periodic forcing. Here, the variation of the normalized maximum decrease in the amplitude of limit cycle oscillations, $(P_0 - P_{min})/P_0$, is plotted as a function of the normalized forcing frequency (f_f/f_{n0}). Here, P_0 indicates the rms of the unforced limit cycle oscillations and P_{min} indicates the minimum amplitude of limit cycle oscillations attained due to forcing when the forcing amplitude is varied at a fixed forcing frequency (see Fig. 12.11a). We observe that forcing of limit cycle oscillations at frequencies lower than the natural frequency leads to a maximum decrease in their amplitude. This decrease is almost 80% of the amplitude of unforced limit cycle oscillations. Mondal et al. [71] referred to this state of quenching of thermoacoustic oscillations due to forcing as asynchronous quenching . In contrast, we observe the amplification (resonance) of the limit cycle oscillations due to forcing of the limit cycle oscillations at frequencies larger than the natural frequency of the system. Thus, we notice asymmetry in the occurrence of asynchronous quenching of limit cycle oscillations across the natural frequency (f_{n0}) in the forced Rijke tube oscillator [71, 111]. However, in some other thermoacoustic systems (e.g., a ducted premixed conical combustor), the behavior of asynchronous quenching is reported on both sides of f_{n0} [42]. Thus, we understood that open-loop control of thermoacoustic instabilities through forcing is effective only in a certain range of the forcing parameters, where the choice of these parameters may be system dependent.

In order to understand the physical reason behind the occurrence of aynchronous quenching of thermoacoustic instability in a system, we need to simultaneously analyze the coupled response of forced acoustic pressure and heat release rate fluctuations. Figure 12.12a, b show the variation of the rms of the acoustic pressure signal (p'_{rms}) and the normalized rms of the heat release rate signal ($\dot{q}'_{\mathrm{rms}}/\overline{q}$) with the forcing frequency (f_f) at a fixed value of the forcing amplitude (A_f), respectively, in a laminar burner having multiple premixed flames (see Fig. 4.1). Here, $\overline{q}$ indicates the temporal mean of the heat release rate signal.

At a lower forcing amplitude ($A_f = 10$ mV), we observe that the amplitude of the acoustic pressure oscillations exhibits a minor variation for the entire range of the forcing frequency (Fig. 12.12a). As the forcing amplitude is increased to 30 mV, we notice a large decrease of p'_{rms} in a small range of the frequency ratio around $f_f/f_{n0} \sim 0.8$. However, at frequencies far from this frequency range, we notice a very small the change in p'_{rms} due to forcing. A further decrease in p'_{rms} is observed around $f_f/f_{n0} \sim 0.8$ as the forcing amplitude is increased to 50 and 70 mV. This indicates that the increase in the forcing amplitude correspondingly decreases the amplitude of acoustic pressure fluctuations only in a specific range of the forcing frequency, and we can refer to this range of forcing frequency as a region of 'asynchronous quenching' in the system. In contrast, the variation of the amplitude of heat release rate fluctuations shows a significant amplification in the region of asynchronous quenching (i.e., around $f_f/f_{n0} \sim 0.8$) at higher values of the forcing amplitude (Fig. 12.12b).

In Fig. 12.13, we present variation of the acoustic power production ($\mathcal{P}$), quantified from the Rayleigh index [101], with a change in forcing frequency. This plot helps us unravel the reason behind the asynchronous quenching of limit cycle oscillation in a specific range of f_f/f_{n0}. Here, $\mathcal{P}$ represents the time-averaged product of the instantaneous values of the acoustic pressure and the heat release rate fluctuations [108]. The Rayleigh index quantifies the acoustic power production (or dissipation), i.e., the acoustic driving or damping, in the system. A positive value of $\mathcal{P}$ indicates acoustic driving and a negative value of $\mathcal{P}$ represents acoustic damping

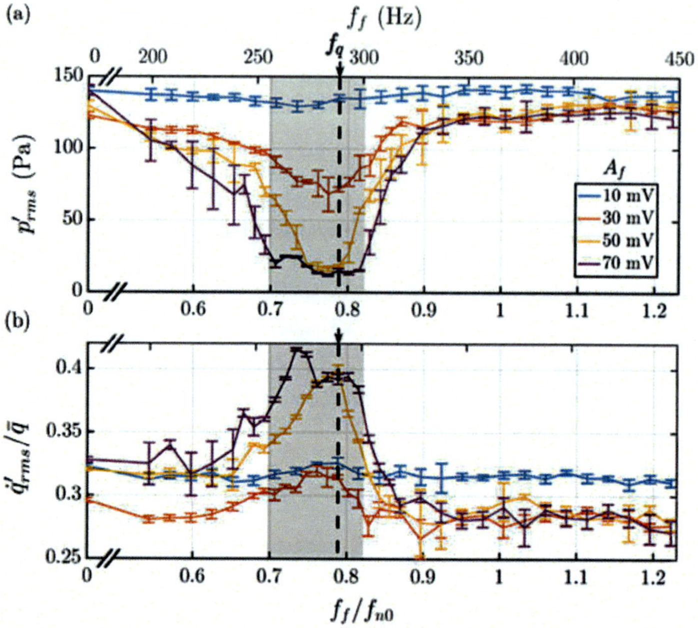

Fig. 12.12 The forced response of thermoacoustic instability in a laminar burner with multiple premixed flames in terms of variation in (**a**) the acoustic pressure fluctuations (p'_{rms}) and (**b**) the normalized heat release rate fluctuations ($\dot{q}'_{\text{rms}}/\overline{q}$) with the normalized forcing frequency (f_f/f_{n0}) for different amplitudes of forcing (A_f). The shaded portion of the plots for $A_f > 30$ mV represents the region of asynchronous quenching. Here, $f_q = 290$ Hz is the frequency of the preferred mode of the heat release rate fluctuations and $f_{n0} = 368$ Hz is the natural frequency of limit cycle oscillations in the unforced state. Reproduced with permission from Roy et al. [108]

in the system. During the occurrence of acoustic driving, the acoustic fluctuations (p') and the heat release rate ($\dot{q}'$) fluctuations exhibit an in-phase relation with each other (i.e., $-\pi/2 < \Delta\phi_{p'\dot{q}'} < \pi/2$, where $\Delta\phi_{p'\dot{q}'}$ is the instantaneous phase difference between the signals).

In Fig. 12.13, we observe a decrease in the value of the acoustic power production to a near zero value in the region of asynchronous quenching (i.e., around $f_f/f_{n0} \sim 0.8$), while a very high value at other ranges of forcing frequencies. Thus, a complete diminishment of the acoustic driving is observed during asynchronous quenching of thermoacoustic instabilities in the system. The temporal variation of $\Delta\phi_{p'\dot{q}'}$ at different instances of forcing frequencies are shown in the insets of Fig. 12.13. In regions I and VII of Fig. 12.13 (i.e., outside the region of asynchronous quenching), we observe that both p' and $\dot{q}'$ signals are locked with a nearly zero rad of phase shift (Fig. 12.13a, i). As a result, the acoustic power production has a very high value in these regions. Once p' and $\dot{q}'$ signals are desynchronized in time, i.e., $\Delta\phi_{p'\dot{q}'}$ shows the behavior of intermittent phase-locking (Fig. 12.13b, h) or phase drifting (Fig. 12.13c, f, g), we notice a decrease in the value of the acoustic power production

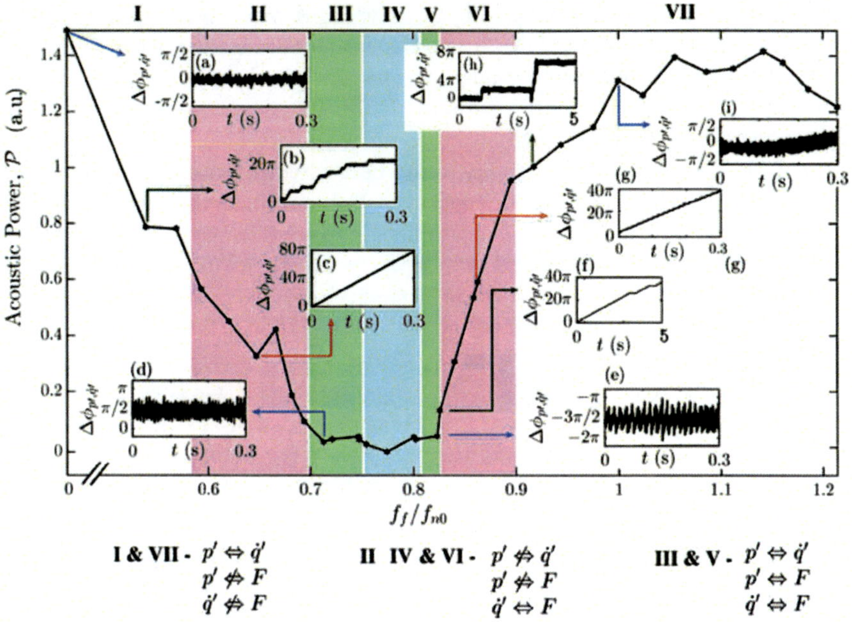

Fig. 12.13 The variation of time-averaged acoustic power production ($\mathcal{P}$) with the normalized forcing frequency (f_f/f_{n0}) at a fixed value of A_f. **(a)–(i)** Insets show the time series of the relative phase between the acoustic pressure and the heat release rate fluctuations ($\Delta\phi_{p'\dot{q}'}$) for different values of forcing parameters. The regions in the plot are highlighted based on the synchronization behavior of these oscillations. We observe that in the asynchronous quenching region, $\Delta\phi_{p'\dot{q}'}$ shows fluctuations around $\pi/2$ or $3\pi/2$ rad, causing acoustic power production $\mathcal{P}$ to approach zero. Reproduced with permission from Roy et al. [108]

(regions II and IV in Fig. 12.13). In contrast, in the frequency range of asynchronous quenching (i.e., minimum acoustic power production), the temporal variation of p' and $\dot{q}'$ exhibits a locking around a phase shift of $\pi/2$ rad (Fig. 12.13d) or $3\pi/2$ rad (Fig. 12.13e). This suggests that an out-of-phase locking of the acoustic pressure and the heat release rate fluctuations is responsible for the occurrence of asynchronous quenching of limit cycle oscillations in the system.

Furthermore, from the comparison of Figs. 12.12a and 12.13 with Fig. 12.12b, we notice that the frequency range corresponding to the quenching of the acoustic pressure oscillations and the decrease of acoustic power in the system matches the frequency range corresponding to the amplification of the heat release rate fluctuations in the system. Roy et al. [108] observed that such amplification of the heat release rate field could be the result of the presence of a preferred mode in the reacting flow field (or the flame) of the burner. Therefore, if the forcing frequency is near the preferred mode of the reacting field, the heat release rate oscillations lock with the forcing signal by breaking their inherent coupling with the acoustic pressure fluctuations developed during the state of thermoacoustic instability. This further results in quenching of thermoacoustic instabilities in the system.

In contrast with the independent application of methodologies based on mutual coupling of two thermoacoustic systems or forcing of a single thermoacoustic system, Sahay et al. [111] studied the combined effect of these methodologies to mitigate thermoacoustic instability, using both experiments and a model. They investigated the dynamics of thermoacoustically unstable two mutually coupled Rijke tube systems when subjected to asymmetric forcing (i.e., the forcing is applied only to one system). Through this study, they found a larger parametric region of oscillation quenching of thermoacoustic instability when the two methodologies are used concurrently, as compared to the individual application of these methodologies. Furthermore, they observed that when two thermoacoustic systems having different natural frequencies (i.e., nonidentical oscillators) are coupled, the effect of forcing is insignificant on synchronization and quenching of oscillations in the oscillator that is not directly forced. We may need to extend this approach in the future to study quenching of thermoacoustic instabilities in coupled can-type or can-annular type combustors through open-loop forcing. Next, we discuss a method of smart passive control to mitigate thermoacoustic in turbulent combustors.

12.4 Identification of Critical Regions in the Spatial Reacting Field

As we discussed in Sec. 12.1, passive control of thermoacoustic instability is achieved through simple modifications in the combustor geometry in the design stage itself. These design modifications are usually selected through several trials that are mostly based on a hit-or-miss approach. To give an example, during the development of F1-engine used in the first Apollo program; engineers had to perform nearly 3200 full-scale tests, among these tests around 2000 tests were just needed to address the problem of thermoacoustic instability [85]. Thus, the control of thermoacoustic instability through such a strategy is an expensive and a challenging task.

To remedy this problem, Krishnan et al. [58] suggested a simple approach to improve the effectiveness of passive controls. They recommended that one has to identify the optimal locations in the reacting field that controls the spatiotemporal dynamics of a combustor during the state of thermoacoustic instability. Once such optimal regions are located, we need to design controls to disturb the flow field at these regions and, thus, mitigate thermoacoustic instability [59, 73, 109, 127]. We can refer to these optimal locations of the reacting field as 'critical regions'.

In order to identify these critical regions, Krishnan et al. [58, 59] employed a complex network approach, similar to that we discussed in Chap. 11, for constructing spatial networks. In this approach, we first need to calculate several network properties, i.e., degree, closeness centrality, betweenness centrality, etc. [10, 84]

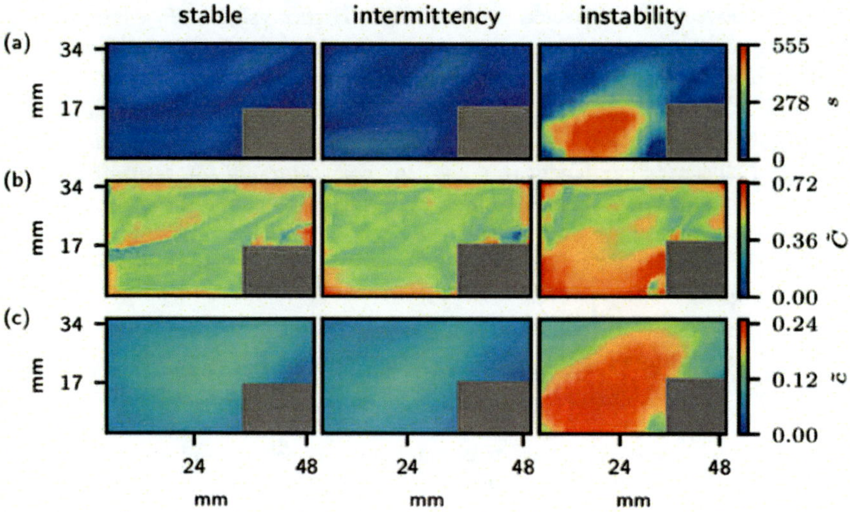

Fig. 12.14 (**a–c**) Weighted spatial networks constructed using different network properties such as the strength of a node (s), local clustering coefficient ($\overline{C}$), and closeness centrality ($\overline{c}$), respectively, during the states of combustion noise, intermittency, and thermoacoustic instability. The grey rectangle in the lower right corner is used to mask the bluff body. High values of network measures during the state of thermoacoustic instability indicate the presence of critical regions in the reaction field of the combustor. Reproduced with permission from Krishnan et al. [58]

from the spatial data of a thermoacoustic system. Then, we need identify the regions corresponding to high values of these network properties as critical regions of the reacting field. This approach of network construction is also used in uncovering spatial patterns in brain networks [34], climate networks [31, 136], and fluid flows [35]. Further, these network properties are successfully used in indicating optimal locations of fire spread [110], the spread of mobile malware [131], and vortex interactions in turbulent flows [130].

In Fig. 12.14, we show weighted spatial correlation networks, constructed from the data of high-speed velocity field obtained from PIV measurements, for the states of combustion noise, intermittency, and thermoacoustic instability [58]. The value of Pearson's correlation coefficient between two nodes (say, i and j) we can obtain as,

$$R_{i,j} = \frac{\langle V(i,t)V(j,t)\rangle - \langle V(i,t)\rangle\langle V(j,t)\rangle}{\sigma(V(i))\sigma(V(j))} \tag{12.3}$$

where $V(t)$ is the signal of total velocity $V(t) = \sqrt{V_x^2(t) + V_y^2(t)}$, $\sigma^2(V(i)) = \langle V(i,t)^2\rangle - \langle V(i,t)\rangle^2$, and $\langle . \rangle$ indicates the time average of the entire signal.

We can connect two nodes in the network if $R_{i,j}$ is greater than a particular value (say, $R_{i,j} \geq 0.5$ [58]). We can then assign the value of $R_{i,j}$ as a weight ($w_{i,j}$) for the link between the connected pairs of nodes. Higher value of $R_{i,j}$ indicates the presence of a stronger connection between the nodes at different locations in the spatial field of the combustor.

Figure 12.14a–c show spatial networks obtained using three measures that highlight the property of nodes in a network such as the strength of a node (s), local clustering coefficient ($\overline{C}$), and closeness centrality ($\overline{c}$), respectively. The mathematical details of these measures are discussed in Chap. 11. In this analysis, the networks are constructed in the upper-half of a bluff-body stabilized dump combustor as shown in Fig. 12.15a, also refer to Fig. 7.3 in Chap. 7 for more details on the experimental setup.

During the states of combustion noise and intermittency, we observe a near uniform distribution of these properties in the flow field (Fig. 12.14). In contrast, during the state of thermoacoustic instability, we notice higher values of these properties in a specific region behind the bluff-body. Higher values of s and $\overline{C}$ indicate the presence of higher correlations between the velocity fluctuations observed at different locations in the system. These high correlations are the result of the periodic formation of the large-scale vortical structures from the dump plane during the state of thermoacoustic instability (as we discussed in Chap. 7). On the other hand, higher values of $\overline{c}$ suggest that perturbations given in the region behind the bluff-body can quickly reach other parts of the flow field. Thus, the region corresponding to higher values of these properties emerges as the most critical region in the combustor flow field. We can implement a smart passive control strategy at this region to mitigate thermoacoustic instability.

In Fig. 12.15a, we show an arrangement of smart passive controls, in the form of local air micro-jet injectors along the combustor casing. The combustor flow field is divided into three regions: region A corresponds to upstream of the bluff-body, region B is present above the bluff-body, and region C exists downstream of the bluff-body. The air flow rate supplied through micro-jets is very small as compared to that supplied in the main combustor through the inlet duct.

During these control experiments, we first need to establish the state of thermoacoustic instability in the system and measure the acoustic response of the system at every instance of the injection of micro-jets from different ports of the combustor. The effect of this micro-jet injection can be represented in the form of the variation of the normalized amplitude of the acoustic pressure fluctuations ($p'_{\mathrm{rms}}/p'_{\mathrm{rmsB}}$) against the total momentum flux ratio $(v_j/v_a)^2$ (Fig. 12.15b). Here, p'_{rms} and p'_{rmsB} are the root mean square of the acoustic pressure signal after and before the injection of micro-jets, v_j represents the air velocity in the micro-jet, and v_a indicates the air velocity in the main combustor.

In Fig. 12.15b, we observe that the injection of the secondary air through micro-jets located at the side wall and the top wall of the combustor in region A leads to almost 86 and 83% reduction in the amplitude of thermoacoustic oscillations, respectively. These values are very close to the amplitude of the acoustic pressure fluctuations observed during the state of combustion noise. This, in turn, indicates the transition of the system behavior from unstable operation to stable operation

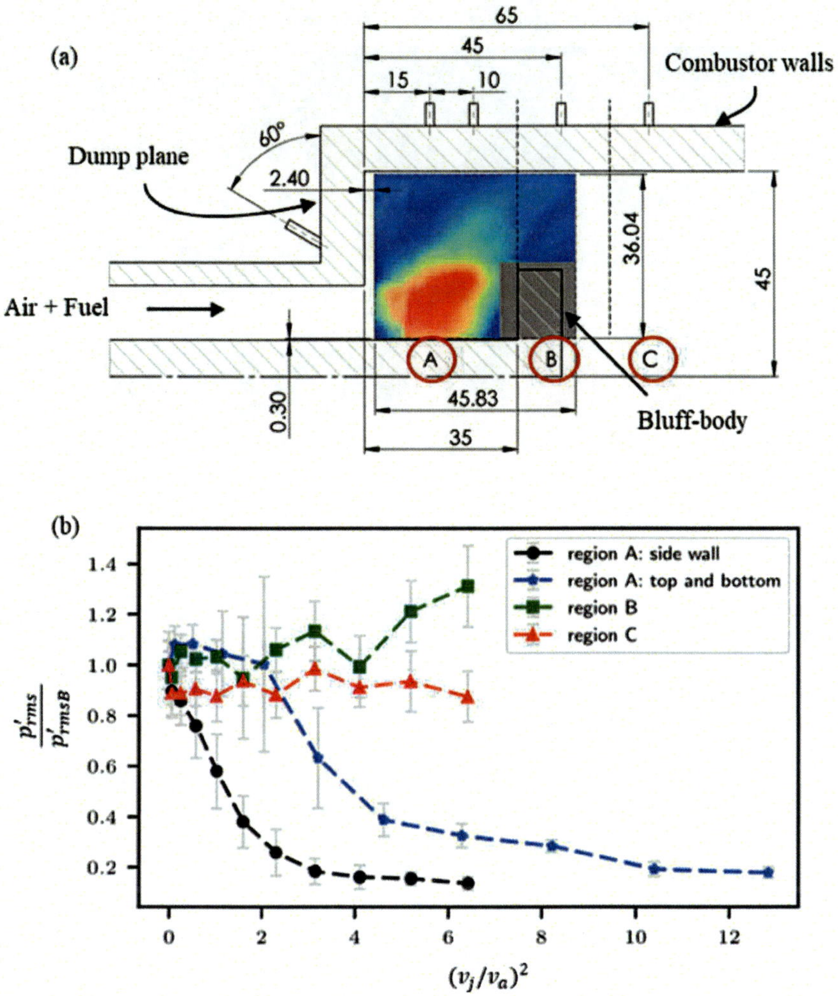

Fig. 12.15 **(a)** The cross-section of the upper half of the bluff-body stabilized dump combustor, showing the arrangement of micro-jet injections along the combustor casing. The spatial distribution of the strength of a node (s) indicates the presence of a critical region in the reaction field during the state of thermoacoustic instability. A, B, C indicate different regions of the combustor. **(b)** Variation of the normalized amplitude of the acoustic pressure fluctuations with the momentum flux ratio of micro-jet injections at different regions of the combustor flow field. The application of micro-jets only at critical regions leads to suppression of thermoacoustic instability in the system. Adapted with permission from Krishnan et al. [58]

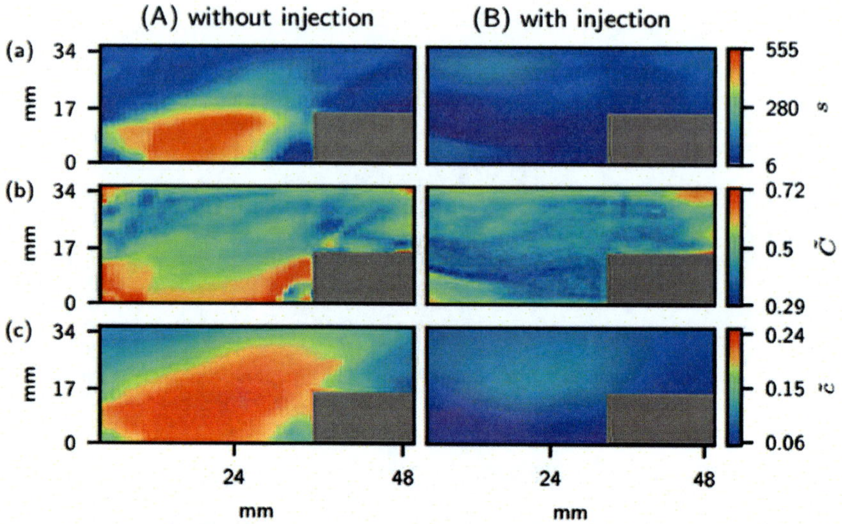

Fig. 12.16 (**a–c**) Weighted spatial networks constructed using different network properties such as the strength of a node (s), local clustering coefficient ($\overline{C}$), and closeness centrality ($\overline{c}$), respectively, (A) during the state of thermoacoustic instability and (B) during the suppression of thermoacoustic instability when micro-jets are injected from the dump plane. We observe that due to the destruction of critical regions in the reaction field, the network measures achieve a lower value. Reproduced with permission from Krishnan et al. [58]

due to the mitigation of thermoacoustic instability in the system. In contrast, the injection of micro-jets in regions B and C does not lead to any reduction in the amplitude of thermoacoustic oscillations. This critical dependence of the reduction in the amplitude of pressure oscillations on the location of micro-jet injectors further proves the hypothesis that, the critical regions identified from weighted correlation networks of the reacting field (Fig. 12.14) correspond to effective locations for the implementation of the passive controls in the combustor.

Figure 12.16 shows the comparison of network properties before and after the suppression of thermoacoustic instability through a smart passive control. We notice that due to micro-jet injection, the critical regions in the combustor flow field are completely suppressed; thus, the mitigation of thermoacoustic instability is achieved. The network properties have a nearly uniform distribution, implying there is no critical region. Such a distribution of network properties is similar to that observed for the state of combustion noise (stable operation) in Fig. 12.15.

In addition to the method identifying critical regions in the combustor using the complex network approach, we can also detect such critical regions through several other methods based on dynamical systems approach, synchronization approach [127] or multifractal approach [109].

To summarize, in this chapter, we discussed methodologies that have been developed to mitigate and evade the occurrence of undesired thermoacoustic instability in practical thermoacoustic systems. We introduced many early warning measures that can help us in forewarning the impending thermoacoustic instability and avoiding the subsequent consequences that could arise due to the presence of these instabilities. We also focused on the detailed description of promising control strategies that are based on various tools from dynamical system theory and complex system theory. We introduced novel phenomena such as amplitude death and asynchronous quenching, and the method of identifying critical regions in the reaction flow field that can be used to quench thermoacoustic instabilities. These technologies may further reduce the development timescales of practical thermoacoustic systems, extend the life of fielded combustors, prevent immediate shutdown of gas turbines, etc. Therefore, we believe that the use of dynamical systems theory and complex system theory provides an elegant, simple, and cost-effective solutions for the control of thermoacoustic instability, and these technologies should be implemented both in fielded combustors and also during the development of new combustors.

In the next chapter, we will discuss different types of oscillatory instabilities existing in various turbulent flow systems (such as aeroacoustic and aeroelastic, and thermoacoustic systems) and present the universality in the dynamical transition to such instabilities observed across these systems.

References

1. Abel, M., Ahnert, K., Bergweiler, S.: Synchronization of sound sources. Phys. Rev. Lett. **103**(11), 114301 (2009)
2. Abel, M., Bergweiler, S., Gerhard-Multhaupt, R.: Synchronization of organ pipes: experimental observations and modeling. J. Acoust. Soc. Am. **119**(4), 2467–2475 (2006)
3. Annaswamy, A.M., Ghoniem, A.F.: Active control in combustion systems. IEEE Control Syst. Mag. **15**(6), 49–63 (1995)
4. Atay, F.M.: Total and partial amplitude death in networks of diffusively coupled oscillators. Phys. D **183**(1–2), 1–18 (2003)
5. Balanov, A., Janson, N., Postnov, D., Sosnovtseva, O.: Synchronization: From Simple to Complex, vol. 17. Springer, Berlin (2009)
6. Balasubramanian, K., Sujith, R.I.: Thermoacoustic instability in a Rijke tube: non-normality and nonlinearity. Phys. Fluids **20**(4), 044103 (2008)
7. Balusamy, S., Li, L.K.B., Han, Z., Juniper, M.P., Hochgreb, S.: Nonlinear dynamics of a self-excited thermoacoustic system subjected to acoustic forcing. Proc. Combust. Inst. **35**(3), 3229–3236 (2015)
8. Banaszuk, A., Mehta, P.G., Hagen, G.: The role of control in design: from fixing problems to the design of dynamics. Control Eng. Pract. **15**(10), 1292–1305 (2007)
9. Bar-Eli, K.: Coupling of chemical oscillators. J. Phys. Chem. A **88**(16), 3616–3622 (1984)
10. Barthélemy, M.: Spatial networks. Phys. Rep. **499**(1–3), 1–101 (2011)
11. Bellows, B.D., Hreiz, A., Lieuwen, T.: Nonlinear interactions between forced and self-excited acoustic oscillations in premixed combustor. J. Propuls. Power **24**(3), 628–630 (2008)
12. Bellucci, V., Flohr, P., Paschereit, C.O., Magni, F.: On the use of Helmholtz resonators for damping acoustic pulsations in industrial gas turbines. J. Eng. Gas Turbines Power **126**(2), 271–275 (2004)

13. Bhattacharya, C., O'Connor, J., Ray, A.: Data-driven detection and early prediction of thermoacoustic instability in a multi-nozzle combustor. Combust. Sci. Tech. 1–32 (2020)
14. Biwa, T., Tozuka, S., Yazaki, T.: Amplitude death in coupled thermoacoustic oscillators. Phys. Rev. Appl. **3**(3), 034006 (2015)
15. Bogoliubovov, N.N., Mitropol'skii, I.A., Mitropol'skij, J.A., Mitropolsky, Y.A.: Asymptotic Methods in the Theory of Non-linear Oscillations, vol. 10. CRC Press, Boca Raton (1961)
16. Bulat, G., McMillan, R.: Fuel ratio control in a combustion apparatus with multiple fuel supply lines. US Patent US8313324B2; Priority date, January 19 (2006)
17. Bulat, G., Skipper, D., McMillan, R., Syed, K.: Active control of fuel splits in gas turbine DLE combustion systems. In: Proceedings of the ASME Turbo Expo 2007: Power for Land, Sea, and Air, GT2007-27266, pp. 135–144 (2007)
18. Candel, S.: Combustion dynamics and control: progress and challenges. Proc. Combust. Inst. **29**(1), 1–28 (2002)
19. Chaudhari, R.R., Sahu, R.P., Ghosh, S., Mukhopadhyay, A., Sen, S.: Flame color as a lean blowout predictor. J. Spray Combust. Dyn. **5**(1), 49–65 (2013)
20. Chiocchini, S., Pagliaroli, T., Camussi, R., Giacomazzi, E.: Chaotic and linear statistics analysis in thermoacoustic instability detection. J. Propul. Power **34**(1), 15–16 (2018)
21. Ćosić, B., Bobusch, B.C., Moeck, J.P., Paschereit, C.O.: Open-loop control of combustion instabilities and the role of the flame response to two-frequency forcing. J. Eng. Gas Turbines Power **134**(6), 061502 (2012)
22. Crowley, M.F., Field, R.J.: Nonlinear Phenomena in Chemical Dynamics, vol. 12. Springer, Berlin (1981)
23. Culick, F.E.C.: Unsteady motions in combustion chambers for propulsion systems. Tech. rep., AGARDograph, NATO/RTO-AG-AVT-039 (2006)
24. Dange, S., Manoj, K., Banerjee, S., Pawar, S.A., Mondal, S., Sujith, R.I.: Oscillation quenching and phase-flip bifurcation in coupled thermoacoustic systems. Chaos **29**(9), 093135 (2019)
25. Daw, C.S., Finney, C.E.A., Tracy, E.R.: A review of symbolic analysis of experimental data. Rev. Sci. Instrum. **74**(2), 915–930 (2003)
26. De, S., Bhattacharya, A., Mondal, S., Mukhopadhyay, A., Sen, S.: Investigation of flame behavior and dynamics prior to lean blowout in a combustor with varying mixedness of reactants for the early detection of lean blowout. J. Spray Combust. Dyn. **11**, 1–20 (2019)
27. De, S., Biswas, A., Bhattacharya, A., Mukhopadhyay, A., Sen, S.: Use of flame color and chemiluminescence for early detection of lean blowout in gas turbine combustors at different levels of fuel–air premixing. Combust. Sci. Tech. **192**(5), 933–957 (2019)
28. Dewan, E.: Harmonic entrainment of van der pol oscillations: phase locking and asynchronous quenching. IEEE Trans. Autom. Control **17**(5), 655–663 (1972)
29. Dey, D., Chaudhari, R.R., Mukhopadhyay, A., Sen, S., Chakravorti, S.: A cross-wavelet transform aided rule based approach for early prediction of lean blow-out in swirl-stabilized dump combustor. J. Spray Combust. Dyn. **7**(1), 69–90 (2015)
30. Domen, S., Gotoda, H., Kuriyama, T., Okuno, Y., Tachibana, S.: Detection and prevention of blowout in a lean premixed gas-turbine model combustor using the concept of dynamical system theory. Proc. Combust. Inst. **35**(3), 3245–3253 (2015)
31. Donges, J.F., Zou, Y., Marwan, N., Kurths, J.: Complex networks in climate dynamics. Eur. Phys. J. Spec. Top. **174**(1), 157–179 (2009)
32. Dowling, A.P., Morgans, A.S.: Feedback control of combustion oscillations. Annu. Rev. Fluid Mech. **37**, 151–182 (2005)
33. Eldredge, J.D., Dowling, A.P.: The absorption of axial acoustic waves by a perforated liner with bias flow. J. Fluid Mech. **485**, 307—–335 (2003)
34. Finn, E.S., Shen, X., Scheinost, D., Rosenberg, M.D., Huang, J., Chun, M.M., Papademetris, X., Constable, R.T.: Functional connectome fingerprinting: identifying individuals using patterns of brain connectivity. Nat. Neurosci. **18**(11), 1664–1671 (2015)
35. Gao, Z., Jin, N.: Flow-pattern identification and nonlinear dynamics of gas-liquid two-phase flow in complex networks. Phys. Rev. E **79**(6), 066303 (2009)

36. Godavarthi, V., Unni, V.R., Gopalakrishnan, E.A., Sujith, R.I.: Recurrence networks to study dynamical transitions in a turbulent combustor. Chaos **27**(6), 063113 (2017)
37. Gopalakrishnan, E.A., Sharma, Y., John, T., Dutta, P.S., Sujith, R.I.: Early warning signals for critical transitions in a thermoacoustic system. Sci. Rep. **6**, 35310 (2016)
38. Gopalakrishnan, E.A., Sujith, R.I.: Influence of system parameters on the hysteresis characteristics of a horizontal Rijke tube. J. Spray Combust. Dyn. **6**(3), 293–316 (2014)
39. Gotoda, H., Hayashi, K., Tsujimoto, R., Domen, S., Tachibana, S.: Dynamical properties of combustion instability in a laboratory-scale gas-turbine model combustor. J. Eng. Gas Turbines Power **139**(4), 041509 (2017)
40. Gotoda, H., Kinugawa, H., Tsujimoto, R., Domen, S., Okuno, Y.: Characterization of combustion dynamics, detection, and prevention of an unstable combustion state based on a complex-network theory. Phys. Rev. Appl. **7**(4), 044027 (2017)
41. Gotoda, H., Shinoda, Y., Kobayashi, M., Okuno, Y., Tachibana, S.: Detection and control of combustion instability based on the concept of dynamical system theory. Phys. Rev. E **89**(2), 022910 (2014)
42. Guan, Y., Gupta, V., Kashinath, K., Li, L.K.B.: Open-loop control of periodic thermoacoustic oscillations: experiments and low-order modelling in a synchronization framework. Proc. Combust. Inst. **37**(4), 5315–5323 (2019)
43. Guan, Y., Gupta, V., Wan, M., Li, L.K.B.: Forced synchronization of quasiperiodic oscillations in a thermoacoustic system. J. Fluid Mech. **879**, 390–421 (2019)
44. Guan, Y., He, W., Murugesan, M., Li, Q., Liu, P., Li, L.K.B.: Control of self-excited thermoacoustic oscillations using transient forcing, hysteresis and mode switching. Combust. Flame **202**, 262–275 (2019)
45. Guan, Y., Murugesan, M., Li, L.K.B.: Strange nonchaotic and chaotic attractors in a self-excited thermoacoustic oscillator subjected to external periodic forcing. Chaos **28**(9), 093109 (2018)
46. Gutmark, E., Parr, T.P., Hanson-Parr, D.M., Schadow, K.C.: On the role of large and small-scale structures in combustion control. Combust. Sci. Tech. **66**(1–3), 107–126 (1989)
47. Gysling, D.L., Copeland, G.S., McCormick, D.C., Proscia, W.M.: Combustion system damping augmentation with Helmholtz resonators. J. Eng. Gas Turbines Power **122**(2), 269–274 (2000)
48. Hashimoto, T., Shibuya, H., Gotoda, H., Ohmichi, Y., Matsuyama, S.: Spatiotemporal dynamics and early detection of thermoacoustic combustion instability in a model rocket combustor. Phys. Rev. E **99**(3), 032208 (2019)
49. Hobson, D.E., Fackrell, J.E., Hewitt, G.: Combustion instabilities in industrial gas turbines—measurements on operating plant and thermoacoustic modeling. J. Eng. Gas Turbines Power **122**(3), 420–428 (2000)
50. Huang, Y., Yang, V.: Dynamics and stability of lean-premixed swirl-stabilized combustion. Prog. Energy Combust. Sci. **35**(4), 293–364 (2009)
51. Hyodo, H., Iwasaki, M., Biwa, T.: Suppression of Rijke tube oscillations by delay coupling. J. Appl. Phys. **128**(9), 094902 (2020)
52. Jegal, H., Moon, K., Gu, J., Li, L.K.B., Kim, K.T.: Mutual synchronization of two lean-premixed gas turbine combustors: phase locking and amplitude death. Combust. Flame **206**, 424–437 (2019)
53. Kantz, H., Schreiber, T.: Nonlinear Time Series Analysis, vol. 7. Cambridge University Press, Cambridge (2004)
54. Kashinath, K., Li, L.K.B., Juniper, M.P.: Forced synchronization of periodic and aperiodic thermoacoustic oscillations: lock-in, bifurcations and open-loop control. J. Fluid Mech. **838**, 690–714 (2018)
55. Keen, B.E., Fletcher, W.H.W.: Suppression of a plasma instability by the method of "asynchronous quenching". Phys. Rev. Lett. **24**(4), 130–134 (1970)
56. Kobayashi, T., Murayama, S., Hachijo, T., Gotoda, H.: Early detection of thermoacoustic combustion instability using a methodology combining complex networks and machine learning. Phys. Rev. Appl. **11**(6), 064034 (2019)

57. Koseska, A., Volkov, E., Kurths, J.: Oscillation quenching mechanisms: amplitude vs. oscillation death. Phys. Rep. **531**(4), 173–199 (2013)
58. Krishnan, A., Manikandan, R., Midhun, P.R., Reeja, K.V., Unni, V.R., Sujith, R.I., Marwan, N., Kurths, J.: Mitigation of oscillatory instability in turbulent reactive flows: a novel approach using complex networks. Europhys. Lett. **128**(1), 14003 (2019)
59. Krishnan, A., Sujith, R.I., Marwan, N., Kurths, J.: Suppression of thermoacoustic instability by targeting the hubs of the turbulent networks in a bluff body stabilized combustor. J. Fluid Mech. **916**, A20 (2021)
60. Lieuwen, T.: Online combustor stability margin assessment using dynamic pressure data. J. Eng. Gas Turbines Power **127**(3), 478–482 (2005)
61. Lieuwen, T., Neumeier, Y.: Nonlinear pressure-heat release transfer function measurements in a premixed combustor. Proc. Combust. Inst. **29**(1), 99–105 (2002)
62. Lieuwen, T.C., Yang, V.: Combustion Instabilities in Gas Turbine Engines (Operational Experience, Fundamental Mechanisms and Modeling), vol. 210. Progress in Astronautics and Aeronautics, AIAA, Reston (2005)
63. Liu, W., Xiao, J., Yang, J.: Partial amplitude death in coupled chaotic oscillators. Phys. Rev. E **72**(5), 057201 (2005)
64. Lubarsky, E., Shcherbik, D., Zinn, B., McManus, K., Fric, T., Srinivasan, S.: Active control of combustion oscillations by non-coherent fuel flow modulation. In: 9th AIAA/CEAS Aeroacoustics Conference and Exhibit, pp. 2003–3180, AIAA, Reston (2003)
65. Manoj, K., Pawar, S.A., Dange, S., Mondal, S., Sujith, R.I., Surovyatkina, E., Kurths, J.: Synchronization route to weak chimera in four candle-flame oscillators. Phys. Rev. E **100**(6), 062204 (2019)
66. Manoj, K., Pawar, S.A., Sujith, R.I.: Experimental evidence of amplitude death and phase-flip bifurcation between in-phase and anti-phase synchronization. Sci. Rep. **8**(1), 1–7 (2018)
67. Manoj, K., Pawar, S.A., Sujith, R.I.: Experimental investigation on the susceptibility of minimal networks to a change in topology and number of oscillators. Phys. Rev. E **103**(2), 022207 (2021)
68. McManus, K.R., Poinsot, T., Candel, S.M.: A review of active control of combustion instabilities. Prog. Energy Combust. Sci. **19**(1), 1–29 (1993)
69. McManus, K.R., Vandsburger, U., Bowman, C.T.: Combustor performance enhancement through direct shear layer excitation. Combust. Flame **82**(1), 75–92 (1990)
70. Mondal, S., Ghalyan, N.F., Ray, A., Mukhopadhyay, A.: Early detection of thermoacoustic instabilities using hidden Markov models. Combust. Sci. Tech. **191**(8), 1309–1336 (2018)
71. Mondal, S., Pawar, S.A., Sujith, R.I.: Forced synchronization and asynchronous quenching of periodic oscillations in a thermoacoustic system. J. Fluid Mech. **864**, 73–96 (2019)
72. Mukhopadhyay, A., Chaudhari, R.R., Paul, T., Sen, S., Ray, A.: Lean blow-out prediction in gas turbine combustors using symbolic time series analysis. J. Propuls. Power **29**(4), 950–960 (2013)
73. Murayama, S., Gotoda, H.: Attenuation behavior of thermoacoustic combustion instability analyzed by a complex-network-and synchronization-based approach. Phys. Rev. E **99**(5), 052222 (2019)
74. Murayama, S., Kaku, K., Funatsu, M., Gotoda, H.: Characterization of dynamic behavior of combustion noise and detection of blowout in a laboratory-scale gas-turbine model combustor. Proc. Combust. Inst. **37**(4), 5271–5278 (2019)
75. Murayama, S., Kinugawa, H., Tokuda, I.T., Gotoda, H.: Characterization and detection of thermoacoustic combustion oscillations based on statistical complexity and complex-network theory. Phys. Rev. E **97**(2), 022223 (2018)
76. Muruganandam, T.M., Nair, S., Scarborough, D., Neumeier, Y., Jagoda, J., Lieuwen, T., Seitzman, J., Zinn, B.: Active control of lean blowout for turbine engine combustors. J. Propuls. Power **21**(5), 807–814 (2005)
77. Murugesan, M., Sujith, R.I.: Detecting the onset of an impending thermoacoustic instability using complex networks. J. Propuls. Power **32**(3), 707–712 (2016)

78. Murugesan, M., Unni, V.R., Nair, V.V., Sujith, R.I.: Devices and methods for early prediction of impending instabilities of a system. US Patent US10337414B2, Priority date: April 29 (2014)
79. Nair, S., Lieuwen, T.: Acoustic detection of blowout in premixed flames. J. Propuls. Power **21**(1), 32–39 (2005)
80. Nair, V., Sujith, R.I.: Multifractality in combustion noise: predicting an impending combustion instability. J. Fluid Mech. **747**, 635–655 (2014)
81. Nair, V., Thampi, G., Karuppusamy, S., Gopalan, S., Sujith, R.I.: Loss of chaos in combustion noise as a precursor of impending combustion instability. J. Spray Combust. Dyn. **5**(4), 273–290 (2013)
82. Nair, V., Thampi, G., Sujith, R.I.: Intermittency route to thermoacoustic instability in turbulent combustors. J. Fluid Mech. **756**, 470–487 (2014)
83. Nair, V.V., Thampi, G., Karuppusamy, S., Gopalan, S., Sujith, R.I.: System and method for predetermining the onset of impending oscillatory instabilities in practical devices. US Patent: Application number 16/663, Publication number US9804054B2, Priority date: October 1 (2012)
84. Newman, M.: Networks. Oxford University Press, Oxford (2010)
85. Oefelein, J.C., Yang, V.: Comprehensive review of liquid-propellant combustion instabilities in F-1 engines. J. Propuls. Power **9**(5), 657–677 (1993)
86. Ohe, K., Takeda, S.: Asynchronous quenching and resonance excitation of ionization waves in positive columns. Contrib. Plasma Phys. **14**(2), 55–65 (1974)
87. Paschereit, C., Gutmark, E.: Control of thermoacoustic instabilities in a premixed combustor by fuel modulation. In: 37th Aerospace Sciences Meeting and Exhibit, 1999-711. AIAA, Reston (1999)
88. Pavithran, I., Sujith, R.I.: Effect of rate of change of parameter on early warning signals for critical transitions. Chaos **31**(1), 013116 (2021)
89. Pavithran, I., Unni, V.R., Sujith, R.I.: Critical transitions and their early warning signals in thermoacoustic systems. Eur. Phys. J. Spec. Top. 1–22 (2021)
90. Pavithran, I., Unni, V.R., Varghese, A.J., Kasthuri, P., Sujith, R.I.: System and method for determining the amplitude of oscillatory instabilities in fluid mechanical devices. US Patent: Application number 16/633658, Publication number US20210131910A1, Priority date: July 25 (2018)
91. Pavithran, I., Unni, V.R., Varghese, A.J., Sujith, R.I., Saha, A., Marwan, N., Kurths, J.: Universality in the emergence of oscillatory instabilities in turbulent flows. Europhys. Lett. **129**(2), 24004 (2020)
92. Pavithran, I., Unni, V.R., Varghese, A.J., Sujith, R.I., Saha, A., Marwan, N., Kurths, J.: Predicting the amplitude of thermoacoustic instability using universal scaling behaviour. In: Proceedings of ASME Turbo Expo 2021: Turbomachinery Technical Conference and Exposition, GT2021-60074 (2021)
93. Pikovsky, A., Rosenblum, M., Kurths, J.: Synchronization: A Universal Concept in Nonlinear Sciences, vol. 12. Cambridge University Press, Cambridge (2003)
94. Poinsot, T.: Prediction and control of combustion instabilities in real engines. Proc. Combust. Inst. **36**(1), 1–28 (2017)
95. Prasad, A., Dana, S.K., Karnatak, R., Kurths, J., Blasius, B., Ramaswamy, R.: Universal occurrence of the phase-flip bifurcation in time-delay coupled systems. Chaos **18**(2), 023111 (2008)
96. Prasad, A., Kurths, J., Dana, S.K., Ramaswamy, R.: Phase-flip bifurcation induced by time delay. Phys. Rev. E **74**(3), 035204 (2006)
97. Prasad, A., Lai, Y.C., Gavrielides, A., Kovanis, V.: Amplitude modulation in a pair of time-delay coupled external-cavity semiconductor lasers. Phys. Lett. A **318**(1–2), 71–77 (2003)
98. Prasanth, R.K., Annaswamy, A.M., Hathout, J.P., Ghoniem, A.F.: When do open-loop strategies for combustion control work? J. Propuls. Power **18**(3), 658–668 (2002)
99. Premraj, D., Manoj, K., Pawar, S.A., Sujith, R.I.: Effect of amplitude and frequency of limit cycle oscillators on their coupled and forced dynamics. Nonlinear Dyn. **103**(2), 1439–1452 (2021)

100. Premraj, D., Suresh, K., Pawar, S.A., Kabiraj, L., Prasad, A., Sujith, R.I.: Dragon-king extreme events as precursors for catastrophic transition. Europhys. Lett. **134**(3), 34006 (2021)
101. Putnam, A.A.: Combustion Driven Oscillations in Industry. Elsevier, Amsterdam (1971)
102. Ramanan, V., Chakravarthy, S.R., Sarkar, S., Ray, A.: Investigation of combustion instability in a swirl-stabilized combustor using symbolic time series analysis. In: Proceedings of the ASME 2014 Gas Turbine India Conference, GTIndia 2014-8280. American Society of Mechanical Engineers, New York (2014)
103. Ray, A.: Symbolic dynamic analysis of complex systems for anomaly detection. Signal Process. **84**(7), 1115–1130 (2004)
104. Rayleigh, J.W.S.: The Theory of Sound. Dover, New York (1945)
105. Reddy, D.V.R., Sen, A., Johnston, G.L.: Experimental evidence of time-delay-induced death in coupled limit-cycle oscillators. Phys. Rev. Lett. **85**(16), 3381–3384 (2000)
106. Richards, G.A., Straub, D.L., Robey, E.H.: Passive control of combustion dynamics in stationary gas turbines. J. Propuls. Power **19**(5), 795–810 (2003)
107. Rouwenhorst, D., Hermann, J., Polifke, W.: Online monitoring of thermoacoustic eigenmodes in annular combustion systems based on a state space model. J. Eng. Gas Turbines Power **139**(2), 021502 (2017)
108. Roy, A., Mondal, S., Pawar, S.A., Sujith, R.I.: On the mechanism of open-loop control of thermoacoustic instability in a laminar premixed combustor. J. Fluid Mech. **884**, A2 (2020)
109. Roy, A., Premchand, C.P., Raghunathan, M., Krishnan, A., Nair, V., Sujith, R.I.: Critical region in the spatiotemporal dynamics of a turbulent thermoacoustic system and smart passive control. Combust. Flame **226**, 274–284 (2021)
110. Russo, L., Russo, P., Siettos, C.I.: A complex network theory approach for the spatial distribution of fire breaks in heterogeneous forest landscapes for the control of wildland fires. PLoS One **11**(10), e0163226 (2016)
111. Sahay, A., Roy, A., Pawar, S.A., Sujith, R.I.: Dynamics of coupled thermoacoustic oscillators under asymmetric forcing. Phys. Rev. Appl. **15**(4), 044011 (2021)
112. Samarasinghe, J., Culler, W., Quay, B.D., Santavicca, D.A., O'connor, J.: The effect of fuel staging on the structure and instability characteristics of swirl-stabilized flames in a lean premixed multinozzle can combustor. J. Eng. Gas Turbines Power **139**(12), 121504 (2017)
113. Sarkar, S., Chakravarthy, S.R., Ramanan, V., Ray, A.: Dynamic data-driven prediction of instability in a swirl-stabilized combustor. J. Spray Combust. Dyn. **8**(4), 235–253 (2016)
114. Sarkar, S., Jha, D.K., Lore, K.G., Sarkar, S., Ray, A.: Multimodal spatiotemporal information fusion using neural-symbolic modeling for early detection of combustion instabilities. In: 2016 American Control Conference (ACC), pp. 4918–4923. IEEE, Piscataway (2016)
115. Sarkar, S., Lore, K.G., Sarkar, S., Ramanan, V., Chakravarthy, S.R., Phoha, S., Ray, A.: Early detection of combustion instability from high-speed flame images via deep learning and symbolic time series analysis. In: Annual Conference of the Prognostics and Health Management, pp. 353–362. Prognostics and Health Management Society (2015)
116. Sarkar, S., Ray, A., Mukhopadhyay, A., Chaudhari, R.R., Sen, S.: Early detection of lean blow out (LBO) via generalized d-markov machine construction. In: 2014 American Control Conference, pp. 3041–3046. IEEE, Piscataway (2014)
117. Sarkar, S., Ray, A., Mukhopadhyay, A., Sen, S.: Dynamic data-driven prediction of lean blowout in a swirl-stabilized combustor. J. Spray Combust. Dyn. **7**(3), 209–241 (2015)
118. Sato, M., Hyodo, H., Biwa, T., Delage, R.: Synchronization of thermoacoustic quasiperiodic oscillation by periodic external force. Chaos **30**(6), 063130 (2020)
119. Saxena, G., Prasad, A., Ramaswamy, R.: Amplitude death: the emergence of stationarity in coupled nonlinear systems. Phys. Rep. **521**(5), 205–228 (2012)
120. Scheffer, M., Bascompte, J., Brock, W.A., Brovkin, V., Carpenter, S.R., Dakos, V., Held, H., Van Nes, E.H., Rietkerk, M., Sugihara, G.: Early-warning signals for critical transitions. Nature **461**(7260), 53–59 (2009)
121. Scheffer, M., Carpenter, S., Foley, J.A., Folke, C., Walker, B.: Catastrophic shifts in ecosystems. Nature **413**(6856), 591–596 (2001)

122. Scheffer, M., Carpenter, S.R., Lenton, T.M., Bascompte, J., Brock, W., Dakos, V., Van de Koppel, J., Van de Leemput, I.A., Levin, S.A., Van Nes, E.H., et al.: Anticipating critical transitions. Science **338**(6105), 344–348 (2012)
123. Sen, S., Chaudhari, R.R., Mukhopadhyay, A.: Lean blowout detection techniques for partially premixed flames in a dump combustor. In: Novel Combustion Concepts for Sustainable Energy Development, pp. 199–232. Springer, Berlin (2014)
124. Sen, S., Sarkar, S., Chaudhari, R.R., Mukhopadhyay, A., Ray, A.: Lean blowout (LBO) prediction through symbolic time series analysis. In: Combustion for Power Generation and Transportation, pp. 153–167. Springer, Berlin (2017)
125. Srikanth, S., Pawar, S.A., Manoj, K., Sujith, R.I.: Dynamical States and Bifurcations in Coupled Thermoacoustic Oscillators. (2021). arXiv preprint arXiv:2109.09600
126. Staubli, T.: Entrainment of self-sustained flow oscillations: phase locking or asynchronous quenching? J. Appl. Mech. **54**(3), 706–712 (1987)
127. Sujith, R.I., Manoj, K., Pawar, S.A., Krishnan, A., Raghunath, M.P., Reeja, V.K., Raghunathan, M., Roy, A.: Systems and methods for suppressing thermo-acoustic instabilities in a combustor. US Patent: Application number 17/061298, Publication number US20210095848A1, Priority date: October 1 (2019)
128. Surovyatkina, E.: Prebifurcation noise amplification and noise-dependent hysteresis as indicators of bifurcations in nonlinear geophysical systems. Nonlinear Process. Geophys. **12**, 25–29 (2005)
129. Swift, G.W.: Thermoacoustic engines. J. Acoust. Soc. Am. **84**(4), 1145–1180 (1988)
130. Taira, K., Nair, A.G., Brunton, S.L.: Network structure of two-dimensional decaying isotropic turbulence. J. Fluid Mech. **795**, R2 (2016)
131. Tang, J., Mascolo, C., Musolesi, M., Latora, V.: Exploiting temporal complex network metrics in mobile malware containment. In: 2011 IEEE International Symposium on a World of Wireless, Mobile and Multimedia Networks, pp. 1–9. IEEE, Piscataway (2011)
132. Thomas, N., Mondal, S., Pawar, S.A., Sujith, R.I.: Effect of noise amplification during the transition to amplitude death in coupled thermoacoustic oscillators. Chaos **28**(9), 093116 (2018)
133. Thomas, N., Mondal, S., Pawar, S.A., Sujith, R.I.: Effect of time-delay and dissipative coupling on amplitude death in coupled thermoacoustic oscillators. Chaos **28**(3), 033119 (2018)
134. Thounaojam, U.S., Sharma, A.: Phase-flip mediated partial amplitude death in relay oscillators. Chaos Solitons Fract. **124**, 97–104 (2019)
135. Tredicce, J.R., Lippi, G.L., Mandel, P., Charasse, B., Chevalier, A., Picqué, B.: Critical slowing down at a bifurcation. Am. J. Phys. **72**(6), 799–809 (2004)
136. Tsonis, A.A., Roebber, P.J.: The architecture of the climate network. Phys. A **333**, 497–504 (2004)
137. Uhm, J.H., Acharya, S.: Low-bandwidth open-loop control of combustion instability. Combust. Flame **142**(4), 348–363 (2005)
138. Unni, V.R., Mukhopadhyay, A., Sujith, R.I.: Online detection of impending instability in a combustion system using tools from symbolic time series analysis. J. Spray Combust. Dyn. **7**(3), 243–255 (2015)
139. Unni, V.R., Nair, V.V., Sujith, R.I.: System and method for detecting precursors to control blowout in combustion systems. US Patent US10267519B2, Priority date: November 8 (2013)
140. Unni, V.R., Nair, V.V., Sujith, R.I., Mukhopadhyay, A.: System and method for controlling oscillatory instabilities in a device. US Patent US10095247B2, Priority date: December 4 (2013)
141. Unni, V.R., Sujith, R.I.: Multifractal characteristics of combustor dynamics close to lean blowout. J. Fluid Mech. **784**, 30–50 (2015)
142. Wei, M.D., Lun, J.C.: Amplitude death in coupled chaotic solid-state lasers with cavity-configuration-dependent instabilities. Appl. Phys. Lett. **91**(6), 061121 (2007)
143. Wei-Qing, L., Jun-Zhong, Y., Jing-Hua, X.: Experimental observation of partial amplitude death in coupled chaotic oscillators. Chin. Phys. **15**(10), 2260 (2006)

144. Xiong, S., Li, J., Ray, A.: Bayesian nonparametric modeling of markov chains for detection of thermoacoustic instabilities. In: 2017 American Control Conference (ACC), pp. 3758–3763. IEEE, Piscataway (2017)
145. Yi, T., Gutmark, E.J.: Real-time prediction of incipient lean blowout in gas turbine combustors. AIAA J. **45**(7), 1734–1739 (2007)
146. Zhao, D., Li, X.Y.: A review of acoustic dampers applied to combustion chambers in aerospace industry. Prog. Aerosp. Sci. **74**, 114–130 (2015)
147. Zou, W., Senthilkumar, D.V., Zhan, M., Kurths, J.: Quenching, aging, and reviving in coupled dynamical networks. Phys. Rep. **931**, 1–72 (2021)

Chapter 13
Oscillatory Instabilities in Other Turbulent Flow Systems

In the previous chapters, we discussed the dynamics of different combustion systems during the onset of thermoacoustic instability, and explored the prediction and control of such instabilities. We showed that, in systems with turbulent flows, the occurrence of thermoacoustic instability is preceded by a state of intermittency. During the occurrence of intermittency, bursts of large amplitude periodic oscillations emerge in between epochs of low amplitude chaotic fluctuations. However, we note that the presence of intermittency is not limited to combustion systems alone. Such a state can also occur prior to oscillatory instabilities in other turbulent flow systems—even if these systems do not involve combustion. Here, we refer to large amplitude self-sustained oscillations occurring in turbulent systems due to the nonlinear interaction between their subsystems as oscillatory instabilities. The presence of such instabilities can have catastrophic consequences on the structural integrity and safety of various technological systems and, hence, needs to be avoided. In this chapter, we will briefly discuss two other turbulent systems—aeroacoustic and aeroelastic systems—where the onset of oscillatory instability is preceded by intermittency. We will show that the intermittency route to oscillatory instability in these systems exhibits a similar scaling behavior. This further reasserts the generality of the existence of intermittent bursting dynamics prior to oscillatory instability across a range of turbulent systems.

Intermittency is a common phenomenon in turbulent flows [6, 32, 33]. It can be observed prior to the transition of a laminar flow to a fully developed turbulent flow. In turbulent flows, eddies of different time and length scales exist. The energy transfer across such eddies at different scales happens through different cascade processes [12, 28]. The interaction of such eddies due to feedback mechanism can cause order to emergence from chaos through the process of self-organization [15]. During the state of ordered oscillations, large coherent structures are formed in the system [4].

The coupled interaction of turbulent flows with other subsystems, at the appropriate operating conditions, can give rise to oscillatory instability. For example, in

R. I. Sujith, S. A. Pawar, *Thermoacoustic Instability*, Springer Series in Synergetics,
https://doi.org/10.1007/978-3-030-81135-8_13

the case of a thermoacoustic system, the mutual interaction between the turbulent reacting flow and the acoustic field leads to the onset of thermoacoustic instability. Similarly, in aeroacoustic systems, the feedback between vortex shedding and the acoustic field causes the generation of aeroacoustic instability or whistling oscillations [5]. In a similar manner, the constructive interaction between turbulent flows and structural elements leads to aeroelastic instability or flutter in aeroelastic systems [10]. In Chap. 6, we have already discussed the onset of thermoacoustic instability in turbulent combustion systems. In the following sections, we turn our attention to the onset of aeroacoustic and aeroelastic instabilities. The aim is to establish the intermittency route to oscillatory instability and the generality of this phenomenon in most turbulent systems.

13.1 Aeroacoustic Instabilities

Aeroacoustic instabilities are a class of oscillatory instabilities that occur as a result of the interaction of unsteady fluctuations in turbulent flows with the acoustic field in a system. Such systems where the flow-induced sound is generated are called aeroacoustic systems. The unsteady fluctuations in the flow create distributed acoustic sources throughout the flow. In a confined system, such distributed acoustic energy accumulates at the resonant modes of the system [29]. In such a situation, the acoustic fluctuations affect the flow fluctuations and, thus, develop a feedback loop. When the timescale of unsteady local hydrodynamic (flow) fluctuations matches that of the acoustic fluctuations, the generation of self-sustained ordered (periodic) oscillations, often called whistling, takes place in the system. The sound generated through aeroacoustic instabilities has been studied in various industrial applications, including driven cavities [18, 30, 31], edge tones [25], howling ejectors [16], jet noise [19], pipe whistling [1], gas pipe networks [3], etc.

In this section, we will discuss the dynamic properties of pressure fluctuations generated by flow turbulence (pseudo-sound) in an aeroacoustic system consisting of a long circular pipe terminated by a circular orifice plate (Fig. 13.1). Figure 13.2 shows time traces of the pressure fluctuations observed for different jet flow velocities at the orifice. When the flow velocity is low (Fig. 13.2a), the pressure signal shows only low amplitude noisy oscillations. With an increase in the

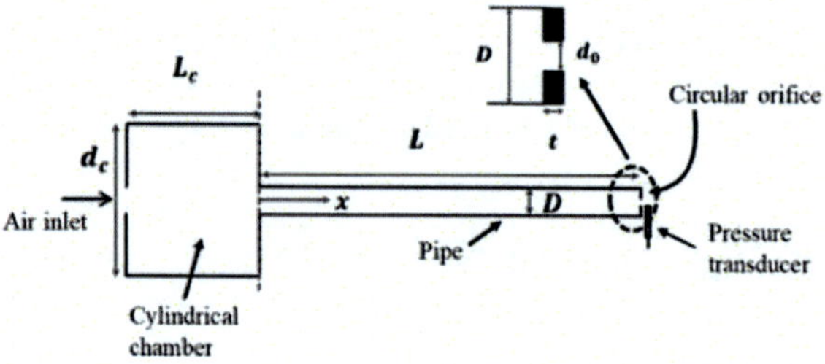

Fig. 13.1 Schematic of the experimental setup of an orifice tube aeroacoustic system. Adapted with permission from Nair and Sujith [17]

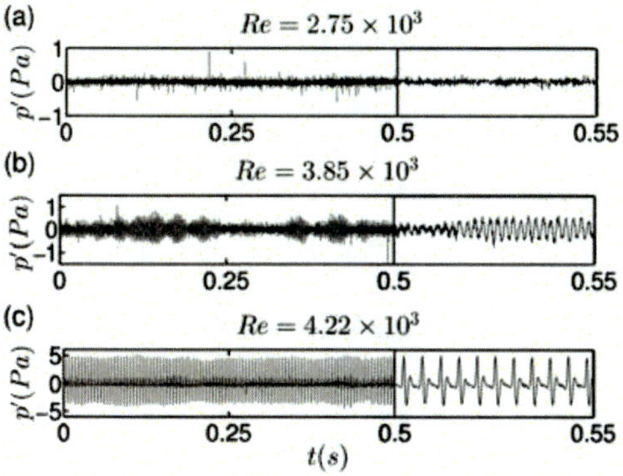

Fig. 13.2 Time series of the pressure fluctuations (p') obtained outside the orifice tube of an aeroacoustic system for different flow velocities corresponding to the states of (**a**) low amplitude noisy fluctuations, (**b**) intermittency, and (**c**) whistling sound. The pressure response of the system shows the intermittency route to whistling instability. Adapted with permission from Nair and Sujith [17]

flow velocity (Fig. 13.2b), bursts of large amplitude periodic oscillations occur intermittently amidst epochs of noisy background fluctuations at apparently random time intervals. In Chap. 6, we discussed the existence of such a state of intermittent pressure oscillations occurring in turbulent thermoacoustic systems. When the flow velocity is increased above a critical value (Fig. 13.2c), the large amplitude periodic oscillations arise, generating a sharp whistling tone in the system. In this case, such pressure oscillations take the form of period-2 behavior. This state of self-sustained periodic oscillations in the pressure signal is referred to as aeroacoustic instability. Thus, the presence of intermittency route to the onset of aeroacoustic instability is observed in an orifice-pipe system. Intermittency is also reported in an aeroacoustic system with a deep cavity subjected to turbulent grazing flow [2].

The presence of an intermittency regime can help to develop early warning signals (or precursors) for aeroacoustic instabilities, in much the same way as for thermoacoustic instabilities (Chap. 12). In Fig. 13.3, we show the variation of different measures calculated from the pressure signal as the flow velocity is varied during the transition from noisy oscillations to self-sustained periodic oscillations via intermittency. For reliable early detection, it is helpful to use measures whose values vary smoothly well in advance as the system transitions from a stable state (noisy fluctuations) to aeroacoustic instability (periodic oscillations).

Figure 13.3a shows the variation of the root mean square of pressure fluctuations (P_{rms}) with changes in the flow velocity (u). This measure is often used as an indicator for sound levels. We notice only a small increase in the amplitude of the pressure signal prior to its sharp rise at the onset of aeroacoustic instability. By contrast, the Hurst exponent H (a fractal quantifier of a signal, as discussed in Chap. 10) changes gradually with increasing flow velocity during the onset of aeroacoustic instability (Fig. 13.3b). This gradual change is due to the presence of

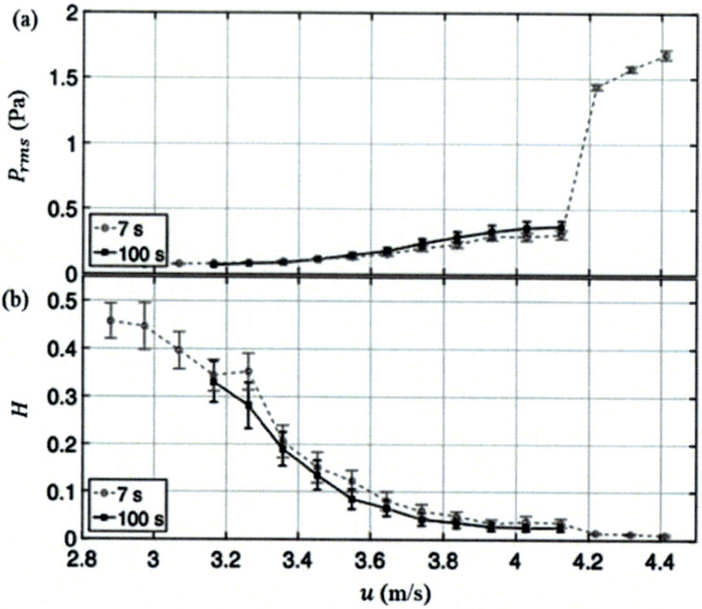

Fig. 13.3 Variation of the pressure response of the aeroacoustic system in terms of (**a**) the root mean square value (P_{rms}) and (**b**) the Hurst exponent (H) for increasing values of the flow velocity (u). The value of H decays well before the value of P_{rms} exhibits a sudden jump; thus, H can be used as a precursor of aeroacoustic instability. Reproduced with permission from Nair and Sujith [17]

intermittency in the system dynamics. In the absence of intermittency, we would observe a sudden jump in the values of this measure during the transition from noisy fluctuations to self-sustained periodic oscillations. We notice that this measure approaches a definite value near zero during the state of the aeroacoustic instability. Furthermore, this measure shows a clear downwards trend well before the condition where the pressure amplitude shows a significant rise. We note that the amplitude of pressure fluctuations is a system dependent property, which can take on any value (low to high) during the onset of aeroacoustic instability. However, the Hurst exponent is a bounded measure, and can be used as a precursor for the occurrence of such instability, as its variation is independent of the system geometry.

13.2 Aeroelastic Instabilities

Another class of oscillatory instabilities is aeroelastic instability. It arises from the nonlinear interaction between the aerodynamic, inertial, and elastic forces acting on a movable body immersed in air flows [7]. Such instabilities are also referred to as flutter. They manifest as self-sustained periodic oscillations in the structure

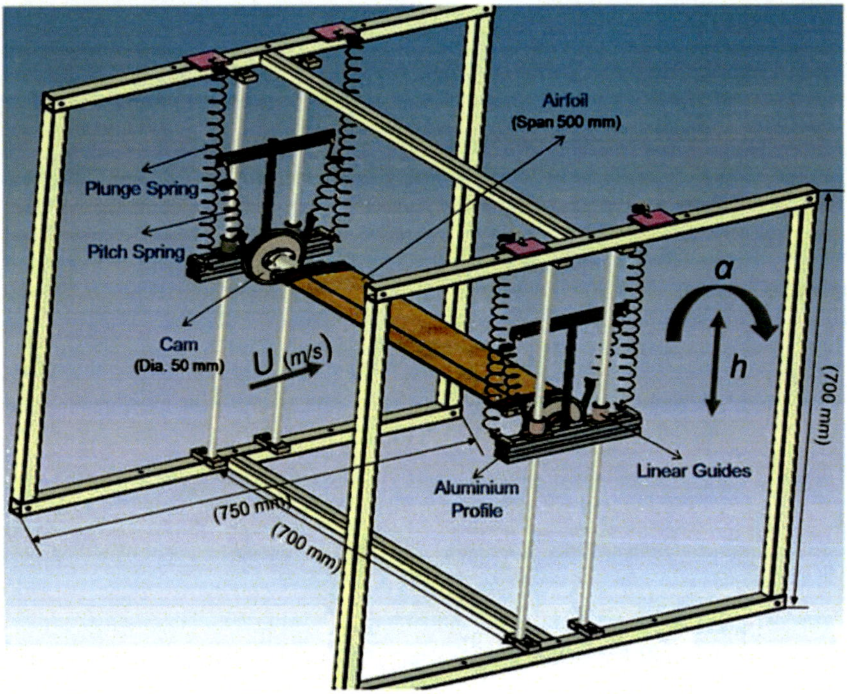

Fig. 13.4 Schematic of the experimental setup of pitch-plunge aeroelastic model of an airfoil. Reproduced with permission from Venkatramani et al. [35]

of the body. Aeroelastic instabilities can potentially afflict different parts of an aircraft, such as its wings, tail, turbine or compressor blades, etc.[11, 14, 38]. The existence of aeroelastic flutter is undesirable as it can lead to structural fatigue and eventual failure. Therefore, for safety and reliability, aeroelastic instabilities need to be avoided. In order to avert such instabilities, we first need to know the dynamical behavior of the system during the onset of aeroelastic instability (or flutter) and then develop precursors that can provide early warning signals prior to their onset [8, 9, 35–37].

In this section, we discuss the dynamic response of a simple aeroelastic system, i.e., NACA 0012 airfoil, subjected to fluctuating flows of different velocities in a low-speed wind tunnel [35], see Fig. 13.4. Figure 13.5 shows the temporal dynamics of the system measured in terms of the plunging (bending) motion of the airfoil for four different wind speeds. When the flow velocity is low, we observe low amplitude aperiodic fluctuations in the plunge response of the system (Fig. 13.5a). When the flow velocity is gradually increased, the response of the system shows intermittency, i.e., the occurrence of bursts of large amplitude periodic oscillations amidst epochs of low amplitude aperiodic fluctuations (Fig. 13.5b). The appearance of bursts in the signal is seemingly random in time. A further increase in flow

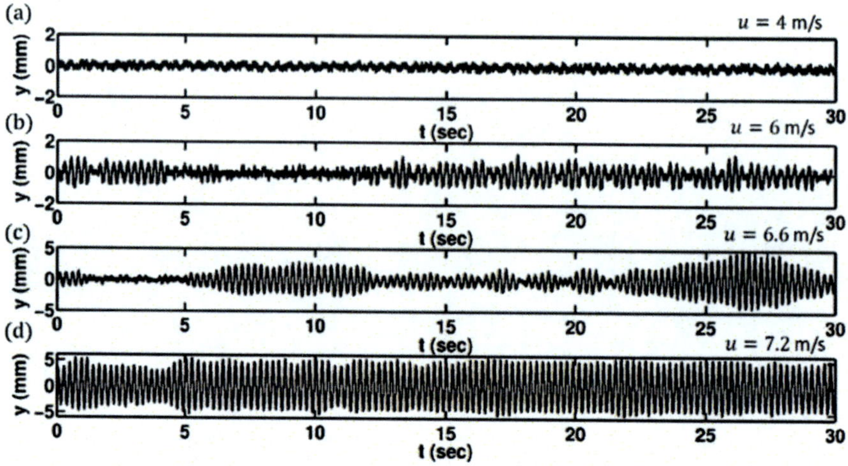

Fig. 13.5 The experimental time series of plunge response of the airfoil for different flow velocities (u) corresponding to the states of (**a**) low amplitude aperiodic fluctuations, (**b**, **c**) intermittency, and (**d**) flutter. Plunge response shows the intermittency route to flutter. Adapted with permission from Venkatramani et al. [35]

velocity increases the duration of periodicity observed in the signal (Fig. 13.5c). Finally, when the flow velocity is above a critical value, the system transitions to a state of aeroelastic flutter, which is characterized by large amplitude limit cycle oscillations. This dynamical transition to a flutter phenomenon suggests that the onset of flutter is preceded by a state of intermittency [34, 35]. This dynamical behavior of an aeroelastic system is similar to what we discussed in Sect. 13.1 for an aeroacoustic system and in Chap. 6 for a turbulent thermoacoustic system.

Venkatramani et al. [35] showed that the pitch (torsion) response of the aeroelastic system (Fig. 13.4) also exhibits the intermittency route to flutter, similar to that observed for its plunge response (Fig. 13.5). Furthermore, Raj et al. [27] modeled the coupled interaction of the pitch and plunge responses of this system, in the presence of a fluctuating flow field, during the intermittency route to flutter instability. They showed that the system undergoes a synchronization transition, wherein it transitions from a desynchronized state to complete phase synchronization via intermittent phase synchronization. Kumar et al. [13] observed the existence of "on-off" intermittency prior to limit cycle oscillations (fluid–elastic dynamic instabilities) in a numerical model capturing the effect of the fluctuating flow over a flexible cantilever plate.

Further, we can use any suitable measure that can characterize the system behavior for early detection of flutter. In Fig. 13.6, we present the variation of the plunge response in terms of the root mean square value (y_{rms}) in Fig. 13.6a and the Hurst exponent in Fig. 13.6b as a function of the flow velocity (u). We notice a significant change in y_{rms} only after the onset of flutter ($u > 6.5$ m/s). As a result, we do not get any indication of the occurrence of flutter before its onset.

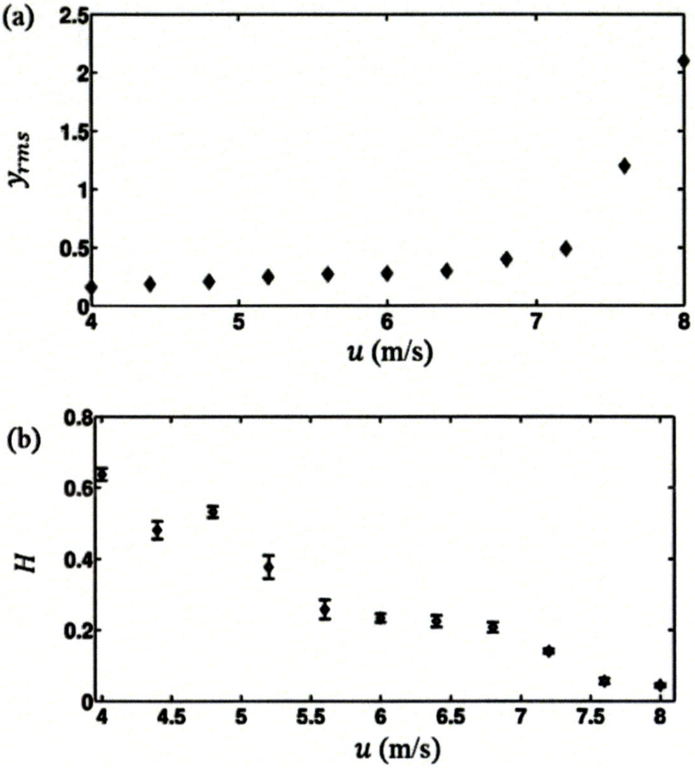

Fig. 13.6 Variation of the plunge response in terms of (**a**) the root mean square value (y_{rms}) and (**b**) the Hurst exponent (H) of the airfoil for increasing values of the flow velocity (u). The value of H displays a decrease much before the value of y_{rms} exhibits a significant growth; hence, H can be used as a precursor of the flutter phenomenon. Adapted with permission from Vennkatramani et al. [36]

However, the Hurst exponent (H) shows a gradual fall with increasing flow speed (Fig. 13.6b). This decrease in H is a result of an increase in the periodicity of the system dynamics. At the onset of flutter, H approaches zero as the signal is dominated by periodic oscillations. Therefore, we can obtain an early warning for the onset of flutter using the Hurst exponent.

Venkatramani et al. [37] have developed several other precursors for flutter based on various measures from dynamical systems theory, such as approximate entropy, sample entropy, Lempel-Ziv complexity, and recurrence quantification. Similarly, Hachijo et al. [9] used methodology combining symbolic dynamics, dynamical systems, and machine learning to develop precursors for the occurrence of cascade flutter in a combustor with low-pressure turbine blades. Furthermore, in another study, they [8] employed measures of complex networks (node strength) and collective synchronization (order parameter) to provide precursors for a cascade flutter phenomenon observed in a model aircraft turbine. Using these methodologies, their study also confirmed the existence of noisy limit cycle oscillations during the state of flutter. From the mitigation perspective, Raj et al. [26] numerically showed that flutter instability observed in two aeroelastic systems (e.g., airfoils) can be simultaneously quenched through the phenomenon of amplitude death.

13.3 Scaling Behavior During the Onset of Oscillatory Instability

From Figs. 13.2 and 13.5 in the present chapter and Fig. 6.3 in Chap. 6, we found that three different classes of turbulent systems (i.e., aeroacoustic, aeroelastic, and thermoacoustic, respectively), each with different interacting subsystems and physical mechanisms for oscillatory instabilities, share a common route to self-sustained oscillatory instability, one that involves intermittency. Pavithran et al. [22, 23] analyzed the properties of the dynamic variables of these systems (i.e., acoustic pressure fluctuations for both thermoacoustic and aeroacoustic systems, and strain fluctuations for an aeroelastic system) during the onset of oscillatory instability via intermittency. The was aim was to characterize this transition to oscillatory instability through a universal scaling relation across different turbulent systems (Fig. 13.7).

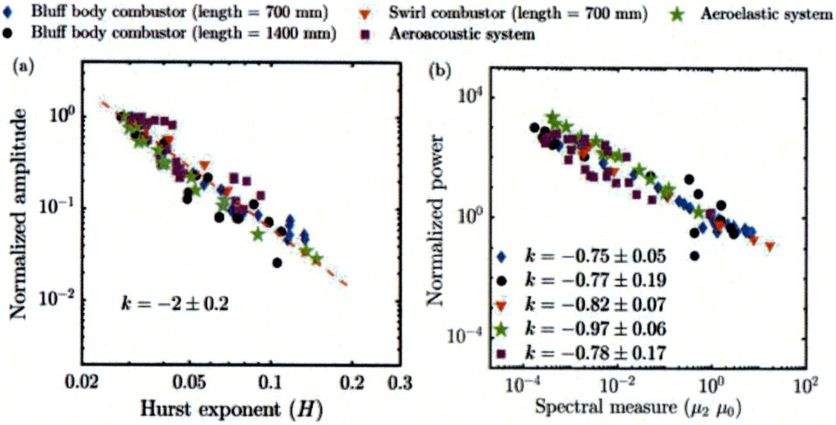

Fig. 13.7 Two-parameter diagrams between (**a**) the spectral amplitude of the dominant mode (A/A_I) and the Hurst exponent (H) of a signal and (**b**) the normalized spectral power and the spectral measure $[\mu_2\ \mu_0]$ of a signal on a log-log scale, for different turbulent systems exhibiting intermittency route to oscillatory instabilities. Both plots show universal power-law scaling for all the systems. (**a**) Adapted with permission from Pavithran et al. [23] and (**b**) adapted with permission from Pavithran et al. [22]

In Fig. 13.7a, we show the variation of the normalized spectral amplitude (A/A_I) as a function of the Hurst exponent (H) on a log-log plot for different turbulent systems. Here, A_I is the amplitude of the dominant frequency observed during the saturated (full-blown) state of oscillatory instability. We notice that the data points for all the systems collapse along a single straight line having a slope of nearly -2 (i.e., $A \propto H^{-2}$). Thus, we notice a striking similarity in the power-law behavior for different systems during the onset of oscillatory instability.

Furthermore, Pavithran et al. [22] analyzed the changes in spectral properties of all the aforementioned turbulent systems. They found the transition of the broad distribution of spectral power to a single sharp peak centered at the dominant mode during the onset of oscillatory instability in all the systems, which they called "spectral condensation". They developed spectral measures that quantifies the sharpening of frequency in the power spectrum of these systems during the onset of oscillatory instability. The spectral measure is considered as the products of moments of the power spectrum raised to integer powers and its general expression can be given as,

$$[\mu_m^x \ \mu_n^y] = \left[\int_{-\delta F}^{+\delta F} \frac{P(F)}{P_0}\left|\frac{F}{f_0}\right|^m dF\right]^x \times \left[\int_{-\delta F}^{+\delta F} \frac{P(F)}{P_0}\left|\frac{F}{f_0}\right|^n dF\right]^y \qquad (13.1)$$

where μ_m indicates the mth moment of the power spectrum, $P(F)$ is the power corresponding to the modified frequency $F = f - f_0$, f is a variable, f_0 is the dominant frequency in the power spectrum, $P_0 = P(f_0)$, indices m, n, x, and y are positive integers. The value of the spectral measures $[\mu_m^x \ \mu_n^y]$ is computed in the neighborhood of f_0 within a width of δF, where $\delta F = f_0/5$.

As per the definition of the spectral measures, the value of $[\mu_m^x \ \mu_n^y]$ decreases as the spectral peak gets sharper. This sharpening property can be easily detected from measures, $[\mu_2]$, $[\mu_2 \ \mu_0]$, and $[\mu_4 \ \mu_4]$. Here, $[\mu_2]$ is the second moment of the power spectrum and, $[\mu_2 \ \mu_0]$ and $[\mu_4 \ \mu_4]$ are the products of higher moments of the distribution. Higher moments give more weightage to the tail ends of the spectrum and thus its variation indicates how the broad tails diminish [22].

In Fig. 13.7b, we show the variation of a spectral measure $[\mu_2 \ \mu_0]$ with the power corresponding to the dominant peak $P(f_0)$ for different turbulent systems during the transition from low amplitude aperiodic oscillations (stable operation) to large amplitude periodic oscillation (oscillatory instability). We notice the collapse of plots for all the systems and the existence of an inverse power law relation with nearly the same scaling exponent (k). Similar inverse power law behavior with different scaling exponents can be observed for $[\mu_0]$ and $[\mu_4 \ \mu_4]$. This scaling relation further indicates that the systems that undergo spectral condensation have these universal scaling relations in spite of having different physical properties. For more details on the universality of scaling relations across different systems exhibiting the emergence of order, a reader may refer to [22–24].

As we discussed before, the emergence of oscillatory instabilities can cause catastrophic damages to system components. Further, we showed that the changes in the spectral and dynamical properties of systems undergoing oscillatory instability are closely related and exhibit a power law relation (see Fig. 13.7). Thus, using such an inverse power law relationship between different properties of the system, we can a priory predict the amplitude of such instabilities that may occur in a system. Such a priori estimation of the amplitude may help in developing strategies for mitigating these instabilities. From Fig. 13.7a, we can plot the power law relation of the amplitude of the dominant mode of oscillations and H using only the pressure measurements acquired during stable operation. Then, we can estimate the amplitude of limit cycle oscillations by extrapolating the scaling towards H tending to zero, i.e., towards oscillatory instability [20]. In Fig. 13.7b, variation of the scaling of the spectral measure with the spectral amplitude provides a methodology to estimate the amplitudes of different modes of oscillations separately, which can be achieved by extrapolating the scaling of spectral measures calculated for different peaks in the amplitude spectrum [21].

In summary, in this chapter, we discussed the dynamical transition that occurs during the onset of oscillatory instability in different turbulent systems, i.e., aeroacoustic, aeroelastic, and thermoacoustic systems. We showed that despite their different physical mechanisms of instabilities, these systems exhibit a similar temporal pattern of the existence of intermittency prior to the onset of oscillatory instability. Furthermore, the universal power-law relations are discovered between different dynamical and spectral properties of signals during the occurrence of oscillatory instability across these turbulent systems. Thus, we note that the dynamical behavior of turbulent systems is unique and that the state of intermittency is very much common before the transition to full-blown instability (i.e., during the emergence of order).

In the next chapter, we provide a summary of the developments so far and our outlook for the future developments in this subject.

References

1. Anderson, A.B.C.: Structure and velocity of the periodic vortex-ring flow pattern of a primary pfeifenton (pipe tone) jet. J. Acoust. Soc. Am. **27**(6), 1048–1053 (1955)
2. Bourquard, C., Faure-Beaulieu, A., Noiray, N.: Whistling of deep cavities subject to turbulent grazing flow: intermittently unstable aeroacoustic feedback. J. Fluid Mech. **909**, A19 (2021)
3. Bruggeman, J.C., Hirschberg, A., van Dongen, M.E.H., Wijnands, A.P.J., Gorter, J.: Self-sustained aero-acoustic pulsations in gas transport systems: experimental study of the influence of closed side branches. J. Sound Vib. **150**(3), 371–393 (1991)
4. Fiedler, H.E.: Coherent structures. In: Advances in Turbulence, pp. 320–336. Springer, Berlin (1987)
5. Flandro, G.A., Majdalani, J.: Aeroacoustic instability in rockets. AIAA J. **41**(3), 485–497 (2003)
6. Frisch, U., Parisi, G.: Fully developed turbulence and intermittency. In: Turbulence and Predictability in Geophysical Fluid Dynamics and Climate Dynamics, pp. 84–88. North-Holland, Amsterdam (1985)
7. Fung, Y.C.: An Introduction to the Theory of Aeroelasticity. Wiley, New York (1955)

8. Hachijo, T., Gotoda, H., Nishizawa, T., Kazawa, J.: Early detection of cascade flutter in a model aircraft turbine using a methodology combining complex networks and synchronization. Phys. Rev. Appl. **14**(1), 014093 (2020)
9. Hachijo, T., Gotoda, H., Nishizawa, T., Kazawa, J.: Experimental study on early detection of cascade flutter in turbo jet fans using combined methodology of symbolic dynamics, dynamical systems theory, and machine learning. J. Appl. Phys. **127**(23), 234901 (2020)
10. Hansen, M.H.: Aeroelastic instability problems for wind turbines. Wind Energy **10**(6), 551–577 (2007)
11. Kehoe, M.W., et al.: A historical overview of flight flutter testing. Tech. rep., NASA Technical Memorandum (1995)
12. Kraichnan, R.H.: Inertial ranges in two-dimensional turbulence. Phys. Fluids **10**(7), 1417–1423 (1967)
13. Kumar, S.K., Sarkar, S., Gupta, S.: Intermittency in a cantilever plate in a randomly fluctuating fluid flow. J. Fluids Struct. **93**, 102855 (2020)
14. Lobitz, D.W.: Aeroelastic stability predictions for a MW-sized blade. Wind Energy **7**(3), 211–224 (2004)
15. Marov, M.Y., Kolesnichenko, A.V.: Turbulence and Self-organization: Modeling Astrophysical Objects, vol. 389. Springer Science & Business Media, New York (2013)
16. Middleton, D.: Theoretical and experimental investigations into the acoustic output from ejector flows. J. Sound Vib. **11**(4), 447–473 (1970)
17. Nair, V., Sujith, R.I.: Precursors to self-sustained oscillations in aeroacoustic systems. Int. J. Aeroacoust. **15**(3), 312–323 (2016)
18. Nelson, P.A., Halliwell, N.A., Doak, P.E.: Fluid dynamics of a flow excited resonance, Part II: flow acoustic interaction. J. Sound Vib. **91**(3), 375–402 (1983)
19. Panda, J.: An experimental investigation of screech noise generation. J. Fluid Mech. **378**, 71–96 (1999)
20. Pavithran, I., Unni, V.R., Sujith, R.I.: System and method for predetermining the onset of impending oscillatory instabilities in practical devices. US Patent US20210131663A1, Priority date: 23 July 2018
21. Pavithran, I., Unni, V.R., Varghese, A.J., Kasthuri, P., Sujith, R.I.: System and method for determining the amplitude of oscillatory instabilities in fluid mechanical devices. US Patent: Application number 16/633658, Publication number US20210131910A1, Priority date: 25 July 2018
22. Pavithran, I., Unni, V.R., Varghese, A.J., Premraj, D., Sujith, R.I., Vijayan, C., Saha, A., Marwan, N., Kurths, J.: Universality in spectral condensation. Sci. Rep. **10**(1), 1–8 (2020)
23. Pavithran, I., Unni, V.R., Varghese, A.J., Sujith, R.I., Saha, A., Marwan, N., Kurths, J.: Universality in the emergence of oscillatory instabilities in turbulent flows. Europhys. Lett. **129**(2), 24004 (2020)
24. Pavithran, I., Unni, V.R., Varghese, A.J., Sujith, R.I., Saha, A., Marwan, N., Kurths, J.: Predicting the amplitude of thermoacoustic instability using universal scaling behaviour. In: Proceedings of ASME Turbo Expo 2021: Turbomachinery Technical Conference and Exposition, GT2021-60074 (2021)
25. Powell, A.: On the edgetone. J. Acoust. Soc. Am. **33**(4), 395–409 (1961)
26. Raaj, A., Mondal, S., Jagdish, V.: Investigating amplitude death in a coupled nonlinear aeroelastic system. Int. J. Nonlin. Mech. **129**, 103659 (2021)
27. Raaj, A., Venkatramani, J., Mondal, S.: Synchronization of pitch and plunge motions during intermittency route to aeroelastic flutter. Chaos **29**(4), 043129 (2019)
28. Richardson, L.F.: Atmospheric diffusion shown on a distance-neighbour graph. Proc. R. Soc. Lond. **110**(756), 709–737 (1926)
29. Rienstra, S.W., Hirschberg, A.: An introduction to acoustics. Tech. rep., Eindhoven University of Technology, IWDE 92-06 (2004)
30. Rockwell, D., Naudascher, E.: Review—self-sustaining oscillations of flow past cavities. Trans. ASME J. Fluids Eng. **100**, 152–165 (1978)

31. Rowley, C.W., Williams, D.R.: Dynamics and control of high-Reynolds-number flow over open cavities. Annu. Rev. Fluid Mech. **38**, 251–276 (2006)
32. Sreenivasan, K.R.: On the fine-scale intermittency of turbulence. J. Fluid Mech. **151**(1), 81–103 (1985)
33. Vassilicos, J.C., et al.: Intermittency in Turbulent Flows. Cambridge University Press, Cambridge (2001)
34. Venkatramani, J., Kumar, S.K., Sarkar, S., Gupta, S.: Physical mechanism of intermittency route to aeroelastic flutter. J. Fluids Struct. **75**, 9–26 (2017)
35. Venkatramani, J., Nair, V., Sujith, R.I., Gupta, S., Sarkar, S.: Precursors to flutter instability by an intermittency route: a model free approach. J. Fluids Struct. **61**, 376–391 (2016)
36. Venkatramani, J., Nair, V., Sujith, R.I., Gupta, S., Sarkar, S.: Multi-fractality in aeroelastic response as a precursor to flutter. J. Sound Vib. **386**, 390–406 (2017)
37. Venkatramani, J., Sarkar, S., Gupta, S.: Investigations on precursor measures for aeroelastic flutter. J. Sound Vib. **419**, 318–336 (2018)
38. Zhang, P., Huang, S.: Review of aeroelasticity for wind turbine: current status, research focus and future perspectives. Front. Energy **5**(4), 419–434 (2011)

Chapter 14
Summary and Perspective

As we mentioned in the Chap. 1, the first papers on nonlinear thermoacoustics were published in the sixties, around the time when the subject of nonlinear dynamics was at its infancy. Nonlinear dynamics developed rapidly, followed by an explosive development of complex systems theory. These subjects influenced the development of not just physics, but many fields including chemistry, biology, medicine, and social sciences. After half a century of lull, an approach based on dynamical systems theory and complex systems theory is rapidly transforming thermoacoustics as well.

14.1 Temporal Analysis

The onset of thermoacoustic instability in thermoacoustic systems was characterized as Hopf bifurcation [1, 31]. Both supercritical and subcritical bifurcations occur in thermoacoustic systems [10, 31]. Triggering instability and hysteresis were explained in the framework of bistability in thermoacoustic systems [2, 3, 14, 15, 19]. This was followed by work on noise-induced transition [18] and rate-dependent transitions [33] to thermoacoustic instability.

Traditionally, the asymptotic state of a thermoacoustic system was thought to be a fixed point or limit cycle. Subramanian et al. [53], Kabiraj et al. [20–23], Kashinath et al. [24], Kasthuri et al. [25], Premraj et al. [48], and Guan et al. [17] showed that limit cycle is just one of the asymptotic states that are possible, and states such as quasiperiodicity, intermittency, mixed-mode oscillations, strange non-chaos (SNA), and chaos are possible even in simple laminar systems.

Unlike the case of laminar combustors, in the case of turbulent combustors, sound is produced even during the stable operation. This incoherent, broadband sound produced by turbulent reacting flow is called combustion noise [52]. The traditional approach treats the stable operation as a fixed point, and models turbulence as white noise [31]. Combustion noise is then considered as the outcome of the dynamical

R. I. Sujith, S. A. Pawar, *Thermoacoustic Instability*, Springer Series in Synergetics,
https://doi.org/10.1007/978-3-030-81135-8_14

system (in this case thermoacoustic system) being forced by noise [7, 30]. By modeling turbulence as additive white noise, Noiray [43] analyzed subcritical and supercritical situations. They extracted quantities such as linear growth rate of the thermoacoustic system. In addition to the net growth rate, recently, Boujo et al. [6] could extract separately the decay rate due to damping and growth rate due to the flame. He also explained the intermittent switching between spinning and standing modes in annular combustors [43]. The suppression of the bistable zone at high levels of noise was also analyzed.

Application of determinism tests on time series data from turbulent combustors have shown that state of combustion noise has features of high-dimensional deterministic chaos, contaminated with white and colored noise [55]. Thus, the transition to thermoacoustic instability (periodicity) is really a loss of chaos [41]. The onset of thermoacoustic instability is portended by bursts of periodic oscillations that are interspersed between epochs of aperiodicity in an near-random manner [38, 42]. As the proximity to the onset decreases, these intermittent bursts last longer, and eventually at the onset, periodicity pervades all through. Intermittency represents the pockets of order seen amidst disorder; a characteristic often seen in complex systems. This intermittency route from chaos to periodic oscillations is seen irrespective of the flame holding mechanism used in the combustor.

Intermittency is often represented and analyzed using recurrence plots [38, 42]. A number of statistical measures such as recurrence rate, determinism, Shannon entropy, and trapping time can be derived from recurrence plots. These measures characterize the proximity of the thermoacoustic system to the onset of an impending instability, and can be used as precursors. Further, using a kicked-oscillator model of the intermittency route to thermoacoustic instability, Seshadri et al. [50] was able to predict the amplitudes at the onset of thermoacoustic instability by examining the signals during stable operation and intermittency.

Complexity often manifests as fractal and multifractal behavior. The time series of acoustic pressure corresponding to combustion noise is self-similar and can be analyzed using the framework of fractals. A time series $a(t)$ is self-similar if the signal $a(ct) = a(t)/c^H$ has the same statistics, where H is the Hurst exponent. The fractal dimension D is then equal to $2 - H$. The oscillations lose their self-similar nature and become periodic at the onset of thermoacoustic instability. The Hurst exponent drops towards zero, well before the amplitude of instability rises, enabling it to be used as a precursor. Further Pavithran et al. [44] report an inverse square law between the amplitude of the oscillations and the Hurst exponent ($A \propto H^{-2}$) for the class of non-equilibrium systems that involve oscillatory instabilities arising in turbulent flows. This scaling enables a prior estimation of the amplitude of oscillations at the onset of oscillatory instability using time series data acquired during stable operation. Knowing the amplitude of the instability at the onset is critical to devising strategies and countermeasures to limit the damages.

Often, a single fractal dimension is inadequate to characterize the complexity in the signal. Fluctuations with different amplitudes can have different scaling; such behavior is called multifractal. The time series obtained during combustion noise displays multifractal characteristics [39]. This multifractality is lost as the signal transforms to periodic oscillations at the onset of thermoacoustic instability. Thus, the onset of thermoacoustic instability is indeed a loss of complexity. This has an interesting parallel with viewing disease and ageing as loss of complexity [32, 51].

As a complex system contains multiple interacting entities, complex networks are ideally suited for investigating such systems and the emergence of nontrivial and nonlinear collective behavior that originate from the interactions. Using visibility algorithm, Murugesan and Sujith [35] showed that the complex networks obtained from the time series of acoustic pressure during the state of combustion noise were scale-free. The scale-invariance during combustion noise manifests as the power-law distribution of connections. This scale-free behavior disappears at the onset of thermoacoustic instability, and the network transitions to a regular network. Statistical measures such as characteristic path length and clustering coefficient that characterize complex networks could be used as precursors as they show the proximity to the onset of thermoacoustic instability [36]. These findings are confirmed by other studies based on recurrence networks [13, 16].

Pawar et al. [45] described that the onset of thermoacoustic instability can be viewed as a mutual synchronization between the acoustic pressure and the heat release rate signals. In a system without any unstable hydrodynamic modes, they showed that, during combustion noise, the acoustic pressure and heat release rate are chaotic signals that are desynchronized. During thermoacoustic instability, these signals are synchronized. During intermittency, these signals are synchronized during the periodic epochs and desynchronized during the aperiodic epochs. Furthermore, they showed that during the state of phase synchronization, the thermoacoustic system exhibits weakly correlated limit cycle oscillations, and during the state of generalized synchronization, the system exhibits strongly correlated limit cycle oscillations. Measures derived from joint and cross recurrence plots and also joint and cross recurrence networks were used to study the interaction between the acoustic pressure and heat release rate fluctuations during the synchronization transition to thermoacoustic instability [12]. Using the measures derived from the cross recurrence networks, Godavarthi et al. [12] identified the directional dependence between the heat release rate and acoustic pressure fluctuations during the onset of thermoacoustic instability.

More interestingly, the intermittency route to periodicity seems to be a universal character of self-organization in turbulent flows, as shown in experiments in aeroacoustic [40] and aeroelastic [60] instabilities, compressor surge [26] and hydrodynamic instabilities [9, 54]. These ideas were met with resistance in the beginning. However, a spate of publications on intermittency in the last 5 years by different research groups across the globe confirms the ubiquity of this route in turbulent systems that transition to oscillatory instabilities. It is also heartening to know that these physics-based precursors that are obtained using the various nonlinear time series analysis have made their way to the industry.

These results have significant implications related to validating multi-fidelity CFD solvers that are used to capture thermoacoustic instability in turbulent combustors. Too often, the rms of pressure fluctuations is used as the measure for validating these solvers. However, this does not ensure that all the flow features are captured, and that the solver can make further predictions correctly. A much more stronger and confidence building validation will consist of:

1. capturing the multifractal spectrum during the stable operation and its loss during the transition to thermoacoustic instability,
2. capturing the intermittency route to thermoacoustic instability,
3. capturing the scale-free network during stable operation and its transition to regular network during the transition from stable operation to thermoacoustic instability.

14.2 Spatiotemporal Analysis

Thermoacoustic systems are spatially extended systems. Rapid developments in the application of time series analysis to the onset of thermoacoustic instability are now being followed by spatiotemporal analysis. Progress in spatiotemporal analysis is slow, as high-quality data at high acquisition rate is not easily available. Further, the analysis techniques are computationally intensive.

There have been a few studies on the flow and flame dynamics during intermittency [37, 49, 58]. What is interesting is that the flow physics are different from situation to situation; nevertheless, the route taken during the transition from chaos to periodic oscillations is universal—and seen even in cases such as aeroacoustic and aeroelastic instabilities, where combustion is not involved!

George et al. [11] proposed that the intertwined and highly intricate interactions between a wide range of spatiotemporal scales in the flow, flame, and acoustic field occur through pattern formation. Small scale vortices collectively interact, leading to sustained periodic emergence of large-scale structures during the onset of thermoacoustic instability. These large coherent structures, where substantial magnitude of instantaneous local acoustic power production is observed, drive self-sustained periodic pressure oscillations. Each subsystem has a direct effect on the emerging spatial and temporal patterns. As a consequence, the subsystems interact through these patterns. The interaction between the small-scale vortices lead to order and pattern formation at the large scale while interacting with the flame and acoustics. The emergence of order at the large scale due to small scale interactions results in the mutual enhancement of coupling across the subsystems. These interactions between the subsystems affect the small scale interactions, leading to self-organization as we approach thermoacoustic instability.

Examining the spatiotemporal dynamics of the acoustic pressure and the heat release rate fluctuations reveals interesting features. We think of the turbulent reacting flow as consisting of a large number of local heat release rate oscillators [34]. During the stable operation, these heat release rate oscillators display aperiodic fluctuations and are asynchronous with the acoustic pressure fluctuations. The fluctuations become phase synchronous and large amplitude oscillations emerge, along with large regions of coherence. The transition happens though an intermediate state where pockets of synchrony co-exist with pockets of asynchrony, resembling a chimera state observed in a network of identical oscillators.

Complex network theory reveals interesting features of the transition. Small fragments of acoustic power sources scattered throughout the reacting flow domain are observed during combustion noise. During the transition to thermoacoustic instability, these fragments nucleate, coalesce and grow in size, to form large clusters [28]. During the transition, the spatiotemporal dynamics of acoustic power sources can be quantified using measures from complex network theory. Examining network centrality measures on the spatial networks constructed for each of the three dynamical regimes namely combustion noise, intermittency, and thermoacoustic instability can help identify critical regions in the flow field [57]. The knowledge of such critical regions in the flow field helps us devise smart passive control strategies [27, 59].

Complex network approach is a very promising approach. The construction of multilayer networks using the acoustic driving ($p'\dot{Q}'$) and vorticity network may help obtain a deeper understanding of the connection between the large coherent structures that emerge during thermoacoustic instability and the emergence of coherent acoustic power. Implementing this is indeed a challenging task, as the resolution of PIV data (velocity vector field) and chemiluminescence or PLIF data (heat release rate field) are often very different. This situation is far more challenging than the ones described in the literature.

Lagrangian coherent structure analysis (LCS) has been performed on the velocity data obtained during the three states of combustion noise, intermittency and limit cycle [46, 47]. During the intermittent regime, when the acoustic and hydrodynamic mode are phase synchronized, bursts of periodic oscillations appear. Using a combination of LCS and dynamic mode decomposition (DMD) analysis revealed that the roll up of shear layers from the dump plane and the leading edge of the bluff-body and the subsequent impingement on combustor walls are the dominant features of the flow during the periodic regimes of intermittency and during thermoacoustic instability.

So far, the spatiotemporal analysis has been performed primarily using experimental data. In future, we envisage that groups working on high fidelity CFD simulations, detailed laser diagnostics, and complex systems theory will work together, to obtain a deeper understanding of thermoacoustic instability.

14.3 Mitigation Strategies

Finally, the proof of the pudding is in the eating. Complex systems theory has given us some novel mitigation strategies that are not available from traditional linear thermoacoustics.

14.3.1 Evasion

A number of precursors based on chaos and fractal theory, intermittency statistics, complex network theory, symbolic logic, etc. can provide us measures of proximity to the onset of thermoacoustic instability. These measures change well before the pressure amplitude rises. Further, we can predict the amplitudes at the onset using techniques based on the above measures, just by examining the time series data of acoustic pressure during stable operation or intermittency. These precursor measures are robust, and can be used to enable automatic machine decision making. In future, these measures from complexity may be combined with artificial intelligence (AI) based precursors or hybrid measures using AI and complex systems theory.

14.3.2 Strategies Based on the Framework of Synchronization Theory

Asynchronous quenching using open-loop forcing has been proved to be effective in quenching oscillations in laminar thermoacoustic systems. Further, amplitude death has been shown to be a very attractive strategy for mitigating instabilities [8]. Amplitude death does not need any actuators; it involves just coupling two thermoacoustic systems appropriately. The effectiveness of AD has been demonstrated in laminar thermoacoustic systems.

The efficacy of these methodologies in mitigating thermoacoustic instability in turbulent combustors is worth exploring. Implementing these methodologies in turbulent systems will indeed bring a host of fresh challenges.

14.3.3 Smart Passive Control

Passive control is a very popular means for controlling thermoacoustic instability. However, passive control strategies are often ad hoc and based on the engineer's gut feeling. They often involve a large number of trials that are expensive and time consuming. In contrast, complex systems theory provides a smart and effective way to implement passive control. The central idea is to detect critical locations that are highly sensitive for the establishment of thermoacoustic instability, and to disrupt them. This disruption can be accomplished by, for example, "tweaking" the flow field (modifying the velocity profiles) by distributing the cooling air to break up the critical regions, by introducing delta wings, or by small tweaks of the fuel distribution. In the case of liquid rockets, we can make slight changes in the injector location to break up the self-organization. Rather than trying these strategies ad hoc as is currently the practice, we can implement these strategies in a "smart" manner, by using methods based on dynamical systems theory and complex systems theory to make the necessary changes [27, 29]. Determination of critical points is possible by using complex network theory, fractal theory, recurrence analysis, and using Lagrangian coherent structures.

14.4 Future Issues

In this section, we list a number of open issues that needs to be addressed in the future. This list is not in any particular order and is not complete, but is just a sample of the exciting work that awaits us.

1. Currently, nonlinear thermoacoustic instability is modeled using the describing function approach. This approach does not capture the chaos to periodic transition via intermittency. Modeling thermoacoustic instability as a synchronization of oscillators can model this transition, and may eventually provide an alternative to the describing function approach.
2. Intermittency is seen before the onset of thermoacoustic instability. It is also seen before blowout. What are the common features between these two occurrences of intermittency and what are the differences?
3. Complex systems theory has been applied primarily to axial instabilities. Applying it to transverse instabilities such as those in annular burners, rockets, or afterburners will be challenging, and will bring out interesting results.
4. Further, most studies are concentrated on thermoacoustic systems that do not have preheated air. The stability of the hydrodynamic flow field in the wake changes significantly as we go from non-preheated (globally stable wake) to a system with preheated air (globally unstable wake). It would be really interesting to unravel the flow physics in preheated systems, using the tools from nonlinear time series analysis and complex systems theory.
5. Most of this work has been concentrated on gas turbine combustors. Application to signals from rockets will bring fresh challenges.
6. Phenomenological models that capture the intermittency route to thermoacoustic instability have been developed. These models need to be improved to capture the multifractal characteristics of the transition. Can we construct simple models that capture the spatiotemporal features of the thermoacoustic transition? Further, we expect that models will be constructed using techniques from AI that can extract the governing equations from data. Further, combining AI with physics may lead to construction of better models.
7. Application of multilayer networks to spatiotemporal PIV and acoustic power data may help unravel the relation between the emergence of coherent structures and onset of thermoacoustic instability.
8. Can we derive amplitude equations such as complex Ginzberg Landau equation (CGLE) from Navier Stokes equations that can capture the behavior of thermoacoustic systems?

9. Can we find one non-dimensional parameter combining equivalence ratio, Reynolds number, damping, that can comprehensively capture the onset of thermoacoustic instability?
10. Recent studies have shown that the thermoacoustic transitions depend not just on the parameter, but on the rate of change of parameter as well [4, 33, 55]. The transitions in a rate dependent situation can be quite different from that of the case where the parameter is varied in a quasi-static manner. Further, we may be able to use this phenomenon to dodge bifurcations [5].
11. Cellular automata have been used successfully to capture blowout [56]. Can we use cellular automata to capture thermoacoustic instabilities?
12. Can we construct a unified theory for the onset of oscillatory instabilities in turbulent flows across all systems?

14.5 Final Thoughts

Traditionally, papers in thermoacoustic instabilities are published in traditional power, propulsion, and aerospace engineering journals such as *ASME Journal of Gas Turbine Power and Engineering, Journal of Propulsion and Power, AIAA Journal, International Journal of Spray and Combustion Dynamics*, and in acoustics journals such as *Journal of Sound and Vibration.*

In the last 5 years, there has been a surge in the number of papers published in more broad based journals such as *Physical Review E, Chaos, Nonlinear Dynamics, International Journal of Nonlinear Dynamics, Nonlinear Mechanics, Proceedings of Royal Society, Scientific Reports, Physical Review Applied, Europhysics Letters, Journal of Fluid Mechanics,* and *Physics of Fluids*. Posing thermoacoustics as a self-organization occurring in turbulent flows is of interest to the wider community that includes physicists, complex system theorists, and fluid mechanics researchers who are interested in studying the problem of order emerging from chaos. Thus, thermoacoustic instability in turbulent flows has emerged as a platform for studying non-equilibrium statistical mechanics and complex systems theory. Researchers in these subjects have developed a large number of tools. Turbulent reacting flow with its large number of degrees of freedom is indeed high in complexity and is the ideal system to test the mettle of these tools.

Linear theory and reductionism are simple and convenient. However, we miss out on a lot of phenomena by looking at the thermoacoustic system through the glasses of linearity and reductionism. Six decades back, computing was hard, and nonlinear dynamics was at its infancy; these were pressing reasons to embrace linearity. However, the human mind has inertia, and the thermoacoustic community is still continuing this embrace of linearity. The tools from nonlinear dynamics and complex systems theory are now very advanced. Computing itself has become very powerful, and the researchers have very powerful machines on their desktops, which could not be even dreamt of in the sixties. It is now time to let go of the embrace of linearity; it is indeed time to embrace nonlinearity and complexity.

References

1. Ananthkrishnan, N., Deo, S., Culick, F.E.C.: Reduced-order modeling and dynamics of nonlinear acoustic waves in a combustion chamber. Combust. Sci. Tech. **177**(2), 221–248 (2005)
2. Balasubramanian, K., Sujith, R.I.: Non-normality and nonlinearity in combustion–acoustic interaction in diffusion flames. J. Fluid Mech. **594**, 29–57 (2008)
3. Balasubramanian, K., Sujith, R.I.: Thermoacoustic instability in a Rijke tube: non-normality and nonlinearity. Phys. Fluids **20**(4), 044103 (2008)
4. Bonciolini, G., Ebi, D., Boujo, E., Noiray, N.: Experiments and modelling of rate-dependent transition delay in a stochastic subcritical bifurcation. R. Soc. Open Sci. **5**(3), 172078 (2018)
5. Bonciolini, G., Noiray, N.: Bifurcation dodge: avoidance of a thermoacoustic instability under transient operation. Nonlinear Dyn. **96**(1), 703–716 (2019)
6. Boujo, E., Denisov, A., Schuermans, B., Noiray, N.: Quantifying acoustic damping using flame chemiluminescence. J. Fluid Mech. **808**, 245–257 (2016)
7. Clavin, P., Kim, J.S., Williams, F.A.: Turbulence-induced noise effects on high-frequency combustion instabilities. Combust. Sci. Tech. **96**(1–3), 61–84 (1994)
8. Dange, S., Manoj, K., Banerjee, S., Pawar, S.A., Mondal, S., Sujith, R.I.: Oscillation quenching and phase-flip bifurcation in coupled thermoacoustic systems. Chaos **29**(9), 093135 (2019)
9. Emerson, B., O'Connor, J., Juniper, M., Lieuwen, T.: Density ratio effects on reacting bluff-body flow field characteristics. J. Fluid Mech. **706**, 219–250 (2012)
10. Etikyala, S., Sujith, R.I.: Change of criticality in a prototypical thermoacoustic system. Chaos **27**(2), 023106 (2017)
11. George, N.B., Unni, V.R., Raghunathan, M., Sujith, R.I.: Pattern formation during transition from combustion noise to thermoacoustic instability via intermittency. J. Fluid Mech. **849**, 615–644 (2018)
12. Godavarthi, V., Pawar, S.A., Unni, V.R., Sujith, R.I., Marwan, N., Kurths, J.: Coupled interaction between unsteady flame dynamics and acoustic field in a turbulent combustor. Chaos **28**(11), 113111 (2018)
13. Godavarthi, V., Unni, V.R., Gopalakrishnan, E.A., Sujith, R.I.: Recurrence networks to study dynamical transitions in a turbulent combustor. Chaos **27**(6), 063113 (2017)
14. Gopalakrishnan, E.A., Sujith, R.I.: Influence of system parameters on the hysteresis characteristics of a horizontal Rijke tube. J. Spray Combust. Dyn. **6**(3), 293–316 (2014)
15. Gopalakrishnan, E.A., Sujith, R.I.: Effect of external noise on the hysteresis characteristics of a thermoacoustic system. J. Fluid Mech. **776**, 334–353 (2015)
16. Gotoda, H., Kinugawa, H., Tsujimoto, R., Domen, S., Okuno, Y.: Characterization of combustion dynamics, detection, and prevention of an unstable combustion state based on a complex-network theory. Phys. Rev. Appl. **7**(4), 044027 (2017)
17. Guan, Y., Gupta, V., Li, L.K.B.: Intermittency route to self-excited chaotic thermoacoustic oscillations. J. Fluid Mech. **894**, R3 (2020)
18. Jegadeesan, V., Sujith, R.I.: Experimental investigation of noise induced triggering in thermoacoustic systems. Proc. Combust. Inst. **34**(2), 3175–3183 (2013)
19. Juniper, M.P.: Triggering in the horizontal Rijke tube: non-normality, transient growth and bypass transition. J. Fluid Mech. **667**, 272–308 (2011)
20. Kabiraj, L., Saurabh, A., Wahi, P., Sujith, R.I.: Route to chaos for combustion instability in ducted laminar premixed flames. Chaos **22**(2), 023129 (2012)
21. Kabiraj, L., Sujith, R.I.: Nonlinear self-excited thermoacoustic oscillations: intermittency and flame blowout. J. Fluid Mech. **713**, 376–397 (2012)
22. Kabiraj, L., Sujith, R.I., Wahi, P.: Bifurcations of self-excited ducted laminar premixed flames. J. Eng. Gas Turbines Power **134**(3), 031502 (2012)
23. Kabiraj, L., Sujith, R.I., Wahi, P.: Investigating the dynamics of combustion-driven oscillations leading to lean blowout. Fluid Dyn. Res. **44**(3), 031408 (2012)

24. Kashinath, K., Waugh, I.C., Juniper, M.P.: Nonlinear self-excited thermoacoustic oscillations of a ducted premixed flame: bifurcations and routes to chaos. J. Fluid Mech. **761**, 399–430 (2014)
25. Kasthuri, P., Unni, V.R., Sujith, R.I.: Bursting and mixed mode oscillations during the transition to limit cycle oscillations in a matrix burner. Chaos **29**(4), 043117 (2019)
26. Kerres, B., Nair, V., Cronhjort, A., Mihaescu, M.: Analysis of the turbocharger compressor surge margin using a Hurst-exponent-based criterion. SAE Int. J. Engines **9**(3), 1795–1806 (2016)
27. Krishnan, A., Manikandan, R., Midhun, P.R., Reeja, K.V., Unni, V.R., Sujith, R.I., Marwan, N., Kurths, J.: Mitigation of oscillatory instability in turbulent reactive flows: a novel approach using complex networks. Europhys. Lett. **128**(1), 14003 (2019)
28. Krishnan, A., Sujith, R.I., Marwan, N., Kurths, J.: On the emergence of large clusters of acoustic power sources at the onset of thermoacoustic instability in a turbulent combustor. J. Fluid Mech. **874**, 455–482 (2019)
29. Krishnan, A., Sujith, R.I., Marwan, N., Kurths, J.: Suppression of thermoacoustic instability by targeting the hubs of the turbulent networks in a bluff body stabilized combustor. J. Fluid Mech. **916**, A20 (2021)
30. Lieuwen, T., Banaszuk, A.: Background noise effects on combustor stability. J. Propuls. Power **21**(1), 25–31 (2005)
31. Lieuwen, T.C.: Experimental investigation of limit-cycle oscillations in an unstable gas turbine combustor. J. Propuls. Power **18**(1), 61–67 (2002)
32. Lipsitz, L.A., Goldberger, A.L.: Loss of "complexity" and aging: potential applications of fractals and chaos theory to senescence. J. Am. Med. Assoc. **267**(13), 1806–1809 (1992)
33. Manikandan, S., Sujith, R.I.: Rate dependent transition to thermoacoustic instability via intermittency in a turbulent afterburner. Exp. Therm. Fluid Sci. **114**, 110046 (2020)
34. Mondal, S., Unni, V.R., Sujith, R.I.: Onset of thermoacoustic instability in turbulent combustors: an emergence of synchronized periodicity through formation of chimera-like states. J. Fluid Mech. **811**, 659–681 (2017)
35. Murugesan, M., Sujith, R.I.: Combustion noise is scale-free: transition from scale-free to order at the onset of thermoacoustic instability. J. Fluid Mech. **772**, 225–245 (2015)
36. Murugesan, M., Sujith, R.I.: Detecting the onset of an impending thermoacoustic instability using complex networks. J. Propuls. Power **32**(3), 707–712 (2016)
37. Murugesan, M., Sujith, R.I.: Physical mechanisms that cause intermittency that presages combustion instability and blowout in a turbulent lifted jet flame combustor. Combust. Sci. Tech. **190**(2), 312–335 (2018)
38. Nair, V., Sujith, R.I.: Identifying homoclinic orbits in the dynamics of intermittent signals through recurrence quantification. Chaos **23**(3), 033136 (2013)
39. Nair, V., Sujith, R.I.: Multifractality in combustion noise: predicting an impending combustion instability. J. Fluid Mech. **747**, 635–655 (2014)
40. Nair, V., Sujith, R.I.: Precursors to self-sustained oscillations in aeroacoustic systems. Int. J. Aeroacoust. **15**(3), 312–323 (2016)
41. Nair, V., Thampi, G., Karuppusamy, S., Gopalan, S., Sujith, R.I.: Loss of chaos in combustion noise as a precursor of impending combustion instability. J. Spray Combust. Dyn. **5**(4), 273–290 (2013)
42. Nair, V., Thampi, G., Sujith, R.I.: Intermittency route to thermoacoustic instability in turbulent combustors. J. Fluid Mech. **756**, 470–487 (2014)
43. Noiray, N.: Linear growth rate estimation from dynamics and statistics of acoustic signal envelope in turbulent combustors. J. Eng. Gas Turbines Power **139**(4), 041503 (2017)
44. Pavithran, I., Unni, V.R., Varghese, A.J., Sujith, R.I., Saha, A., Marwan, N., Kurths, J.: Universality in the emergence of oscillatory instabilities in turbulent flows. Europhys. Lett. **129**(2), 24004 (2020)
45. Pawar, S.A., Seshadri, A., Unni, V.R., Sujith, R.I.: Thermoacoustic instability as mutual synchronization between the acoustic field of the confinement and turbulent reactive flow. J. Fluid Mech. **827**, 664–693 (2017)

46. Premchand, C.P., George, N.B., Raghunathan, M., Unni, V.R., Sujith, R.I., Nair, V.: Lagrangian analysis of intermittent sound sources in the flow-field of a bluff-body stabilized combustor. Phys. Fluids **31**(2), 025115 (2019)
47. Premchand, C.P., George, N.B., Raghunathan, M., Unni, V.R., Sujith, R.I., Nair, V.: Lagrangian analysis of flame dynamics in the flow field of a bluff body-stabilized combustor. J. Eng. Gas Turbines Power **142**(1), 011015 (2020)
48. Premraj, D., Pawar, S.A., Kabiraj, L., Sujith, R.I.: Strange nonchaos in self-excited singing flames. Europhys. Lett. **128**(5), 54005 (2020)
49. Sampath, R., Mathur, M., Chakravarthy, S.R.: Lagrangian coherent structures during combustion instability in a premixed-flame backward-step combustor. Phys. Rev. E **94**(6), 062209 (2016)
50. Seshadri, A., Pavithran, I., Unni, V.R., Sujith, R.I.: Predicting the amplitude of limit-cycle oscillations in thermoacoustic systems with vortex shedding. AIAA J. **56**(9), 3507–3514 (2018)
51. Sleimen-Malkoun, R., Temprado, J.J., Hong, S.L.: Aging induced loss of complexity and dedifferentiation: consequences for coordination dynamics within and between brain, muscular and behavioral levels. Front. Aging Neurosci. **6**, 140 (2014)
52. Strahle, W.C.: Combustion noise. Prog. Energy Combust. Sci. **4**(3), 157–176 (1978)
53. Subramanian, P., Mariappan, S., Sujith, R.I., Wahi, P.: Bifurcation analysis of thermoacoustic instability in a horizontal Rijke tube. J. Spray Combust. Dyn. **2**(4), 325–355 (2010)
54. Suresha, S., Sujith, R.I., Emerson, B., Lieuwen, T.: Nonlinear dynamics and intermittency in a turbulent reacting wake with density ratio as bifurcation parameter. Phys. Rev. E **94**(4), 042206 (2016)
55. Tony, J., Gopalakrishnan, E.A., Sreelekha, E., Sujith, R.I.: Detecting deterministic nature of pressure measurements from a turbulent combustor. Phys. Rev. E **92**(6), 062902 (2015)
56. Unni, V.R., Chaudhuri, S., Sujith, R.I.: Flame blowout: transition to an absorbing phase. Chaos **28**(11), 113121 (2018)
57. Unni, V.R., Krishnan, A., Manikandan, R., George, N.B., Sujith, R.I., Marwan, N., Kurths, J.: On the emergence of critical regions at the onset of thermoacoustic instability in a turbulent combustor. Chaos **28**(6), 063125 (2018)
58. Unni, V.R., Sujith, R.I.: Flame dynamics during intermittency in a turbulent combustor. Proc. Combust. Inst. **36**(3), 3791–3798 (2017)
59. Unni, V.R., Sujith, R.I., Krishnan, A., Marwan, N., Kurths, J.: System and method for optimizing passive control of oscillatory instabilities in turbulent flows. US Patent US10895382B2, Priority date: 27 Feb 2018
60. Venkatramani, J., Nair, V., Sujith, R.I., Gupta, S., Sarkar, S.: Precursors to flutter instability by an intermittency route: a model free approach. J. Fluids Struct. **61**, 376–391 (2016)

Appendix A
Governing Equations for the One-Dimensional Fluid Flow

Fluid flow is governed by three basic principles: conservation of mass, conservation of momentum, and conservation of energy. We can mathematically formulate these principles of the fluid flow into corresponding governing equations as continuity equation, momentum equation, and energy equation, respectively. In this appendix, we provide derivation of these equations from first principles, by considering a fluid flow through an infinitesimal control volume fixed in space [1–4]. The derivation of these conservation equations is based on several assumptions. We consider a one-dimensional, isentropic, calorifically perfect, inviscid, non-conducting, compressible flow at low Mach number. Body forces are ignored in the analysis. The variables and their units (in brackets) involved in the derivation of the conservation equations are: m is the mass (kg), ρ is the density (kg/m^3), p is the pressure (Pa), u is the velocity (m/s), $\dot{Q}$ is the heat release rate (W), A is the area (m^2), t is the time (s), and x, y are the spatial coordinates (m).

A.1 Continuity Equation

The continuity equation is based on the principle of conservation of mass in a control volume. The law of conservation of mass states that mass can neither be created nor be destroyed. When a fluid is passing through the control volume, the net mass balance across the control surfaces is given as,

$$\begin{aligned}&\text{Rate of mass flow entering the control volume}\\&-\text{ Rate of mass flow leaving the control volume}\\&=\text{ Rate of change of mass inside the control volume}\end{aligned} \tag{A.1}$$

R. I. Sujith, S. A. Pawar, *Thermoacoustic Instability*, Springer Series in Synergetics,
https://doi.org/10.1007/978-3-030-81135-8

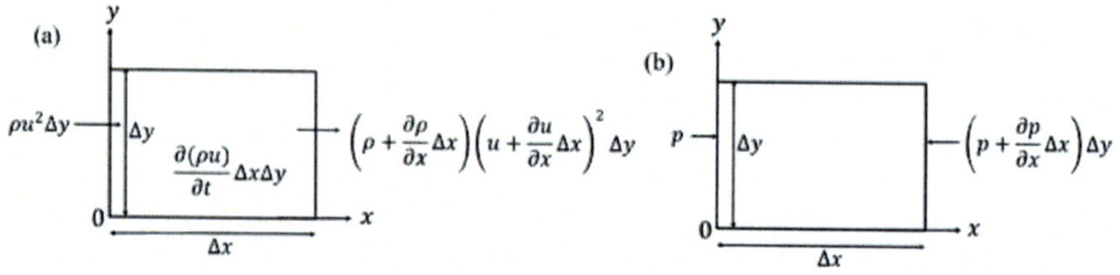

Fig. A.1 Schematic showing the balance of mass fluxes across the control volume and the accumulation of mass inside the control volume

Consider an infinitesimally small, stationary, rectangular control volume of sides Δx and Δy with unit depth located at the origin. Therefore, the cross-sectional area of the side boundary is $A = \Delta y.1 = \Delta y$. The fluid flow is one-dimensional, i.e., in x-direction (Fig. A.1).

The mass of the fluid entering the control volume per unit time at $x = 0$ is,

$$\rho u A|_{x=0} = \rho u \Delta y \tag{A.2}$$

The mass of the fluid leaving the control volume per unit time at $x = \Delta x$ is,

$$\rho u A|_{x=\Delta x} = \left(\rho + \frac{\partial \rho}{\partial x}\Delta x\right)\left(u + \frac{\partial u}{\partial x}\Delta x\right)\Delta y \tag{A.3}$$

The rate of mass accumulation of the fluid in the control volume is,

$$\frac{\partial m}{\partial t} = \frac{\partial \rho}{\partial t}\Delta x \Delta y \tag{A.4}$$

As there is no generation of mass in the control volume, we can obtain the continuity equation (A.1) in x-direction as,

$$\rho u \Delta y - \left(\rho + \frac{\partial \rho}{\partial x}\Delta x\right)\left(u + \frac{\partial u}{\partial x}\Delta x\right)\Delta y = \frac{\partial \rho}{\partial t}\Delta x \Delta y \tag{A.5}$$

By ignoring terms higher than the first order and rearranging the equation, we get,

$$\frac{\partial \rho}{\partial t} + \frac{\partial(\rho u)}{\partial x} = 0 \tag{A.6}$$

Thus, (A.6) represents the mass conservation (continuity) equation for a one-dimensional compressible fluid flow.

A.2 Momentum Equation

The law of conservation of momentum is based on Newton's second law of motion. This law states that the net rate of change of momentum in a control volume is balanced by the momentum fluxes and the net force acting on the control volume. Further, this momentum change takes place in the direction of the net force. Therefore, we can obtain the momentum equation as follows,

$$\begin{aligned}&\text{Rate of change of momentum inside the control volume}\\&= \text{Rate of momentum flow entering the control volume}\\&- \text{Rate of momentum flow leaving the control volume}\\&+ \text{Net force acting on the control volume}\end{aligned} \tag{A.7}$$

Let us consider a fluid flowing through an infinitesimally small, stationary, rectangular control volume of size Δx and Δy with unit depth (see Fig. A.2).
Then, we get the mass flow rate of the fluid entering the control volume as,

$$\rho A u = \rho \Delta y u \tag{A.8}$$

The rate of momentum of the fluid entering the control volume therefore becomes,

$$\text{mass flow rate} \times \text{ velocity} = (\rho \Delta y u) u = \rho u^2 \Delta y \tag{A.9}$$

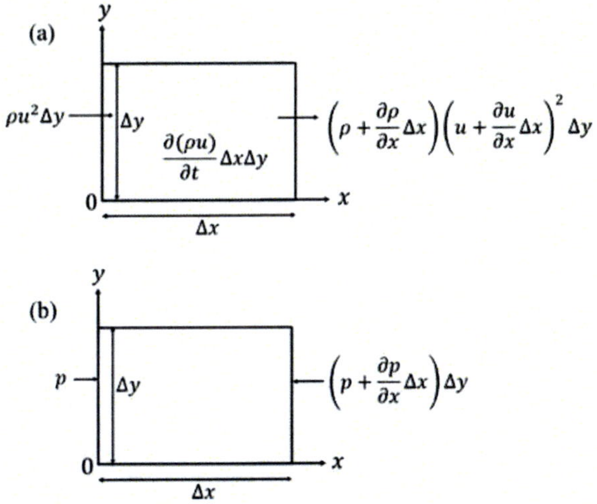

Fig. A.2 Schematics showing (**a**) the momentum fluxes across the control volume and the accumulation of momentum inside the control volume, and (**b**) a force balance on the control surface

Similarly, the rate of momentum of the fluid leaving the control volume is given as,

$$\left(\rho + \frac{\partial \rho}{\partial x}\Delta x\right)\left(u + \frac{\partial u}{\partial x}\Delta x\right)^2 \Delta y \tag{A.10}$$

Thus, we get the net change of momentum flow across the control volume as,

$$\begin{aligned}
&= \text{Rate of momentum flow entering the control volume} \\
&- \text{Rate of momentum flow leaving the control volume} \\
&= \rho u^2 \Delta y - \left(\rho + \frac{\partial \rho}{\partial x}\Delta x\right)\left(u + \frac{\partial u}{\partial x}\Delta x\right)^2 \Delta y
\end{aligned} \tag{A.11}$$

Further, the rate of momentum accumulation inside the control volume is,

$$= \frac{\partial(\rho u)}{\partial t}\Delta x \Delta y \tag{A.12}$$

We know that there are two types of external forces that act on the fluid element: body forces (e.g., gravity, electromagnetic field, etc.) and surface forces (e.g., normal pressure, shear stress, etc.). We ignore the body forces acting on the fluid element. Since the flow is inviscid, we also neglect the surface forces arising from the shear stress.

Hence, the net force acting on the fluid element is the force due to the pressure (Fig. A.2b), which we can express as,

$$\begin{aligned}
F_x &= pA - \left(p + \frac{\partial p}{\partial x}\Delta x\right) A, \\
&= p\Delta y - \left(p + \frac{\partial p}{\partial x}\Delta x\right) \Delta y, \\
&= -\frac{\partial p}{\partial x}\Delta x \Delta y
\end{aligned} \tag{A.13}$$

Substituting (A.11)–(A.13) into (A.7), we get the final momentum equation in x-direction as,

$$\frac{\partial(\rho u)}{\partial t}\Delta x \Delta y = \rho u^2 \Delta y - \left(\rho + \frac{\partial \rho}{\partial x}\Delta x\right)\left(u + \frac{\partial u}{\partial x}\Delta x\right)^2 \Delta y - \frac{\partial p}{\partial x}\Delta x \Delta y$$

By ignoring the higher-order terms and rearranging the remaining equation, we get,

$$\rho \frac{\partial u}{\partial t} + \rho u \frac{\partial u}{\partial x} + u\left(\frac{\partial \rho}{\partial t} + \frac{\partial(\rho u)}{\partial x}\right) = -\frac{\partial p}{\partial x}$$

The term in the parenthesis on the left hand side of the above equation becomes zero due to the conservation of mass equation (A.6). With this simplification, we obtain,

$$\rho \frac{\partial u}{\partial t} + \rho u \frac{\partial u}{\partial x} = -\frac{\partial p}{\partial x} \tag{A.14}$$

Therefore, (A.14) represents the momentum conservation equation for one-dimensional, inviscid, compressible fluid flow through a control volume.

A.3 Energy Equation

The principle of conservation of energy is based on the first law of thermodynamics. It states that energy can be neither created nor destroyed; however, it can be converted from one form to another. In other words, the net energy transfer to and from a system (control volume) is equal to the net change of internal energy in the system. Mathematically, we can express the energy equation as,

Rate of change of internal energy (E) inside the control volume

= Rate of energy flow entering the control volume due to fluid flow

− Rate of energy flow leaving the control volume due to fluid flow (A.15)

+ Work done ($\dot{W}$) by stresses on the control volume

+ Rate of heat addition ($\dot{Q}$) to the control volume.

The total energy of the flow in the control volume equals the sum of internal energy, potential energy, and kinetic energy (Fig. A.3a).

Therefore, we obtain the total energy per unit mass as,

$$E = e + gy + u^2/2$$

where e is the specific internal energy, g is the gravitational acceleration, y is the distance of the fluid element above a reference level.

Since we are ignoring body forces (including gravitational force), we modify the above equation as,

$$E = e + u^2/2 \tag{A.16}$$

Thus, we obtain the rate of energy flowing into the control volume as,

$$\rho A u \left(e + \frac{u^2}{2}\right) = \rho u \left(e + \frac{u^2}{2}\right) \Delta y \tag{A.17}$$

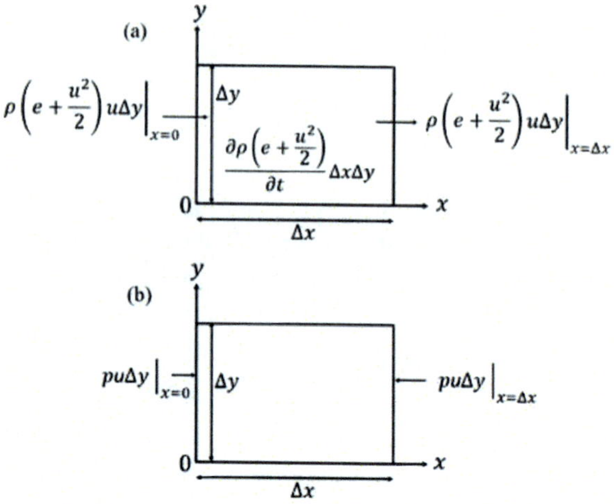

Fig. A.3 Schematics showing (**a**) the energy fluxes across the control volume and the accumulation of energy inside the control volume, and (**b**) the work done by pressure forces on the control surface

Similarly, we can express the rate of energy flowing out of the control volume as,

$$\left(\rho + \frac{\partial \rho}{\partial x}\Delta x\right)\left(u + \frac{\partial u}{\partial x}\Delta x\right)\left[\left(e + \frac{\partial e}{\partial x}\Delta x\right) + \frac{1}{2}\left(u + \frac{\partial u}{\partial x}\Delta x\right)^2\right]\Delta y \tag{A.18}$$

Further, we can write the rate of energy accumulation inside the control volume as,

$$\frac{\partial\left[\rho\left(e + u^2/2\right)\right]}{\partial t}\Delta x \Delta y \tag{A.19}$$

Consider the rate of heat addition per unit volume in the control volume be, $\dot{Q}$ (A.20)

The work done on a system can be of different forms, which include the work done by a rotating (shaft) or translating (piston) element of the system, work done by the pressure on the control surface, work done by the viscous forces on the control surface, and work done by other forces such as electric force, magnetic force, surface tension force, etc. Since we assume that there are no body forces and that the flow is inviscid, we consider only the work done by pressure forces in x-direction on the control surface as follows (Fig. A.3b),

$$\begin{aligned} \dot{W} &= pAu - \left(p + \frac{\partial p}{\partial x}\Delta x\right) A \left(u + \frac{\partial u}{\partial x}\Delta x\right), \\ &= \left(-u\frac{\partial p}{\partial x} - p\frac{\partial u}{\partial x}\right)\Delta x \Delta y \end{aligned} \tag{A.21}$$

Therefore, by substituting (A.17)–(A.21) into (A.15), we get the final energy equation as,

$$\begin{aligned}&\frac{\partial\left[\rho\left(e+u^2/2\right)\right]}{\partial t}\Delta x\Delta y\\&\qquad=\rho u\left(e+\frac{u^2}{2}\right)\Delta y\\&\qquad-\left(\rho+\frac{\partial\rho}{\partial x}\Delta x\right)\left(u+\frac{\partial u}{\partial x}\Delta x\right)\left[\left(e+\frac{\partial e}{\partial x}\Delta x\right)+\frac{1}{2}\left(u+\frac{\partial u}{\partial x}\Delta x\right)^2\right]\Delta y\\&\qquad-\left(u\frac{\partial p}{\partial x}+p\frac{\partial u}{\partial x}\right)\Delta x\Delta y+\dot{Q}\Delta x\Delta y\end{aligned} \tag{A.22}$$

On simplification,

$$\begin{aligned}&\rho\frac{\partial e}{\partial t}+\rho u\frac{\partial e}{\partial x}\\&\qquad=-e\left(\frac{\partial\rho}{\partial t}+\rho\frac{\partial u}{\partial x}+u\frac{\partial\rho}{\partial x}\right)-\frac{u^2}{2}\left(\frac{\partial\rho}{\partial t}+u\frac{\partial\rho}{\partial x}+\rho\frac{\partial u}{\partial x}\right),\\&\qquad-\left(\rho\frac{\partial(u^2/2)}{\partial t}+\rho u^2\frac{\partial u}{\partial x}+u\frac{\partial p}{\partial x}\right)-p\frac{\partial u}{\partial x}+\dot{Q}\end{aligned} \tag{A.23}$$

Now, multiply the momentum equation (A.14) by u and moving the pressure gradient term to the left hand side, we get,

$$\rho\frac{\partial(u^2/2)}{\partial t}+\rho u^2\frac{\partial u}{\partial x}+u\frac{\partial p}{\partial x}=0 \tag{A.24}$$

Further, the continuity equation (A.6) of the fluid flow can be expressed as,

$$\frac{\partial\rho}{\partial t}+\rho\frac{\partial u}{\partial x}+u\frac{\partial\rho}{\partial x}=0 \tag{A.25}$$

Substituting (A.24) and (A.25) into (A.23), we get,

$$\rho\frac{\partial e}{\partial t}+u\rho\frac{\partial e}{\partial x}=\dot{Q}-p\frac{\partial u}{\partial x} \tag{A.26}$$

Thus, (A.26) indicates the conservation of energy equation for one-dimensional, compressible fluid flow through a control volume.

The governing equations derived in this section are nonlinear. Obtaining analytical solutions for these equations is difficult. Furthermore, numerical simulation of these equations is still computationally expensive. Hence, we often resorts to linearization of these equations to gain insights into many fluid-flow problems encountered in practical applications.

A.4 Linearized Governing Equations for the Acoustic Field

In this section, we will derive linearized governing equations for a fluid (i.e., gas or air) flow in one-dimension [5, 6]. In Chap. 9, we have used these linearized equations for modeling different thermoacoustic phenomena in a turbulent combustor. To derive the acoustic equations from the governing equations of the compressible fluid flow, we linearize the governing equations as small perturbations about a base flow. When an acoustic wave passes through a fluid, it produces local compressions and rarefactions in the flow. This results in local fluctuations in the density and the pressure. We assume that the disturbances induced by the propagation of the acoustic wave are very small as compared to the mean quantities of the fluid flow.

We also consider that the mean flow is quiescent (i.e., $\overline{u} = 0$), and the flow properties are steady in time and uniform in space (i.e., $\overline{p}(x)$ = constant and $\overline{\rho}(x)$ = constant).

Hence, we get

$$\frac{\partial \overline{p}}{\partial t} = 0, \quad \frac{\partial \overline{\rho}}{\partial t} = 0, \quad \frac{\partial \overline{p}}{\partial x} = 0, \quad \frac{\partial \overline{\rho}}{\partial x} = 0.$$

Further, we represent the acoustic variables by prime, such as ρ', p', and u', and the mean variables by overbar, viz., $\overline{\rho}$, $\overline{p}$, and $\overline{u}$. We also assume that the fluctuating quantities (ρ', p') are very small as compare to their mean values [7]. Therefore, the total value of a particular variable is expressed as the sum of its mean and fluctuating quantities as follows [7, 8],

$$p = p' + \overline{p}, \quad \rho = \rho' + \overline{\rho}, \quad u = u' + \overline{u} = u', \tag{A.27}$$

where

$$\overline{p} = \lim_{T\to\infty} \frac{1}{T} \int_t^{t+T} p(t)dt,$$

and

$$\lim_{T\to\infty} \frac{1}{T} \int_t^{t+T} p'(t)dt = 0.$$

Next, we linearize the governing equations of conservation of mass, momentum, and energy (indicated by (A.6), (A.14), and (A.26), respectively) by ignoring terms of higher order than the first order from these equations, as the perturbations are of small magnitude.

A.4.1 Linearized Conservation of Mass Equation

The continuity equation (A.6) derived before is given as,

$$\frac{\partial \rho}{\partial t} + \frac{\partial (\rho u)}{\partial x} = 0.$$

By substituting linearized expressions of all state variables (see (A.27)) in the continuity equation, we get,

$$\frac{\partial (\rho' + \overline{\rho})}{\partial t} + \frac{\partial (\rho' + \overline{\rho}) u'}{\partial x} = 0$$

In other words,

$$\frac{\partial (\rho' + \overline{\rho})}{\partial t} + \frac{\partial (\rho' u' + \overline{\rho} u')}{\partial x} = 0$$

Neglecting the second-order terms resulting from the product of fluctuating quantities, and noting that the time and space derivatives of mean quantities are zero, we get,

$$\frac{\partial \rho'}{\partial t} + \overline{\rho} \frac{\partial u'}{\partial x} = 0 \tag{A.28}$$

Therefore, (A.28) represents the linearized form of the continuity equation.

A.4.2 Linearized Momentum Equation

As we derived before, the momentum equation (A.14) of the fluid flow is given as,

$$\rho \frac{\partial u}{\partial t} + u \rho \frac{\partial u}{\partial x} + \frac{\partial p}{\partial x} = 0$$

By substituting the linearized expressions of all state variables (see (A.27)) in the momentum equation, we get,

$$(\rho' + \overline{\rho}) \frac{\partial u'}{\partial t} + u' (\rho' + \overline{\rho}) \frac{\partial u'}{\partial x} + \frac{\partial (p' + \overline{p})}{\partial x} = 0$$

By ignoring higher order terms of fluctuating quantities and by considering constant mean flow quantities, we get,

$$\bar{\rho}\frac{\partial u'}{\partial t} + \frac{\partial p'}{\partial x} = 0 \tag{A.29}$$

Thus, (A.29) denotes the linearized form of the momentum equation.

A.4.3 Linearized Energy Equation

From the ideal gas equation, we know,

$$p = \rho R T \tag{A.30}$$

We can express the specific internal energy of an ideal, calorifically perfect gas as

$$e = c_v T = \frac{1}{\gamma - 1} R T$$

From (A.30), we get,

$$e = \frac{1}{\gamma - 1}\frac{p}{\rho}$$

or

$$e\rho = \frac{1}{\gamma - 1} p \tag{A.31}$$

where c_v is the specific heat at constant volume (J/kg·K), T is the temperature (K), R is the universal gas constant (J/mol·K), and γ is the ratio of specific heats. Since,

$$\frac{\partial(e\rho)}{\partial t} = \rho\frac{\partial e}{\partial t} + e\frac{\partial \rho}{\partial t} \tag{A.32}$$

We can write,

$$\rho\frac{\partial e}{\partial t} = \frac{\partial(e\rho)}{\partial t} - e\frac{\partial \rho}{\partial t} \tag{A.33}$$

Similarly, we can get,

$$\rho\frac{\partial e}{\partial x} = \frac{\partial(e\rho)}{\partial x} - e\frac{\partial \rho}{\partial x} \tag{A.34}$$

By substituting (A.31) into (A.33), we obtain,

$$\rho\frac{\partial e}{\partial t}=\frac{1}{\gamma-1}\frac{\partial p}{\partial t}-\frac{1}{\gamma-1}\frac{p}{\rho}\frac{\partial \rho}{\partial t} \tag{A.35}$$

Now, on multiplying (A.34) by u and then substituting (A.31), we get,

$$u\rho\frac{\partial e}{\partial x}=u\left[\frac{\partial(e\rho)}{\partial x}-e\frac{\partial \rho}{\partial x}\right]$$

$$u\rho\frac{\partial e}{\partial x}=u\left[\frac{1}{\gamma-1}\frac{\partial p}{\partial x}-\frac{1}{\gamma-1}\frac{p}{\rho}\frac{\partial \rho}{\partial x}\right] \tag{A.36}$$

Thus, by substituting (A.35) and (A.36) into energy equation (A.26), we get,

$$\frac{1}{\gamma-1}\left[\frac{\partial p}{\partial t}-\frac{p}{\rho}\frac{\partial \rho}{\partial t}\right]+\frac{1}{\gamma-1}u\left[\frac{\partial p}{\partial x}-\frac{p}{\rho}\frac{\partial \rho}{\partial x}\right]=\dot{Q}-p\frac{\partial u}{\partial x}$$

$$\left[\frac{\partial p}{\partial t}-\frac{p}{\rho}\frac{\partial \rho}{\partial t}\right]+u\left[\frac{\partial p}{\partial x}-\frac{p}{\rho}\frac{\partial \rho}{\partial x}\right]=(\gamma-1)\rho\dot{Q}-(\gamma-1)p\frac{\partial u}{\partial x}$$

Multiply the above equation by ρ and by simplifying terms, we get,

$$\rho\frac{\partial p}{\partial t}-p\frac{\partial \rho}{\partial t}+u\rho\frac{\partial p}{\partial x}-pu\frac{\partial \rho}{\partial x}=(\gamma-1)\rho\dot{Q}-(\gamma-1)p\rho\frac{\partial u}{\partial x} \tag{A.37}$$

Now, we multiply the continuity equation (A.6) with p and on simplification, we obtain,

$$p\frac{\partial \rho}{\partial t}+p\frac{\partial(\rho u)}{\partial x}=0$$

$$p\frac{\partial \rho}{\partial t}=-p\rho\frac{\partial u}{\partial x}-pu\frac{\partial \rho}{\partial x} \tag{A.38}$$

Substituting (A.38) into (A.37), we get,

$$\rho\frac{\partial p}{\partial t}+p\rho\frac{\partial u}{\partial x}+pu\frac{\partial \rho}{\partial x}+u\rho\frac{\partial p}{\partial x}-pu\frac{\partial \rho}{\partial x}=(\gamma-1)\rho\dot{Q}-\gamma p\rho\frac{\partial u}{\partial x}+p\rho\frac{\partial u}{\partial x}$$

On simplification,

$$\frac{\partial p}{\partial t}+u\frac{\partial p}{\partial x}+\gamma p\frac{\partial u}{\partial x}=(\gamma-1)\dot{Q} \tag{A.39}$$

By substituting linearized expressions of all state variables (from (A.27)) in the above (A.39), we get,

$$\frac{\partial(p'+\overline{p})}{\partial t}+u'\frac{\partial(p'+\overline{p})}{\partial x}+\gamma(p'+\overline{p})\frac{\partial u'}{\partial x}=(\gamma-1)(\dot{Q}'+\overline{\dot{Q}})$$

On ignoring higher-order terms, using constant mean quantities, and substituting $\overline{\dot{Q}}=0$, we get,

$$\frac{\partial p'}{\partial t}+\gamma\overline{p}\frac{\partial u'}{\partial x}=(\gamma-1)\dot{Q}' \quad (A.40)$$

This (A.40) represents the linearized form of the energy equation.

To summarize, the linearized continuity, momentum, and energy equations are as follows [9]:

Continuity equation:

$$\frac{\partial \rho'}{\partial t}+\overline{\rho}\frac{\partial u'}{\partial x}=0 \quad (A.28)$$

Momentum equation:

$$\frac{\partial u'}{\partial t}+\frac{1}{\overline{\rho}}\frac{\partial p'}{\partial x}=0 \quad (A.29)$$

Energy equation:

$$\frac{\partial p'}{\partial t}+\gamma\overline{p}\frac{\partial u'}{\partial x}=(\gamma-1)\dot{Q}' \quad (A.40)$$

Note that the aforementioned linearized conservation equations can be used for low Mach number flows, i.e., when the mean flow velocity is much smaller than the speed of sound, i.e., $\overline{u} \ll c$ [9]. Further, these equations form the basis of our further analysis on studying the transition to thermoacoustic instability from the state of combustion noise in a turbulent combustor (as elaborated in Chap. 9). By substituting an appropriate model for the heat release rate in the linearized energy equation, we will qualitatively capture the dynamical transitions observed in experiments (as discussed in Chaps. 6 and 8) during the intermittency route to thermoacoustic instability in such systems.

References

1. Anderson, J.D., Wendt, J.: Computational Fluid Dynamics, vol. 206. Springer, Berlin (1995)
2. Cengel, Y.A.: Fluid Mechanics. Tata McGraw-Hill Education, New York (2010)

3. Biswas, G.: Introduction to Fluid Mechanics and Fluid Machines. Tata McGraw-Hill Education, New York (2003)
4. Cantwell, B.: Fundamentals of compressible flow. AA210, Department of Aeronautics and Astronautics, Stanford University (1996)
5. Sujith, R.I.: Lecture 2: Introduction to Acoustics Part II : Conservation Equations (2012). https://nptel.ac.in/courses/101/106/101106031/
6. Sujith, R.I.: Lecture 3: Wave Equation and Its Solution in Time Domain (2012). https://nptel.ac.in/courses/101/106/101106031/
7. Kuttruff, H.: Acoustics: An Introduction. CRC Press, Boca Raton (2007)
8. Tennekes, H., Lumley, J.L.: A First Course in Turbulence. MIT Press, Cambridge (2018)
9. Nicoud, F., Wieczorek, K.: J. Spray Combust. Dyn. **1**(1), 67 (2009)

Index

R. I. Sujith, S. A. Pawar, *Thermoacoustic Instability*, Springer Series in Synergetics,
https://doi.org/10.1007/978-3-030-81135-8